International Series in Operations Research & Management Science

Volume 320

The book series **International Series in Operations Research and Management Science** encompasses the various areas of operations research and management science. Both theoretical and applied books are included. It describes current advances anywhere in the world that are at the cutting edge of the field. The series is aimed especially at researchers, advanced graduate students, and sophisticated practitioners.

The series features three types of books:

- Advanced expository books that extend and unify our understanding of particular areas.
- Research monographs that make substantial contributions to knowledge.
- Handbooks that define the new state of the art in particular areas. Each handbook will be edited by a leading authority in the area who will organize a team of experts on various aspects of the topic to write individual chapters. A handbook may emphasize expository surveys or completely new advances (either research or applications) or a combination of both.

The series emphasizes the following four areas:

Mathematical Programming: Including linear programming, integer programming, nonlinear programming, interior point methods, game theory, network optimization models, combinatorics, equilibrium programming, complementarity theory, multiobjective optimization, dynamic programming, stochastic programming, complexity theory, etc.

Applied Probability: Including queuing theory, simulation, renewal theory, Brownian motion and diffusion processes, decision analysis, Markov decision processes, reliability theory, forecasting, other stochastic processes motivated by applications, etc.

Production and Operations Management: Including inventory theory, production scheduling, capacity planning, facility location, supply chain management, distribution systems, materials requirements planning, just-in-time systems, flexible manufacturing systems, design of production lines, logistical planning, strategic issues, etc.

Applications of Operations Research and Management Science: Including telecommunications, health care, capital budgeting and finance, economics, marketing, public policy, military operations research, humanitarian relief and disaster mitigation, service operations, transportation systems, etc.

This book series is indexed in Scopus.

More information about this series at https://link.springer.com/bookseries/6161

Said Ali Hassan • Ali Wagdy Mohamed •
Khalid Abdulaziz Alnowibet
Editors

Decision Sciences for COVID-19

Learning Through Case Studies

Springer

Editors
Said Ali Hassan
Department of Operations Research and Decision Support, Faculty of Computers and Artificial Intelligence
Cairo University
Giza, Egypt

Ali Wagdy Mohamed
Department of Operations Research, Faculty of Graduate studies for Statistical Research
Cairo University
Giza, Egypt

Khalid Abdulaziz Alnowibet
Department of Statistics and Operations Research
King Saud University
Riyadh, Saudi Arabia

ISSN 0884-8289 ISSN 2214-7934 (electronic)
International Series in Operations Research & Management Science
ISBN 978-3-030-87018-8 ISBN 978-3-030-87019-5 (eBook)
https://doi.org/10.1007/978-3-030-87019-5

This Springer imprint is published by the registered company Springer Nature Switzerland AG.
The registered company address is: Gewerbestrasse 11, 6330 Cham, Switzerland

Preface

Managers, researchers, and interested readers in all disciplines especially those related to the transportation of people, materials, and products from one place to another face an endless list of complex issues every day. They must make the appropriate and most beneficial decisions about each activity they make. One of the essential managerial skills is the ability to allocate and utilize limited resources appropriately for achieving the optimal performance efficiently and, ultimately, to use the most appropriate scientific tools in various disciplines to achieve the desired objectives. That is why "decision sciences" appear at the top of the list that everyone interested in this field needs.

"Decision sciences" is a collaborative approach involving mathematical formulae, business tactics, technological applications, and behavioral sciences to help senior management and practitioners make scientific-driven decisions. Such approaches along with up-to-date technology will help businesses execute various processes in a smoother and better way. These approaches also help in day-to-day operations and projections; they will make it easier for business leaders to make smarter and more efficient decisions that will improve their work performance and the functioning of the overall marketplace. Ultimately, the decision sciences are about enhancing individual, group, organizational, societal, and national decision making in support of a better future.

This book aims to provide relevant theoretical and practical frameworks and the latest research findings in the domain of decision sciences and applications for disaster risk reduction of COVID-19. It is written for researchers, executives, and practitioners who want to enrich their scientific and practical knowledge and improve their understanding of the decision-making process in facing real-world problems in reducing the emerging disaster risk of COVID-19.

The used modern and up-to-date decision sciences techniques will make it easier for business leaders and managers in the government and in the health sector to make smarter and more efficient decisions that will improve the firms' functioning and performance.

The target audience of this book are composed of researchers, professionals, leaders, managers, and executives working in the field of decision sciences and applications in most of the application areas especially in the field of COVID-19 and similar pandemics. The book is appropriate and helpful for individual, group, organizational, and societal decision makers seeking for making better decisions and achieving optimum solutions to real-life problems. The diversity of the contained chapters related to multidisciplinary decision sciences and applications topics will benefit researchers and practitioners facing real-world problems who seek to optimize their decision-making process and enhance the performance of their works for a better future.

Giza, Egypt Said Ali Hassan
Giza, Egypt Ali Wagdy Mohamed
Riyadh, Saudi Arabia Khalid Abdulaziz Alnowibet

Acknowledgment

The editors would like to acknowledge the help of each one of the respected authors involved in this book who contributed their time and expertise to this valuable work.

The editors would also like to acknowledge the valuable contributions of all the reviewers regarding the improvement of quality, intelligibility, and content presentation of the enclosed chapters.

Without the support of all mentioned persons, this book would not have become a reality; we highly appreciate their contribution.

Introduction

In December 2019, Wuhan, China was the origin of a pneumonia of unknown cause. Cases of COVID-19 were not limited to this city, and by January, confirmed cases were detected outside Wuhan. Currently, the entire world is suffering from a global epidemic of COVID-19 that has infected thousands of people in almost all countries.

Currently, the entire world is suffering from a global epidemic of COVID-19 that has infected millions of people in almost all countries. The new coronavirus (COVID-19) put humans in all countries in front of a huge danger. The major signs of COVID-19 are declared so that any individual can discover whether or not he has such symptoms.

Updated statistics indicate clearly that the current and the near future situations are not quite right in such a way that foreshadows a catastrophe for many countries of the world, if the situation continues as it is now and if strict measures are not taken at the level of governments and people.

The pandemic of coronavirus has become the talk of the whole world, and it will remain so in the coming period because of the imminent danger it poses to humanity in all countries of the world.

It is very important that this worldwide focus draw the attention of scientists and researchers in the field of optimization to the effective contribution in the optimal utilization of available resources and optimal decision making to help in reducing the emerging disaster risk of COVID-19.

In this book, authors of chapters utilize the applications of decision science techniques and tools in combating against the epidemic of COVID-19 to stand up and face the various challenges facing the whole world.

The new applications of decision science techniques can be regarded as a continuously emerging field in all areas of interest including this very important area of coronavirus pandemic which has become the focus of attention of the whole world. Managers in all disciplines face an endless list of complex issues every day. They must make decisions about each activity they make. One of the essential managerial skills is the ability to allocate and utilize limited resources appropriately in the efforts of achieving the optimal performance efficiently.

Decision sciences use a number of mathematical techniques to solve problems in all fields of study. It is helpful to aid in making decisions to achieve desired objectives and help managers do their jobs more effectively. These techniques help managers allocate resources more effectively and enable them to better optimize the performance of their works. They allow people to analyze a greater number of alternatives and constraints which results in greater confidence in the optimal choice.

The advancements in decision sciences theory and applications can be regarded as a continuously emerging field in all areas of interest including technology, industry, energy, healthcare, education, agriculture, social sciences, and others. Managers in all disciplines face an endless list of complex issues every day; they must make decisions about each activity they make. One of the essential managerial skills is the ability to allocate and utilize limited resources appropriately in the efforts of achieving the optimal performance efficiently.

Quantitative approaches use a number of mathematical techniques to solve problems in all fields of study. It is helpful to aid in making decisions to achieve desired objectives and help managers do their jobs more effectively. These techniques help managers allocate resources more effectively and enable them to better optimize the performance of their works. The mathematical models allow people to analyze a greater number of alternatives and constraints which results in greater confidence in the optimal choice.

Quantitative techniques are the most widely applied methods; they play an important role in the operation of problem solving and this importance is likely to increase, creating the opportunity to play an even greater role. With the recent dramatic improvements in the real-time availability of computer speed and evolving of new exact and approximate solution techniques, this role will increase.

It is very necessary that this draws the attention of scientists and researchers in the fields of decision sciences and supporting decision making to the effective contribution in the optimal utilization of available resources and optimal decision making to help in contributing to the theory and in facing the real-world problems.

List of topics of decision sciences includes mathematical modeling; linear, nonlinear, integer, binary programming models; multi-criteria decision making; network optimization models; project management; game theory models; queueing theory models; inventory theory models; stochastic models; decision making under uncertainty and risk; decision support systems; modeling and simulation; statistical models; forecasting techniques; metaheuristic techniques; heuristic techniques; artificial intelligence and machine learning; data analytics; operations management; crisis management; risk management; supply chain management; business decision-making problems; leadership decision making; strategic planning; innovation and creativity; and behavioral decisions.

To this end, we included high-quality original submissions to this book that reflect the unprecedented momentum garnered by this research area.

Organization of the Book

The book is organized into seven parts with 25 chapters. Parts are as follows:

Part I: Artificial Intelligence
Part II: Forecasting Techniques
Part III: Social Sciences
Part IV: Optimization Techniques
Part V: Data Science
Part VI: COVID-19 Detection
Part VII: Economy

Contents

Part I Artificial Intelligence

1 Application of Artificial Intelligence and Big Data for Fighting COVID-19 Pandemic 3
Joseph Bamidele Awotunde, Sakinat Oluwabukonla, Chinmay Chakraborty, Akash Kumar Bhoi, and Gbemisola Janet Ajamu

2 An IOT-Based COVID-19 Detector Using *K*-Nearest Neighbor 27
T. M. Okediran, O. R. Vincent, A. A. Abayomi-Alli, and O. J. Adeniran

3 Predictive Analytics for Early Detection of COVID-19 by Fuzzy Logic 45
V. Kakulapati, R. Sai Sandeep, and V. Kranthikumar

4 Role of Artificial Intelligence in Diagnosis of Covid-19 Using CT-Scan 67
Karim Sherif, Yousef Emad Gadallah, Khalil Ahmed, Salma ELsayed, and Ali Wagdy Mohamed

5 Predicting the Pandemic Effect of COVID 19 on the Nigeria Economic, Crude Oil as a Measure Parameter Using Machine Learning 79
Cecilia Ajowho Adenusi, Olufunke Rebecca Vincent, Jesufunbi Bolarinwa, and Ologunye Oluwakemi

Part II Forecasting Techniques

6 Time Series Analysis and Forecast of COVID-19 Pandemic 97
Pawan Thapa

7 **A Multi-Step Predictive Model for COVID-19 Cases in Nigeria Using Machine Learning** 107
Sakinat Oluwabukonla Folorunso, Ezekiel Adebayo Ogundepo, Joseph Bamidele Awotunde, Femi Emmanuel Ayo, Oluwatobi Oluwaseyi Banjo, and Abass Ishola Taiwo

8 **Nigerian COVID-19 Incidence Modeling and Forecasting with Univariate Time Series Model** 137
Abass Ishola Taiwo, Adedayo Funmi Adedotun, and Timothy Olabisi Olatayo

9 **Predicting the Spread of COVID-19 in Africa Using Facebook Prophet and Polynomial Regression** 151
Cecilia Ajowho Adenusi, Olufunke Rebecca Vincent, Abiodun Folurera Ajayi, and Bukola Taibat Adebiyi

10 **Comparing Predictive Accuracy of COVID-19 Prediction Models: A Case Study** 165
Dmitriy Klyushin

Part III Social Sciences

11 **AIC Algorithm for Entrepreneurial Intention in Covid19 Pandemic** 183
Nguyen Thi Ngan and Bui Huy Khoi

12 **Sentiment Analysis for COVID Vaccinations Using Twitter: Text Clustering of Positive and Negative Sentiments** 195
Dwijendra Nath Dwivedi and Shailendra Pathak

13 **Participation and Active Contribution of Private Universities in the Prevention of the Covid-19 Pandemic Transmission** 205
Mahyudin Ritonga, Ahmad Lahmi, Sandra Ayu, Yufi Latmini Lasari, Sri Wahyuni, Desminar, Armalena, and Fitria Sartika

14 **Effects on Mental Health by the Coronavirus Disease 2019 (COVID-19) Pandemic Outbreak** 217
Sanjay Kumar Rout and Duryodhan Sahu

15 **Multimodal Analysis of Cognitive and Social Psychology Effects of COVID-19 Victims** 247
V. Kakulapati, S. Mahender Reddy, and Sriman Naini

16 **Public Transport Passenger's Density Estimation Tool for Supporting Policy Responses for COVID-19** 271
Nilton A. Henao-Calle, Mateo Arroyave-Quintero, Semaria Ruiz-Alvarez, and Danny A. J. Gómez-Ramírez

Part IV Optimization Techniques

17 **A Generalized Model for Scheduling Multi-Objective Multiple Shuttle Ambulance Vehicles to Evacuate COVID-19 Quarantine Cases** 287
Said Ali Hassan and Ali Wagdy Mohamed

18 **Hyperparameters Optimization of Deep Convolutional Neural Network for Detecting COVID-19 Using Differential Evolution** . . . 305
Abdelrahman Ezzeldin Nagib, Mostafa Mohamed Saeed, Shereen Fathy El-Feky, and Ali Khater Mohamed

Part V Data Science

19 **Quality Design for the COVID-19 Pandemic: Use of a Web Scraping Technique on Text Comments and Quality Ratings from Multiple Online Sources** 329
Praowpan Tansitpong

20 **Data Science Models for Short-Term Forecast of COVID-19 Spread in Nigeria** 343
Ijegwa David Acheme, Olufunke Rebecca Vincent, and Olaniyi Mathew Olayiwola

Part VI COVID-19 Detection

21 **Attention-Based Residual Learning Network for COVID-19 Detection Using Chest CT Images** 367
R. Karthik, R. Menaka, Sundar Anand, Annie Johnson, and Kanagala Srilakshmi

22 **COVID-19 Face Mask Detection Using CNN and Transfer Learning** 393
Cecilia Ajowho Adenusi, Olufunke Rebecca Vincent, Jesufunbi Abodunrin, and Bukola Taibat Adebiyi

23 **LASSO-DT Based Classification Technique for Discovery of COVID-19 Disease Using Chest X-Ray Images** 407
Roseline Oluwaseun Ogundokun, Joseph Bamidele Awotunde, Paul Onawola, and Taye Oladele Aro

Part VII Economy

24 **Economic Policies for the COVID-19 Pandemic: Lessons from the Great Recession** 425
Guillermo Peña

25 **QOL Barometer for the Well-being of Citizens: Leverages during Critical Emergencies and Pandemic Disasters** 457
Arindam Chakrabarty, Uday Sankar Das, and Saket Kushwaha

About the Editors

Said Ali Hassan received his Ph.D. degree in Informatics from National Institute Polytechnic, Toulouse, France, 1981. He received two M.Sc. degrees in Operations Research Applications, Military Technical College, Cairo, Egypt, 1974, and in Management Development, Euro-Arab Management School, Granada, Spain, 2000. He received his B.Sc. degree from the Military Technical College in Cairo, Egypt, in 1971.

He is a full professor in the Department of Operations Research and Decision Support, Faculty of Computers and Artificial Intelligence, Cairo University, Egypt.

Prior to his current position, he was a staff member in the Military Technical College, Cairo, from 1971 to 1993. He was the Chairman of Operations Research Department in Maritime Research and Consultation Center (MRCC), Arab Academy for Science and Technology, Alexandria, Egypt (1993–1994). He was the Chairman of Operations Research and Decision Support Department in the Faculty of Computers and Artificial Intelligence, Cairo University (2000–2003). He was the Vice Dean, Arab Development Institute, Jeddah, Kingdom of Saudi Arabia (2004–2005). He was professor at the Industrial Engineering Dept., Faculty of Engineering, King Abdulaziz University, Jeddah, Kingdom of Saudi Arabia (2005–2018).

He has field experience in working as Scientific Manager, International Center for Systems (ICS), Zamalek, Cairo, Egypt (1986–1996); Member of the Board of Directors of the Higher Institute of Computer Science and Information Systems, 6th of October City,

Egypt, 1998; Member of the Board of Directors of the Faculty of Information Systems and Computer Science, 6th of October City, Egypt, 1999; Training Manager, Pyramids Computer Center (PCC), Nasr City, Cairo, Egypt (2001–2004); Member of the Founding Committee of Engineering Faculty, Northern Borders University, Saudi Arabia, 2009; and Member of the Academic Accreditation Committee, College of Engineering, King Abdulaziz University, 2009.

He serves as an editor and as a reviewer for many international accredited journals; he published more than 110 research papers in national and international journals and conferences, supervised more than 60 M. Sc. and Ph.D. theses in Egypt and Saudi Arabia, and participated as a PI and Co-PI in more than 30 funded projects.

His research interests lie broadly in operations research, optimization, decision sciences, forecasting, management science, strategic management, and modeling and solving of real-life applications.

Ali Wagdy Mohamed received his B.Sc., M.Sc., and Ph.D. degrees from Cairo University, in 2000, 2004, and 2010, respectively. Ali Wagdy is an Associate Professor at Operations Research Department, Faculty of Graduate Studies for Statistical Research, Cairo University, Egypt. Currently, he is an Associate Professor of Statistics at Wireless Intelligent Networks Center (WINC), Faculty of Engineering and Applied Sciences, Nile University. Recently, he has been recognized among the top 2% scientists according to Stanford University report 2019. He serves as a reviewer of more than 70 international accredited top-tier journals and has been awarded the Publons Peer Review Awards 2018, for placing in the top 1% of reviewers worldwide in assorted fields. He is an associate editor with *Swarm and Evolutionary Computation Journal*, Elsevier. He is editor of more than 10 journals of information sciences, applied mathematics, Engineering, system science, and operations research. He has presented and participated in more than 5 international conferences. He participated as a member of the reviewer committee for 35 different conferences sponsored by Springer and IEEE. He has obtained Rank 3 in CEC'17 competitions on single objective bound constrained real-parameter numerical optimization in Proc of IEEE Congress on Evolutionary Computation, IEEE-CEC 2017, San Sebastián, Spain. Besides, he obtained Rank 3 and Rank 2 in CEC'18 competitions on single objective bound constrained real-parameter numerical optimization and competition on large scale global optimization, in Proc of IEEE Congress on Evolutionary Computation, IEEE-CEC 2017, Sao Paolo, Brazil. He has obtained Rank 2 in CEC'20 competitions on single objective bound constrained real-parameter numerical optimization in Proc of IEEE Congress on Evolutionary Computation, IEEE-CEC 2020, UK. He has published more than 70 papers in reputed and high impact journals like *Information Sciences, Swarm and Evolutionary Computation, Computers & Industrial Engineering, Intelligent Manufacturing, Soft Computing,* and *International Journal of Machine Learning and Cybernetics*. He is interested in mathematical and statistical modeling, stochastic and deterministic optimization, swarm intelligence, and evolutionary computation. Additionally, he is also interested in real-world problems such as industrial, transportation, manufacturing, education, and capital investment problems. 2 Ph.D. and 1 master's degrees have been completed under his supervision.

Khalid Abdulaziz Alnowibet was born on 1971, Riyadh, Saudi Arabia. Mr. Alnowibet got his B.-Sc. degree in Operations Research in 1992 from the College of Sciences, King Saud University, Riyadh, Kingdom of Saudi Arabia. In 1994, Mr. Alnowibet moved to the USA to get his M.Sc. degree in Operations Research from the School of Engineering and Applied Sciences, George Washington University, Washington DC, USA. In 2004, Mr. Alnowibet got his Ph.D. degree in Operations Research, School of Engineering, North Carolina State University, North Carolina State, USA. His dissertation title was "Nonstationary Loss Queues and Network of Nonstationary Loss Queues." Mr. Alnowibet published several papers in several research areas related to operations research such as Performance Evaluation Using Stochastic Modeling, Queueing Networks Applications, Queueing Theory and Applications, Stochastic Processes Theory and

Applications, and Modeling Communication Networks. Mr. Alnowibet now is an associate professor at King Saud University, Department of Operations Research, for more than 15 years. During his service in the university, Mr. Alnowibet was appointed to several key administrative positions in the university. He also supervised graduate students and participated in defending many theses.

Part I
Artificial Intelligence

Chapter 1
Application of Artificial Intelligence and Big Data for Fighting COVID-19 Pandemic

Joseph Bamidele Awotunde, Sakinat Oluwabukonla, Chinmay Chakraborty, Akash Kumar Bhoi, and Gbemisola Janet Ajamu

Abstract The coronavirus (COVID-19) pandemic is playing sensitive havoc in socio-communal systems, humanity and creates economic crises worldwide. Many strategies have been used to managed and curtailed the COVID-19 outbreak, but many countries are still helpless in fighting and containing the outbreak. In an increasingly knowledge-driven, healthcare innovation, and linked society, fighting COVID-19 becomes easier. The Big Data drives the digital revolution by providing solutions focused on big data analytics empowered by Artificial Intelligence (AI) to reduce the difficulty and cognitive burden of accessing and processing large quantities of data. Hence, big data and AI can have been applied in fighting COVID-19 pandemic since the use of both technologies empowered Big Data Analytics (BDA) and yielded imaginable results in combating infectious diseases globally. Therefore, this paper reviews the applicability and importance of AI and Big Data methods to data produced from the countless ubiquitously connected healthcare devices that produced entrenched and distributed information handling capabilities in fighting

J. B. Awotunde (✉)
Department of Computer Science, University of Ilorin, Ilorin, Nigeria
e-mail: awotunde.jb@unilorin.edu.ng

S. O. Folorunso
Department of Mathematical Sciences, Olabisi Onabanjo University, Ago-Iwoye, Nigeria
e-mail: sakinat.folorunso@oouagoiwoye.edu.ng

C. Chakraborty
Department of Computer Science and Engineering, Birla Institute of Technology, Mesra, Jharkhand, India
e-mail: cchakrabarty@bitmesra.ac.in

A. K. Bhoi
Department of Electrical and Electronics Engineering, Sikkim Manipal Institute of Technology (SMIT), Sikkim Manipal University (SMU), Majitar, Sikkim, India

G. J. Ajamu
Department Agricultural Extension and Rural Development, Landmark University, Omu-Aran, Nigeria
e-mail: ajamu.gbemisola@lmu.edu.ng

S. A. Hassan et al. (eds.), *Decision Sciences for COVID-19*, International Series in Operations Research & Management Science 320,
https://doi.org/10.1007/978-3-030-87019-5_1

COVID-19 outbreak. In the area of managing big data for real-time diagnosing, monitoring, and treating COVID-19 patients, AI enabled with big data analytics has shown tremendous potential. The technologies can also be used in the development of drugs and vaccines within the shortest of time, more than ever before.

Keywords Artificial intelligence · Coronavirus · Big data analytics · Healthcare system · Decision science

1.1 Introduction

The world has changed significantly due to the spread of the Coronavirus Disease-19 (COVID-19) outbreak; this is not only affect the healthcare industry but the whole aspects of human life like education, economics, transportations, politics, among many others (Pham et al. 2020a). For instance, many school was closed globally due to this deadly pandemic, various companies and organization has to send their staff home because of the fear of COVID-19 transfer from one person to another. Several countries have to change their date of elections because of the COVID-19 outbreak, various other activities of human endeavors were also affected by the pandemic. As at the time of writing this chapter, there have been 127,877,462 confirmed cases, with 2,796,561 deaths affecting 215 countries around the world, as recorded on World Health Organization website as of 29 March 2021, and a total of 520,540,106 vaccine doses have been administered.

The transmissions of COVID-19 in human-to-human makes it much more treacherous, deadly, and easily spread than other infectious disease families. The National Health Commission of China has identified droplets, direct exposure, feces, respiratory aspirates, and aerosols transmission as a carrier of COVID-19 pandemic (Liu et al. 2020a). According to the recent report, ages between 30 and 79 years, approximately 86.60% are susceptible to COVID-19 pandemic of all patients infected so far, thus have a median age of 47 years (Guan et al. 2020). Though researches have not reported significant gender preponderance, the men are noted to exhibit a higher propensity for this outbreak (Wang et al. 2020a). Although research has not provided concrete evidence about sporadic reports about the vertical transmission of SARS-CoV-2 it has been shown that these also can be a possible way of spread COVID-19 (Kannan et al. 2020).

The efforts, assistance, and collaboration between various nations and several organizations are significantly reducing the risks and the spreading COVID-19 outbreak. For instance, Google as a search engine giant launched a COVID-19 website (ww.google.com/covid19) use for gathering useful information like frequently asked questions on COVID-19 pandemic, latest statistics, and COVID-19 map. Also, the effort of the White House in collaborations with organizations like Google, Amazon, IBM, and Microsoft designed a supercomputing system on importance researches for the COVID-19 pandemic (Wong et al. 2019). Also, various giant publishers like Elsevier and Springers offer free access to related articles on the

COVID-19-like pandemic with technical standards and other related documents on the outbreak and archival websites like bioRxiv, medRxiv, and arXiv enabled fast access to all preprints researches on COVID-19 pandemic (Lu et al. 2018).

The introduction of modern technologies has been a blessing to various fields of studies like Artificial Intelligence (AI) and Big Data (BD) in the healthcare system, AI-banking, AI-agriculture, AI-computer science, and AI-transportations, etc. The advancement of AI with various integration of devices and systems has achieved significant success in dealing with the challenges of infectious diseases like influenzas (Jang et al. 2016), SARS (Kim et al. 2016), and MERS (Brown et al. 2015). Hence, AI with Big Data can play prominent role in the global fight against the COVID-19 outbreak. In building a robust system that can be used for diagnosis, monitoring, forecast, and providing the vaccine for fighting the COVID-19 outbreak, AI plays a crucial role. The integration of several systems and devices enabled with AI mechanisms in designing various applications would require AI techniques like computational modeling, data analytics, and most importantly, machine learning models a major powerful subfield of AI. Different efforts have been made in developing systems and applications that use AI models for combatting COVID-19 outbreaks. AI makes use of the present and past data to predict patterns, actions, and activity in the future; hence, the prediction can be rendered using statistical analysis, automated AI, and quantitative questions (Adeniyi et al. 2021a). The use of AI in BDA has resulted in a knowledge-based system that transformed Big Data into big Knowledge with new approaches and visions to provide people with better understanding and information driven results (Awotunde et al. 2021c).

The Internet of Things has found it way into the healthcare systems, and the medical version is the Internet of Medical Things (IoMT). The IoMT-based devices are used to gathered huge BD, and this collected data can be used to predict and diagnose any type of disease provided that the capture data are related to the disease in question. The use of AI instruments from data collected using IoMT-based devices will create a meaningful result for both medical experts and their patients. This has resulted in the development of the wearable application that progressively adds an AI engine inside medical applications using wearable tools. Furthermore, AI aided the creation of an intelligent medical care platform using collected smartphones, IoT gadgets, websites, and wearables devices to capture data and provide fascinating decision-making for medical experts and their patients (Ahmed et al. 2020; Rodgers 2020).

The population-based researches enabled with data computing, and AI gives easy access to the big data generated in the healthcare systems; the capture data in this regard are Big Data (Priyanka and Kulennavar 2014). For researches in clinical and public health to make a great impact, there is a need for an improvement in healthcare researches for better and oriented results that will surpass the traditional methods in the healthcare system. The BD operation is very diverse from conventional treatment-oriented and narrow statistical analysis-based clinical data. Hence, it is now becoming expedient to solely depend on BD for accurate results in a heterogeneous population in the healthcare system (Abiodun et al. 2021). The use of AI on the generated Big Data has swift research environment with quick advancement in computing technology combine collected data for making useful decision-making, and to reduce their complexity (Oussous et al. 2018).

The virtualization approach provided by cloud computing allows effective data processing on the Big Data generated from medical devices, thus turn the medical sectors into a BD generations production. Nevertheless, the great improvement in modern technologies coupled with the generation of data from various sources has greatly helped in the creation of BD in the healthcare system in recent years. The healthcare industries are in the era of BD, touching every aspect of human life, especially in the areas of patient diagnosis, monitoring, and treatment in this twenty-first century (Olaronke and Oluwaseun 2016). The moving away from conventional medical records to Electronic Health Record (EHR) systems has resulted in an unprecedented increase in generated data in the healthcare system (Hulsen et al. 2019).

To make evidence-driven decisions-making in enhancing patient diagnosis, monitoring, and treatment, BD has been proved very useful and offers the great prospect for consultants, epidemiologists, and specialists in medical fields (Austin and Kusumoto 2016). In the search for new knowledge by medical experts, the BD paradigm is an imperative weapon that needs grasped and seen as a modern reality in the healthcare industry (Galetsi et al. 2019). Therefore, this chapter discusses the possible application of AI and Big Data to tackle the COVID-19 pandemic by improving conventional approaches using to diagnose, control, manage, and identify infected patients (Awotunde et al. 2021d). The various applicability and challenges of using AI and BD were also discussed in this chapter. The use of AI and Big Data has been identified as a weapon to fight the COVID-19 outbreak globally (Awotunde et al 2021d).

Overall, AI is used to classify, monitor, and anticipate occurrences, as well as assist in virus diagnosis. It is used in the analysis of health services. Drones and robots are used to distribute food and drugs, as well as to disinfect public spaces. Using supercomputers, AI is assisting in the development of drugs and a coronavirus vaccine (Lacroix 2019). AI is also designing detection technology, such as surveillance wristbands, to aid in the identification of people who are breaking the prohibition law. Fever and contagious patients can also be detected using mobile devices and Automation thermal cameras (Bullock et al. 2020a). Several scientists are using AI to discover new treatments and therapies for cures, with some software engineers concentrating on identifying contagious patients using biomedical imaging techniques like X-rays and CT scans (Nguyen 2020). Countries such as Taiwan have integrated their public healthcare insurance database with data from immigration and customs, revealing COVID-19 patients based on their travel history and symptoms (Bhattacharya et al. 2021). To fight the COVID-19 outbreak, AI focuses primarily on patient and virus identification, radiology, disease monitoring, and assessment. On the other hand, it also involves internet-based alerting, public knowledge, and social regulation.

The contribution of the chapter is as follows: The chapter presents the effectiveness use of AI and BD in combatting COVID-19 outbreak; related state-of-the-art papers that used these two technologies are discussed. The challenges and associated issues with AI and BD techniques are presented to understand their applications better. Some papers from several sources like Elsevier, Springers, IEEE Xplore, and some preprints archived websites are reviewed for a better understanding of their applications. The remaining part of this chapter is organized thus: Sect. 1.2 discusses the application AI to fight COVID-19 Pandemic. Section 1.3 presents applications of BD and analytics to battle the COVID-19 pandemic. Section 1.4 discusses the problems of using AI and BD to combat the COVID-19 Pandemic. Finally, Sect. 1.5 concludes the chapter.

1.2 Applications of Artificial Intelligence in Combat COVID-19 Pandemic

In the battle against the COVID-19 outbreak, AI has been proved useful and allows computer-based models for diagnosis, monitoring, treatment, and used for optimization utilizing Machine Learning, machine vision, natural language processing, robotics, and automation for the development of the useful applications. Also, AI is very useful in the areas of pattern recognition, forecasting, description, and classification of medical data. COVID-19 infections are being predicted, diagnose, forecasted, treat, and optimized to generate useful decision-making by the medical experts, government, and other policymaking by AI and data science researchers. AI cannot substitute scientific knowledge, but it can help with COVID-19 patient detection, predicting, preventing, and monitoring outbreaks, sterilizing environments, assisting in the production of immunotherapies and therapies, health administration, business, and trade, greater accountability and policymaking, and more. Figure 1.1 shown some areas of applicability of AI-based models during the COVID-19 pandemic.

1.2.1 The Diagnosis and Prediction of COVID-19 Outbreak Using Artificial Intelligence

The early diagnosis and prediction of the COVID-19 outbreak are one of the effective and prominent ways of combating the pandemic. The reverse transcription polymerase chain reaction (RT-PCR) detection technique is currently the gold standard for classifying respiratory viruses. There have been various works in response to the COVID-19 outbreak, and some work has gone into improving this strategy (Rao and Vazquez 2020) as well as looking for other options (Wang et al. 2020b). These methods are characterized by time consuming, very costly, imprecision, require specific equipment, instruments, and materials, coupled with low true positive rate. Also, there are insufficient and lack of testing kits in various countries due to limited resources and techniques. Hence, there is need for a standard

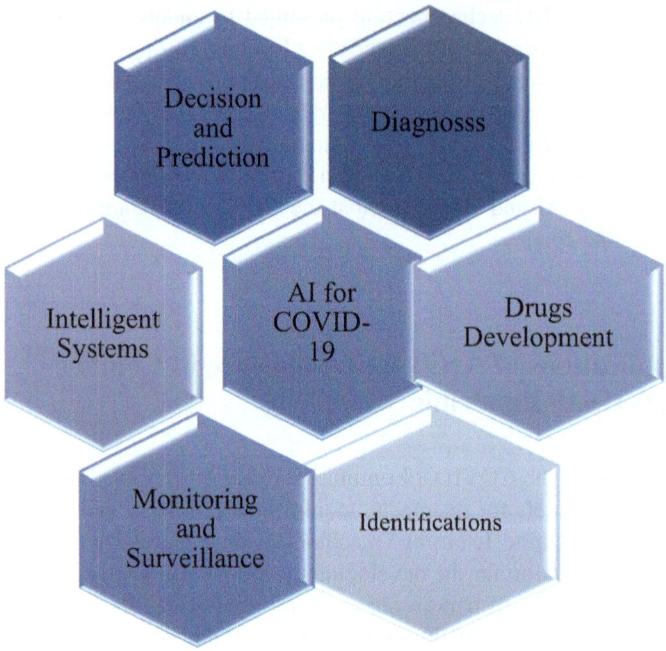

Fig. 1.1 Some basic roles of artificial intelligence in fighting COVID-19 pandemic

technique to meet the needs for a fast diagnosis and monitoring during the COVID-19 pandemic since the present methods cannot meet these requirements. The use of smart devices enabled with AI models can provide a low-cost and simple solution for COVID-19 diagnosis and prediction (Fomsgaard and Rosenstierne 2020; Rao and Vazquez 2020), referred to as mHealth technologies in the literature (Silva et al. 2015). The use of smart devices is very useful in this direction since they are using daily for various purposes. The emergence of edge and cloud computing will reduce the limitation of mobile devices in the area of computing capacities, storage, and battery lifespan (Pham et al. 2020b).

There are various AI-based models for the diagnosis, monitor, and treatment of COVID-19 cases using clinical signs and radiological images and several methods have been suggested in these directions (Li et al. 2020) and many others. An AI-based architecture based on a cell phone-based survey was proposed to classify COVID-19 cases faster and more automatically (Apostolopoulos and Mpesiana 2020). In terms of diagnosis and prediction, radiological images like X-ray with AI-based models have been proposed (Nguyen 2020; Apostolopoulos and Mpesiana 2020) that have been published in the literature. CT scan (Li et al. 2020), which uses a variety of various computational techniques. There have been systematic reviews on the diagnosis and prediction of the COVID-19 pandemic recently (Awotunde et al. 2020).

The application of AI-based models on COVID-19 datasets for diagnosis, prediction, monitoring, and forecasting is a work in progress in a larger population. The use of X-ray, CT, and MRI has been used to generated medical images and has made AI the prominent method in the fight against COVID-19 outbreaks (Wynants et al. 2020). The CT scan reports revealed the presence of the COVID-19 virus in the lungs (Bernheim et al. 2020), as well as claims that the CT scan reports could diagnose COVID-19 faster than current RT-PCR tests. It is worth noting that COVID-19's imaging characteristics were somewhat close to those of other forms of pneumonia, making a proper diagnosis complicated. The study's main aim was to study current AI algorithms, use significant features incorporating medical image processing applications from COVID-19 CT scan images, and then minimize prediction errors to improve the accuracy of potential estimations in order to reduce this challenge.

The use of AI in the COVID-19 outbreak is a fundamental instrument to recognize devices proficient that can offer accuracy in the diagnosis of these infectious diseases. AI techniques can provide useful policymaking for medical experts, government, and the patients themselves by breaking systematic difficulties capture data into useful information. The techniques can still be used to examine the performance of diverse AI models for the categorization of COVID-19 samples (Albahri et al. 2020). Diverse categorization performance capacities have been employed for the assessment of the performance of CI approaches. Six commonly used CI approaches are as follows: Artificial Neural Network (ANN), Support Vector Machine (SVM), linear regression, *k*-Nearest Neighbor (*k*-NN), multilayer perceptron and classification algorithm. By analyzing medical imaging, AI can help develop COVID-19 detection, provide various methods to study disease development using appropriate instruments, and produce infection spread opportunities to expand on complex predictive analysis, including electronic health information. AI can be used for epidemiological exploration modeling, experiential statistics, and forecasting the number of cases based on various public policy options from a social perspective (Awotunde et al. 2020).

Through collecting and investigating all past records relevant to this epidemic, AI plays a crucial role in determining the collection of cases and forecasting the place where this disease will spread in the future. This technology can identify disease characteristics and predict its existence based on information gathered from social media and other digital or print media channels, such as the potential for contagion and the nature of its spread. In addition, based on relevant data and knowledge, it can forecast the number of infected cases and deaths in any given region (Vaishya et al. 2020). Magdon-Ismail (2020) presents an ML study of the COVID-19 pandemic that is based on data from the early stages of infection dynamics. The aim is to extract useful information for a better understanding of community health. The likelihood of disease spread, the rate at which a mild infection becomes a severe infection, asymptomatic approximations for infections, and predictions of new infections over time are all examples of these understandings. The AI models were widely used in prediction and diagnosis, filtering social networks, and computational biology, hence, considered to be effective techniques for analyzing COVID-19 outbreak. AI can use to scrutinize the abnormal symptoms as well as other "obvious signs" swiftly and alerts patients and medical officials (Haleem et al. 2020a).

1.2.2 Artificial Intelligence in COVID-19 Outbreak Epidemiology

The AI is very important in the epidemiology of COVID-19 outbreaks like viral forecasting, spread dynamics and control of the pandemic and other related infectious diseases. The epidemiological research is critical for stopping the spread by tracing infectious trails and identifying link chains that contribute to the rapid spread. The COVID-19 outbreak, on the other hand, was aided by the high movement of citizens during China's most important traditional carnival, which exacerbated the virus's spread while also greatly amplifying the epidemiological investigation's challenges (Zhong et al. 2020). When making decisions on how to cope with a health crisis, it is important to keep an eye on and predict the progression of the outbreak. In this case, the AI model has gotten more attention and consideration in epidemiological studies. Forecasting and predicting the spread of COVID-19 will enable the nation, healthcare providers, businesses, and individuals plan, prepare, and handle the pandemic.

AI techniques can be used to forecast and model the spread of the COVID-19 outbreak in real-time. AI-based model can be used to trained available demographic and COVID-19 outbreak data to predict the spread of the infectious disease. But, due to the lack of huge data and various noisy and outlier data, AI-based predictions of viral spread are not yet very accurate and reliable (Elmousalami and Hassanien 2020). As a result, most of the models used to monitor and forecast the COVID-19 outbreak, such as SIR epidemiological models (Ogundokun et al. 2020), SEIR (Cui et al. 2021), SIRD (Gunawan 2021), expanded SIR model integrating quarantine protocols like isolation policies, and community-level micro inspection steps (Wang et al. 2020c), and ARIMA model (Joseph et al. 2021).

There are various AI-based models that have been reported and applied by several researchers, some of them are multilayer feedforward ANN for COVID-19 outbreak forecast (Ballı 2021); for transmission dynamics a modified stacked autoencoder was used (Raza 2020); the affected, deaths, and recovered was predicted using non-linear hybrid cellular automata classifier (Ayo et al. 2020a); for the prediction of growth and containment strategy of COVID-19 outbreak, an agent-based artificial intelligence simulation platform (EnerPol) was used (Kafieh et al. 2021); to predict the cumulative number of confirm cases, a multi-input deep CNN was on COVID-19 dataset (Khan et al. 2020a); to generate similarity map of transmission dynamics, topological autoencoder was used (Osman et al. 2021); the viral spread dynamics was predicted using Differential Evolution as parameter estimator with SEIR-SD (with social distancing) model (Rica and Ruz 2020); an hybrid AI was used for the forecasting of COVID-19 outbreak using Support Vector Regression (SVR),

Polynomial Regression (PR), deep feedforward neural network, and Long Short Term Memory (LSTM) network (Khan et al. 2020b).

A hybridized deep learning with fuzzy rule method was used to forecast COVID-19 outbreak included limited data in early Composite Monte-Carlo (Fong et al. 2020), and the prediction of the COVID-19 peaks and size of the outbreak was performed using modified SEIR-LSTM model (Yang et al. 2020). The major issue in using AI in the COVID-19 outbreak globally is the limitation of available data and datasets, but as huge relevant data is available, the accuracy of AI-based models will greatly improve. There is no doubt that AI place prominent roles in combatting the COVID-19 pandemic in heightening several strategies used, making the vital decision, and in medical diagnosis, however, the most critical step now is to forecast to know whether or how the second wave of this pandemic will hit China, or other parts of the nations will still have suffered COVID-19 pandemic.

1.2.3 Artificial Intelligence in the Vaccine Development for COVID-19

The most effective way to combat the COVID-19 outbreak is developing a vaccine or therapeutic agents, and AI is one of the most appropriate and faster methods in the development of these drugs. In the development of an effective solution for novel drug candidate-specific against COVID-19 pandemic, AI has shown an encouraging solution (Ortega et al. 2020). The application of AI and in silico has been applied to the repositioning of existing atazanavir drugs for fighting and treating COVID-19 patients (Beck et al. 2020). Also, to design good vaccine contestants against other immunogenic protein fragments and most especially SARSCoV-2 spike protein AI and immunology models have being used. To properly understand the 3Dstructure of protein with the COVID-19, Google's DeepMind developed AlphaFold using AI models, and in proposing a new drug design, the models provided advantageous information. Benevolent AI-based system a UK-based start-up is another queue AI technique to generate and design drug for the treatment of COVID-19 outbreak (Stebbing et al. 2020). The foundation and definition for COVID-19 vaccine and drugs have been provided (Ong et al. 2020) with the establishment of Vaxigen Reverse Vaccinology equipment embedded into AL methods.

The process in developing vaccines was time consuming, challenging, and expensive, but in a situation like this, there is an urgent need for vaccine and drugs production for COVID-19 within the shortest possible time for their development. Current game-changing developments like AI, and Big Data, and IoT, combined with a deep understanding of molecular and structural biology, basic virology, and immunoinformatics, have aided in the development of vaccines for the COVID-19 pandemic. AI has been proved to be one of the fastest methods that can aid the development of vaccines and drugs during this urgent period of the pandemic (Bharadwaj et al. 2021).

Since the production of a new vaccine takes too long, drug repositioning is the best and expedient way to treat COVID-19 infected patients. On the other hand, the development of long-term drug targets is to find inhibitors that target the replication processes linked to this disease and inhibit main coronavirus proteins, as well as what could be used as starting points for drug development (Liu et al. 2020b). Several studies have used AI to establish a COVID-19 therapeutic strategy and drug design. According to Savioli (Savioli 2020), glycosylated spike (S) protein extracted from the Heptad Repeat 1 (HR1) has the least flexibility, making it a promising target. The used of a Siamese Neural Network (SNN) to train the entire 2019-nCoV protein sequence resulted in the discovering that virus proteins had knowledge of peptide linkage. A huge proportion of there was examined against the particular region of HR1 in 2019-nCoV, with the peptidyl-prolyl cis-trans isomerase (PPlase) peptide exhibiting a high affinity for HR1, potentially opening up new research avenues.

Chenthamarakshan et al. (2020) proposed CogMol, a DL multiplicative model system, to developed customized particular protein sequence target drug candidates. This was used on three 2019-nCoV proteins, namely: replicase the primary protease, S protein's receptor-binding domain (RBD), and non-structural protein 9 (NSP9). Beck et al. (2020) used a Molecular Transformer-Drug Target Interaction (MT-DTI) based on DL-based drug-target interaction model to reposition drugs used in treating HIV and discovered that atazanavir is an antiviral drug that has the best inhibitory potency against 2019-nCoV 3C-like proteinase, and their findings suggest that several antiviral drugs, such as Kaletra (lopinavir/ritonavir), can bind to the replication complex components of 2019-nCoV with sufficient inhibitory potency. AI-based models have also been used by other researchers for drug repositioning (Mohapatra et al. 2020), and the models have been proved very useful in this direction. Table 1.1 discusses the different approaches of ML and DL and their key findings in some existing researches discussed in this study.

The repurposing of proven licensed drugs, such as ritonavir, lopinavir, and others, to treat COVID-19 is ineffective. As a consequence, discovering novel chemical compounds to fight this deadly virus is crucial and urgent. Tang et al. (2020) built an advanced deep Q-learning network with fragment-based drug design (ADQN-FBDD) to identify possible lead compounds for 2019-nCoV's 3CL. They presented work lead compounds that were completely developed using the ADQN-FBDD model. To create novel drug-like compounds, a deep learning-based drug discovery platform has been created (Zhavoronkov et al. 2020). Other reviews based AI-based assisted drug designing and production for COVID-19 can be find in (Ogundokun et al. 2021). In previous research works, AI has been used in medical and in the biomedical field (Ayo et al. 2020b; Oladele et al. 2020), for the involvement of heart disease and diabetes (Oladipo et al. 2020), Berglund et al. (2006) investigated diabetes proteins. Academics have used ANN, SVM, Fuzzy Logic Systems, K-means classifier, and many other AI methods (Oladipo et al. 2021).

Table 1.1 Summary of ML and DL only approaches

S/N	Study	Dataset, classes	Classes	ML and DL	Key findings
1	Jiang et al. (2020)	Text	ARDS, COVID-19	L R, *k*-NN ($k = 5$), DT, RF and SVM	SVM and *k*-NN performed best with 80% accuracy on the classification of COVID-19.
2	Brunese et al. (2020)	CXR	COVID-19, TB, Healthy	VGG-16 on ImageNet dataset	Average COVID-19 detection time was 2.5 s with 97% accuracy
3	Sharifrazi et al. (2021)	CT scan	COVID-19, pneumonia and normal	SVM	ResNet+SVM obtained the best metric based on accuracy, sensitivity, FPR and F1-score of 95.33%, 95.33%, 2.33% and 95.34% respectively
4	Barstugan et al. (2020)	CT scan	coronavirus/non-coronavirus	SVM	Accuracy of 99.68% with the GLSZM.
5	Tartaglione et al. (2020)	CXR	CORDA, CORDA&CXR, CORDA&RSNA, CORDA&COVID-CXR, COVID-CXR	ResNet-18, ResNet-50 and COVID-Net	Resnet 18+ COVID-19 + CXR obtained an accuracy of 100%
6	Abbas et al. (2021)	CXR	COVID-19, normal and SARS	ImageNet and DeTraC on ResNet 18, PCA	DeTraC+ ResNet18 achieved the highest Accuracy, Sensitivity, and Specificity 95.12%, 97.91% and 91.87% respectively.
7	Pereira et al. (2020)	CXR	Normal, COVID-19, MERS, SARS, Varicella, Streptococcus, and Pneumocystis	*k*-NN, SVM, MLP, DT, and RF	Multi-classification: F1-Score of 0.65 with MLP + LBP Hierarchical Classification: F1-Score of 0.89 with clustering+BSIF +EQP + LPQ+ SMOTE +TL.
8	Folorunso et al. (2022)	CXR	Normal, PTB, and COVID-19	HOG, PCA, Extra Trees	ROC value of 0.97, Precision value of 0.81

1.3 Applications of Big Data in Fighting COVID-19 Pandemic

Various technologies have been used in fighting COVID-19 pandemic, and big data has one of those techniques shown its capability to support combat infectious diseases (Awotunde et al. 2021a). The application of the techniques has a potential number of solutions to fight the outbreak. The combination of BD and AI analytics helps in outbreak diagnosis, vaccine development, monitoring, pandemic structure and strategies dynamics (Hussain et al. 2021). For instance, AI-based models with associated big data can be used to design complex simulations using COVID-19 data for outbreak diagnosis, monitoring, and pandemic estimation. The AI enabled with big data can be used to support a future forecast of COVID-19 outbreak based on data aggregation capability to leverage on the huge amount of data for early diagnosis. Thus, help healthcare professionals in tracking the spread of the COVID-19 pandemic, and preparing better preventative measures against the outbreak (Buckee 2020). AI-enabled with big data analytics using various real-time collected data from infected patients will help to implement real-time COVI-19 research to design inclusive accuracy with high reliability for patient treatment (Bragazzi et al. 2020).

This will provide a better response to the various diagnosis and treatments, and healthcare experts will understand better the nature of the virus development. AI and BD will assist in the handling of vast quantities of data generated from public health surveillance, real-time disease outbreak tracking, event now-casting/forecasting, daily circumstance summary and reports from regulatory entities and species, and healthcare facilities usage data. When faced with a huge crisis like the COVID-19 outbreak, policymakers must consider why a threat is developing, how the threat scenario evolves, and if the general public embraces containment steps. Governments and research organizations need data in order to gain insight into these issues, with real-time data being especially appealing (Zwitter and Gstrein 2020). Figure 1.2 displayed the framework for an AI-based enabled big data model for combatting the COVID-19 pandemic.

Technological advances can promote and allow for a more detailed definition and review of health conditions, which can contribute to promote the demographics' wellbeing (Abdel-Basset et al. 2021). Precision can help in the health sector because there are more precise criteria for determining illness, toxins, exposure, and behavior. To meet personalized disease treatment for individual, there is a need for various types of data like genetic, physiological, and environmental in the healthcare system (Johnson et al. 2021). The capture data using various devices resulting in BD can be used for diagnosis, monitoring, and treatment of COVID-19 patients, and enabling researchers to learn more about the SARS-CoV-2 virus. The BD can be used to store relevant information like infected, recovered, and expired cases of COVID-19 outbreak (Haleem et al. 2020b). Big data is known as one of the most precise innovations for fighting the current COVID-19 disease outbreak (Bragazzi et al. 2020; Benke and Benke 2018). The advent of BD technologies helps to enhance

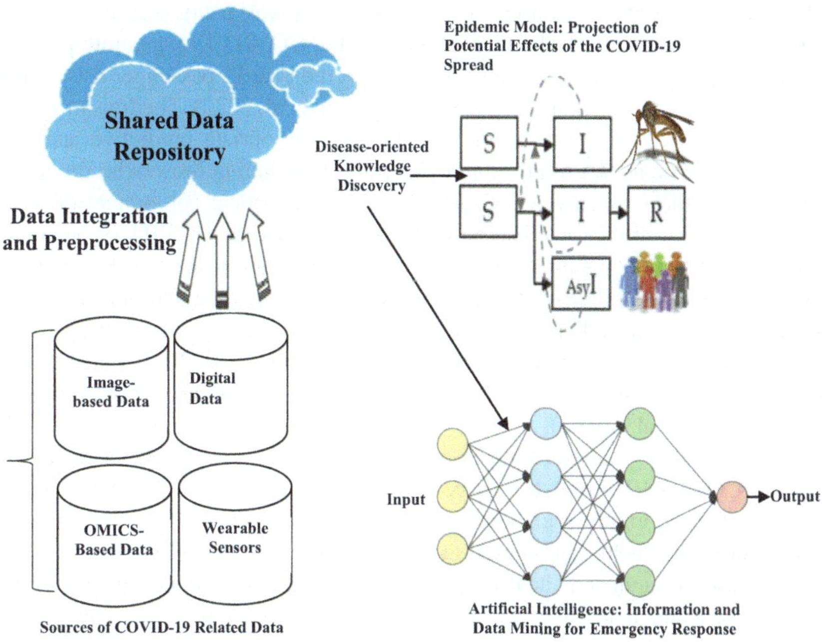

Fig. 1.2 Framework for artificial intelligence enabled with big data for COVID-19 pandemic

patient facilities through the use of Electronic Health Records (EHR) in healthcare sectors. Medical data like doctors and nurses' records, physician files, X-Ray documents, and information about the COVID-19 outbreak area are needed for managing the COVID-19 pandemic (Pham et al. 2020a).

Handling the COVID-19 by policymaker makes use of BD because it helps in the identification of new events, whether they are suspected or confirmed. The task of BD in the COVID-19 outbreak is to estimate the amount of instances so that clinician management on healthcare systems, such as having beds and ventilators, can be done more effectively (Agbehadji et al. 2020). The time series model has been used to estimate the number of infected cases COVID-19 spread to its height with a geographical area (Mahalle et al. 2020). AI and cloud computing were used to forecast epidemiological models for the COVID-19 pandemic using big data, thus, aids in determining the magnitude of COVID-19's distribution across the globe. Big data is used in the management of COVID-19 because it helps in the identification of new events, whether they are suspected or confirmed. Medical signs are used to diagnose cases in individuals such as coughing, pneumonia, and fever. The COVID-19 pandemic could spread across the globe if data analysis is done correctly (Jiang et al. 2020). Individuals, communities, and regions can all use spatial analysis, and the results can be contrasted across measures to determine spatial risk, medical needs, and transportation capability. With spatial identification, it is hoped to avoid a COVID-19 pandemic that has a detrimental effect on the economy, education, and other facets of society (Zhou et al. 2020).

Big data can help in handling of COVID-19 outbreak using data from various data sources, thus useful in the management of the COVID-19 pandemic problem. Several meta-data enabled tags on social media, credit cards, passenger lists, among others, can be used to track the populace (Agbehadji et al. 2020). The transmit power of COVID-19 outbreaks can be measured using appropriate predictive analytics, which can provide model monitoring to all people, including those who do not use smart devices like a wristwatch. Contact tracing is a critical component of public health response to infectious disease outbreaks, such as the COVID-19 pandemic, particularly in the early stages when treatment options are restricted. Big data will guarantee the accessibility of data about the spread of disease and harvest in real-time via an active surveillance system, where the data architecture must be considered to make an accurate outbreak decision (Bragazzi et al. 2020). Contact tracing will also aid in recognition of the virus's structure and timely care for those who are infected (Pham et al. 2020a).

Big data analysis can store huge data about the people who have been infected with the COVID-19 virus. It assists in a detailed understanding of the virus's existence. The information gathered can be used to train new preventive methods in the future. This technique is used to keep records from all forms of COVID-19-affected cases (diagnosed, healed, and discontinued). This data can be used to classify cases and help distribute resources for improved public health security (Anshari et al. 2019; MacLaren et al. 2020). Patient location, proximity, patient-reported travel, co-morbidity, patient physiology, and current symptoms are just a few examples of digital data that can be digitized and used to produce actionable insights at the population and demography levels. Table 1.1 shows how big data is being used in the COVID-19 outbreak.

Big data provides scientists, health workers, and epidemiologists with a vast amount of knowledge, enabling better decisions in the battle against the COVID-19 virus. These data can continually monitor the virus on a global scale and to spur technological development (Xia et al. 2020). It will assist in predicting the effect of COVID-19 on a specific area and the entire population. It contributes to the expansion and study of novel therapeutic procedures. Big data might provide citizens with potential sources and resources, assisting them in dealing with difficult situations. In general, this system enables data for the spread of disease, movement, and health monitoring and prevention systems to be analyzed. Big data can aid in the advanced analytics of Orthopedic and Mental distress surgery using existing evidence in the surgical specialty of Orthopaedics. Trauma mortality and morbidity are similar to the current pandemic, and ideas from it could cross-fertilize. This technology is useful for keeping track of orthopedic patients, which can lead to more informed and positive clinical decisions. As a result of a better understanding of the treatment protocols derived from the appropriate collected data may improve efficiency. Big data is a boon for conducting clinical trials and reviewing medical information to speed up the treatment process. Table 1.2 displays the general applicability of big data during the COVID-19 pandemic.

Big Data Analytics systems take a large amount of time to provide consumers with input and advice, although only a few applications (Singh and Reddy 2015) can

Table 1.2 The applications of big data in COVID-19 pandemic

Applications	Importance during COVID-19 outbreak
COVID-19 outbreak symptoms	COVID-19 patients will receive prompt and reliable care thanks to big data's ability to store patient symptoms due to their variety, length, and velocity attributes. Big data will keep track of a patient's fever and other symptoms and suggest whether or not medical treatment is needed. With the right data, it aids in the identification of suspicious cases and other misconceptions.
Fast moving pathogens must be detected and examined.	Thanks to the velocity characteristics of big data, it is able to analyze fast-moving diseases effectively, which aids in timely early detection.
Detection of COVID-19 pandemic at an early stage	Speedily assists in the early identification of the infected COVID-19 patient and aids in the analysis and identification of people who might become infected with this virus in the future.
Cases that have been contaminated are identified.	Because of the density attributes that enable the storing of huge volumes of data, big data will fully store the medical history of COVID-19 patients. This aids in the identification and analysis of threats. Efficiency in order to store a large amount of data, it can store the full medical history of all patients. This enables efficient in detection of affected individuals and the further examination of the level of danger by providing the collected data.
Travel history	Used to keep track of passenger travel background in order to determine risk. Aids in the identification of individuals who might have come into contact with a virus-infected patient.
Treatments for medical conditions are being established at a faster pace.	Big data will include a history of a patient's medication, allowing them to receive care as soon as possible. Medical professionals benefit from the attributes of big data speed and diversity in ensuring reliable treatment.
Identified a person who enters or leaves an infected area	Due to velocity characteristics, big data can help evaluate and classify opportunities for viruses to infect humans in a given environment. Since COVID-19 spreads rapidly, data analytics is required to help real-time epidemiological databases.
Details about the lockdown	Because of the diversity and frequency of moving objects during the lockout, big data can track and control them. This is due to the fact that big data technology can store huge volumes of data from a variety of sources, allowing for human monitoring in large populations with high mobility.

handle massive data sets within a short processing period. By comparison, the greatest of the outstanding approaches employ a complex trial-and-error approach to tackle large volumes of data sets and data complexity (Siddiqa et al. 2016). There are big data analytics platforms, for instance, the Investigative Data Examination System (Steed et al. 2013) is a big data graphic analytics program that is employed to scrutinize complicated earth structure models of massive quantities of datasets. Big data volume is enormous, and thus conventional applications for Database Management cannot be used to accumulate and investigate huge data. The resolution comes from modern warehousing databases such as Apache Hadoop that support the analysis of distributed data.

1.4 Application Challenges of Artificial Intelligence and Big Data for Fight COVID-19 Pandemic

AI has the ability to become an instrument in fighting COVID-19 and related infectious diseases. Conversely, AI programs are only at a foundational level, and the effects of such AI interventions will take some time to be evident (Kent 2020). In its analysis of the use of CI toward COVID-19, Bullock et al. (2020b) concluded that at this point, very little of the AI programs tested are operationally mature. In reality, AI is limited by the lack of data and by much more data. There are (yet) inadequate available data to train AI models, inadequate open datasets and approaches to work on, yet also possible problems of big data hubris, non-adjustment of methods, and outer data and a deluge of scientific results, all of which have to be switched.

Valuable as these are, the concern is that after the epidemic is over, data protection degradation will not be pushed back and policymakers will continue to help their increased capacity to monitor their populations- and use the data collected for other purposes in the battle with COVID-19. As advises, "though if coronavirus diseases are almost being eliminated, some information-hungry policymakers may claim they wanted to keep biometric tracking mechanisms in place because they are afraid of a second wave of the diseases, or because a new Ebola outbreak is emerging in the central region of Africa, or because you have the concept. The use of AI for COVID-19 prediction and diagnosis is hindered by deficiency of relevant training samples, AI resources such as image recognition and automation are not.

While it has been stated that a number of Chinese clinics have implemented "AI-Based" radiology innovation, the promise of AI is still not implemented. Radiologists elsewhere have raised concerns that inadequate data are available to train AI models, that most of the accessible COVID-19 samples emerge from Chinese medical centers and may suffer from sample preference, and that the use of CT scans and X-rays may pollute instruments and further transmit the infection. The use of CT scans in European medical centers has progressively decreased after the disease outbreak has ended. Owing to a lack of evidence, crowded social platforms and outlier data, big data hubris, and computational complexities, COVID-19 spread AI predictions are still not very reliable and valid (Maier and Brockmann 2020). The majority of frameworks used for monitoring and forecasting so far do not use AI approaches. Rather, most analysts consider epidemiologically proven models the so-called SIR models.

The Robert Koch Institute in Berlin, for example, uses an immunological SIR method that captures into account government control steps, such as lockdowns, quarantines, and social distancing recommendations. This method was established in China to demonstrate that containment would succeed in limiting the incidence to slower than the (Maier and Brockmann 2020) exponential value. While working with Big Data, data scientists face a lot of obstacles (Awotunde et al. 2021d). One obstacle is the compilation, integration, and storage of tremendous datasets generated from disseminated sources, with far less needed hardware and software (Awotunde et al. 2021b). The management of big data results in another huge problem. Effective big data management is essential to promote the abstraction of accurate information to maximize investment. Proper management of data is the basis for BDA (Awotunde et al. 2021d).

Another difficulty is to synchronize with an organization's internal infrastructures external data sources, and Big Data dispersed systems (sensors, apps, databases, networks, etc.) (Abiodun et al. 2021). Most of the time, an analysis of the data produced within organizations is not sufficient. We need to go one step further to integrate internal data with external data sources to gain useful insight and information. External data may include the sources from third parties, market fluctuations details, weather forecasts and traffic situations, social network data, comments from clients, and resident input. This will aid to optimize the power of analytical models used in the analysis, for example.

Big data and data stream uses the concept of collaborative and incremental learning. They deal with numerous challenges that arise during the learning of data such as data accessibility and restricted resources. They make use of User profiling and stock forecasting applications. The application of incremental learning to data produces a faster means of prediction when dealing with the new data stream. Hence, using the incremental algorithm is highly recommended when the drift concept is not available. On the other hand, it is highly recommended to use ensemble algorithms to obtain accurate outcomes with large drift concepts. Also, when dealing with a modest data stream or real-time processing, the incremental algorithm should be considered for use. Conversely, in the case of a complex data stream, an ensemble algorithm is considered the best.

Another big problem is analogical reasoning: fundamentally, disease forecasting is unpredictable. In the narrative on Big Data and CI, there is often an underlying assumption about the statistical covariates or cell phone data and difficult models of simulation that avoid the need to gather basic epidemiological details. Nevertheless, for evolving COVID-19 epidemics emphasizing this opinion still lack a reliable fact on situation counts and biotic processes that drive an outbreak of the disease, let only the behavioral reactions of infected individuals, and make it difficult to rapidly adjust or understand precise, intricate prototypes on spatiotemporal scales applicable in making a verdict. Hopefully, the best effective systems would continue to be simple (Viboud et al. 2018) equally, out of necessity for versatile modules that provide quick responses given a great controversy surrounding emergency epidemiological studies and since simple models are available easier to interpret and communicate (Kahn et al. 2019).

Clear knowledge of both the importance and weaknesses of typical outputs is a precondition for their successful implementation but is often absent. Because policymakers typically have not yet reached in-depth modeling experience, lack of consistent message threatens two destructive results: assume that the method would be misinformed without skepticism and choices, or ignore off-hand modeling and neglect the proof that we must control the epidemics most efficiently. During epidemics, Verdicts have to be taken rapidly, on a patchy basis, and inaccurate data models could be effective tools for guiding them. Considering the above obstacles, computing capacity, novel methods, and novel data are advancing steadily sources give authentic courage for improved monitoring and valuable predicting systems. Novel database available to us comprises not only inactively experimental large data streams from cell phones but also comprehensive ecological data and local sensor information from embedded devices, internet search information, pathogenic genomic data that can be produced quickly informing response during an outbreak, Adeniyi et al. (2021b) and crowd-sourced methods to track emergencies changing rapidly (Schiavo et al. 2014).

Information sharing systems and structured aggregation strategies are being established to secure personal data privacy (De-Montjoye et al. 2018), and through internet access enables rapid data transmission and collaboration between geographically distant respondent teams. Methodologically, effective modeling methods that combine multiple predictions to reduce uncertainty are being built (Ray and Reich 2018). In reaction to the outbreak of COVID-19 between academic organizations, an innovative, collaborative approach has unfolded: for example, Twitter and other channels to exchange, evaluate, and publicly debate the effects of new research as it arises. (Ironically, the control of the misinformation that is now proliferating on social networking during a crisis is likely to become one of the most significant concerns for containing the disease).

Unless the above problems are resolved, these technologies will remain dislocated and impracticable. It is promising that all three problems may be strengthened by transferring most of the funding and knowledge focus to those communities most helpless to epidemics. The unpredictability of the COVID-19 outbreak and the progressively technological mechanisms of these methods will require agile, dispersed groups of people covering the systematic and functioning dimensions of the epidemic response using novel Big Data methods to supplement and explain the limited spread, socio-geographical data that are necessary towards the progress of important predictions. Provincial or small groups with investigative experience and current connections with government and industry stakeholders can contribute to scalable modeling strategies that exploit broad curated and international scientific database skills, building on partnerships formed if no new or ongoing outbreaks occur. This strategy may likewise be of help to mitigate some policy concerns related to data sharing, particularly though sensitive information cannot be shared publicly.

1.5 Conclusion and Future Research Directions

A good method in detecting COVID-19 in infested patients is significant and necessary, and Computational Intelligence has been proved useful in this direction. This study had discussed recent works that had applied CI towards the control of COVID-19. We found out that most of the works reviewed were done on diagnosis, prediction, and monitoring of the patient with COVID-19 pandemic, and image classification with CXR images and CTs for prognosis and diagnosis purposes. While most of the diagnostic approaches discussed are achieving good results, there is, however, no production-ready solution, and the scientific community is still working assiduously to improve the sensitivity of the AI models. An AI model can be very useful in diagnosing COVID-19 from the available dataset. Therefore, the current level of the diagnostic efforts is already contributing to the control of the COVID-19 pandemic. While it is granted that AI has not yet played a critical role in fighting against COVID-19 from the viewpoints of pharmaceuticals, it is achieving some results in other areas while struggling to solve the constraint of a lack of historical data, which may be overcome soon. Following the generalized results from this study, researchers, and practitioners may be guided on some of the things that are scientifically possible in exploiting AI approaches for the control of COVID-19. The COVID-19 predictor can be used to forecast COVID-19's future trend, but it has its own limitations and cannot account for other variables that can influence the trend, such as flexibility, environment, and so on. Our future work will focus on developing a more robust model that can forecast trends with a variety of features. Big data analytics for the COVID-19 outbreak, such as spread detection, virus spread monitoring, diagnosis and treatment, and vaccine and drug discovery, have also been presented. Furthermore, we addressed the barriers that must be resolved in order for AI and big data to be successful in combating the COVID-19 pandemic. Finally, we have outlined some key takeaways and suggestions for policymakers and researchers.

References

Abbas A, Abdelsamea MM, Gaber MM (2021) Classification of COVID-19 in chest X-ray images using DeTraC deep convolutional neural network. Appl Intell 51(2):854–864

Abdel-Basset M, Chang V, Nabeeh NA (2021) An intelligent framework using disruptive technologies for COVID-19 analysis. Technol Forecast Soc Chang 163:120431

Abiodun MK, Awotunde JB, Ogundokun RO et al (2021) Cloud and big data: a mutual benefit for organization development. J Phys Conf Ser 1767(1):012020

Adeniyi EA, Ogundokun RO, Awotunde JB (2021a) IoMT-based wearable body sensors network healthcare monitoring system. Stud Comput Intell 933:103–121

Adeniyi EA, Ogundokun RO, Awotunde JB, (2021b) IoMT-Based wearable body sensors network healthcare monitoring system. In IoT in healthcare and ambient assisted living (pp. 103–121). Springer, Singapore

Agbehadji IE, Awuzie BO, Ngowi AB, Millham RC (2020) Review of big data analytics, artificial intelligence and nature-inspired computing models towards accurate detection of COVID-19 pandemic cases and contact tracing. Int J Environ Res Public Health 17(15):5330

Ahmed MB, Boudhir AA, Santos D, El Aroussi M, Karas IR, (Eds.). (2020) Innovations in smart cities applications edition 3: the proceedings of the 4th international conference on smart city applications. Springer Nature

Albahri OS, Zaidan AA, Albahri AS et al (2020) Systematic review of artificial intelligence techniques in the detection and classification of COVID-19 medical images in terms of evaluation and benchmarking: Taxonomy analysis, challenges, future solutions and methodological aspects. J Infect Public Health 13(10):1381–1396

Anshari M, Almunawar MN, Lim SA, Al-Mudimigh A (2019) Customer relationship management and big data enabled: Personalization & customization of services. Appl Comput Inform 15 (2):94–101

Apostolopoulos ID, Mpesiana TA (2020) Covid-19: automatic detection from x-ray images utilizing transfer learning with convolutional neural networks. Phys Eng Sci Med 43(2):635–640

Austin C, Kusumoto F (2016) The application of Big Data in medicine: current implications and future directions. J Interv Card Electrophysiol 47(1):51–59

Awotunde JB, Ayo FE, Jimoh RG, et al (2020) Prediction and classification of diabetes mellitus using genomic data. Intelligent IoT Systems in Personalized Health Care, pp 235–292

Awotunde JB, Jimoh RG, Oladipo ID, Abdulraheem M (2021a) Prediction of malaria fever using long-short-term memory and big data. Commun Comput Inform Sci 1350:41–53

Awotunde JB, Adeniyi AE, Ogundokun RO et al (2021b) MIoT-based big data analytics architecture, opportunities and challenges for enhanced telemedicine systems. Stud Fuzziness Soft Comput 2021(410):199–220

Awotunde JB, Folorunso SO, Jimoh RG et al (2021c) Application of artificial intelligence for COVID-19 epidemic: an exploratory study, opportunities, challenges, and future prospects. In: Studies in systems, decision and control, vol 358, pp 47–61

Awotunde JB, Jimoh RG, Oladipo ID et al (2021d) Big data and data analytics for an enhanced COVID-19 epidemic management. In: Studies in systems, decision and control, vol 358, pp 11–29

Ayo FE, Ogundokun RO, Awotunde JB, et al (2020a, July) Severe acne skin disease: a fuzzy-based method for diagnosis. Lecture Notes in Computer Science (including subseries Lecture Notes in Artificial Intelligence and Lecture Notes in Bioinformatics), 2020, 12254 LNCS, pp 320–334

Ayo FE, Awotunde JB, Ogundokun RO et al (2020b) A decision support system for multi-target disease diagnosis: a bioinformatics approach. Heliyon 6(3):e03657

Ballı S (2021) Data analysis of Covid-19 pandemic and short-term cumulative case forecasting using machine learning time series methods. Chaos, Solitons Fractals 142:110512

Barstugan M, Ozkaya U, Ozturk S, (2020) Coronavirus (covid-19) classification using ct images by machine learning methods. arXiv preprint arXiv:2003.09424

Beck BR, Shin B, Choi Y et al (2020) Predicting commercially available antiviral drugs that may act on the novel coronavirus (SARS-CoV-2) through a drug-target interaction deep learning model. Comput Struct Biotechnol J 18:784–790

Benke K, Benke G (2018) Artificial intelligence and big data in public health. Int J Environ Res Public Health 15(12):2796

Berglund E, Sitte J (2006) The parameterless self-organizing map algorithm. IEEE Trans Neural Netw 17(2):305–316

Bernheim A, Mei X, Huang M et al (2020) Chest CT findings in coronavirus disease-19 (COVID-19): relationship to duration of infection. Radiology:200463

Bharadwaj KK, Srivastava A, Panda MK et al (2021) Computational intelligence in vaccine design against COVID-19. In: Computational intelligence methods in COVID-19: surveillance, prevention, prediction and diagnosis. Springer, Singapore, pp 311–329

Bhattacharya S, Maddikunta PKR, Pham QV et al (2021) Deep learning and medical image processing for coronavirus (COVID-19) pandemic: a survey. Sustain Cities Soc 65:102589

Bragazzi NL, Dai H, Damiani G et al (2020) How big data and artificial intelligence can help better manage the COVID-19 pandemic. Int J Environ Res Public Health 17(9):3176
Brown DE, Abbasi A, Lau RY (2015) Predictive analytics: predictive modeling at the micro level. IEEE Intell Syst 30(3):6–8
Brunese L, Mercaldo F, Reginelli A, Santone A (2020) Explainable deep learning for pulmonary disease and coronavirus COVID-19 detection from X-rays. Comput Methods Prog Biomed 196: 105608
Buckee C (2020) Improving epidemic surveillance and response: big data is dead, long live big data. Lancet Digital Health 2(5):e218–e220
Bullock J, Luccioni A, Pham KH et al (2020a) Mapping the landscape of artificial intelligence applications against COVID-19. J Artif Intell Res 69:807–845
Bullock J, Pham KH, Lam CSN, Luengo-Oroz M, (2020b) Mapping the landscape of artificial intelligence applications against COVID-19. arXiv preprint arXiv:2003.11336
Chenthamarakshan V, Das P, Padhi I, et al (2020) Target-specific and selective drug design for covid-19 using deep generative models. arXiv preprint arXiv:2004.01215
Cui Y, Ni S, Shen S (2021) A network-based model to explore the role of testing in the epidemiological control of the COVID-19 pandemic. BMC Infect Dis 21(1):1–12
De-Montjoye YA, Gambs S, Blondel V et al (2018) On the privacy-conscientious use of mobile phone data. Sci Data 5(1):1–6
Elmousalami HH, Hassanien AE (2020) Day level forecasting for Coronavirus Disease (COVID-19) spread: analysis, modeling and recommendations. arXiv preprint arXiv:2003.07778
Folorunso SO, Awotunde JB, Adeboye NO, Matiluko OE (2022) Data classification model for COVID-19 pandemic. In: Studies in systems, decision and control, vol 378, pp 93–118
Fomsgaard AS, Rosenstierne MW (2020) An alternative workflow for molecular detection of SARS-CoV-2–escape from the NA extraction kit-shortage, Copenhagen, Denmark, March 2020. Eurosurveillance 25(14):2000398
Fong SJ, Li G, Dey N et al (2020) Composite Monte Carlo decision making under high uncertainty of novel coronavirus epidemic using hybridized deep learning and fuzzy rule induction. Appl Soft Comput 93:106282
Galetsi P, Katsaliaki K, Kumar S (2019) Values, challenges and future directions of big data analytics in healthcare: a systematic review. Soc Sci Med 241:112533
Guan WJ, Ni ZY, Hu Y et al (2020) Clinical characteristics of coronavirus disease in 2019 in China. N Engl J Med 382(18):1708–1720
Gunawan AA (2021) Forecasting Social Distancing impact on COVID-19 in Jakarta using SIRD Model. Proc Comput Sci 179:662–669
Haleem A, Vaishya R, Javaid M, Khan IH (2020a) Artificial Intelligence (AI) applications in orthopaedics: an innovative technology to embrace. J Clin Orthop Trauma 11:S80–S81
Haleem A, Javaid M, Khan IH, Vaishya R (2020b) Significant applications of big data in COVID-19 pandemic. Indian J Orthop 54:526–528
Hulsen T, Jamuar SS, Moody A et al (2019) From big data to precision medicine. Front Med 6:34
Hussain AA, Dawood BA, Al-Turjman F (2021) Application of AI techniques for COVID-19 in IoT and big data era: a survey. Artificial intelligence and machine learning for COVID-19, pp 175–211
Jang S, Lee S, Choi SM, et al (2016) Comparison between SARS CoV and MERS CoV using apriori algorithm, decision tree, SVM. In MATEC Web of Conferences (vol 49, p 08001). EDP Sciences
Jiang X, Coffee M, Bari A et al (2020) Towards an artificial intelligence framework for data-driven prediction of coronavirus clinical severity. Comput Mater Contin 63(1):537–551
Johnson KB, Wei WQ, Weeraratne D et al (2021) Precision medicine, AI, and the future of personalized health care. Clin Transl Sci 14(1):86–93
Joseph JJ, Paul A, Saijan S et al (2021) Assessment of lockdown effectiveness in the wake of COVID-19 in India using the auto regressive integrated moving average model. Infect Dis Clin Pract 29(1):e13–e19

Kafieh R, Arian R, Saeedizadeh N et al (2021) COVID-19 in Iran: forecasting pandemic using deep learning. Comput Math Methods Med 2021:6927985
Kahn R, Mahmud AS, Schroeder A et al (2019) Rapid forecasting of cholera risk in Mozambique: translational challenges and opportunities. Prehosp Disaster Med 34(5):557–562
Kannan S, Ali PSS, Sheeza A, Hemalatha K (2020) COVID-19 (Novel Coronavirus 2019)-recent trends. Eur Rev Med Pharmacol Sci 24(4):2006–2011
Kent J (2020) Understanding the COVID-19 pandemic as a big data analytics Issue. Erişim Tarihi 11:2020
Khan MB, Zhang Z, Li L et al (2020a) A systematic review of non-contact sensing for developing a platform to contain COVID-19. Micromachines 11(10):912
Khan SD, Alarabi L, Basalamah S (2020b) Toward smart lockdown: a novel approach for COVID-19 hotspots prediction using a deep hybrid neural network. Computers 9(4):99
Kim D, Hong S, Choi S, Yoon T, (2016, March) Analysis of transmission route of MERS coronavirus using decision tree and Apriori algorithm. In 2016 18th international conference on advanced communication technology (ICACT) (pp 559–565). IEEE
Lacroix P (2019) Big data privacy and ethical challenges. In: Big data, big challenges: a healthcare perspective. Springer, Cham, pp 101–111
Li L, Qin L, Xu Z et al (2020) Artificial intelligence distinguishes COVID-19 from community acquired pneumonia on chest CT. Radiology:200905
Liu Y, Gayle AA, Wilder-Smith A, Rocklöv J (2020a) The reproductive number of COVID-19 is higher compared to SARS coronavirus. J Travel Med 27(2):taaa021
Liu C, Zhou Q, Li Y, et al (2020b) Research and development on therapeutic agents and vaccines for COVID-19 and related human coronavirus diseases
Lu FS, Hou S, Baltrusaitis K et al (2018) Accurate influenza monitoring and forecasting using novel internet data streams: a case study in the Boston Metropolis. JMIR Public Health Surveill 4(1): e4
MacLaren G, Fisher D, Brodie D (2020) Preparing for the most critically ill patients with COVID-19: the potential role of extracorporeal membrane oxygenation. JAMA 323(13):1245–1246
Magdon-Ismail M (2020) Machine learning the phenomenology of COVID-19 from early infection dynamics. arXiv preprint arXiv:2003.07602
Mahalle PN, Sable NP, Mahalle NP, Shinde GR, (2020) Data analytics: Covid-19 prediction using multimodal data. In Intelligent systems and methods to combat Covid-19 (pp. 1–10). Springer, Singapore
Maier BF, Brockmann D (2020) Effective containment explains subexponential growth in recent confirmed COVID-19 cases in China. Science 368(6492):742–746
Mohapatra S, Nath P, Chatterjee M et al (2020) Repurposing therapeutics for COVID-19: rapid prediction of commercially available drugs through machine learning and docking. PLoS One 15(11):e0241543
Nguyen TT (2020) Artificial intelligence in the battle against coronavirus (COVID-19): a survey and future research directions. arXiv preprint arXiv:2008.07343
Ogundokun RO, Lukman AF, Kibria GB et al (2020) Predictive modelling of COVID-19 confirmed cases in Nigeria. Infect Dis Model 5:543–548
Ogundokun RO, Awotunde JB, Misra S, Umoru DO (2021) Drug Verification System Using Quick Response Code. Communications in Computer and Information Science 1350:535–545
Oladele TO, Ogundokun RO, Awotunde JB et al (2020, July) Diagmal: a malaria coactive neuro-fuzzy expert system. Lecture Notes in Computer Science (including subseries Lecture Notes in Artificial Intelligence and Lecture Notes in Bioinformatics), 2020, 12254 LNCS, pp 428–441
Oladipo ID, Babatunde AO, Aro TO, Awotunde JB (2020) Enhanced neuro-fuzzy inferential system for diagnosis of diabetes mellitus (DM). Int J Inform Process Commun 8(1):17–25
Oladipo ID, Babatunde AO, Awotunde JB, Abdulraheem M (2021) An improved hybridization in the diagnosis of diabetes mellitus using selected computational intelligence. Commun Comput Inform Sci 2021(1350):272–285

Olaronke I, Oluwaseun O (2016, December) Big data in healthcare: prospects, challenges and resolutions. In 2016 future technologies conference (FTC) (pp 1152–1157). IEEE
Ong E, Wong MU, Huffman A, He Y (2020) COVID-19 coronavirus vaccine design using reverse vaccinology and machine learning. Front Immunol 11:1581
Ortega JT, Serrano ML, Pujol FH, Rangel HR (2020) Role of changes in SARS-CoV-2 spike protein in the interaction with the human ACE2 receptor: an in silico analysis. EXCLI J 19:410
Osman N, Torki M, ElNainay M, AlHaidari A, Nabil E (2021) Artificial intelligence-based model for predicting the effect of governments' measures on community mobility. Alex Eng J 60 (4):3679–3692
Oussous A, Benjelloun FZ, Lahcen AA, Belfkih S (2018) Big data technologies: a survey. J King Saud Univ Comput Inform Sci 30(4):431–448
Pereira RM, Bertolini D, Teixeira LO et al (2020) COVID-19 identification in chest X-ray images on flat and hierarchical classification scenarios. Comput Methods Prog Biomed 194:105532
Pham QV, Nguyen DC, Huynh-The T, et al (2020a) Artificial intelligence (AI) and big data for coronavirus (COVID-19) pandemic: a survey on the state-of-the-arts
Pham QV, Fang F, Ha VN et al (2020b) A survey of multi-access edge computing in 5G and beyond: fundamentals, technology integration, and state-of-the-art. IEEE Access 8:116974–117017
Priyanka K, Kulennavar N (2014) A survey on big data analytics in health care. Int J Comput Sci Inform Technol 5(4):5865–5868
Rao ASS, Vazquez JA (2020) Identification of COVID-19 can be quicker through artificial intelligence framework using a mobile phone–based survey when cities and towns are under quarantine. Infect Control Hosp Epidemiol 41(7):826–830
Ray EL, Reich NG (2018) Prediction of infectious disease epidemics via weighted density ensembles. PLoS Comput Biol 14(2):e1005910
Raza K (2020) Artificial intelligence against COVID-19: a meta-analysis of current research. Big Data analytics and artificial intelligence against COVID-19: innovation vision and approach, pp 165–176
Rica S, Ruz GA (2020, October) Estimating SIR model parameters from data using differential evolution: an application with COVID-19 data. In 2020 IEEE conference on computational intelligence in bioinformatics and computational biology (CIBCB) (pp. 1–6). IEEE
Rodgers W (2020) Artificial intelligence in a throughput model: some major algorithms. CRC Press
Savioli N (2020) One-shot screening of potential peptide ligands on HR1 domain in COVID-19 glycosylated spike (S) protein with deep siamese network. arXiv preprint arXiv:2004.02136
Schiavo R, May Leung M, Brown M (2014) Communicating risk and promoting disease mitigation measures in epidemics and emerging disease settings. Pathog Glob Health 108(2):76–94
Sharifrazi D, Alizadehsani R, Roshanzamir M et al (2021) Fusion of convolution neural network, support vector machine and Sobel filter for accurate detection of COVID-19 patients using X-ray images. Biomed Signal Process Control 68:102622
Siddiqa A, Hashem IAT, Yaqoob I et al (2016) A survey of big data management: taxonomy and state-of-the-art. J Netw Comput Appl 71:151–166
Silva MB, Rodrigues JJ, de la Torre DI et al (2015) Mobile-health: a review of current state in 2015. J Biomed Inform 56:265–272
Singh D, Reddy CK (2015) A survey on platforms for big data analytics. J Big Data 2(1):1–20
Stebbing J, Phelan A, Griffin I et al (2020) COVID-19: combining antiviral and anti-inflammatory treatments. Lancet Infect Dis 20(4):400–402
Steed CA, Ricciuto DM, Shipman G et al (2013) Big data visual analytics for exploratory earth system simulation analysis. Comput Geosci 61:71–82
Tang B, He F, Liu D, et al (2020) AI-aided design of novel targeted covalent inhibitors against SARS-CoV-2. BioRxiv
Tartaglione E, Barbano CA, Berzovini C et al (2020) Unveiling covid-19 from chest x-ray with deep learning: a hurdles race with small data. Int J Environ Res Public Health 17(18):6933

Vaishya R, Javaid M, Khan IH, Haleem A (2020) Artificial Intelligence (AI) applications for COVID-19 pandemic. Diabetes Metab Syndr Clin Res Rev 14(4):337–339

Viboud C, Sun K, Gaffey R et al (2018) The RAPIDD ebola forecasting challenge: synthesis and lessons learnt. Epidemics 22:13–21

Wang D, Hu B, Hu C et al (2020a) Clinical characteristics of 138 hospitalized patients with 2019 novel coronavirus–infected pneumonia in Wuhan, China. JAMA 323(11):1061–1069

Wang CJ, Ng CY, Brook RH (2020b) Response to COVID-19 in Taiwan: big data analytics, new technology, and proactive testing. JAMA 323(14):1341–1342

Wang L, Zhou Y, He J, Zhu B et al (2020c) An epidemiological forecast model and software assessing interventions on the COVID-19 epidemic in China. J Data Sci 18(3):409–432

Wong ZS, Zhou J, Zhang Q (2019) Artificial intelligence for infectious disease big data analytics. Infect Dis Health 24(1):44–48

Wynants L, Van Calster B, Collins GS et al (2020) Prediction models for diagnosis and prognosis of covid-19: systematic review and critical appraisal. BMJ 369:m1328

Xia W, Sanyi T, Yong C, et al (2020) When will be the resumption of work in Wuhan and its surrounding areas during COVID-19 epidemic? A data-driven network modeling analysis. Scientia Sinica Mathematica

Yang Z, Zeng Z, Wang K et al (2020) Modified SEIR and AI prediction of the epidemics trend of COVID-19 in China under public health interventions. J Thorac Dis 12(3):165

Zhavoronkov A, Aladinskiy V, Zhebrak A, et al (2020) Potential COVID-2019 3C-like protease inhibitors designed using generative deep learning approaches. Insilico Medicine Hong Kong Ltd A, 307, E1

Zhong L, Mu L, Li J et al (2020) Early prediction of the 2019 novel coronavirus outbreak in the mainland China based on simple mathematical model. Ieee Access 8:51761–51769

Zhou C, Su F, Pei T et al (2020) COVID-19: challenges to GIS with big data. Geogr Sustain 1 (1):77–87

Zwitter A, Gstrein OJ (2020) Big data, privacy and COVID-19–learning from humanitarian expertise in data protection

Chapter 2
An IOT-Based COVID-19 Detector Using *K*-Nearest Neighbor

T. M. Okediran, O. R. Vincent, A. A. Abayomi-Alli, and O. J. Adeniran

Abstract COVID-19 is rampaging the world, increasing medical emergencies, imposing a high cost on every Nation's expenditure without minding the budget, and causing a continuous rise in the death rate. It has come to stay and live with people in the world. The cure is no longer the case, but how to manage it is now the fact. It is essential to save time, cost of running tests and test kits, cost of purchasing vaccine, create awareness in public places, decongest the isolation centers, save time to get test results, and provide mobility advantage for testing people anywhere and everywhere. This model will quickly detect and report COVID-19 symptoms on patients with cost-effectiveness to bring down the rising curve. It presents an Internet of Things (IOT) based COVID-19 detector that lowers the cost of testing by using machine learning techniques for easy and timely detection of Covid-19 symptoms in a patient. The device integrates an Infrared (IR) camera, Infrared skin thermometer, IR stethoscope, IR sphygmomanometer for blood pressure, a liquid crystal display screen to show the result, a Buzz alarm and its database, hosted in the cloud and connected via the wireless network. The training and test data gives an accuracy of 0.995 and 0.996, respectively, by using the *k*-nearest neighbor's classifier. It shows that the model performs well on training and test data without overfitting. Nine patients were further tested with the model, four were reported positive by this model, five returned negative, and the maximum time taken to complete the check was 28 s, while the minimum time was 20.5 s, which shows that the device is time-efficient and can be used where large numbers of people are expected.

Keywords Internet of things · Cost · IR · Corona virus · Symptoms · *k*-nearest neighbor

---

T. M. Okediran (✉) · O. R. Vincent · A. A. Abayomi-Alli
Department of Computer Science, Federal University of Agriculture, Abeokuta, Nigeria

O. J. Adeniran
Department of Mathematics, Federal University of Agriculture, Abeokuta, Nigeria

S. A. Hassan et al. (eds.), *Decision Sciences for COVID-19*, International Series in Operations Research & Management Science 320,
https://doi.org/10.1007/978-3-030-87019-5_2

2.1 Introduction

The coronavirus, fondly called COVID-19, is one deadly disease that broke out in Wuhan City of China in December 2019 and spreading through the entire nations of the world (Waris et al. 2020; Javaid and Khan 2021) which include Italy, Germany, the United Kingdom, the United States of America, Nigeria, Ghana, Spain, India, and many other nations totaling 223 countries with 170,051,718 confirmed cases in the world as on May, 31, 2021, 02:48 pm CEST, including 3,540,437 deaths and 1,579,416,705 vaccines doses have been administered as reported by the world health organization (WHO). Initially, the virus had been said to have no cure nor any successfully tested vaccines (Singh et al. 2020). However, many proposed vaccines that went through testing and trial have been proven to be adequate as a preventive measure or curative by medical research experts. Nevertheless, COVID-19 is still affecting our wellbeing and now coming with different variants that still requires that research be done to bring down the rise and its occurrence to a halt and reduce the rising cases around the world despite the discovery of several vaccines like the Oxford AstraZeneca vaccine and Johnson & Johnson vaccine. This research work is aimed at providing an Internet of Things (IOT) based device that reduces the cost of running tests and buying testing kits using machine learning to quickly detect and report the symptoms of COVID-19 on any carrier entering a House, an Elevator, Hall, Car, Train, and any other places where more than a person can be found.

The Virus is a class of other viruses that comes with symptoms like pneumonia, dry cough, fever, headache, lung infections and difficulty in breathing, tiredness, sore throat, diarrhea, and nasal congestions, among others (Adhikari et al. 2020a, b; WMHC 2020). Various drugs had been used in the treatment of COVID-19, and the main ones are Chloroquine, Remdesivir, lopinavir/Ritonavir, Oseltamivir, and Favipiravir before the discovery of vaccines. Since the virus affects the whole World, vaccines and new curative antiviral drugs have now been administered to end the pandemic. Many of the existing vaccines still have their side effects on patients of various immune systems and underlying health issues, which does not make the vaccine a total solution (Baden et al. 2021; Huang et al. 2020). From this view, large-scale observational studies are needed. The World Health Organization (WHO) coined the word 2019 novel coronavirus (COVID-19), referring to a coronavirus that infected the lower respiratory tract of patients with pneumonia in Wuhan, China, on December 29, 2019, and announced that the official name of the 2019 novel coronavirus was coronavirus disease (COVID-19) (Bilgin et al. 2020; Nigerian Center For Disease Control (NCDC) 2020; Shi et al. 2020).

The World Health Organization, together with other medical personnel, had stated that the early detection of COVID-19 could be easily managed, and it will reduce the risk of damage done to human organs, reduce the spread, lower death rate, and also bring down the rising curve of the outbreak and its effects in the society. Therefore, the proposed device will be programmed to detect a carrier of the Virus using a trained dataset with machine learning technique, and an alarm system is attached to alert everyone in the environment of the evolving danger by installing the device in any Private and Public transport system, Elevator, Hall, Home, and many more. The general control method popularly accepted had been social distancing, stay at home, face and nose masks, hand washing, and hand sanitizers.

The device will be used for early detection of the disease, which will serve as a preventive measure to curb the spread of the disease (Shen et al. 2020; Mydukuri et al. 2021), because prevention is better than cure and the effect of the aftermath of this pandemic in our societies, economy, educational system, religious activities, and business perspective is not palatable showing that the spread should not be encouraged. Nigeria and other nations of the world are presently facing the considerable cost attached to the purchase of test kits, running tests, limited space in isolation centers, cost of maintaining the infected patient, cost of purchasing vaccines, limited supply of vaccine and time to run the test. Therefore, early detection using electronic and computerized devices will provide a better means of overcoming the challenges imposed on the world's nations by COVID-19.

2.2 Related Work

Researchers all over the world in different fields are devoting and contributing efforts in providing the solution to control the spread of the COVID-19 and provide vaccines to cure the infected people around the world; some of these include: the prevention and control measure proposed by Adhikari et al. (2020a, b), stating prevention, control strategies and methods are reported at three levels: national level, case-related population level, and general population level. Considering the national level, the National Health Commission of the People's Republic of China (NHCPRC 2020) issued the "first announcement" on January 20, 2020, with the official inclusion of coronavirus into the management of class B legal infectious diseases; it also made class A infectious diseases' preventive and control measures implementable (Hamid et al. 2020). Based on the first announcement, the medical institution can use isolation treatment and observation methods to prevent and control the spread of COVID-19 (Jin et al. 2020a, b; Wei and Ren 2020).

The research conducted by Kamal et al. (2020), recommended the use of IOT devices to meet COVID-19 in detection and treatment. The IOT has been a useful tool in the medical field, as it comes up with the ingredients needed to help the countries in minimizing the effect of COVID-19. It has a scalable network with the potential to deal with huge amount of data received from sensors used by a number of applications to fight against COVID-19 and its reliability decreases the delivery time of crucial information which can help in providing timely response during the global pandemic of COVID-19. However, the paper strongly discussed the effectiveness of IOT in combating the outbreak of COVID-19 using several case studies and recommended the use of IOT as digital health care, Telecare and smart ambulance.

Ali (2020) proposed an IOT-based smart solution for early detection of COVID-19 Patients; this work itemizes the risk of COVID-19 in society and the aftermath effect on the affected areas of the world. It uses IOT sensors and smartphones to collect clinical data like heart rate, temperature, and cough sound. The collected data was analyzed using the random forest algorithm.

Mavrikou et al. (2020) conducted medical research and developed a Portable, Ultra-Rapid, and Ultra-Sensitive Cell-based Biosensor that can directly detect SARS-CoV-2 S1 Spike Protein Antigen. The work uses a biosensor based on membrane-engineered mammalian cells to bear the human chimeric spike S1 antibody. It further demonstrated protein attachment to the membrane-bound antibodies resulted in a selective and considerable change in the cellular bioelectric properties measured employing a Bioelectric Recognition Assay. The biosensor provided results in an ultra-rapid manner within a space of three minutes (3 min), having a detection limit of 1 fg/mL, semi-linear range of response between 10 fg and 1 μg/mL.

Chamola et al. (2020) examined the use of IOT technologies to prevent the impact of the Covid-19 outbreak by stating the opportunities of different IOT devices such as smart wearables, medical drones, medical robots, smart thermometers and smart ambulances. The work also described the challenges and all possible solutions of different technologies which can be used in the covid-19 pandemic.

The research work of Jin et al. (2020a, b), called rapid advice guidelines for the diagnosis and treatment of 2019 novel coronavirus (2019-nCoV) infected pneumonia. It was stated that the detection of COVID-19 on the patient is basically on clinical examination, physical examination, imaging examination, which is not a measure to quickly alert neighbors of the presence of someone with COVID-19 symptoms in a House, Car, or any social gathering which is dangerous to everyone.

Hossen and Ali (2014) introduced an automatic heart rate counting system from the fingertip based on a microcontroller. It uses an infrared (IR) light-emitting-diode (LED) and a photodiode sensor unit. The IR diode transmits an infrared light into the fingertip through the sensor unit. The photodiode senses the reflected light, and the intensity of reflected light is proportional to fingertip blood volume. Therefore, the heartbeat rate varies with the amount of reflected infrared light detected by the photodiode. The pulse output magnitude from the photodiode is too small for the microcontroller to sense directly. For this reason, an Operational Amplifier (Op-Amps) is designed with a two-stage high gain, active low pass filter, which can filter and amplify the signal to appropriate voltages so that the pulses can be counted by a microcontroller to determine the heart rate and displayed on the LCD screen.

This work is organized into seven sections: Sect. 2.1 is the introduction, Sect. 2.2 provides related works, while Sect. 2.3 establishes the proposed model and methods. Section 2.4 provides information about *k*-NN algorithm, Sect. 2.5 presents the various classes of COVID-19 symptoms, Sect. 2.6 shows the discussion and results, respectively. Section 2.7 provides the conclusion of this work.

2.3 The Proposed Model Using *k*-Nearest Neighbor

This research will iterate the various COVID-19 symptoms and write an algorithm to detect and report vital signs of the body that can reveal the presence of these symptoms employing an alarm system connected to the device and a digital display screen to show the readings of the vital signs with the database stored in the cloud using wireless internet connection. The model is trained to report a suspected case of COVID-19 if at least two-thirds (2/3) of these symptoms on any patient are detected at any time. Such patients will be isolated immediately and reported to the proper authority for proper testing. The false alarm rate is $\leq 10\%$. This model is made of an Infrared (IR) camera, Infrared skin thermometer, IR stethoscope, IR sphygmomanometer for blood pressure, liquid crystal display screen, and Buzz alarm system connected to a database that is hosted in the cloud (see Figs. 2.1 and 2.2). The value

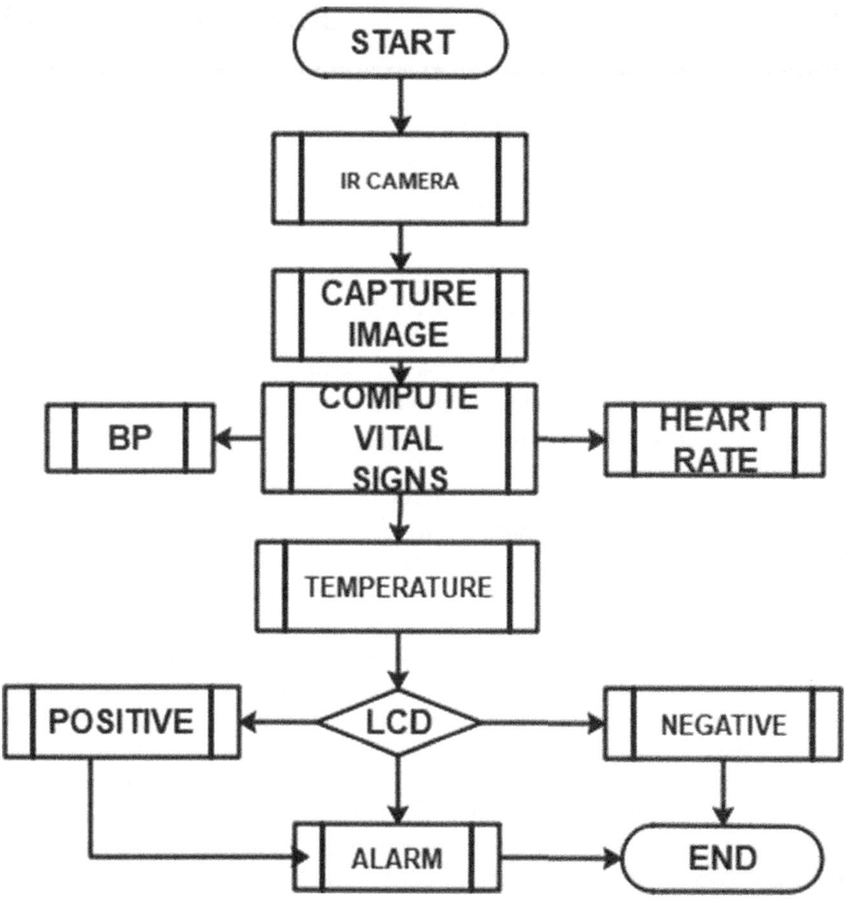

Fig. 2.1 Flowchart for the proposed model

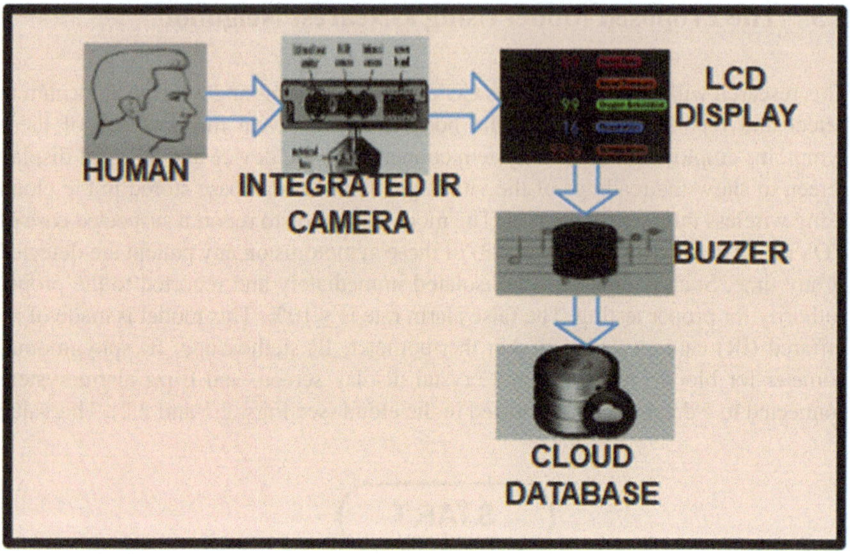

Fig. 2.2 The architecture for the proposed model

of all symptoms is learned and programmed to the range, and the time taken by the model to detect, analyze, and report is also recorded. The model uses a supervised machine learning technique and the *k*-nearest neighbor (*k*-NN) machine learning algorithm to classify, train, and test the data set. The essence of supervised learning is to predict outputs from any given input data, such as vital signs, which predicts the symptoms of COVID-19. The *k*-NN enables the prediction of COVID-19 on patients by looking for the closest training data point to the new data point when patients are scanned, and vital signs are taken, which predicts COVID-19 symptoms before laboratory tests.

The training data set contains five features: Age, Blood Pressure (BP), Body Temperature (T), Pulse, and Heartrate (HR), to build a model that performs well on our test data. The IR camera tracks and detects a patient's face automatically as they come in contact with the camera through a face detection algorithm using sliding windows and gathers temperature information through the IR image. The patient will not have to stay in front of the camera all through the time of analysis because, the camera will capture the face within five seconds (5 s) of close contact and store the image for further processing. The IR sphygmomanometer will monitor blood pressure continuously using an IR lighting source and a luminosity sensor to measure the blood pressure without using a cuff. By the optical pulsatile method, blood pressure measurement is gathered from various parts of the body. Numerical readings of systolic and diastolic blood pressure are recorded and displayed on the screen. Since the heart rate reading is used to determine the heart's soundness and assess the state of the cardiovascular system of humans, a heart rate monitor is designed to take samples of heartbeats and compute the beats per minute (bpm) through an IR signal. The resulting values are used to determine the proper heart condition, as show in Fig. 2.2 of the architecture for the proposed model.

Electrical and optical methods are two ways to build heart rate monitoring systems. The electrical method has proven to return a 1% error, and the optical method is accurately rated to be 15%. The human heart rate at rest is about 70 bpm when the reading is regular for adult males and 75 bpm for adult females, while others may vary based on age, fitness, and genetics. Heart rate can be directly measured using pulse through specialized medical devices or merely pressing a finger against the artery. Generally, listening to heartbeats using a stethoscope is a conventional and more accurate means of measuring the heart rate. There are many other methods to measure heart rates like Phonocardiogram (PCG), ECG, blood pressure waveform, and pulse meters (Thoppil and Thomas 2016).

2.4 The *k*-Nearest Neighbor (*k*-NN)

This classification technique is used for modeling a classification problem in machine learning and data mining applications with outstanding performance, low computational cost and objective to predict the label of the test data point, which is mainly classification (Zhang et al. 2017; Al-Dosary et al. 2019). The label of the test data point will be predicted by the value of *k*, which is most similar in the feature space (Cheng et al. 2017); the value of *k* and the distance between neighbors influence the performance of the *k*-NN algorithm. The problem of missing value is visible in the real world, and data mining or classification problem, *k*-NN also dealt with such missing value problem by providing missing value input. The ideology behind the *k*-NN is near the red, near the blue by your neighbor so that your category can be choosing depending on the distance. To use the *k*-NN algorithm, the dataset should be divided into two subsets: the training dataset, on which the algorithm bases its predictions, and the testing dataset, which is used to test the algorithm's performance on new data set (Imandoust and Bolandraftar 2013). The training dataset is further divided into vectors, while, for each point in the testing dataset, the distance from the data point to its neighbors in the training dataset is calculated using the Euclidean distance equation as shown in Eq. (2.1). When *k* is selected, predictions will be made using Eq. (2.2) based on the average from the outcomes of *k*-NN.

$$S = \sqrt{}(x_0 - y_0)^2 + (x_1 - y_1)^2 + (x_2 - y_2)^2 + \cdots (x_{n-1} - y_{n-1})^2 \tag{2.1}$$

Here, *x* and *y* are different points on the data set known as neighbors, and *S* is the Euclidean distance between the two points.

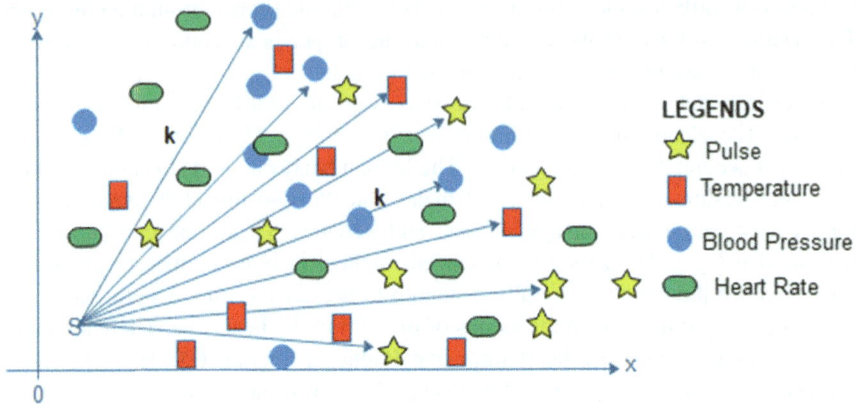

Fig. 2.3 *k*-NN neighbors showing the different vital signs

$$y = \frac{1}{k}\sum_{i=1}^{k} y_i \tag{2.2}$$

where i is the sample data and y is the prediction point.

Therefore, in this work, the temperature, blood pressure, heart rate, and pulse values used to train the data set will be compared to all input data sets to classify a patient category as positive or negative to COVID-19. The value of k used is 5, where k is the number of neighbors. Figure 2.3 shows the distance between neighbors where the neighbors are represented with the vital sign.

2.5 Classification of the COVID-19 Symptoms

According to WHO and the Nigerian Center for Disease Control (NCDC), some prevalent symptoms of COVID-19 include fever, tiredness, high temperature and dry cough. Some patients may have aches and pains, nasal congestion, runny nose, sore throat, or diarrhea (Fisher and Heymann 2020; Satu et al. 2021). These symptoms are usually mild at the early stage and rise steadily; many people get infected and may not develop symptoms. It was also noted that few out of many who get infected with COVID-19 become seriously ill and develop difficulty in breathing. Older people, and those with other critical medical problems like high blood pressure, heart problems, diabetes, are more likely to develop severe illnesses. The time between catching the virus and the patient's time begins to have symptoms of the disease; the patient still moves about. Most estimates of the incubation period for COVID-19 range from 1 to 14 days and some within 5 days (Fisher and Heymann 2020; Vinod and Mahadevaswamy 2018).

Table 2.1 Stages of Covid-19 and vital sign range

Stages	Classes	Degree Celcius (0 °C)	Degree Fahrenheit (0 °F)	Age (years)	Regular heart rate (BPM)
Hypothermia		<35.0	95.0	Above 150	1–2
Middle-stage	Normal	36.5–37.5	97.7–99.5	95–140	2–5
Advance stage	Fever or hyperthermia	>37.5 or 38.3	99.5 or 100.9	80–120	6–12
Chronic stage	Hyperpyrexia	>40.0 or 41.5	104.0 or 106.7	60–100	20–40

The symptoms of COVID-19 can be of different stages based on the patients' reaction and the number of days the infection has stayed on the affected person. The stages are summarized as follows (Wei and Ren 2020; Hashem et al. 2010).

- Infant-Stage: It is a stage where the patient experiences cold-dampness to be obstructing the lung. The manifestations include cold with or without fever, dry cough, dry throat, fatigue, and hydrodynamic oppression in the chest, epigastric fullness, nausea, loose stool, etc.
- Middle-Stage**:** It is a stage when there is an epidemic toxin causing the blockage in the lung. At this stage, the symptoms include persistent fever or alternating cold and heat, cough with less phlegm or yellow phlegm, abdominal distension and constipation; oppression in the chest with annihilation, cough with wheezes, panting on exertion; or red tongue, slimy yellow fur or dry yellow fur, slippery and rapid pulse.
- Advance Stage**:** It is characterized as heat toxin generating stasis with medical diagnosis known as high fever, oppression in the chest with annihilation, purple-black facial complexion, dark swollen lips, obnubilation, crimson tongue, dry yellow fur, surging and slight, rapid string-like pulse.
- Chronic Stage**:** This is characterized as inner blocking, causing collapse. It comes with dyspnea, panting on exertion, or need to assist ventilation, accompanied by coma and agitation, cold limbs and sweating, dark purple tongue, thick dry tongue fur, floating, and rootless pulse. Table 2.1 present a temperature level of the four stages and the Heart rate with conditions.

The heart rate and blood pressure range for every individual is also an indication of the state and health condition of that individual. It can vary based on age, activities, and body weight. Table 2.2 presents the blood pressure range of any patient (Baby et al. 2016; Karambelkar et al. 2017; Somboonkaew et al. 2017; Zneid et al. 2014).

Table 2.2 Blood pressure analysis

Blood pressure category	Systolic (mm Hg)	Diastolic (mm Hg)
Normal	<120	<80
Prehypertension	120–139	80–89
Hypertension (high)	140–159	90–99
Hypertension (very high)	≥160	≥100
Critical hypertensive	≥180	≥110

2.6 Discussion and Result

This work introduces an IOT solution with machine learning to quickly detect a patient with COVID-19 symptoms. It is useable anywhere at any time, especially in the Car, Train, and Elevator, to ensure that the cost of running tests and purchasing test kits is lowered to reduce the expenditure of the government especially, the cost of vaccines in the recent time. To decongest the various isolation centers because they are limited, especially in Nigeria where the discovered cases are still rising due to the movement of people in and out of the country with the low level of compliance with covid-19 government travel rule. It is also essential to provide test result as quick as possible because there are so many patients to be tested, and the testing centers are limited in number, resources, and staff thereby increasing the time to carry out the test and produce a result that can be resolved using IOT smart devices. The model is divided into four stages, which include detections of IR images by the IR camera. The images will be examined to get vital signs, which is very important as a determinant of fever and some other body functions, although the person will not have to stay in front of the camera until image analysis is done since it has been captured; it will further report on the LCD screen the result obtained from various vital signs such as Temperature, Blood Pressure, Heart Rate, Pulse among others as shown in Fig. 2.2. The Alarm is activated when the report is positive to show that the patient is suspected of COVID-19 based on the fact that two-third of the symptoms are found positive on such a patient. Results are stored in the database and hosted in the cloud. There is a need to isolate the patient for further laboratory examinations. There are five features and 230 observations on the training dataset (230 rows × 5 columns) which contains the vital signs of patients. Data was collected at Goodness Medical Center in Lagos, Nigeria. The data was standardized and splitted into the training (X_train, y_traing) and testing (X_test, y_test) dataset with dimensions (172, 5) (28, 5), respectively, as shown in Fig. 2.4.

The model was built with default parameters of the *k*-NN algorithm from the sklearn neighbors and setting $k = 5$, fitted the model and check the accuracy level of the training and testing for our model. It was observed that the training accuracy was 0.995, and test accuracy was 0.996. Although, there was no overfitting on this model, showing that the model performed so well on the training dataset and was able to also perform on the test data. It was able to memorize the pattern and compare an input during testing, meaning that our model can perform well for the prediction of the presence of any patient having the following symptoms based on vital signs.

```
In [12]: import pandas as pd
         from sklearn.neighbors import KNeighborsClassifier
         df= pd.read_excel(r"e:\TESTDATA.xlsx")
         df

Out[12]:
```

	TEMPERATURE (0°C)	SYS. B.P	DIAL B.P	HEART RATE	PULSE
0	38.6	150.0	100.0	24	70
1	35.4	120.0	100.0	20	60
2	36.8	140.0	100.0	18	60
3	37.2	130.0	100.0	20	50
4	36.8	140.0	100.0	20	60
...	...	...	...	...	...
225	36.0	120.0	98.0	20	80
226	36.0	100.0	79.0	20	80
227	36.0	100.0	80.0	20	80
228	36.6	110.0	110.0	20	80
229	36.1	1409.0	100.0	20	80

Fig. 2.4 Dataset of patients for the proposed model

When two out of three of the various symptoms appeared, the implication is that the patient would require a medical check to ascertain the infection of COVID-19 by taking blood samples to conduct a laboratory test. The model detects vital signs, such as body temperature, respiration rate, and resting heart rate, among others, and the temperature range is between 32.7 °C and 37 °C (Somboonkaew et al. 2017; Ghassemi et al. 2018; Ibrahim et al. 2021). If the temperature exceeds this specified range, it confirms fever's presence, and the alarm is turned on. The appropriate respiration rate for an adult at rest is 12–20 breaths per minute. A respiration rate under 12 or over 25 breaths per minute at rest is considered abnormal, and an average resting heart rate for adults ranges from 60 to 100 beats per minute. Generally, a lower heart rate at rest implies more efficient heart function and better cardiovascular fitness (Mardani et al. 2020; Dakappa et al. 2017; Hossen and Ali 2014). When the result reveals at least 2/3 of all the symptoms to be positive, the alarm is turned on. Otherwise, the process is terminated.

The number of cases is still increasing rapidly as on February 1, 2021, across the various states in Nigeria, as shown in Table 2.3, and that is why different measures must be taken to curb the spread of COVID-19 because it is affecting other nations like India even with the discovery of the vaccine, we now have the second wave with scientist still predicting the third wave of COVID-19. The graph is plotted to depict the effects and increasing curve of the spread of COVID-19. The use of the proposed

Table 2.3 A glance view of COVID-19 condition in Nigeria as at first February, 2021

Total number of 1,302,410 samples tested	
Total number of confirmed cases	131,242
Total number of active cases	24,667
Number of discharged cases	104,989
Number of death	1586

Table 2.4 Result of our model

S/N	Age	Gender	Time to complete check (seconds)	Results
1	23	Male	22.0	Positive
2	35	Female	25.1	Positive
3	42	Male	20.5	Negative
4	46	Female	28.0	Positive
5	52	Male	21.0	Negative
6	40	Female	22.3	Negative
7	53	Male	26.2	Positive
8	30	Female	24.4	Negative
9	61	Male	26.1	Negative

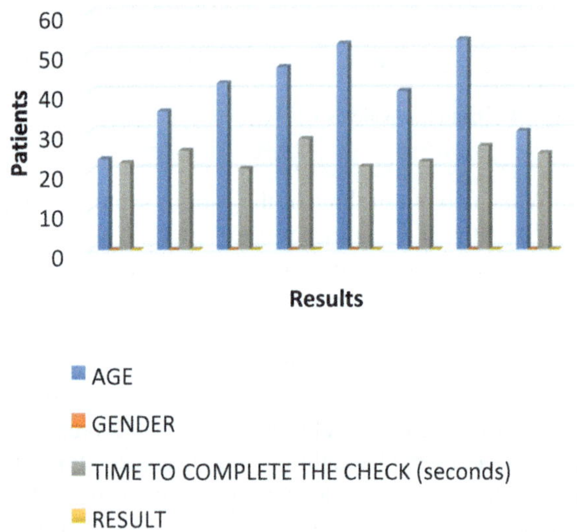

Fig. 2.5 Showing the result of the tested patient using the proposed model

model to alert people in a place shows that the spread of this virus can be controlled easily, and the death rate will reduce because of early detection and control measures that can be aided by the system. It was observed that Lagos had the highest number of cases (active, discharged, and death), which is 50% of the total cases in Nigeria. Table 2.4 shows the result of our model and the graph shown in Fig. 2.5. The graph in Fig. 2.6 shows the training and testing accuracy for the model, while Fig. 2.7 is the histogram of the visualization for the model.

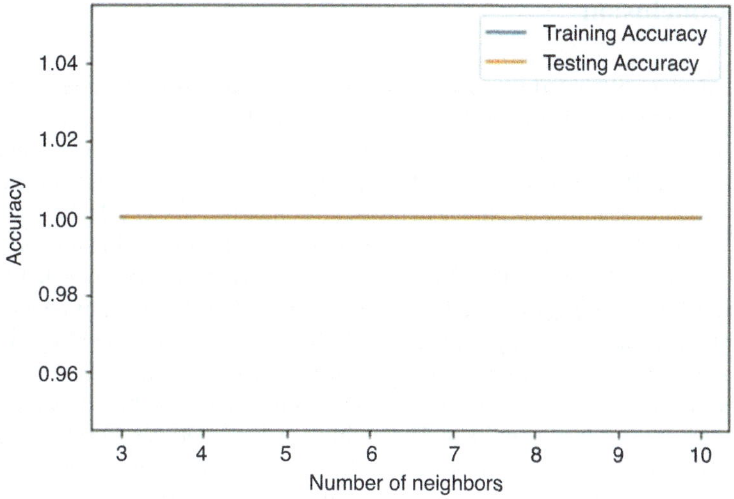

Fig. 2.6 Showing accuracy of the model with 10 neighbors

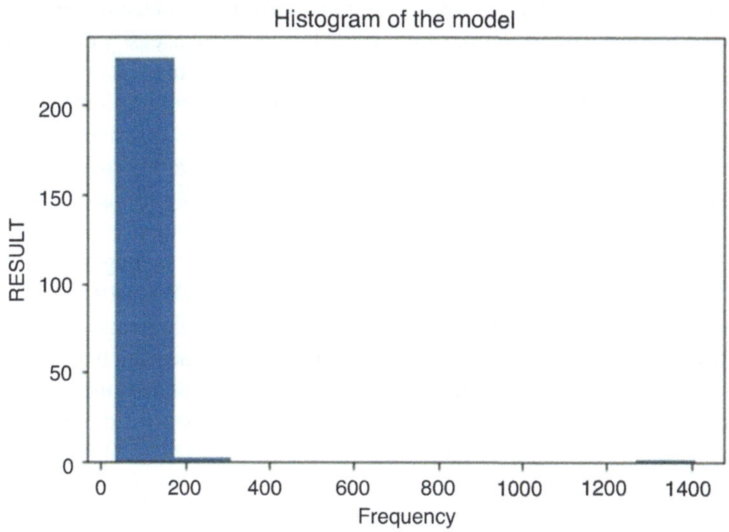

Fig. 2.7 Histogram of the model for training and testing data

The proposed model had revealed that the control of this pandemic would effectively provide a result that can stop the spread of COVID-19 in our society based on data obtained. The result obtained was able to report the presence of fever on four patients out of nine that were tested for COVID-19 using the symptoms stated in this work, and the analysis is given in Table 2.4. The four patients were further tested using the COVID-19 test kit in the laboratory by taking blood samples.

2.7 Conclusion

This study has presented a model called an IOT-based cost-effective COVID-19 detector using a machine learning technique as a technological tool for easy detection of COVID-19 symptoms to control the spread of the virus in Nigeria and other nations. The method was an easy and fastest way to bring down the rising curve by using a machine learning technique to predict the presence of COVID-19 in a patient at any public places such as School, Church, Train, Elevator, and many more. The model uses the *k*-NN algorithm for prediction, and samples were collected to train the model. The method introduced in this work showed that four out of nine samples tested positive based on body temperature, heartbeat rate, pulse, and blood pressure readings that rise above average. The abnormal body temperature revealed fever, abnormal heartbeat rate also calls for medical attention, and the high blood pressure shows a critical medical condition. The training and test data produced an accuracy of 0.995 and 0.996, respectively, by using the *K*-nearest neighbors. It shows that our model performs well on training and test data without overfitting. The maximum time taken to complete the check was within 28 s, and the minimum time was 20.5 s, which showed that the device is time-efficient and can be used where large numbers of people are expected. The device can be moved from one place to another, and it can also be fixed in a particular position where the camera can easily detect an image and must be within range. The system operation is in five main stages, the camera detects image through the IR camera, captures the image, computes vital signs, show the result on the LCD screen, and turns on the alarm if it is positive. Otherwise, it terminates and resets. The advantages of this work are: to safe the cost of purchasing the test kit and running the test, high cost and limited supply of vaccine due to the financial situation of the country, especially in Nigeria where people pay about seventy thousand naira ($140) to run the COVID-19 test in Lagos state on one patient. To quickly and easily detect the symptoms of COVID-19 in any patient where ever we go and to combat the spread in Nigeria and other nations of the world motivated this research. However, it can be managed, and knowing that the virus has come to stay in the world, awareness must continue using different methods. Several works done on COVID-19 did not rely on IOT to combat the spread of the virus and machine learning techniques to model the symptoms as provided in this work, thus, making this model better research. To further improve this work, the model can be built into a machine installed in a car, train, automated doors, and more. However, this work's challenges were access to the test kit, the hostility of patients for data collection, test laboratory where the blood samples of all the 230 patients can be further examined for COVID-19 and the cost of running the test because there was no research grant or funding for this research. Therefore, only nine (9) samples were further examined.

References

Adhikari SP, Meng S, Wu Y, Mao Y, Ye R, Wang Q, Zhou H (2020a) A literature review of 2019 Novel Coronavirus during the early outbreak period: Epidemiology, causes, clinical manifestation and diagnosis, prevention, and control 7(6):35–42

Adhikari SP, Meng S, Wu YJ, Mao YP, Ye RX, Wang QZ, Sun C, Sylvia S, Rozelle S, Raat H, Zhou H (2020b) Epidemiology, causes, clinical manifestation and diagnosis, prevention and control of coronavirus disease (COVID-19) during the early outbreak period: a scoping review. Infect Dis Poverty 9(1):1–12

Al-Dosary NMN, Al-Hamed SA, Aboukarima, a.m. (2019) K-nearest neighbors method for prediction of fuel consumption in tractor-chisel plow systems. Engenharia Agrícola 39 (6):729–736

Ali M (2020) IoT-based smart solution to early detect COVID-19 patients. Masters of Computer Science and Engineering, United International University

Baby A, Sulochana B, Latha T (2016) Comparison of body temperature of elderly with adult in South India. Manipal J Nurs Health Sci 2(1):25–30

Baden LR, El Sahly HM, Essink B, Kotloff K, Frey S, Novak R, Diemert D, Spector SA, Rouphael N, Creech CB, McGettigan J (2021) Efficacy and safety of the mRNA-1273 SARS-CoV-2 vaccine. N Engl J Med 384(5):403–416

Bilgin S, Kurtkulagi O, Kahveci GB, Duman TT, Tel BMA (2020) Millennium pandemic: a review of coronavirus disease (COVID-19). Exp Biomed Res 3(2):117–125

Chamola V, Hassija V, Gupta V, Guizani M (2020) A comprehensive review of the COVID-19 pandemic and the role of IoT, drones, AI, blockchain, and 5G in managing its impact. Ieee Access 8:90225–90265

Cheng D, Zhang S, Liu X, Sun K, Zong M (2017) Feature selection by combining subspace learning with sparse representation. Multimedia Systems 23(3):285–291

Dakappa PH, Prasad K, Rao SB, Bolumbu G, Bhat GK, Mahabala C (2017) A predictive model to classify undifferentiated fever cases based on twenty-four-hour continuous tympanic temperature recording. J Healthcare Eng 2017:5707162

Fisher D, Heymann D (2020) Q&A: the novel coronavirus outbreak causing COVID-19. BMC Med 18(1):1–3

Ghassemi P, Pfefer TJ, Casamento JP, Simpson R, Wang Q (2018) Best practices for standardized performance testing of infrared thermographs intended for fever screening. PLoS One 13(9): e0203302

Hamid S, Mir MY, Rohela GK (2020) Noval coronavirus disease (COVID-19): a pandemic (Epidemiology, Pathogenesis and potential therapeutics). New microbes and new infections, p 100679

Hashem MMA, Shams R, Kader MA,Sayed MA (2010, May) Design and development of a heart rate measuring device using fingertip. In International conference on computer and communication engineering (ICCCE'10) (pp 1–5). IEEE

Hossen MB, Ali MZ (2014) A microcontroller-based automatic heart rate counting system from fingertip. J Theor Appl Inf Technol 62:3

Huang AT, Garcia-Carreras B, Hitchings MD, Yang B, Katzelnick LC, Rattigan SM, Cummings DA (2020) A systematic review of antibody mediated immunity to coronaviruses: kinetics, correlates of protection, and association with severity. Nat Commun 11(1):1–16

Ibrahim AU, Ozsoz M, Serte S, Al-Turjman F, Yakoi PS (2021) Pneumonia classification using deep learning from chest X-ray images during COVID-19. Cogn Comput:1–13

Imandoust SB, Bolandraftar M (2013) Application of k-nearest neighbor (knn) approach for predicting economic events: theoretical background. Int J Eng Res Appl 3(5):605–610

Javaid M, Khan IH (2021) Internet of Things (IoT) enabled healthcare helps to take the challenges of COVID-19 Pandemic. J Oral Biol Craniofac Res 11(2):209–214

Jin H, Hong C, Chen S, Zhou Y, Wang Y, Mao L, Li Y, He Q, Li M, Su Y, Wang D (2020a) Consensus for prevention and management of coronavirus disease 2019 (COVID-19) for neurologists. Stroke Vasc Neurol 5(2):146–151

Jin YH, Cai L, Cheng ZS, Cheng H, Deng T, Fan YP, Fang C, Huang D, Huang LQ, Huang Q, Han Y (2020b) A rapid advice guideline for the diagnosis and treatment of 2019 novel coronavirus (2019-nCoV) infected pneumonia (standard version). Mil Med Res 7(1):1–23

Kamal M, Aljohani A, Alanazi E, (2020) IoT meets COVID-19: status, challenges, and opportunities. arXiv preprint arXiv:2007.12268

Karambelkar K, Balkawade S, Kapgate J, Adoni K (2017) Non-invasive optical blood pressure sensing and measurement. Int J Eng Trends Technol 46(5)

Mardani R, Vasmehjani AA, Zali F, Gholami A, Nasab SDM, Kaghazian H, Kaviani M, Ahmadi N (2020) Laboratory parameters in detection of COVID-19 patients with positive RT-PCR; a diagnostic accuracy study. Arch Acad Emerg Med 8(1)

Mavrikou S, Moschopoulou G, Tsekouras V, Kintzios S (2020) Development of a portable, ultra-rapid and ultra-sensitive cell-based biosensor for the direct detection of the SARS-CoV-2 S1 spike protein antigen. Sensors 20(11):3121

Mydukuri RV, Kallam S, Patan R, Al-Turjman F, Ramachandran M (2021) Deming least square regressed feature selection and Gaussian neuro-fuzzy multi-layered data classifier for early COVID prediction. Exp Syst:e12694

National Health Commission of People's Republic of China (2020) Notice on prevention and control of novel coronavirus infection pneumonia in the elderly people. http://www.nhc.gov.cn/lljks/tggg/202001/96e82ba8a14d41b283da990d39771493.shtml. Accessed 31 Jan 2020

Nigerian Centre for Disease Control (2020) Symptoms and causes of COVID-19. https://covid19.ncdc.gov.ng

Satu M, Howlader KC, Mahmud M, Kaiser MS, Shariful Islam SM, Quinn JM, Alyami SA, Moni MA (2021) Short-term prediction of COVID-19 cases using machine learning models. Appl Sci 11(9):4266

Shen K, Yang Y, Wang T, Zhao D, Jiang Y, Jin R, Zheng Y, Xu B, Xie Z, Lin L, Shang Y (2020) Diagnosis, treatment, and prevention of 2019 novel coronavirus infection in children: experts' consensus statement. World J Pediatr 16(3):223–231

Shi H, Han X, Jiang N, Cao Y, Alwalid O, Gu J, Fan Y, Zheng C (2020) Radiological findings from 81 patients with COVID-19 pneumonia in Wuhan, China: a descriptive study. Lancet Infect Dis 20(4):425–434

Singh SK, Jain SK, Sharma SN (2020) Prevention and control of COVID-19 in Indian perspective. Epidemiol Int (E-ISSN: 2455-7048) 5(1):32–38

Somboonkaew A, Prempree P, Vuttivong S, Wetcharungsri J, Porntheeraphat S, Chanhorm S, Sumriddetchkajorn S. (2017). Mobile-platform for automatic fever screening system based on infrared forehead temperature. In 2017 Opto-electronics and communications conference (OECC) and photonics global conference (PGC), pp 1–4

Thoppil VT, Thomas A (2016) A non invasive heart rate and blood pressure remote monitoring system. Int J Recent Innov Trends Comput Commun 4(6):582–585

Vinod KP, Mahadevaswamy UB (2018) Unilateral vital signs monitoring systems on IoT. Int J Eng Manuf 8(1):53

Waris A, Khan AU, Ali M, Ali A, Baset A (2020) COVID-19 outbreak: current scenario of Pakistan. New microbes and new infections, p 100681

Wei Q, Ren Z (2020) Disinfection measures for pneumonia foci infected by novel coronavirus in 2019. Chin J Disinfect 37:59–62

WMHC (2020) Wuhan Municipal Health and Health Commission's Briefing on the Current Pneumonia Epidemic Situation in Our City. http://wjw.wuhan.gov.cn/front/web/showDetail/2019123108989. Accessed 1 Feb 2020

Zhang S, Li X, Zong M, Zhu X, Cheng D (2017) Learning k for knn classification. ACM Trans Intell Syst Technol (TIST) 8(3):1–19

Zneid BA, Al-Zidi M, Al-Kharazi T (2014, April) Non-invasive blood pressure remote monitoring instrument based microcontroller. In 2014 IEEE region 10 Symposium. IEEE, pp 248–253

Chapter 3
Predictive Analytics for Early Detection of COVID-19 by Fuzzy Logic

V. Kakulapati, R. Sai Sandeep, and V. Kranthikumar

Abstract Today entire world is struggling, and significant cases are rising due to Coronavirus, namely COVID-19. Healthcare providers are busy in clinical trials to investigate the vaccine for this pandemic. If this virus attacks the person, nobody can know that person is going to be tested positive. This virus is spreading through the droplets of one person or dirty hands. The primary task of healthcare providers is to provide diagnostic product services at low costs and accurately diagnose patients. Machine learning methods can use for disease identification because they mainly apply to data and prioritize specific tasks' outcomes. In this work, a multistage fuzzy rule-based algorithm for detection and CART algorithm is utilizing to produce the fuzzy rules. Implementation results exhibit that the proposed method differentiated the development of the disease prediction accuracy. The integration of these two techniques, multistage fuzzy rules and CART algorithms with unrelated data removal methods, could help predict disease. The proposed system can be helpful for healthcare providers in predicting the early stages of COVID-19.

Keywords Corona · CART · Multistage · Fuzzy · Prediction · Accuracy · Technique · Rules

3.1 Introduction

The majority of people diagnosed with COVID-19 are affected as on June 5, 2021, which is shown in Table 3.1 with moderate to severe lung infections and recover without particular medication. Exceedingly probable of becoming severe disease seem to be older generations and those with underlying health conditions such as lung disease, diabetes, Acute breathing Disease, and malignancy.

V. Kakulapati (✉) · R. S. Sandeep · V. Kranthikumar
Sreenidhi Institute of Science and Technology, Yamnampet, Ghatkesar, Hyderabad, Telangana, India

S. A. Hassan et al. (eds.), *Decision Sciences for COVID-19*, International Series in Operations Research & Management Science 320,
https://doi.org/10.1007/978-3-030-87019-5_3

Table 3.1 Statistics of Novel Coronavirus confirmed, recover, and critical cases

#	Country, other	Total cases	Total deaths	Total recovered	Serious, critical
	World	174,209,885	3,747,389	157,594,354	87,022
1	USA	34,211,687	612,386	28,123,111	5280
2	India	28,974,152	350,631	27,282,022	8944
3	Brazil	16,947,062	473,495	15,342,286	8318
4	France	5,712,753	109,998	5,409,110	2527
5	Turkey	5,293,627	48,255	5,167,350	1060
6	Russia	5,135,866	124,117	4.743,202	2300
7	UK	4,522,476	127,841	4.277,098	134
8	Italy	4,233,698	126,588	3,918,657	759
9	Argentina	3,955,439	81,214	3,529,033	7776
10	Germany	3,709,268	89,872	3,542,700	203

A reasonable awareness of the virus COVID-19, the autoimmune disease, and how it accelerates is using to avoid and relatively slow transmission. Secure themself as well as from contamination by cleaning hands or by consistently and use an alcohol scrub and do not touch mouth or nose. Whenever an affected person coughed, the COVID-20 virus is spreading mainly by spit particles and via the nose. But you will need breathing etiquette, for instance, by sneezing in a stretched elbow.

A variety of investigations focused on artificial intelligence (AI) intend to tackle the epidemic to assess suspected cases and anticipate the disease. Many such approaches, combined with high reliance on specific unpredictable model parameters and inconsistently observed variables, could decrease efficiency and precision as Covid-19 propagates. The usage of DL approaches for detecting cases of COVID-19 is mostly focused on X-ray images (Ozturk et al. 2020; Narin et al. 2020; Khan et al. 2020; Hashem et al. 2020), with some initiatives of been developed to use some of the valuable, informative, and relevant data in the electronic medical records process for patients. Observing that any such constraints may lead to inconsistent predicted consequences, which will often turn to harmful and inaccurate outcomes negatively influences healthcare plans' strategy and effectiveness. Recent investigations provide positive alternative detection modeling approaches to address the drawbacks of observational and AI models, acquire more precise consequences, and reduce existing constraints by incorporating CART and Fuzzy-based fast-detection strategies to analyze the emerging epidemic.

Novel coronavirus (Xu et al. 2020) can protect people from early and efficient detection to ensure the patient is directed at a reasonable time to quarantine, efficiently transmit severe cases to advanced intensive care units, and prevent disease transmission. Even if the assessment is such a rapid process, the expense of diagnostics in the nations with healthcare systems of treatment or utilization of healthcare systems is suitable for unaffordable costs to impact Governments and individuals.

During these crisis periods, researchers and scientists have received significant assistance from investigators using ML and DL algorithms to help society classify the virus inside humans. Fuzzy set theory is a classification model that has provided us a statistical approach for cognitive computing to differentiate between different forms of inconsistencies. The performance to demonstrate cognitive information in grammar is producing by brief reasoning.

A variety of methods of trying to diagnose infections (Dhiman et al. 2019) have been using to detect symptoms and outbreaks by many medical professionals. The algorithm determines whether a vaccine and disease will target the virus in the future based on the extraction parameters. The CART algorithm utilized in the classification model predicts symptoms' the probable symptoms based on training data (https://iris.paho.org/bitstream/handle/10665.2/52276/PAHOEIHISCOVID-19200007_eng.pdf?sequence=5&isAllowed=y).

Data analytics is a mathematical analysis that used the empirical calculation method to recognize behavioral trends and insights to anticipate possibilities involving knowledge extraction, machine learning, and optimization. Even if forecasting is a well-known diagnostic method, the availability of vast volumes of data, enhanced interpret of the data collected, and advanced diagnostic approaches have improved.

This work aims to forecast early detection (Kakulapati et al. 2019) of a pandemic's symptoms to avoid disease. In this, analyzed the fuzzy model to estimate the accuracy of COVID-19. Expected values are assessed with experimental values for precision to be producing. This analysis provides novel coronavirus with a professional diagnosis and medical care with better outcomes.

The remaining manuscript is describing as a review of literature on this epidemic is discussed in related works. The present state about the diagnosis of the infection is described in this section using analytics. Section 3.4 discusses the fuzzy inference and integration of the cart algorithm to forecast the classification of COVID-19 patients. Section 3.5 outlines the potential study of the COVID-19 application and early identification. Section 3.6 addresses the need to detect COVID-19 early and makes conclusions and recommendations in the subsequent sections. Section 3.8 also includes functional, analytical focus areas.

3.2 Related Work

At the critical stage of the illness, COVID-19 can recognize. In contrast, coronavirus patients isolated from SARS in the third week of disease (Chan et al. 2004) and Diagnosis of coronavirus can occur after this duration due to the apparent continuous transmission of immutable infections in these kinds of patients and the existence in bacterial species of detecting autoantibodies, avoiding disease progression in biological samples.

Infections were showing to be persistent in clinical signs of SARS-Cov-2. Covid-19 manifestations also differ slightly among regions. Intestinal symptoms are becoming more common than in China in the USA (Holshue et al. 2020). Various investigations have demonstrated symptomless, moderate, and extreme symptoms (Lu et al. 2020). No therapies are recommending for symptomless or mild conditions. Early signs were initially fever above 37.5 ° C and sore-throats and can worsen to severe hospitalized patients. In minor symptoms, temperature, sore-throat, abdominal discomfort, and inaccurate blood inflammatory markers have been recording. Extreme instances of breathing deficiency, chest pain and tachypnea and hydraulic ventilation (Guan et al. 2020) are required. The associated symptoms of medical issues or pre-existing diseases were recurrent cough, fever, and fatigue, not related to cardiovascular disorders or asthma—plasma pO_2 levels reduction. Lymphocytopenia, thrombopenia, and excessive monoterpene have been producing by plasma therapies in mild and extreme cases. In severe cases, immune cells worsened, and the ventilation system was necessary. The core characteristics in persistent with pre-existing respirational problems are constant temperature and symptom persistent cough—to begin with, dry for few days preceded by persistent coughing—and some indications have been geographically variable (Gudbjartsson et al. 2020). This research focused on the disease progression of Covid-19, which was investigated by utilizing clinical symptoms to establish a feasible method for the Diagnosis and assessment of the infection stage by using appropriate detecting and adequate measures.

ANN, K-Means Clustering and Fuzzy C Means (Chitra et al. 2013) for heart disease prediction have been classifying. The exactness of these methodologies is 85%, 88%, and 92%, respectively. Fuzzy C Means Clustering probably had the best performance that is 4% better than others at least. The neural network (NN) and fuzzy logic (Rashedur et al. 2013) have been utilizing to anticipate cardiovascular risk. NN and Fuzzy Logic precision were 79.19% and 83.85%, respectively. The Decision tree (DT), the naïve Bayes (NB) Classification (Patel et al. 2014), recommended a forecast of cardiac arrest. The performance of these methodologies is 99.2%, 96.5%, and 88.3%, respectively. Decision Tree obtained at least 2.7% more than other evaluations. CART methodology for classification for foreseeing heart attack. The method is a precision of 84.49%, and the duration of that algorithm takes 0.23 s, and the mean error is 0.3 (Chaurasia et al. 2013).

NN and Genetic algorithms (Nilakshi et al. 2014) are establishing for predicting heart disease, and accuracy is 98%, respectively. Adaptive-Fuzzy Framework for the risk of cardiac diagnostic techniques. A fuzzified, genetic algorithm-based optimization approach is utilizing to identify CAD. Maximize the model parameters, and the optimization algorithm is using and evaluated by CAD and has 88.79 percent accuracy. The prediction of cardiovascular disease is describing by the CART classifier (Suganya et al. 2016) and distinguishes data between different classes at a minimal distance. The performance of the algorithm of the CART is 83%.

Fuzzification is describing as a robust methodology for modeling uncertainty in clinical care (Zadeh 2008). In a healthcare setting, some health interpretations are vague (Tuncer et al. 2019). Such factors usually are complex to understand and evaluate (Vieira et al. 2019). Fuzzy reasoning is deciding a misinterpretation, ambiguity, and vagueness setting. Fuzzified strategies protect classes of ambiguity and complex boundaries (Pereira et al. 2007). The healthcare sector was among the first domains to incorporate fuzzy theory (Reyna et al. 2003) and a degree of Membership between 0 and 1.

The presence of fuzzy logic and classification techniques (Benamina et al. 2018) increases the latency and precision of case-based generalization in knowledge fields. The Fuzzy CBR is composed of two complementary parts: one per fuzzy decision tree classification segment and the other per case reasoning segment. A fuzzy inference system's role is to lower the uncertainty of determining the extent of similarities among patients. The results produced by the fuzzy decision-making tree are most efficient in improving the precise rating of disease and thus improve the retrieval stage of CBR rationalization.

To enhance the adaptive neuro-fuzzy inference system performance accuracy by applying two optimization metaheuristic techniques (Al-Qaness et al. 2020). In 10 days, this method was used to approximate and predict the number of confirmed cases of a new coronavirus that had been identified in China.

Numerous models of neural networks for the forecast of the COVID19 time series are designed with fuzzy aggregation (Melin et al. 2020). ANOVA, developed by Fisher (Kahraman et al. 2016), and was dealing with testing the significance of many samples, is a well-known data analysis technique. If the data is not understood, the methods depend on fuzzy systems that are consolidated into the ANOVA decision-making method (Doane et al. 2016). ANOVA using a set of statistical methods (Gil et al. 2006) for variance, using a bootstrap methodology to FANOVA Fuzzy (FANOVA), was described (Buckley 2006) and pessimistic and optimistic level of FANOVA triangular fuzzy data focused on Zadeh's delay principle (Kalpanapriya et al. 2012; Gonzalez et al. 2012). A single-way ANOVA assessment for imprecise analyses has been developed (Parchami et al. 2017), where the vague experts are interpreted as functional Hilbert space outcomes. ANOVA with two factors using Trapezoidal Fuzzy Numbers (Parthiban et al. 2016) is considered. IFANOVA was developed (Kalpanapriya et al. 2017) on two channels by transforming IFSs into fuzzy sets. In an evaluation of more approximate numbers like the one-way and double-way IFANOVA that can perform on methodological individualism, fuzzy data (IF), based on Intuitionistic Fuzzy and Index matrices, enhance the conventional ANOVA to one-way and two-ways in their approach.

3.3 Need for Early Detection of Covid-19

Also, early treatment of Coronavirus provides possibilities for accurate Diagnosis by relying on patients' early intervention. g Late-stage presentations, particularly in lower resource conditions and vulnerable communities, are typical to uncertainties in COVID-19 diagnosis. Low chances of healing, the higher mortality rate of Diagnosis, and increased healthcare costs lead to preventable deaths, and coronavirus deficiency is implicated in the consequences of poor or inadequate medical care. Early Diagnosis strengthens the results of COVID-19 by proper Diagnosis and is a critical strategy in all contexts of public health.

The COVID-19 test can diagnose the pathogen SARS-CoV-2 and usually involves identifying the existence of the virus and immune cells processed for infection. Serological testing can be used for both treatment and demographic tracking. Confirmatory testing affects how a person struggled, such as patients with mild or symptomlessness. The resulting data can evaluate an actual infection mortality rate and livestock antibodies in the community. Consequently, it remains unresolved how long and how efficiently this antibody response will be, and the percentages of false positive and false negatives need to be interpreted appropriately.

For analysis of COVID-19 disease testing and antimicrobial Research Center, 2020, two categories of methods are available (Google COVID-19 Tests 2020). There is a wide variety of tests available. An infection testing indicates that the patient had an outbreak, and if the patient had an outbreak previously, a clinical diagnosis suggests. If the patient has an infectious disease, an anticorder examination cannot notify, and this will require 1–3 weeks to develop anticorps. Till now, nobody knows whether antibodies to the virus will prevent anybody from acquiring the virus-infected afterward and how far it can last. To recognize if infectious experiments are being used to evaluate pathogens. However, this assessment is not necessary for all of us. Numerous people suffer from mild disease and can recover without treatment at residence and do not need testing. The Department of health gives guidelines on who can be evaluated, but the state and local systems or health care facilities are making the choices on testing. Contact the health professional (CDC Centre of Disease Control and Prevention 2020; SAEM Academic Emergency Medicine 2020) first if anyone has COVID-19 symptoms and would like to get screened.

Accurate Diagnosis by an early and effective screening of COVID-19 (https://economictimes.indiatimes.com/news/politics-and-nation/early-detection-timely-testing-played-pivotal-role-in-fighting-covid-19government/articleshow/77763576.cms?utm_source=contentofinterest&utm_medium=text&utm_campaign=cppst) diseases played a crucial role. Early assessments give the possibility to identify or hospitalize appropriate cases effectively in advance. 'Early identification is not necessarily an essential consideration.

Early analyses of threats, Covid-19 screening, and control caused issues, especially to healthcare providers in developing nations who cannot deal with the problem. While epidemic, beginning moderate indications are accurate in detecting Covid-19 earlier than usual.

The Coronavirus occurs with individual antibiotic-resistant contact because of lung infections such as pneumonia. A comprehensive analysis of novel Coronavirus features called SARSCov-2 due to similar symptoms of ARS (acute respiratory syndrome) caused by recently noticed coronaviruses (Zhu et al. 2020). In comparison with influenza viruses, SARS-Cov-2 described a longer incubation duration. The prolonged onset of symptoms will benefit from premature assessment and therapy during Covid-19 pathogenesis (Morens et al. 2020). The essential and useful knowledge on the pathogenesis of Covid-19 even before the severe assault of SARS-Cov-2 potentially triggering the emergence of new covid-19 moderate symptoms (Li et al. 2020). The time between mild symptoms in acute Covid-19 attacks gives a chance to evaluate prematurely shown in Fig. 3.1 by detecting typical clinical signs without severe covid-19 threats. The initial stage of contamination with Covid-19 is

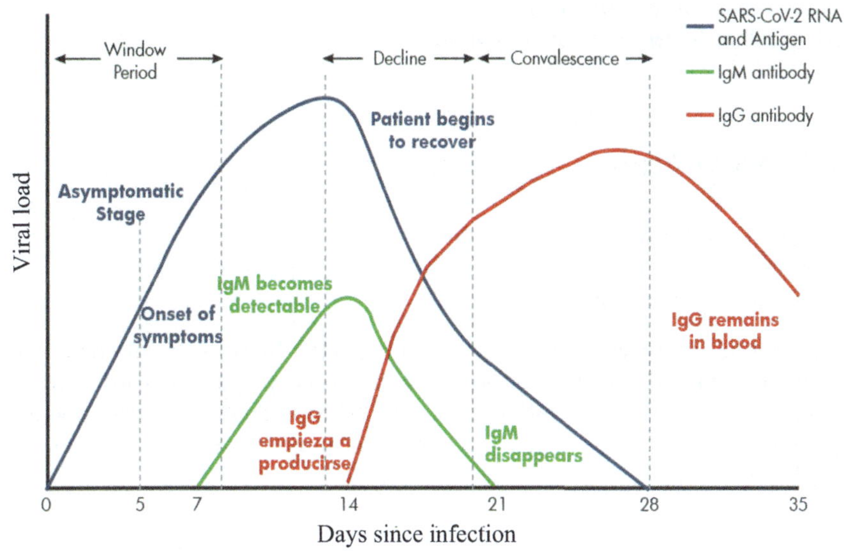

Fig. 3.1 SARS-CoV-2 during infection (https://www.hcmarbella.com/en/techniques-for-an-accurate-and-early-diagnosis-of-covid-19)

likely to lead to interjectory treatments. The significance of the time interval and the progression of signs provide relevant insights into prevent Covid-19, adding value to a comprehensive understanding of Covid-19 pathogenesis (Fauci et al. 2020).

COVID-19 impacts individuals in diverse types. Many individuals who are affected will experience moderate to severe symptoms.

3.3.1 *More Frequent Signs*

- Pneumonia
- cough
- Fatigue

3.3.2 *Common Signs*

- headaches and agony
- stinging throat
- Diarrhea
- Eye problems

- loss of sense of taste or smell
- skin rash or finger or toe discoloration

3.3.3 Severe Signs

- Breathing problems or loss of breathing
- Cardiovascular issues
- Speaking or movement loss

If a patient having severe symptoms, consult a doctor. Individuals with moderate, healthy symptoms should treat at home. In this work, propose an early warning method for Coronavirus, an early warning method for s To improve detection and precision, CART uses an implicit function segment using various techniques and fuzzy logic.

3.4 About Fuzzy Logic

Fuzzy logic was developing to model natural language unsafeness as a mathematical concept. "Fuzzy logic, developed in the mid-1960s on a prior study on fuzzy set theory, was invented by the Lotfi Zadeh, of the University of California, in Berkeley. He has also proposed a fuzzy control concept that makes it possible to employ a minimal set of 'perceptive principles' for the performance of electronic equipment (Kakulapati et al. 2020).

It is a variable processing strategy to process multiple values via the same variable. Fuzzy logic tries to fix issues using an open, imperfect data spectrum and heuristics, enabling accurate conclusions to be attaining.

This logic is imminent to solve issues through the assessment and inference of all relevant data. It comes from the statistical modeling of conceptual ideas that also incorporates membrane data sets. When about complex notions and ambiguous analysis, researchers might use a variety of words. These words are broadly categorized and precisely as vague semantics (https://www.investopedia.com/terms/f/fuzzy-logic.asp).

"Fuzzy logic is a methodology of ambiguous data analysis and interpretation. Every fact or statement must be either true or false, even under the conventional set theory. However, most of the knowledge that individuals use as to the environment is uncertain. Much like the concept of probabilities, varying logic assigns numerical values between 0 and 1 for every proposal to signify vulnerability. While the idea of probabilistic analyses the probability of a valid argument, heuristics evaluate precisely accurate, the proposition is (https://www.scientificamerican.com/article/what-is-fuzzy-logic-are-t).

The methodology of the Fuzzy set is applying for the medical applications of diagnosis and treatment expert systems, control, and monitoring were integrating with machine learning to control healthcare systems. Considerable efforts have been directed in many AI systems by the use of diagnosis sets into management practices. Using uncertain data sets and if the rules are inadequately recognized, this methodology was extremely effective.

Fuzzy logic can be utilized in cases where basic logic technology is not successful. For instance, systems and equipment cannot precisely describe computer simulations, methods with considerable uncertainties or inconsistent variables, and technologies or scenarios operated by language. As Lotfi Zadeh originally said, faulty logic will not replace existing logic or methodologies but instead enhance these conditions in which traditional methods do not help solve the problem.

3.4.1 The Fuzzy Framework Contains

- RULE BASE: It incorporates the guidelines established and the IF-THEN constraints generated by the analysts, upon on sample size was based evidence, to regulate the decision-making process. Significant advancements in fuzzy set theory employ several effective methods for control system modeling and control. In a lot of instances, the rule set is reduced.
- FUZZIFICATION: It is being used for converting inputs to fuzzy sets, i.e., precise quantities. Precise entries are essentially units of good inputs sent only through the computation monitoring system, including temperature, stress, etc.
- INFERENCE ENGINE: Every condition influences the amount of the actual fuzzy input and determines which rules are activated by the attribute. Moreover, the triggered rules for system parameters are grouped collectively.

3.4.2 Membership Function

A graph shows that every data object is transformed to memberships of 0 to 1. The weight vector is frequently called a discussion environment, which incorporates all the relevant aspects relevant to any function.

3.4.3 Fuzzifier

It may be characterized as a mapping state of a certain input state of argumentation from an observable input space to vague set labels.

For each variable, the mapping procedure assumes the associated assessment errors. The mapping function is intended to interpret the analyzing set of variables, represented as accurate fuzzy approximations of the corresponding actual value (https://forumautomation.com/t/what-are-fuzzifier-and-defuzzifier/4148).

Fuzzifiers are three types:

- SF (Singleton fuzzifier).
- GF (Gaussian fuzzifier)
- TF (Trapezoidal fuzzifier)

3.4.4 Fuzzy Control

- It is a technique that incorporates conceptual frameworks into a system of control.
- The logic may not be precise but is intended to provide an acceptable reason.
- It can imitate inferential cognitive reasoning, i.e., how humans extract information from actual knowledge.
- Also, with the aid of flippant logic, any ambiguities may readily be addressed.

The framework of the Fuzzy logic system is comparable to that incurred in making solutions. Due to always one decrease being performed, understanding how objectives are met is an inexpensive add feature for fuzzy systems. Nevertheless, several fuzzy systems can illustrate the procedures for extracting outcomes and the visual representation of the membership function. The adaptability of fuzzy systems and their effectiveness in several other domains, especially medicine, differs from the flexibility to execute logical operations (AND, OR, NOT) using several logic types. Progressive levels of certainty in fuzzy logic serve the needs of the treatment plan and human competence and promote health professionals who are used to referencing beliefs, possibility, and probability. The capability to manage with contradicting rules that are not consistent under human reasoning truth with any quasi logic—also has an advantage.

The treatment options of patient care are increasingly using precise logic methodologies. Several other areas of patient care in which observational reasoning is not feasible and which metrics are only partially applicable, significant uses suitable for applying futile logic to pass with almost quantitative to qualitative reasoning, completing the gap.

3.4.5 Fuzzy Set

These are constructed on the concept of set theory and augmentation of conventional sets. The set features are termed elements or set members in a classic set. An item x that is part of a set A must be defined as $X = A$ and an element that is not part μA is defined as $x + A$. An element of the universe U having numerical values of 1 or 0 is a function or a membership. For every $x \in U$.

$$\mu_A(X) = \begin{cases} 1\,\text{for}\,x \in A \\ 0\,\text{for}\,x \in A \end{cases}$$

It can also be stated as $\mu_A(x) \in (0,1)$.

The Membership function for fuzzy sets is given interval values [0, 1]. The 0 to 1 range is called membership or membership degree.

It can be defined as:

$$A = \{(x, \mu_A(x))|x \in A, \mu_A(x) \in [0, 1]\}$$

where μA is the interval of the membership function [0, 1].

3.4.6 *Defuzzification*

It transforms the inadequate fuzzy sets into a feature vector. There are several defuzzification approaches, and with a given expert system, the best one is utilized to minimize the error (https://www.geeksforgeeks.org/fuzzy-logic-introduction/).

The defuzzification concept aims to determine a region in W that matches the fuzzy set C, derived by the expert system. This may be characterized as lowering the set C from a universe of fuzzy controlling measures in the environment of the transmitter output W al R to an area of explicit transient response $z^* \in W$.

There are other possibilities to generate this precise optimal threshold z^*. Nonetheless, throughout the implementation of a defuzzification process, the following factors should be taken into account:

1. **Acceptability:** Intuitively, point z^* should reflect C; for instance, it may be about the center of supporting of C'' or have a high degree of membership of C''.
2. **Simplicity:** This is especially crucial for fuzzy control, as fluid controllers work in real-time.
3. **Steadiness:** Some little variation of C' need not result in a significant z^* change.
4. **Discrepancy:** It means that the defuzzification method should always produce a unique value for z^*.

3.5 Methodology

This chapter covered a concise description of the techniques utilized and incorporated in this experiment: an overview of the complete Fuzzy logic methodology and cart implementation analysis for early intervention of COVID-19.

The utilized fuzzy logic rules in (Dhiman and Sharma 2020; Muka et al. 2017; Princy and Dhenakaran 2016; Walia et al. 2016) The necessary steps are.

3.5.1 Identification of Parameters

First of all, it is essential to examine all appropriate corrective measurement parameters that reduce the risk during a pandemic. All decisions should prevent considerations of patients and health care settings.

3.5.2 Fuzzification (Membership Function)

For each value of the input parameters, membership functions are identified. The object-level corresponding to a fuzzy set is represented as μ by a membership function ranging from 0 to 1. For each parameter, membership functions, along with their μ values, were defined.

3.5.3 Fuzzy Inference Method and Rules

This is constructed by using input parameters, and the inputs, outputs, and the level of the uncertainty vary from minor-medium to high are then split into 3 linguistic groups. To analyze the information to develop the risk levels, the "if-then" guidelines were used. This method comprises three measures, including the fuzzification of membership functions, the translation and denoting ($\mu = 0$ - 1), linguistic levels or formulation of rules, and defuzzification.

3.5.4 Classification and Regression Trees (CART)

A classification model is a technique with a categorical target variable. The method can be used to determine which "class" is the target variable. There may be simple, binary classifications where only one of two mutually exclusive values can be taken from the classified dependent variable. An algorithm with the destination variable and used to forecast its value is referred to in the regression tree. It performs by segregating the dataset into different sections by making statements on the characteristics. In this case, it has been used as a useful mathematical method for classification, analysis, and data processing due to its successful use in medical science.

3.6 Implementation

Using CART to determine the expected number of individuals infected to execute the regression tree. The regression tree's primary objective would be that if with the variables under analysis, the total global number of infected individuals can be estimated very well. It indicates that the most significant variables for the total number of infected patients, cumulative deaths accompanied by cumulative recovery cases, new cases, and cumulative severe cases, are the nation and the cumulative active cases. Finding the most impressive way to predict accumulated infectious tests is to occur as one of the most significant variables.

Data set is extracted from verified sources such as the Kaggle database (Tuncer et al. 2019). The data set contains confirmed COVID-19 cases, fatality cases, recovered cases, and the disease symptoms for affected patients. The Data set looks as shown in Table 3.2.

Create the frbs structure by classification process and the features in the fuzzy interface processed for the membership function. Triangular is quick and easy to learn and frequently used for diverse applications. The cart rules are used to denote the results in the regulatory editor of the fuzzy rule base showing in Table 3.3. The operator continues to generate the defuzzification yield and can quickly learn if the persistence is infected by Coronavirus or not.

	precision	recall	f1-score	support
0	0.71	0.70	0.70	60406
1	0.49	0.51	0.50	34634
accuracy			0.63	95040
macro avg	0.60	0.60	0.60	95040
weighted avg	0.63	0.63	0.63	95040

Table 3.2 The symptoms of COVID-19 disease

Country	Gender	Symptoms	Experiencing_Symptoms	Severity	Contact
China	Male	Fever, Tiredness, Dry-Cough, Difficulty-in-Breathing, Sore-Throat	Pains, Nasal-Congestion, Runny-Nose, Diarrhea	Mild	Yes
Italy	Female	Fever, Tiredness, Dry-Cough, Difficulty-in-Breathing	Pains, Nasal-Congestion, Runny-Nose	Moderate	No
Iran	Transgender	Fever, Tiredness, Dry-Cough	Pains, Nasal-Congestion	Severe	Dont-Know
Republic of Korean		Fever, Tiredness	Pains	None	
France		Fever	Nasal-Congestion, Runny-Nose, Diarrhea		
Spain		Tiredness, Dry-Cough, Difficulty-in-Breathing, Sore-Throat	Nasal-Congestion, Runny-Nose		

Figure 3.2 depicts the decision tree classifier area under curve and the accuracy is 71%/

Confusion Matrix and measures

```
Prediction 1    2
    1   239   5
    2     0   153

      Accuracy: 98.7%
        95% CI: (0.9709, 0.9959)
  No Information Rate: 0.602

        Kappa: 97.3%

 P-Value: 0.07364

      Sensitivity: 1
      Specificity: 97%
    Optimistic Pred Value: 98%
    Undesirable Pred Value: 1
       Occurrence: 60%
    Rate of Detection: 60%
  Detection Prevalence: 62%
   Well-adjusted Accuracy: 98.4.%
```

Table 3.3 The covid patient having symptoms (1 yes and 0 no)

	Fever	Tiredness	Dry-cough	Difficulty-in-breathing	Sore-throat	None_sympton	Pains	Nasal-congestion	Runny-nose
0	1	1	1	1	1	0	1	1	1
1	1	1	1	1	1	0	1	1	1
2	1	1	1	1	1	0	1	1	1
3	1	1	1	1	1	0	1	1	1
4	1	1	1	1	1	0	1	1	1
...	...	...	...	...	...	...	...	...	...
316795	0	0	0	0	0	1	0	0	0
316796	0	0	0	0	0	1	0	0	0
316797	0	0	0	0	0	1	0	0	0
316798	0	0	0	0	0	1	0	0	0
316799	0	0	0	0	0	1	0	0	0

316800 rows × 9 columns

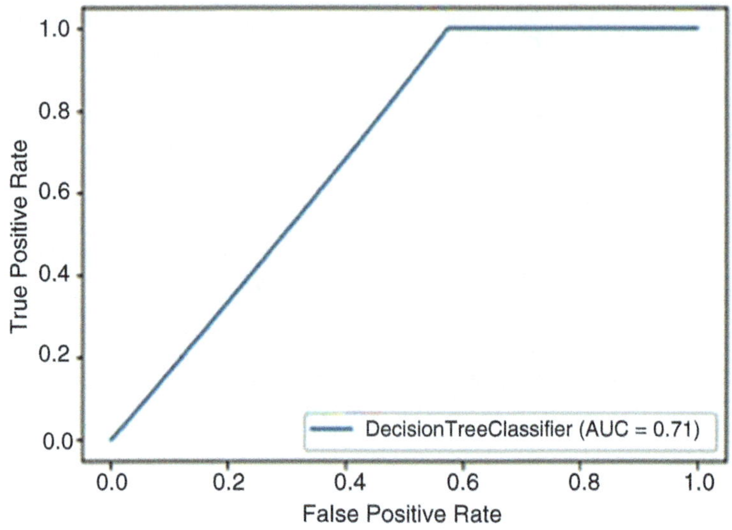

Fig. 3.2 Roc curve for COVID-19 decision tree classifier

```
'Optimistic' Class: 1. The limit values of membership function on the
input variable (regulated):
 v.1_a.1 v.1_a.2 v.2_a.1 v.2_a.2 v.3_a.1 v.3_a.2 v.4_a.1 v.4_a.2
v.5_a.1 v.5_a.2 v.6_a.1 v.6_a.2 v.7_a.1 v.7_a.2
[1,]  2.00  3.00  2.00  3.00  2.00  3.00  2.00  3.00  2.00  3.00  2.00
3.00  2.00  3.00
[2,]  0.00  0.50  0.00  0.50  0.00  0.50  0.00  0.50  0.00  0.50  0.00
0.50  0.00  0.50
[3,]  0.25  0.75  0.25  0.75  0.25  0.75  0.25  0.75  0.25  0.75  0.25
0.75  0.25  0.75
[4,]  0.50  1.00  0.50  1.00  0.50  1.00  0.50  1.00  0.50  1.00  0.50
1.00  0.50  1.00
[5,]   NA   NA   NA   NA   NA   NA   NA   NA   NA   NA   NA   NA   NA   NA
   v.8_a.1 v.8_a.2 v.9_a.1 v.9_a.2 v.10_a.1 v.10_a.2 v.11_a.1 v.11_a.2
v.12_a.1 v.12_a.2 v.13_a.1 v.13_a.2 v.14_a.1
[1,]  2.00  3.00  2.00  3.00  2.00  3.00  2.00  3.00  2.00  3.00  2.00
3.00   2.00
[2,]  0.00  0.50  0.00  0.50  0.00  0.50  0.00  0.50  0.00  0.50  0.00
0.50   0.00
[3,]  0.25  0.75  0.25  0.75  0.25  0.75  0.25  0.75  0.25  0.75  0.25
0.75   0.25
[4,]  0.50  1.00  0.50  1.00  0.50  1.00  0.50  1.00  0.50  1.00  0.50
1.00   0.50
[5,]   NA   NA   NA   NA   NA   NA   NA   NA   NA   NA   NA   NA   NA
   v.14_a.2 v.15_a.1 v.15_a.2
[1,]   3.00   2.00   3.00
[2,]   0.50   0.00   0.50
[3,]   0.75   0.25   0.75
```

```
[4,]   1.00   0.50   1.00
[5,]    NA     NA     NA

The fuzzy IF-THEN rules:
 V1    V2 V3   V4 V5    V6 V7   V8 V9  V10 V11   V12 V13        V14 V15  V16
V17
1 IF Sore.throat is v.1_a.1 and Running.2se is v.2_a.2 and Asthma is
v.3_a.2 and Chronic.Lung.Disease is v.4_a.2 and
2 IF Sore.throat is v.1_a.1 and Running.2se is v.2_a.1 and Asthma is
v.3_a.1 and Chronic.Lung.Disease is v.4_a.2 and
3 IF Sore.throat is v.1_a.1 and Running.2se is v.2_a.2 and Asthma is
v.3_a.2 and Chronic.Lung.Disease is v.4_a.1 and
4 IF Sore.throat is v.1_a.1 and Running.2se is v.2_a.2 and Asthma is
v.3_a.1 and Chronic.Lung.Disease is v.4_a.2 and
5 IF Sore.throat is v.1_a.1 and Running.2se is v.2_a.1 and Asthma is
v.3_a.1 and Chronic.Lung.Disease is v.4_a.2 and
6 IF Sore.throat is v.1_a.1 and Running.2se is v.2_a.2 and Asthma is
v.3_a.2 and Chronic.Lung.Disease is v.4_a.1 and so on.,

The certainty factor:
  [1,] 1.3167702
 [2,] 0.6832298
 [3,] 0.6832298
 [4,] 1.3167702
 [5,] 1.3167702
 [6,] 1.3167702
 [7,] 1.3167702
 [8,] 1.3167702
 [9,] 1.3167702
[10,] 1.3167702
[11,] 0.6832298
[12,] 1.3167702
[13,] 1.3167702
[14,] 1.3167702
[15,] 1.3167702
[16,] 1.3167702
[17,] 0.6832298
[18,] 0.6832298
[19,] 0.6832298
[20,] 0.6832298
[21,] 0.6832298
[22,] 1.3167702
[23,] 1.3167702
[24,] 1.3167702
[25,] 1.3167702

Predicting COVID-19 is attack or not
 [1,]  1,  [2,]  1  [3,]  1  [4,]  1 [5,]  1  [6,]  1  [7,]  2  [8,]  1  [9,]
1
 [10,]  1 [11,]  2 [12,]  2 [13,]  2 [14,]  2 [15,]  1 [16,]  2 [17,]  1
[18,]  1 [19,]  2 [20,]  1 [21,]  1 [22,]  2 [23,]  1 [24,]  1 [25,]  1
Error:  [1] 2.518892
```

Figure 3.3 depicts the fuzzy sets of COVID-19 symptoms classification.

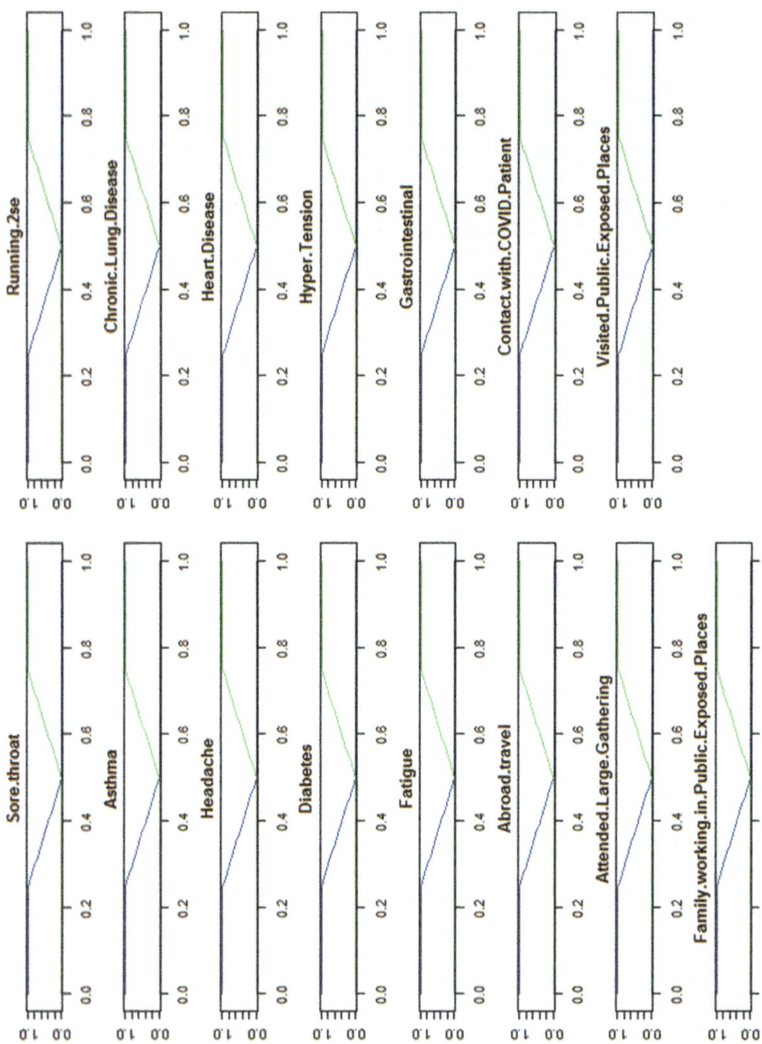

Fig. 3.3 Fuzzy-based classification of symptoms of COVID-19

3.7 Conclusion

To prevent transmitting the disease to numerous individuals, an early prediction of COVID-19 patients is necessary. The primary objective of this research was to target people with mild and severe symptoms around the world. The fuzzy decision tree's significant advantage is the primary classification method of treatment insights that help physicians and hospital staff assess prospective patient interference. This proposed model assimilates to clinician expertise and hospital staff, suggesting a precise approach for specific cases. The assessment's recommended analytical evidence is an appropriate and efficient technique of describing uncertainty assertions or hypotheses towards more comprehension of the membership features and non-membrane features. Sufficient identification can be recommended in patients prematurely. This chapter ends with premature coronavirus awareness and prevention performance metrics.

3.8 Future Work

In the future, the use of coronavirus cases and the behavior analyses of the COVID-19 patient would be further enhanced for improving classification precision. The use of clustering methods to prevent the progression of coronavirus diagnosis and care. For forecasting analysis of disease by a comparative evaluation of the impact of appropriate algorithms for machine learning. To accurately interpret, it can review online patient results and be much more user-friendly and reliable, which can be incorporated into a vast volume of data.

References

Al-Qaness MAA et al (2020) Optimization method for forecasting confirmed cases of COVID-19 in China. J Clin Med 9(3):674

Benamina M et al (2018) Diabetes diagnosis by case-based reasoning and fuzzy logic. IJIMAI 5 (3):72–80

Buckley J (2006) Fuzzy probability and statistics. Springer, Berlin

CDC Centre of Disease Control and Prevention (2020) Coronavirus Disease 2019 (COVID-19), Testing for COVID-19

Chan KH et al (2004) Detection of SARS coronavirus in patients with suspected SARS. Emerg Infect Dis 10:294–299

Chaurasia V et al (2013) Early prediction of heart diseases using data mining techniques. Caribb J Sci Technol 1:208–217

Chitra R et al (2013) Heart disease prediction system using supervised learning classifier. Bonfring Int J Softw Eng Soft Comput 3(1):1–7

Dhiman N, Sharma MK (2020) Fuzzy logic inference system for identification and prevention of coronavirus (COVID-19). Int J Innov Technol Explor Eng 9(6)

Dhiman N et al (2019) Diabetes diagnostic model based on truth-value restriction method using inference of intuitionistic conditional and qualified fuzzy propositions. Int J Eng Adv Technol 9 (2)

Doane D et al (2016) Applied statistics in business and economics. McGraw-Hill Education, New York

Fauci AS et al (2020) Covid-19 – navigating the uncharted. N Engl J Med 382(13):1268–1269. https://doi.org/10.1056/NEJMe2002387

Gil M et al (2006) Bootstrap approach to the multi-sample test of means with imprecise data. Comp Stat Data Anal 51:14816

Gonzalez G et al (2012) Fuzzy data treated as functional data: a one-way ANOVA test approach. Comput Stat Data Anal 56:943–955

Google COVID-19 Tests (2020) Images COVID-19 tests

Guan WJ et al (2020) Clinical characteristics of coronavirus disease 2019 in China. N Engl J Med 382:1708–1720. https://doi.org/10.1056/NEJMoa2002032

Gudbjartsson DF et al (2020) Spread of SARS-CoV-2 in the Icelandic population. N Engl J Med. https://doi.org/10.1101/2020.03.26.20044446

Hashem IAT et al. (2020). A machine learning solution framework for combatting COVID-19 in smart cities from multiple dimensions. medRxiv 2020.05.18.20105577. doi:https://doi.org/10.1101/2020.05.18.20105577

Holshue ML et al (2020) First case of 2019 novel coronavirus in the United States. N Engl J Med 382:929–936. https://doi.org/10.1056/NEJMoa2001191

https://economictimes.indiatimes.com/news/politics-and-nation/early-detection-timely-testing-played-pivotal-role-in-fighting-covid-19government/articleshow/77763576.cms?utm_source=contentofinterest&utm_medium=text&utm_campaign=cppst

https://forumautomation.com/t/what-are-fuzzifier-and-defuzzifier/4148

https://iris.paho.org/bitstream/handle/10665.2/52276/PAHOEIHISCOVID-19200007_eng.pdf?sequence=5&isAllowed=y

https://www.geeksforgeeks.org/fuzzy-logic-introduction/

https://www.hcmarbella.com/en/techniques-for-an-accurate-and-early-diagnosis-of-covid-19

https://www.investopedia.com/terms/f/fuzzy-logic.asp

https://www.scientificamerican.com/article/what-is-fuzzy-logic-are-t

Kahraman C et al (2016) Fuzzy statistical decision-making: theory and applications, studies in fuzziness and soft computing. Springer, Cham

Kakulapati V et al. (2019) "Fuzzy based predictive analytics for early detection of Diabetes" IGI global book "Pre-screening systems for early disease prediction, detection, and prevention", ISBN 1522571310, 9781522571315

Kakulapati V et al. (2020) Fuzzy-based predictive analytics for early detection of disease – a machine learning approach. ICT4SD, 23–24 July, Goa

Kalpanapriya D et al (2012) Fuzzy hypotesis testing of ANOVA model with fuzzy data. Int J Mod Eng Res 2(4):2951–2956

Kalpanapriya D et al (2017) Intuitionistic fuzzy ANOVA and its application using different techniques. In: Advances in algebra and analysis. Trends in mathematics. Birkhuser, Cham, pp 457–468

Khan AI et al (2020) CoroNet: a deep neural network for detection and diagnosis of COVID-19 from chest x-ray images. Comput Methods Prog Biomed 196C(2020):105581. https://doi.org/10.1016/j.cmpb.2020.105581

Li Q, Guan X, Wu P et al (2020) Early transmission dynamics in Wuhan, China, of novel coronavirus–infected pneumonia. N Engl J Med 382:1199–1207. https://doi.org/10.1056/NEJMoa2001316

Lu X et al (2020) Wuhan children's hospital Wuhan, China. N Engl J Med 382:1370–1371

Melin P et al (2020) Multiple ensemble neural network models with fuzzy response aggregation for predicting COVID19 time series: the case of Mexico. Healthcare 8(181):1–13

Morens DM et al (2020) Escaping Pandora's box – another novel coronavirus. N Engl J Med. https://doi.org/10.1056/NEJMp2002106

Muka Z, Maraj E, Kuka S (2017) Rainfall prediction using fuzzy logic. Int J Innov Sci Eng Technol 4(12)

Narin A, et al. (2020) Automatic detection of Coronavirus Disease (COVID-19) using X-ray images and deep convolutional neural networks. Retrieved from https://arxiv.org/ftp/arxiv/papers/2003/2003.10849.pdf

Nilakshi P et al (2014) Genetic neural approach for heart disease prediction. Int J Adv Comput Res 4(3):331–338

Ozturk T et al (2020) Automated detection of COVID-19 cases using deep neural networks with X-ray images. Comput Biol Med. https://doi.org/10.1016/j.compbiomed.2020.103792

Parchami A et al (2017) Analysis of variance in uncertain environments. Complex Intell Syst 3 (3):189–196

Parthiban S et al (2016) A comparative study of Two factor ANOVA model under fuzzy environments using trapezoidal fuzzy numbers. Int J Fuzzy Math Arch 10(1):1–25

Patel SB et al (2014) Predict the diagnosis of heart disease patients using classification mining techniques. IOSR J Agric Vet Sci 4(2):61–64

Pereira CC et al (2007) Profiles of multiple causes of death related to HIV/AIDS in the cities of Sao Paulo and Santos, Brazil, 2001. Cad Saude Publica 23(3):645–655

Princy S, Dhenakaran S (2016) Comparison of triangular and trapezoidal fuzzy membership function. J Comput Sci Eng 2(8)

Rashedur M et al (2013) Comparison of various classification techniques using different data mining tools for diabetes diagnosis. J Softw Eng Appl 6(3):85–97

Reyna VF et al (2003) Fuzzy-trace theory, risk communication, and product labelling in sexually transmitted diseases. Risk Anal 23(2):325–342

SAEM Academic Emergency Medicine (2020) Symptom criteria for COVID-19 testing of heath care workers

Suganya S et al (2016) A proficient heart disease prediction method using fuzzy-cart algorithm. Int J Sci Eng Appl Sci 2(1):1–6

Tuncer T et al (2019) Face recognition with triangular fuzzy set-based local cross patterns in wavelet domain. Symmetry 11(6):787

Vieira L et al (2019) Fuzzy logic and hospital admission due to respiratory diseases using estimated values by mathematical model. Cien Saude Colet 24(3):1083–1090

Walia N, Singh H, Sharm A (2016) Effective analysis of lung infection using fuzzy rules. Int J Artif Intell WHO. 2020. Cleaning and disinfection of environmental surfaces in the context of COVID-19. Interim guidance

Xu Z et al (2020) Pathological findings of COVID19 associated with acute respiratory distress syndrome. Lancet Respir Med. https://doi.org/10.1016/S2213-2600(20)30076-X

Zadeh LA (2008) Is there a need for fuzzy logic? Inf Sci (NY) 178(13):2751–2779

Zhu N et al (2020) A noval corona virus form patient of pneumonia in China, 2019. N Engl J Med 382:727–733

Chapter 4
Role of Artificial Intelligence in Diagnosis of Covid-19 Using CT-Scan

Karim Sherif, Yousef Emad Gadallah, Khalil Ahmed, Salma ELsayed, and Ali Wagdy Mohamed

Abstract Machine learning (ML) and deep learning (DL) have been broadly used in our daily lives in different ways. Early detection of COVID-19 built on chest Computerized tomography CT empowers suitable management of patients and helps control the spread of the disease. We projected an artificial intelligence (AI) system for rapid COVID-19 detection using analysis of CTs of COVID-19 depending on the AI system. We developed and evaluated our system on a large dataset with more than 3000 CT volumes from COVID-19, viral community-acquired pneumonia (CAP) and non-pneumonia subjects—1601 positive cases, 1626 negative cases.

Keywords Covid-19 · CT · Diagnosis · Machine learning · Artificial intelligence · Covid positive CT image

4.1 Introduction

Coronavirus disease has now become a global pandemic, a life-threatening disease with an enormous rate of growth and a process of transmission that is not properly understood. The virus can lead to rapidly progressive pneumonia, which often might be fatal in 2–8% of those who are infected. Every day a significant number of people lose their lives due to this disease. Not only does this disease affect a single country,

---

K. Sherif · Y. E. Gadallah · K. Ahmed · S. ELsayed
School of Information Technology and Computer Science, Nile University, Giza, Egypt
e-mail: Kar.sherif@nu.edu.eg; y.gadallah@nu.edu.eg; kha.ahmed@nu.edu.eg; S.Elsayed@nu.edu.eg

A. W. Mohamed (✉)
Faculty of Graduate Studies for Statistical, Operations Research Department, Cairo University, Giza, Egypt

Wireless Intelligent Networks Center (WINC), School of Engineering and Applied Sciences, Nile University, Giza, Egypt

S. A. Hassan et al. (eds.), *Decision Sciences for COVID-19*, International Series in Operations Research & Management Science 320,
https://doi.org/10.1007/978-3-030-87019-5_4

but also the entire world is suffering from it. Also, many other kinds of viruses such as (SARS, MERS, flu, etc.) came into the picture in the past 10 years, but they were not as dangerous as COVID-19; they have only been standing for a few days or a few months (Ye et al. 2020; Pan et al. 2020). The most crucial part is that while the entire world is affected by this virus, there is no scientist from any single country can prepare a vaccine for it. In the meantime, plasma therapy, X-ray images, CT-scans are used as methods of prediction of COVID-19 in addition to many other methods, but until now, there is no precise remedy to this fatal virus. Due to the high diagnostic cost of this disease in the context of a country, state, patients, not everyone is able to make it (Udugama et al. 2020).

Due to inadequate supply of vital protective equipment and qualified providers, the faster pace of outbreaks has weakened healthcare systems globally, driven in part by varying access to point-of-care testing methodologies, such as reverse transcription polymerase chain reaction methodologies (RT-PCR). As rapid RT-PCR testing becomes more accessible, challenges still exist, such as high false-negative rates, test technique variations, processing delays, and sometimes reported sensitivity as low as 60–70% (Liang et al. 2020).

The common symptoms of COVID-19 are fever, cough, shortness of breath, headache, and sore throat, loss of appetite, loss of smell, and taste (Zhou et al. 2020; Fang et al. 2020). COVID-19 has similar symptoms to that of Influenza-A, Influenza-B, common flu, and other types of common cold (Huang et al. 2020). This high similarity ratio resulted in the spreading of the virus rapidly. The genome sequences of the respiratory samples of affected patients were analyzed. It suggested a new type of beta coronavirus that was similar to Severe Acute Respiratory Syndrome Coronavirus (SARS-CoV) (Zhou et al. 2020; He et al. 2020). Thus, careful and thorough identification processes were required, making the entire process time demanding.

Computed tomography (CT) is a scan that creates a window into pathophysiology, which might put a spotlight on the diagnosis and evolution of several stages of the disease. While still obstacles continue with a fast diagnosis of COVID-19, radiologists disclose an infection pattern which is somewhat characteristic of typical characteristics such as ground-glass opacification in the periphery of the lung, rounded opacities, enlarged intra-infiltrate vessels, and afterward more consolidations which are an indication of a crucial disease progressing. While CT and RT-PCR are quite often identical, early COVID-19 can indeed be discovered by CT in patients with a negative RT-PCR test or in patients who have no symptoms or before symptoms began to appear or after symptoms disappear (Zhao et al. 2020a; Li et al. 2020a).

One of the most uses global technology right now is Artificial Intelligence (AI) that can track the speed and detect the growth rate of the coronavirus, identify the risk and severity of Coronavirus patients. AI can also anticipate the possibility of death by adequately analyzing previous patient data. Artificial intelligence can assist us with battling the virus by testing individuals, medical assistance, data and information, and recommendations about disease control. In order to solve complex problems in our lives, AI is a broad umbrella consisting of many subareas. These subareas include learning, preparation, thinking, representation of information and searching. Machine Learning (ML) and Deep Learning (DL) are a subset of AI areas that consist of several algorithms that mimic human brains and behaviors based on data monitoring to identify or cluster particular tasks (Fang et al. 2020).

4.2 Literature Review

4.2.1 What Is Covid-19 Disease?

Extreme Acute Respiratory Syndrome CoronaVirus 2 (SARS-CoV-2) is as of now an impressive irresistible overall sickness. Covid 2019 (COVID-19) caused by an infection named SARS-CoV-2 was absolute initially informed in Wuhan, China, in December 2019 and later in numerous parts everywhere on the world. On January 03, 2020, the World Health Association announced that COVID-19 is a Public Health Crisis of International Concern (PHEIC) and affirmed it as a plague on March 11, 2020. (Huang et al. 2020). This infection has been represented in 216 nations and districts for what it is worth from May 16, 2020. The infection has spread and provoked groundbreaking results, with 31,668,942 cases that affirmed Covid and 972,437 number of passing on September 22, 2020 (Fang et al. 2020).

4.2.2 New Techniques Used to Diagnose This Disease

The wellbeing business is enthusiastically searching for new advances what is more, methods to track and control the development of the Covid plague in this worldwide wellbeing emergency. One of the most uses worldwide innovations right presently is Artificial Intelligence (Computer-based intelligence) that can follow the speed and recognize the development pace of the Covid, recognize the danger and seriousness of Covid patients. Comp teased intelligence can likewise foresee the chance of passing by sufficiently examining past patient information. Collecting knowledge can help us with fighting the infection by testing people, clinical help, information and data, and proposals about infectious prevention.

4.2.3 How the Machine Learning and Deep Learning Are Used to Examine the Disease?

ML and DL are utilized to look at SARSCoV-2 protein-related hereditary qualities and anticipate novel blends that can be utilized for drug creation and immunization. In expansion, for an enormous scope, COVID-19 case information and social media information, AI astute models dependent on ML and DL figure out how to build illness transmission models that precisely anticipate episodes, transmission way, transmission list, and impacts. ML and DL are likewise inconceivably utilized in pestilence security what is more, public observing, for example, security registration in air terminals, understanding following, and scourge identification. In this study, we present the primary extent of AI centering on ML and DL towards COVID-19 examination fuses the sides of illness conclusion and medication and antibody improvements. Note that because of the quick advancement of the COVID-19 pandemic, we have cited many distributed examination works before a careful examination, where these works really ought to be an overview for their accuracy and quality in friend surveys nationally and internationally, the Coronavirus (COVID-19) outbreak is rising. In the universal battle against COVID-19, for example, medical imaging, X-ray, and computed tomography (CT) play a key role, and the latest AI developments tend to improve the capacity of imaging tools and facilitate healthcare personnel. Medical imaging research is commonly used for the identification of COVID-19 by clinicians. Chest X-ray and lung CT image samples are mostly used in COVID-19 clinical imaging trials. AI innovation plays a significant role in medical imaging testing. It has produced enormous results in image identification, organ recognition, geographic infection classification, and disease classification. It not only decreases the picture diagnostic time of the radiologist but also increases the accuracy and execution of the diagnosis. Through correct diagnostic precision in X-ray and CT imaging, AI can enhance work performance, making it easier to test (Li et al. 2020a).

4.2.4 Chest CT-Image Detection

An esteemed component of the appraisal of patients with dicey SARS-CoV-2 disease is the chest CT picture. There is a developing exploration on the job of COVID-19 imaging for treatment and conclusion. The contamination triggers a gigantic range of CT examine imaging revelations, most ordinarily ground-glass opacities and lung outskirts combinations. Chest CT affectability to analyze Coronavirus has been discovered to be essentially higher and can happen before a positive viral lab test (Soghaier et al. 2015). Thusly, Emergency clinics with enormous amounts of affirmations use CT for the quick evaluation of patients with possible COVID-19 illness in epidemic domains where the essential medical care framework is under tension. In the assessment of COVID-19 patients with extreme and compound respiratory manifestations, chest CT plays an indispensable job. In light of sweeps, it is conceivable to decide how severely the lungs are undermined and how the ailment of the individual advances, which is successful in making clinical decisions. There is a developing comprehension of the abrupt occurrence of lung defects prompted by COVID-19 in CT checks directed for some other clinical signs, such as stomach CT examines for gut issues or patients without respiratory manifestations. In this pandemic, by diminishing the strain on clinicians, the assessment of AI may turn into the most important factor. In spite of the fact that it can require as long as 15 min to decipher a CT check physically, AI can examine the pictures in 10 s in this way, progressed picture handling with Fake neural organization has the likelihood to fundamentally improve the capacity of CT in COVID 19 recognition by permitting a huge extent of patients to distinguish infection without any problem also, quickly with exactness. The continuation of AI-based CT imaging tests typically includes the accompanying advances: provincial division of the Region of Interest (ROI), the expulsion of respiratory tissue, the ID of provincial contamination, and order of Coronavirus. An essential reason for examining AI-based symbolism is the acknowledgment of lung organs and ROIs (Mei et al. 2020).

Zhang et al. (2020b). defined their AI method for the analysis of COVID-19 pneumonia based on chest CT images. The role of the system was as good as to that of practicing radiologists with important clinical practice and could help and develop the work of junior radiologists. Supporting this was the progress of a progressing clinical model based on their AI system using CT imaging data and clinical data. This supports the conflict that AI can be used to augment clinical diagnosis. Built on their database, they could identify a compound score of $\geq$0.5 as the high-risk group in terms of the final progression into marked or serious illness resulting in intensive care unit admission, automated ventilation, or even death. Significantly, an expected time to this progress could also be supported (Zhao et al. 2020b). Some studies reported that artificial intelligence (AI) could play an important role in assisting doctors in making a rapid and accurate diagnosis, and contribute to alleviating the epidemic. (Huang et al. 2020; He et al. 2020; Zhao et al. 2020a) In particular, Abbasian Ardakani et al., further demonstrated the feasibility of AI in the auxiliary diagnosis of COVID-19 based on the CT images (Li et al. 2020a).

4.3 Methodology

Modern research shows that AI algorithms can even reach or exceed the performance of human experts in certain medical image diagnosis tasks and lung diseases. Comparing to other lung diseases, distinguishing COVID-19 from other pneumonias has special difficulty, i.e., the high similarity of pneumonias of different types (especially in early stage) and large variations in different stages of the same type. Hence, developing AI diagnosis algorithm specific to COVID-19 is essential. The AI diagnosis algorithm has the benefits of high efficiency, high repeatability, and easy large-scale arrangement. (Wang et al. 2020; Luo et al. 2020; Li et al. 2020b; Ren et al. 2019). The dataset used and the methodology used are explained in Fig. 4.1.

We used Chest CT-scans of Covid-19 affected, normal and pneumonia. This collected dataset is not meant to claim the diagnostic ability of any Deep Learning model but to research different possible ways of efficiently detecting Coronavirus infections using computer vision techniques. The collected dataset consists of 3227 total chest CT images. This data set is further divided into 1601 normal, and 1626 are Covid. The scans were scaled down 250 $\times$ 250 to aid the fast training of our program. We visualize the dataset and plotting it as shown in Figs. 4.2 and 4.3.

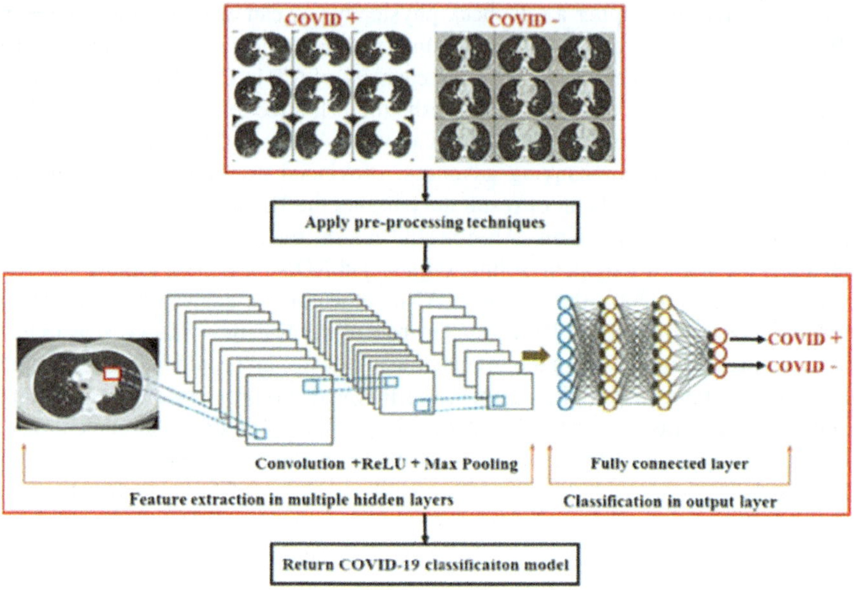

Fig. 4.1 The Covid-19 chest CT dataset and the methodology

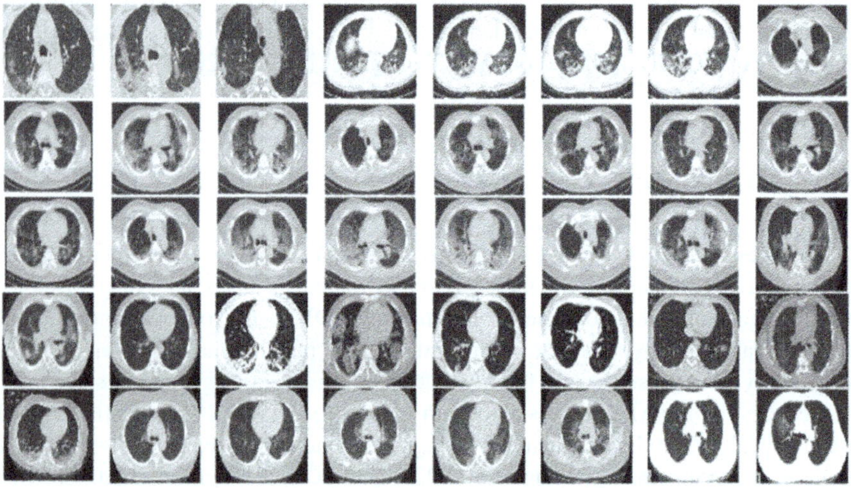

Fig. 4.2 Positive Covid-19 images using CT-scan

Then we split the dataset into train, test, and concatenate the datasets with test_size = 0.2.

We then Retrained our model, such as VGG were used. These models were pertained using the ImageNet data set and were further trained using the CT-scan data set. **VGG** is an innovative object-recognition model that supports up to

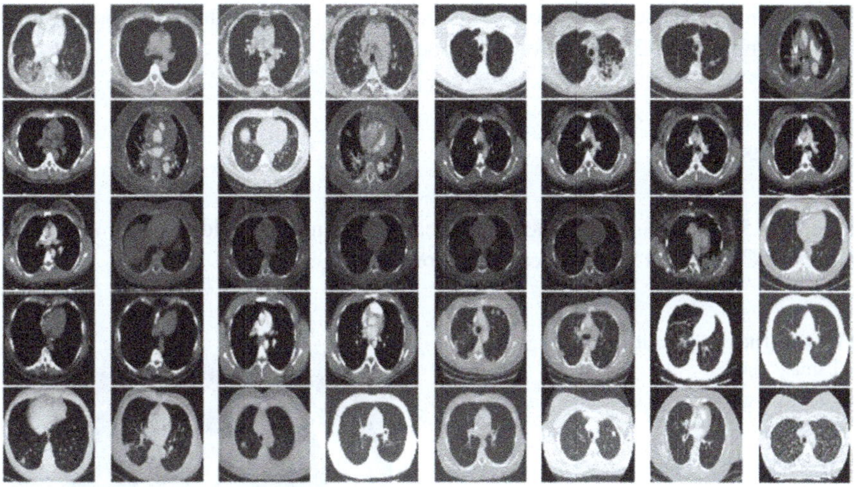

Fig. 4.3 Negative Covid-19 using CT-scans

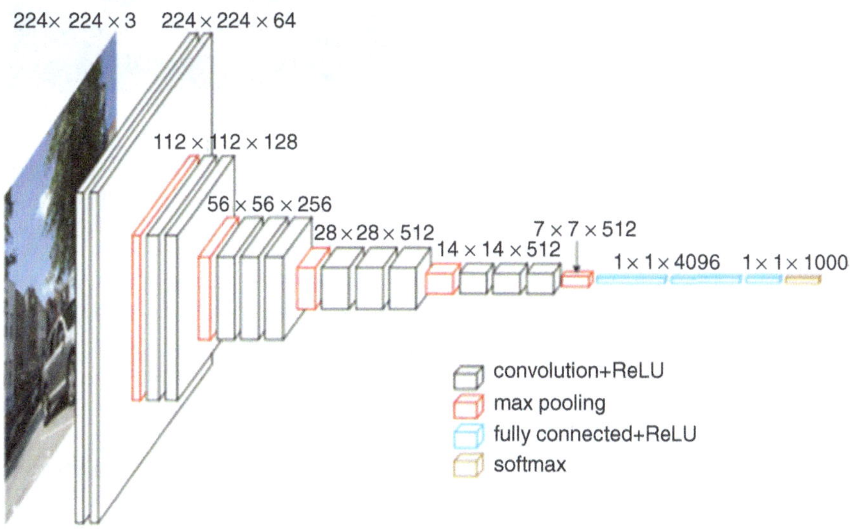

Fig. 4.4 Visual geometry model

19 layers. Built as a deep CNN, **VGG** also outperforms baselines on many tasks and datasets outside of ImageNet, Such as Fig. 4.4.

The input images were fed to the visual geometry group-16 (VGG-16) model with a dimension of 150 × 150 × 3. The model consists of 19 layers, having 5 convolutional blocks, each block consisting of two or three convolutional layers and 5 max-pooling layers, finally ending with 2 fully connected (FC) and a softmax layer. We just replaced the softmax layer with the sigmoid layer for binary classification purposes. The model was trained with root mean square propagation and a learning rare $(2^*e - 5)$ for 100 epochs.

4.4 Results

Figures 4.5 and 4.6 show a model that can differentiate covid positive and covid Negative CT images with an accuracy more than 70%.

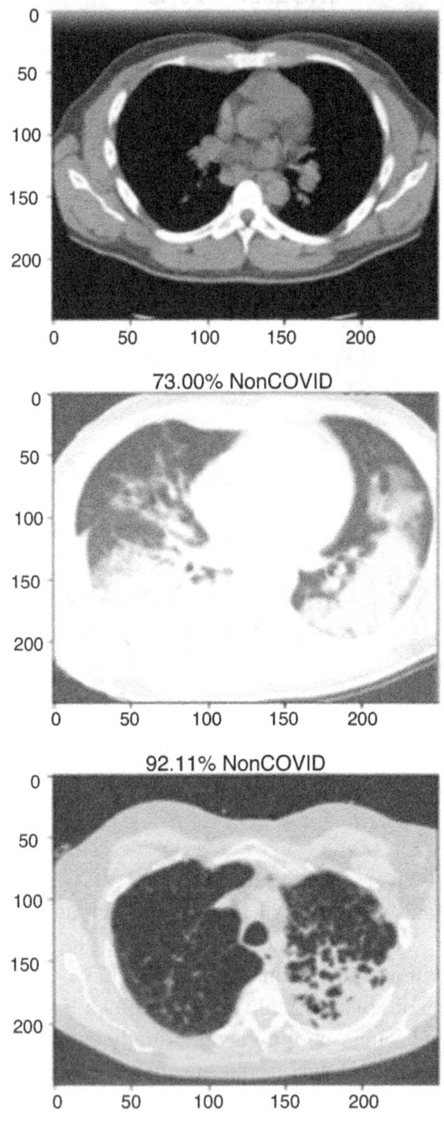

Fig. 4.5 The covid positive CT image model

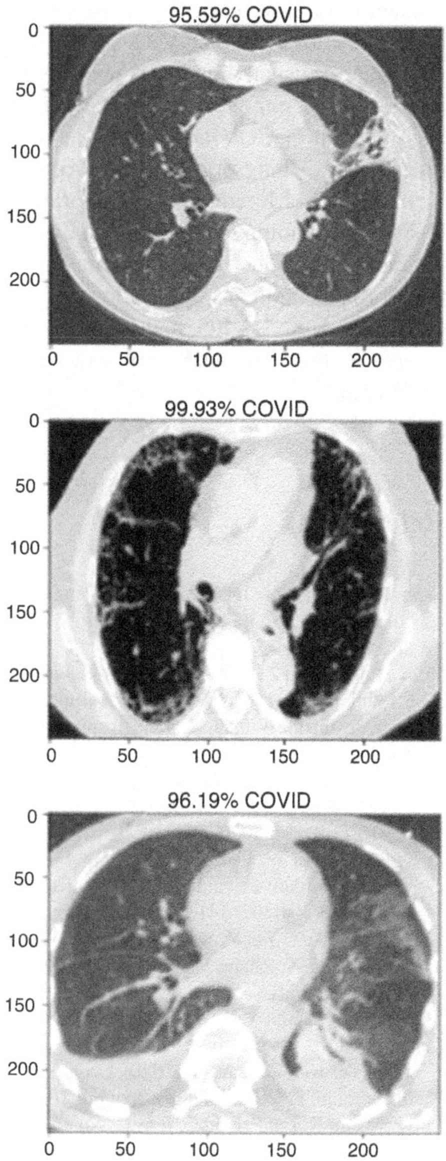

Fig. 4.6 The covid negative CT image model

4.5 Conclusion and Future Work

During the Covid 19 pandemic, the assessment of AI may become the most useful factor by reducing the pressure on doctors. Though it can take up to 15 min to manually read a CT-scan, AI can examine the images in 10 s. Therefore, the use of artificial neural networks for progressive image processing can enable most patients to identify diseases easily, quickly, and accurately, thereby significantly improving the purpose of CT in COVID-19 identification. The continuation of AI-based CT imaging tests usually involves the following steps: regional division of the Region of Interest (ROI), removal of pulmonary tissue, identification of regional infection, and classification of COVID-19. A basic basis for examining AI-based imagery is the recognition of lung organs, lung lobes, bronchopulmonary segments, and regions with infection. Moreover, we are working on Selling this project as a product to the hospitals, Radiology Centers, Clinics, Medical & Research Centers. As a result, upgrade our product by diagnosing more and more diseases PET CT-Scans. Furthermore, creating an application for diagnosing the CT-Scans by sending the CT-Scans, X-rays, EEG, and ECG on the application and the doctors. Clinics will receive it, hospitals analyze it and send the patient's feedback remotely on the application.

References

Fang Y, Zhang H, Xie J, Lin M, Ying L, Pang P, Ji W (2020) Sensitivity of chest CT for COVID-19: comparison to RT-PCR. Radiology 296:E115–E117. https://doi.org/10.1148/radiol.2020200432

He JL, Luo L, Luo ZD, Lyu JX, Ng MY, Shen XP, Wen Z (2020) Diagnostic performance between CT and initial real-time RT-PCR for clinically suspected 2019 coronavirus disease (COVID-19) patients outside Wuhan, China. Respir Med 168:105980. https://doi.org/10.1016/j.rmed.2020.105980

Huang L, Han R, Ai T, Yu P, Kang H, Tao Q, Xia L (2020) Serial quantitative chest CT assessment of COVID-19: deep-learning approach. Radiol Cardiothorac Imaging 2:e200075. https://doi.org/10.1148/ryct.2020200075

Li K, Fang Y, Li W, Pan C, Qin P, Zhong Y, Liu X, Huang M, Liao Y, Li S (2020a) CT image visual quantitative evaluation and clinical classification of coronavirus disease (COVID-19). Eur Radiol 30:4407–4416. https://doi.org/10.1007/s00330-020-06817-6

Li L, Qin L, Xu Z, Yin Y, Wang X, Kong B, Bai J, Lu Y, Fang Z, Song Q, Cao K, Liu D, Wang G, Xu Q, Fang X, Zhang S, Xia J, Xia J (2020b) Artificial intelligence distinguishes COVID-19 from community acquired pneumonia on chest CT. Radiology:200905. https://doi.org/10.1148/radiol.2020200905

Liang T, Liu Z, Wu CC, Jin C, Zhao H, Wang Y, Wang Z, Li F, Zhou J, Cai S, Liang Y, Zhou H, Wang X, Ren Z, Yang J (2020) Evolution of CT findings in patients with mild COVID-19 pneumonia. Eur Radiol. https://doi.org/10.1007/s00330-020-06823-8

Luo Z, Wang N, Liu P, Guo Q, Ran L, Wang F, Tang Y, Li Q (2020) Association between chest CT features and clinical course of Coronavirus Disease 2019. Respir Med 168:105989. https://doi.org/10.1016/j.rmed.2020.105989

Mei X, Lee H, Diao K, Huang M, Lin B (2020) Artificial intelligence–enabled rapid diagnosis of patients with COVID-19. Nat Med 26:1224–1228

Pan Z, Yanbing D, Xia W, Junke L, Yanjun Z, Yiming L (2020) The epidemiology, diagnosis and treatment of COVID-19. Int J Antimicrobial Agents 55:105955. https://doi.org/10.1016/j.ijantimicag.2020.105955

Ren J, Ren R, Green M, Huang X (2019) Defect detection from X-ray images using a three-stage deep learning algorithm. 2019 IEEE canadian conference of electrical and computer engineering (CCECE). doi:https://doi.org/10.1109/ccece.2019.8861944

Soghaier MA, Saeed KMI, Zaman KK (2015) Public health emergency of international concern (pheic) has declared twice in 2014; polio and ebola at the top. AIMS Public Health 2:218
Udugama B, Kadhiresan P, Kozlowski HN, Malekjahani A, Osborne M, Li VYC, Chen H, Mubareka S, Gubbay JB, Chan WCW (2020) Diagnosing COVID-19: the disease and tools for detection. ACS Nano 14:3822–3835. https://doi.org/10.1021/acsnano.0c02624
Wang K, Kang S, Tian R, Zhang X, Zhang X, Wang Y (2020) Imaging manifestations and diagnostic value of chest CT of coronavirus disease 2019 (COVID-19) in the Xiao an area. Clin Radiol 75:341–347. https://doi.org/10.1016/j.crad.2020.03.004
Ye Z, Zhang Y, Wang Y, Huang Z, Song B (2020) Chest CT manifestations of new coronavirus disease 2019 (COVID-19): a pictorial review. Eur Radiol. https://doi.org/10.1007/s00330-020-06801-0
Zhao W, Zhong Z, Xie X, Yu Q, Liu J (2020a) CT Scans of patients with 2019 novel coronavirus (COVID-19) pneumonia. Theranostics 10:4606–4613. https://doi.org/10.7150/thno.45016
Zhao J, Zhang Y, He X, Xie P (2020b) COVID-CT-Dataset: A CT scan dataset about COVID-19. ArXiv preprint arXiv: 2003.13865
Zhou Z, Guo D, Li C, Fang Z, Chen L, Yang R, Li X, Zeng W (2020) Coronavirus disease 2019: initial chest CT findings. Eur Radiol. https://doi.org/10.1007/s00330-020-06816

Chapter 5
Predicting the Pandemic Effect of COVID 19 on the Nigeria Economic, Crude Oil as a Measure Parameter Using Machine Learning

Cecilia Ajowho Adenusi, Olufunke Rebecca Vincent, Jesufunbi Bolarinwa, and Ologunye Oluwakemi

Abstract The spread of the Covid-19 pandemic in Nigeria has had a great impact on the economy. Prior to 2019, the crude oil price (US $/l) ranged from 62 to 79.59 starting in 2018. The year 2019 ranges between 59.1 to 73.65 per barrel. It will be between 14.28 and 66.65 per barrel in 2020. The Covid-19 pandemic will commence around December 2019 in China before being discovered outside China in the year 2020. Because of the aforementioned crude oil prices, crude oil prices increased until the year 2020, when there was a total lockdown on the export of oil products, causing crude oil prices to fall. This study investigates the impact of the pandemic on the Nigerian economy, utilizing crude oil as a good parameter measure. This was achieved by using a machine model (random tree) for prediction, and the results were further compared (Sequential Minimal Optimization (SMOreg), decision table, Random Forest, M5 tree model (M5P) and Gaussian Processes) for better accuracy. The results obtained thus far from the combination of datasets used can be used to manage the Nigerian economy, particularly the crude oil industry, in search of ways to mitigate the damage caused by the pandemic.

Keywords Crude oil · M5P · Prediction · Random tree · SMOreg · NCDC · Decision table · Random forest · CBN · Price · Barrel

---

C. A. Adenusi (✉)
Linux Professional Institute, Nigeria Master Affiliate, Ibadan, Nigeria

O. R. Vincent · J. Bolarinwa
Department of Computer Science, Federal University of Agriculture, Abeokuta, Nigeria

O. Oluwakemi
Redeemer University, Ogun State, Nigeria

S. A. Hassan et al. (eds.), *Decision Sciences for COVID-19*, International Series in Operations Research & Management Science 320,
https://doi.org/10.1007/978-3-030-87019-5_5

5.1 Introduction

This pandemic, dubbed Covid-19, has had a significant negative impact on the African continent, particularly in terms of business and economics. For the first time on December 30, 2019, Jinyintan Hospital in Wuhan, China, detected a sickness. This pandemic had spread to 213 nations as of November 24, 2020, with 4,464,255 cases, 299,410 fatalities, and 1,678,246 recovered cases, including Africa. The five continents that comprise Africa are Western, Eastern, Northern, Middle, and Southern Africa, with the Western half having the most people (206,139,589) and the major companies in Africa being oil and gas, real estate, technology, telecommunications, health care, finance, insurance, media, agriculture, tobacco, transportation, building materials, drinks, metal and mining, The business sector is further classified into technology (19.9%), finance (15%), metals and mining (14.7%), media (10.8%), telecommunications (7.7%), luxury goods (4.7%), retail (4%), insurance (3.4), and all other industries (19.8%). All of these African continents were severely impacted by the epidemic, and as of the writing of this article, there are no licensed vaccinations now being dispersed across the world, with larger parts going to the most severely impacted countries and less going to the least severely impacted. This vaccine is a big breakthrough that is expected to provide additional respite from the country's health problems, which would, in turn, have a rippling impact on the global economy, including in Africa.

The oil and gas industry has been reduced to less than half of what it was before the start of the epidemic, and the crude oil market in the USA has also been revealed to be low. The reduction in crude oil prices, which is now hitting the African economy, has had a significant impact on businesses and initiatives (Nwoba et al. 2017). As a mono-product economy, the African continent remains powerless to advancement and growth in global unrefined fuel costs. Crude oil is an important component of African money and financial strategy, accounting for 80% of governmental revenue, 12% of GDP (Yusuf 2015), and 90–95% of foreign products, which is a percentage of export revenue (Yusuf 2015). In the medical record, time series, computer vision, natural language processing, and image classification, there has been a lot of interest in AI recently, and deep learning approaches have shown excellent results (Cireşan 2013). The dual-channel convolutional neural network (DCCNN) model was suggested by Jianfang Cao (2019) after the two convolutional neural networks (CNN) were combined with various topologies. The results show that DCCNN outperforms CNN in terms of accuracy. As evidenced by the AlexNet model (Krizhevsky et al. 2012), which demonstrated better improvement in picture categorization. Machine learning will be used in this research. (Random Tree) to estimate crude oil prices in Nigeria as a parameter measure for evaluating the impact of Covid-19 on the Nigerian economy, and the findings of the other five models will be compared to the real model used. This study's remaining chapters are divided into separate sections. Section 5.2 provides an overview of previous work. Section 5.3 provides full details on the methods utilized, and the techniques used to carry them out, the dataset, and Sect. 5.4 provide the whole outcome of the models used in this research study. Finally, Sect. 5.5 summarizes our work, including limitations and contributions to knowledge.

5.2 Related Works

The use of Long-term short-term memory (LSTM) for forecasting economic commodity prices, like crude oil prices (98.2 prices) based on the variability of commodity prices, has gotten a lot of attention from researchers and investors over the years. This trend is continuing with the use of Long-term short-term memory (LSTM) for forecasting economic commodity prices, like crude oil prices (98.2 prices) based on the variability of commodity prices. The findings show that the pandemic has had an impact on commodity price variability based on the total number of deaths and confirmed cases. This was done using a hybrid ARIMA-Wavelet model to predict Covid19's spread across the commodity market, which involves many players, such as traders, portfolio managers, brokers, and investors as a whole (Moews and Ibikunle 2020).

Deep learning is a special type of artificial neural network (ANN) that allows for high-level abstraction and numerous processing layers to model data, and it automatically retrieves acceptable input data using a generic learning approach. The recurrent neural network (RNN) and its modifications, time series forecasting for finance, and convolutional neural networks (CNNs) for image identification were all featured in this study. (Chen He 2018) examined CNN and deep learning approaches for forecasting the Chinese stock market's stock prices, and the results demonstrate that deep learning is feasible. Chen et al. (2016) proposed deep convolutional neural networks for predicting the Taiwan Stock Index in place of an improved algorithmic training framework.

Under the effects of the epidemic and the confrontation between Saudi Arabia and Russia, crude oil has undergone unexpected and unexpected developments. As a result, an analysis was performed on the volatilities of Brent, Dubai, and West Texas Intermediate (WTI) on daily crude oil price data spanning from May 2006 to March 2020, using Lyapunov exponent tests and Shanon Entropy, with the test results indicating that oil prices are chaotic. Currently, this study proposes a hybrid modeling methodology based on the Logistic Smooth Transition Autoregressive Generalized Autoregressive Conditional Heteroscedasticity (LSTARGARCH) model and the Logistic Smooth Transition Autoregressive Conditional Heteroscedasticity (LSTM) model for analysis of the volatility of oil prices.

Some researchers or papers have used a genetic algorithm (GA) (Kaboudan 2001), a support vector machine (SVM) (Xie et al. 2006), a neural network (Moshiri and Foroutan 2006), and support vector regression with a meta-heuristic algorithm, while another method used in the literature is the decomposition integration (EMD) method (Zhang et al. 2009). By proposing a class of Neural Network NN models (Donaldson and Kamstra 1997). To anticipate oil prices, Gonzalez Miranda and Burgess et al. (1997) coupled the GARCH model with the neural network. Roh

(2007) combined the GARCH, EWMA, and EGARCH models with a feedforward neural network and discovered that the combination produces the best results.

5.3 Methodology

This entails the methods used for data gathering and collation, analysis, and prediction methods that were executed in this research work.

5.3.1 Dataset

The COVID19 data used was gotten from the Nigeria Center for Disease (NCDC), and it comprises data from January 2019 to December 2020, and the daily crude oil price used in Nigeria was gotten from the central bank of Nigeria website (www.cbn.gov.ng). The oil prices were taken daily up to the time of writing this paper, and the analysis was done using Orange and prediction was done using Random tree via Weka application.

5.3.2 Machine Learning Models

Machine learning is an artificial intelligence branch that may be used for data processing, visualization, mining, and other tasks. This is divided into three (3) categories: supervised learning, unsupervised learning, and reinforcement learning. Classification and regression are two types of supervised learning. The user's random tree falls within the category of supervised learning, and the other models utilized in the comparison are all supervised machine learning models.

5.3.2.1 Random Tree Model

A random tree is defined as an arborescence created by stochastic processes in computer science and mathematics, and its forms include a random binary tree, recursive tree, random forest, Brownian tree, spanning tree, and the like. This always comes with three ways of structuring or representing the model, and its visualized results are always in tree form.

The Algorithm for Random Tree that was made in this study is given below:

```
Begin:
Input: To generate F Classifier,
 For I = 1 to F, do
     Randomly sample training data D with replacement to produce D1
     Create a root node N, containing D1,
     Call BuildTree (N)
 End for
BuildTree (N)
If N contains instances of only one class then,
Return.
 Else:
       Randomly select z% of the possible splitting features in N
       Select the feature F with the highest information gain to split on.
       Create f child nodes of n, n1......nf,
       Where f has f possible values ( f1.....fn)
For i=1 to f do
       Set content of N1 to D1 where D1 is all instances in N that match F1.
       Call BuildTree (N)
 End for
End if.
```

5.3.2.2 Decision Table

Decision tables are a simple visual representation of specifying which actions to take in response to certain circumstances. They are algorithms that produce a set of actions as their output. The information in decision tables might potentially be represented as decision trees or a set of if-then-else and switch-case instructions in a computer language. Each choice is associated with a variable, relation, or predicate, the potential values of which are presented among the condition alternatives. Each item specifies whether (or in what sequence) the action should be executed given the set of condition alternatives that the entry relates to.

5.3.2.3 Random Forest

Random forests, also known as random decision forests, are an ensemble learning method for classification, regression, and other problems that work by training a large number of decision trees. For classification tasks, the random forest's output is the class chosen by the majority of trees. The mean or average prediction of the individual trees is returned for regression tasks.

5.3.2.4 M5P

Quinlan (1992) presented the M5 tree as a decision tree learner for regression situations. By classifying or splitting the entire data space into numerous subspaces, this tree approach assigns linear regression functions to the terminal nodes and applies a multivariate linear regression model to each subspace and subspace. The

M5 tree approach is used to handle continuous classes. Rather than separate classes, problems can be solved, and tasks can be completed, having a high degree of dimensionality. It displays the piecewise information of each linear model built to approximate the data set's nonlinear interactions.

5.3.2.5 Gaussian Processes

A Gaussian process is a stochastic process (a collection of random variables indexed by time or space) that has a multivariate normal distribution for every finite collection of those random variables, i.e., every finite linear combination of them is normally distributed in probability theory and statistics. Gaussian processes benefit from traits inherited from the normal distribution, making them valuable in statistical modeling. The distributions of various derived values can be computed directly if a random process is treated as a Gaussian process, for example.

5.3.2.6 SMOreg

SMOreg is a regression program that uses the support vector machine. Various algorithms can be used to learn the parameters. The RegOptimizer is used to choose the algorithm. Shevade, Keerthi, and others developed the most prevalent method (RegSMOImproved), which is the default RegOptimizer.

5.3.3 Proposed Model Used

Figure 5.1 illustrates the workflow of this research work, ranging from the collation of needed information via the prediction of the crude oil price and it was drawn using edraw max to sketch out how it works. There was a dataset used for this research work. The first is from the NCDC website, which includes Nigeria Covid-19 information, and the second is from the CBN website, which includes crude oil price information. These two data information were put into consideration. This paper also includes analysis and visualization of covid-19 data. Likewise, analysis and prediction of crude oil data were visualized, models developed, prediction done via Random Tree, validation of the model used was done by comparing the model with the other five (3) diverse models and the result gotten was tabulated in Table 5.3.

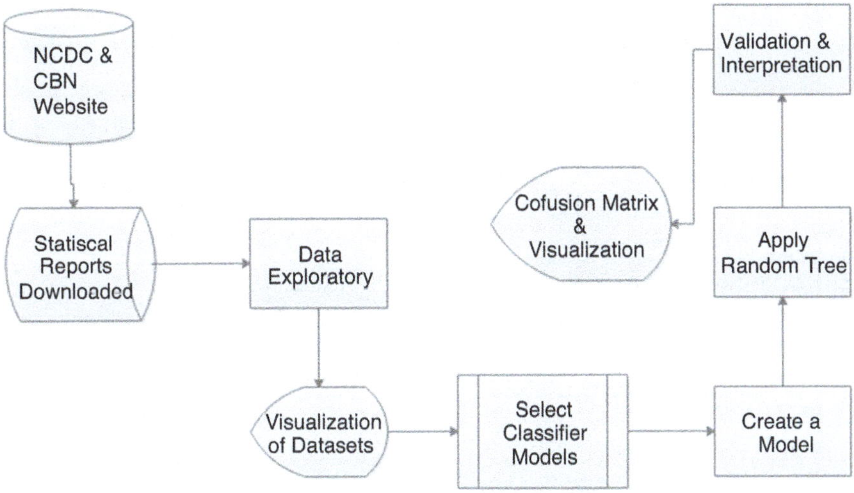

Fig. 5.1 Predicted crude oil model developed

Fig. 5.2 Crude oil price outcome. Source: https://www.cbn.gov.ng/rates/crudeoil.asp

5.3.4 Production, Crude Oil Price and Export from Central Bank of Nigeria

Figure 5.2 depicts the crude oil price per barrel for the years 2019 and 2020. This output was only caught since it falls inside the Covid-19 timeframe, and it was obtained from the above-mentioned source, and crude oil prices are dropping, steady, rising, and declining.

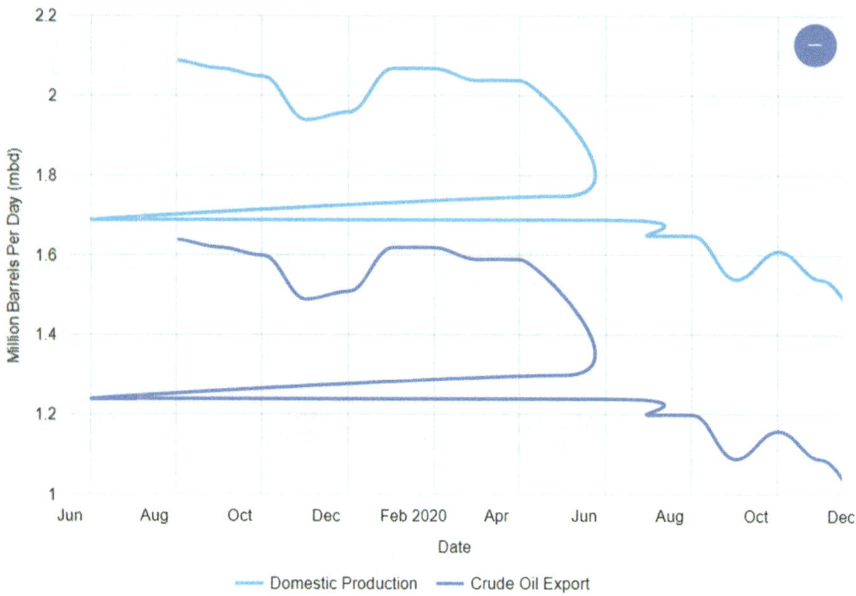

Fig. 5.3 Production outcome. Source: https://www.cbn.gov.ng/rates/crudeoil.asp

Table 5.1 Crude oil export outcome

	Year 2020		
Month	October	November	December
Crude oil price (Bonny)	39.74	42.70	50.33
Domestic Production	1.61	1.54	1.42
Crude Oil Export	1.16	1.09	0.97

Source: Nigerian National Petroleum Corporation (NNPC) and Reuters

Figure 5.3 depicts the crude oil production results for Covid-19, and the graph depicts the falling and rising rates of crude oil price production.

Table 5.1 describes the crude oil that was exported during the pandemic era for the year 2020, and the source of the data is also included above.

5.3.5 *Covid 19 Nigeria Dashboard*

The covid-19 cases in Nigeria are depicted in Fig. 5.4. According to the graph, there was no record of covid-19 in 2019, and only a small amount was identified in weeks 20 and 40 of 2020, while early 2021 had an increase in confirmed cases, but this fell again from week 20 to the present, when we have no record of covid-19 in Nigeria (Fig. 5.5).

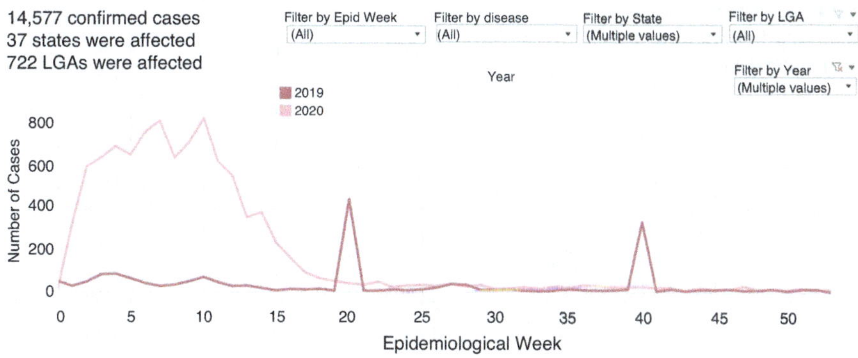

Fig. 5.4 Confirmed Cases of Covid-19. Source: https://ncdc.gov.ng/data

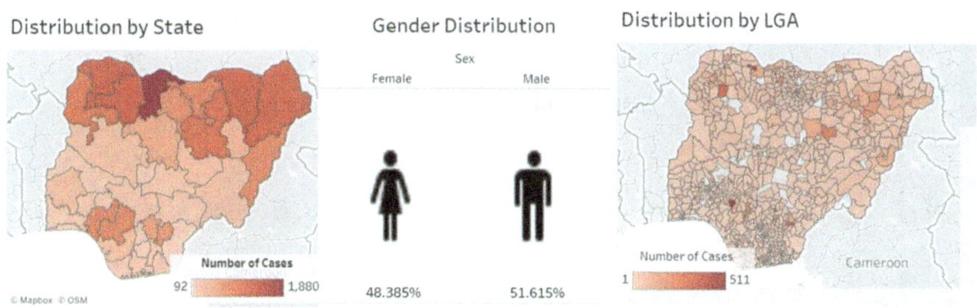

Fig. 5.5 Visualization of COVID-19 by state, gender, and local government areas. Source: https://ncdc.gov.ng/data

5.4 Results and Discussion

Comparison of results was done using five diverse machine learning techniques, which are SMOreg, M5P, Decision Table, Random Forest and Gaussian processes, and the overall performance of the model used is assessed by MAE (Mean Absolute Error), Mean Absolute Scaled Error (MASE) and Root Mean Square Error (RMSE).

Table 5.2 shows the feature data selected from the dataset downloaded from the CBN website and it only contains three distinct variables, which are the year, the month, and the crude oil prices. These tables were used for analysis as well as visualization.

The scatter graph in Fig. 5.6 analyzes monthly crude oil price data. The prices with multiple colors and a regression line of 0.3 were analyzed.

Table 5.2 Crude oil data table with 179 instances and no missing data

	Year	Month	CrudeOilPrice
1	2006	1	63.85
2	2006	2	61.33
3	2006	3	65
4	2006	4	72.09
5	2006	5	71.18
6	2006	6	69.32
7	2006	7	75.13
8	2006	8	75.15
9	2006	9	62.97
10	2006	10	59.49
11	2006	11	59.81
12	2006	12	64.7
13	2007	1	55.57
14	2007	2	59.97
15	2007	3	64.28
16	2007	4	70.46
17	2007	5	70.4

Source: Crude oil dataset on cbn.org.ng

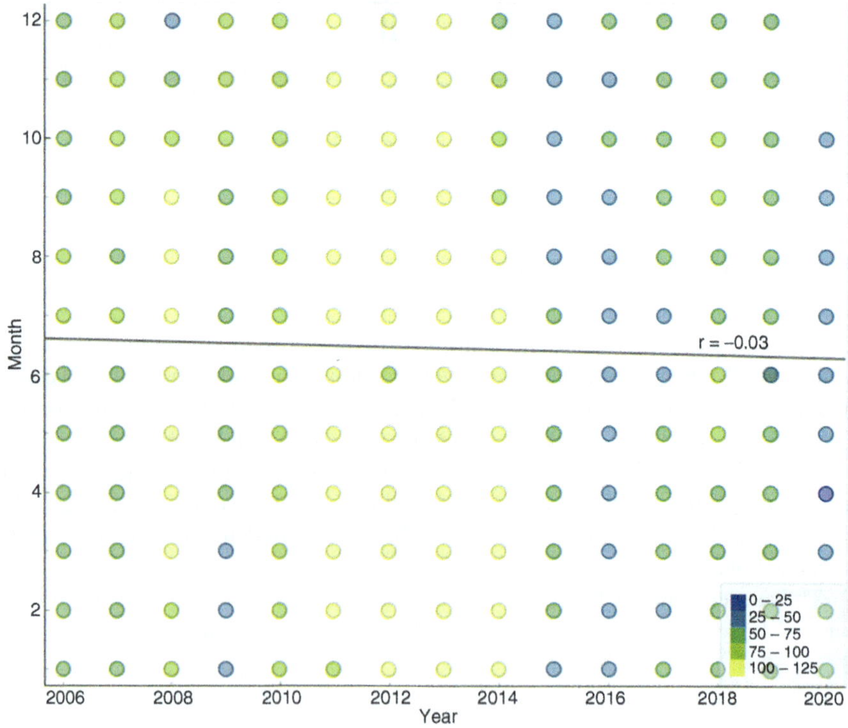

Fig. 5.6 Visualization of crude oil price showing the regression line with scatter plot

5.4.1 *Prediction Results*

Crude oil price predictions were made using prices extracted from the dataset for the years 2019 and 2020 because that was the Covid19 era nationwide and in Nigeria, using the Random Tree model, which had the lowest mean absolute error (1.2161) and Root mean squared error (3.9018).

Table 5.3 shows the result obtained by using a random tree of a machine learning model for predicting the next crude oil price or the probability of increasing or decreasing in prices. Instances are made up of data from the years 2019 (1–12, January–December) and 2020 (13–23, January–November). The actual column entails the real price from the dataset downloaded, and the predicted column entails the derived/predicted crude oil prices from the method used, and the error column entails the percentage error of occurrences of the predicted values.

Figure 5.7 shows the actual values against the predictable values of crude oil prices in Nigeria and, from the results obtained, some new prices were predicted along with some that remain at the same value as the actual values.

Figure 5.8 shows the visualization of the prediction processes in a tree-like format. Lesser values appear on the left side of the tree or nodes, while greater values appear on the right.

Table 5.3 Crude oil prediction result using random tree

	Instances	Actual	Predicted	Error
1	1	60.39	60.39	0
2	2	64.89	64.89	0
3	3	67.67	67.67	0
4	4	73.08	73.365	0.285
5	5	73.65	73.365	-0.285
6	6	66.74	53.52	-13.22
7	7	66.24	66.24	0
8	8	61.05	61.05	0
9	9	65.27	65.27	0
10	10	59.1	59.1	0
11	11	63.56	63.56	0
12	12	68.56	68.56	0
13	13	66.68	66.68	0
14	14	58.45	58.45	0
15	15	32.29	32.29	0
16	16	14.28	14.28	0
17	17	27.9	27.9	0
18	18	40.3	53.52	13.22
19	19	40.3	40.3	0
20	20	44.1	44.58	0.48
21	21	45.06	44.58	-0.48
22	22	40.85	40.85	0
23	23	39.74	39.74	0

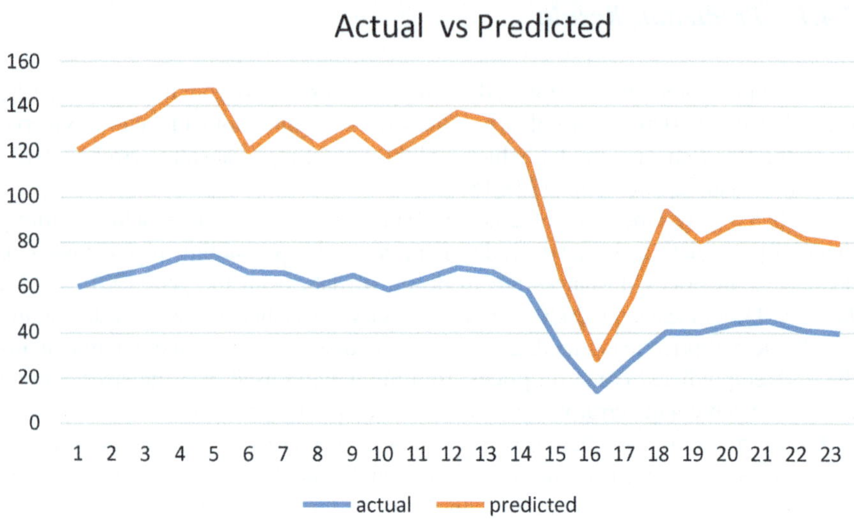

Fig. 5.7 Actual vs. predicted crude oil prices

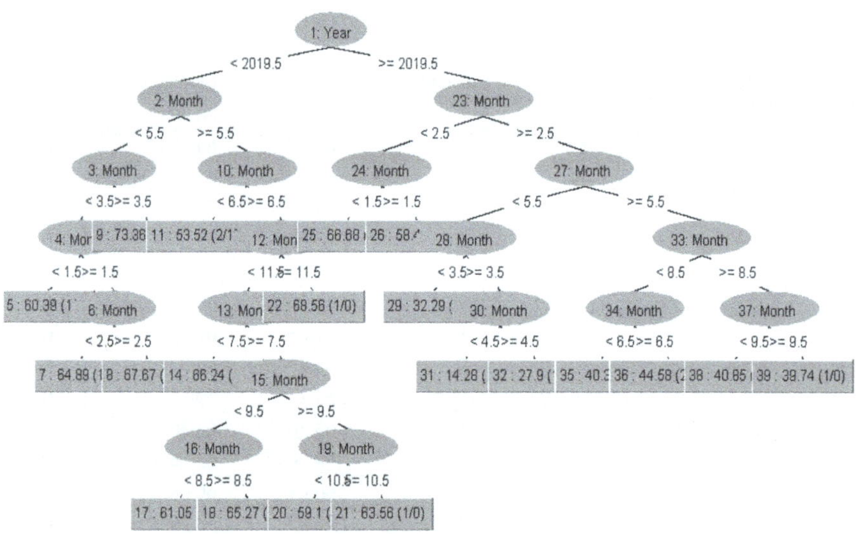

Fig. 5.8 Visualization of random tree model used

The result obtained using the random forest was further compared with other machine learning models like Gaussian Process, SMOreg, Decision Table, M5P, and Random Forest for better accurate results and the lower mean absolute error and root mean squared error were higher than recommended, among others, as shown in Table 5.4.

Table 5.4 Comparative results tested model

Evaluations	Gaussian processes	SMOreg	Decision table	M5P	Random forest	Random tree
Mean absolute error	9.7646	7.2411	7.4307	7.4307	3.5763	1.2161
Root mean squared error	12.4577	11.0617	11.0055	11.0055	5.3609	3.9018
Relative absolute error	70.20%	51.90%	53.26%	53.26%	25.63%	8.72%
Root relative squared error	78.76%	69.93%	69.58%	69.58%	33.89%	24.67%

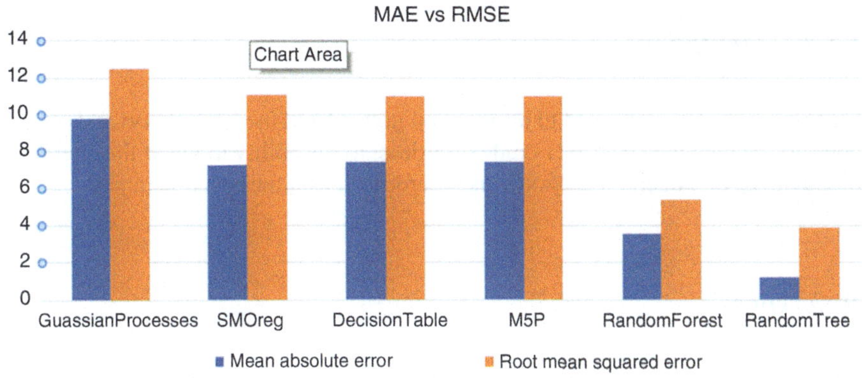

Fig. 5.9 MAE vs. RMSE

Figure 5.9 shows the mean absolute error and root mean squared error of all the machine learning models used, denoting their values using a bar graph. With Gaussian Processes having the highest MAE and RMSE, while the Random Tree has the lowest MAE and RMSE. This is why the Random Tree is used as a crude oil price prediction model.

5.5 Conclusively

In this study, Covid19 data was examined using the Nigeria Center for Disease Control website, and visualizations based on cases, distribution, gender, and local government areas were captured. Likewise, information about the crude oil price, production, and export oil data was captured from the central bank of Nigeria's website under the statistics report tab. The dataset was downloaded and used for visualization and prediction. Visualization was achieved via Orange data mining application and predicted crude oil prices with the Weka java application. Data from 2000 till 2020 was used for the crude oil visualization while data from 2019 to 2020 was used for the prediction, being the covid19 era. The visualization and prediction results are all under the result section of this work.

According to Figs. 5.4 and 5.5 of this research work, in the year 2019, there were few records of cases of covid-19 in Nigeria, and in the early year 2020, there were high cases of covid-19 which led to the total lockdown of the country, and the rate started dropping from week 20 of 2020. To date, there is no record of covid-19. As result of the discussion of this work, it includes information about crude oil prices, beginning with the raw dataset and progressing through feature selection, analysis, visualization, prediction, and performance metrics. The Covid-19 era had a significant impact on crude oil prices (Table 5.2), particularly around the early year 2020.

5.5.1 Limitation

This research work is limited to information gotten from the CBN report and, which does not include the year 2021 crude oil prices, to the ability to know the aftermath of Covid-19 on Nigeria's economy using crude oil as a measured parameter.

References

Cao J, Wu C, Chen L, Cui H, Feng G (2019) An improved convolutional neural network algorithm and its application in multilabel image labeling. Comput Intell Neurosci 1:1–11

Chen S, He H (2018) Stock prediction using convolutional neural network. In: IOP Conference series: materials science and engineering, vol 435. IOP Publishing, p 012026

Chen JF, Chen WL, Huang CP, Huang SH, Chen AP (2016) Financial time-series data analysis using deep convolutional neural networks. In: 2016 7th International conference on cloud computing and big data (CCBD). IEEE, pp 87–92

Cireşan DC, Giusti A, Gambardella LM, Schmidhuber J (2013) Mitosis detection in breast cancer histology images with deep neural networks. In: International conference on medical image computing and computer-assisted intervention. Springer, Berlin, pp 411–418

Donaldson RG, Kamstra M (1997) An artificial neural network–GARCH model for international stock return volatility. J Empir Financ 4:17–46

Gonzalez Miranda F, Burgess N (1997) Modelling market volatilities: the neural network perspective. Eur J Financ 3(2):137–157

https://ncdc.gov.ng/data

https://www.cbn.gov.ng/rates/crudeoil.asp

https://www.offshore-technology.com/investment/covid-19-creates-supply-and-demand-crisis-for-oil-and-gas/ (Accessed on 14 May 2020)

https://www.worldometers.info/coronavirus/ (Accessed on 14 May 2020)

Kaboudan MA. Compumetric forecasting of crude oil prices. In Proceedings of the 2001 congress on evolutionary computation, Seoul, Korea, 27–30 May 2001, vol 1, pp 283–287

Krizhevsky A, Suskever I, Hinton GE (2012) ImageNet classification with deep convolutional neural networks. In: International conference on neural information processing systems, pp 1106–1114

Moews B, Ibikunle G (2020) Predictive intraday correlations in stable and volatile market environments: evidence from deep learning. Physica A: Stat Mech Appl 547:124392

Moshiri S, Foroutan F (2006) Forecasting nonlinear crude oil futures prices. Energy J 27:81–96
Nwoba MO, Nwonu C, Agbaeze EK (2017) Impact of fallen oil prices on the nigeria economy. J Poverty Invest Dev 33:75–82
Roh TH (2007) Forecasting the volatility of stock price index. Expert Syst Appl 33:916–922
Xie W, Yu L, Xu S, Wang S. Anew method for crude oil price forecasting based on support vector machines. In Proceedings of the international conference on computational science, Reading, 28–31 May 2006, pp 444–451
Yusuf M (2015) An analysis of the impact of oil price shocks on the growth of the Nigerian economy: 1970–2011. Afr J Bus Manag 9(3):103–115
Zhang X, Yu L, Wang S, Lai KK (2009) Estimating the impact of extreme events on crude oil price: An EMD-based event analysis method. Energy Econ 31:768–778

Part II
Forecasting Techniques

Chapter 6
Time Series Analysis and Forecast of COVID-19 Pandemic

Pawan Thapa

Abstract Background: The coronavirus has killed over 80 million individuals globally. Thus, the linear regression and autoregressive integrated moving average (ARIMA) model analyze the pattern of COVID-19 and identify the future confirmed cases.

Methods: In this study, the dataset was used from the Johns Hopkins University (JHU CSSE) data repository in COVID-19 analytics package and prophet library. The time series analysis creates a simulating linear regression and ARIMA model for COVID-19 confirmed cases. The best fit model is select by Akaike information criteria (AIC) and predicts short-term issues validated by Ljung-Box Q test using RStudio Cloud.

Results: The linear regression and ARIMA model identifies a best-fit model for time series data. From this model, forecast of more than 300,000 to 1,500,000 from 2020 to 2022. In addition, it depicts a significant increasing trend in the future predictions of confirmed cases.

Conclusion: This forecast can help estimate the number of cases that information can provide control measures for an epidemic outbreak. It can suggest the government plan the policies regarding the control of the spread of the virus.

Keywords Time series · Confirmed cases · Linear regression · ARIMA · COVID-19 · Predication

List of Abbreviations

ARIMA	Auto-Regressive Integrated Moving Average
ACF	Auto Correlation Function
ADF	Augmented Dicky-Fuller test

P. Thapa (✉)
Department of Geomatics Engineering, Kathmandu University, Dhulikhel, Nepal
e-mail: pawan.thapa@ku.edu.np

S. A. Hassan et al. (eds.), *Decision Sciences for COVID-19*, International Series in Operations Research & Management Science 320,
https://doi.org/10.1007/978-3-030-87019-5_6

AIC	Akaike Information Criteria test
PACF	Partial Auto Correlation Function
JHU	Johns Hopkins University
WHO	World Health Organization

6.1 Introduction

The coronavirus cases had spread in every continent ("Coronavirus Spreads to Antarctic Research Station" 2020; Coronavirus Update (Live) n.d.; Thapa 2021). It has impacted every countries people socioeconomically, physically, and mentally. Each country's governments implement partial or complete lockdown to reduce its spread, restriction on mass gathering, social distancing, and face mask use. Still, these measures are unable to control the spread worldwide. Also, a new variant of coronavirus confirmed cases identified in December 2020 in the United Kingdom, France, Italy, Germany, Spain, Sweden, Switzerland, Denmark, the Netherlands, Canada, Australia, Japan, Singapore, India, Lebanon, and the United Arab Emirates (Hauck n.d.; Wise 2020). It has spread rapidly, with 33 countries confirmed cases (Gutierrez et al. 2021). A model recommends an increased transmission of 70% compared to the previous one (Hauck n.d.). Several other experts predict that the number of new virus cases will rise considerably in the upcoming days (20201005-Weekly-Epi-Update-8. Pdf n.d.; Petropoulos and Makridakis 2020; Scudellari 2020).

In Nepal, five passengers from the United Kingdom had tested positive. It raises chances of spreading in the country (Five Arrivals from UK Test Positive, Raising Fears New Coronavirus Variant Is Here n.d.-a; "Nepal Not Sure If It Has New Coronavirus Variant with 5 Arriving from UK Testing Positive" n.d.). They are in hotel quarantine to prevent its spread throughout the country (Five Arrivals from UK Test Positive, Raising Fears New Coronavirus Variant Is Here n.d.-a). As increasing international travel on December 22, the Civil Aviation Authority of Nepal had declared not to bring passengers from those new type confirmed countries. Nepal is among the most vulnerable countries where new variants can plummet cases that might not handle care with limited health facilities.

Several studies have reported that the government lacks proper concern and plans to control its spread and irresponsible activities of people not wearing masks, gathering people for mass protest ("Covid," 2020; Li et al. 2020). However, there have been few studies in Nepal to understand the COVID-19 epidemic spread and its prediction (Basu et al. 2020; Thapa 2021). This research study creates linear regression and autoregressive integrated moving average (ARIMA) model for the spread of COVID-19 of Nepal and forecasts three years, which will help government, policymakers make an informed decision regarding this pandemic. An objective uses time series data for finding trends and forecasts from 2020 to 2022 of COVID-19. This model chosen from Akaike information criteria (AIC) and predicted infected cases validated by Ljung-Box Q test for structure and residuals using R in RStudio Cloud.

6.2 Materials and Methods

6.2.1 Study Area

Nepal's study area has confirmed more than two lakhs sixty thousand cases till January 02, 2021 (Coronavirus Update (Live) n.d.). It shows a gradual decrease in a few days as people self-isolate and stops testing; however, this will significantly increase as people protest without the mask and neglect social distancing. Another will be travelers, migrant workers coming to bring new variant type and spread in the country. Due to travel, mass protest and ignorance of face masks, social distancing, people are at alarming risk of new virus variants. Similarly, the government focuses on vaccines rather than testing, contact tracing, and genome sequence that are costly, time-taking with limited resources (Fig. 6.1).

6.2.2 Data Collection

The study used datasets acquired from different sources are shown in Table 6.1.

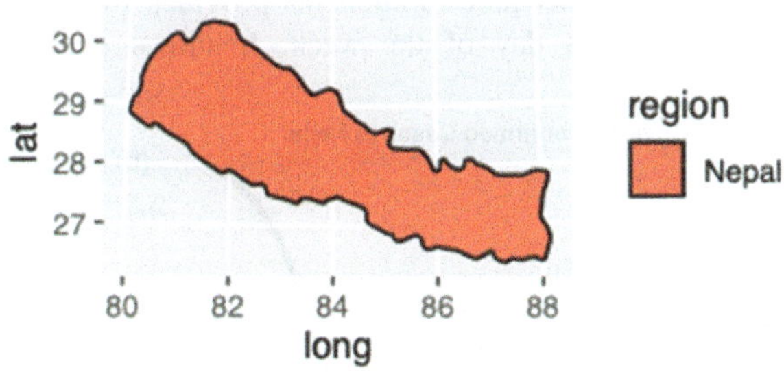

Fig. 6.1 Study area used for predicating COVID-19

Table 6.1 The datasets used for the modeling

Datasets	Data source
COVID-19	Johns Hopkins University Center for Systems Science and Engineering (JHU CSSE) data repository
Prophet	https://github.com/CSSEGISandData/COVID-19 Prophet library https://facebook.github.io/prophet/

6.2.3 Methods

This study used data from Johns Hopkins University (JHU) data repository in COVID-19 analytics package. The data time is from January 22, 2020, to January 1, 2021, which includes confirmed cases, death, date, recovered patients for all countries (COVID-19 n.d.; Ponce and Sandhel 2020). This study analyzes and predicts COVID-19 in Nepal with time series analysis using linear regression and the ARIMA model. It forecast three years (2020–2022) confirmed case scenarios using forecast, prophet library, and validation. Time series is data organized by time order and applied to develop and improve forecasting models. Prophet can handle trend shifts and outliers. The ARIMA model depends upon error lags, that is, the difference between forecasted and actual outcomes. For the ARIMA model, data must be check for stationary autocorrelation. The Auto Arima function will check via autocorrelation function (ACF) and Partial autocorrelation function (PACF). After that, fit the model using Akaike information criteria (AIC), then the best-fitted model is accepted for forecasting; otherwise, repeat the same process. Finally, predicated is valid statistically is checked using the Ljung-Box Q test using R in RStudio Cloud.

6.3 Results

These are the confirmed cases in Nepal, which give cumulative numbers that provide a total number of confirmed cases on a specific date Fig. 6.2a. In Fig. 6.2b, these confirmed cases and dates are stored in y and ds variable, respectively, and forecast using prophet library that shows increasing trends of confirmed cases in the future assuming that this current condition continues.

Using prophet plot components provides two plots of a trend with a date and weekly with a day of the week. It shows a pattern of confirmed cases will

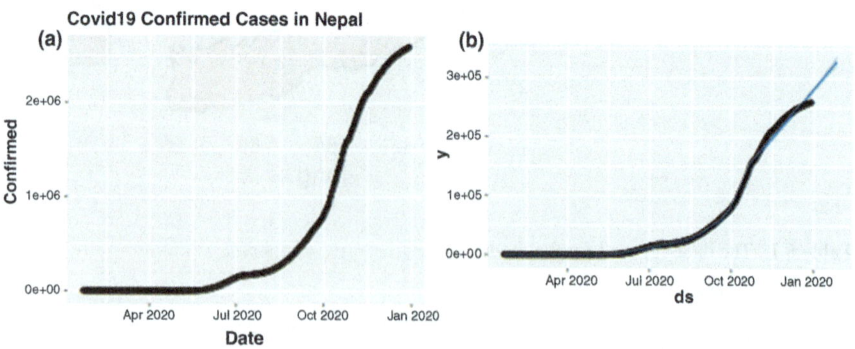

Fig. 6.2 Line graph of daily confirmed cases Coronavirus of Nepal (**a**) and Forecast confirmed cases of Nepal (**b**)

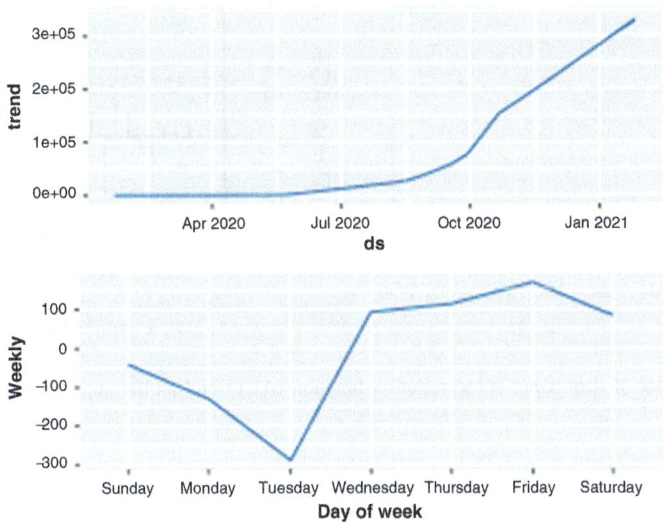

Fig. 6.3 Predicting the trend of confirmed cases monthly and weekly

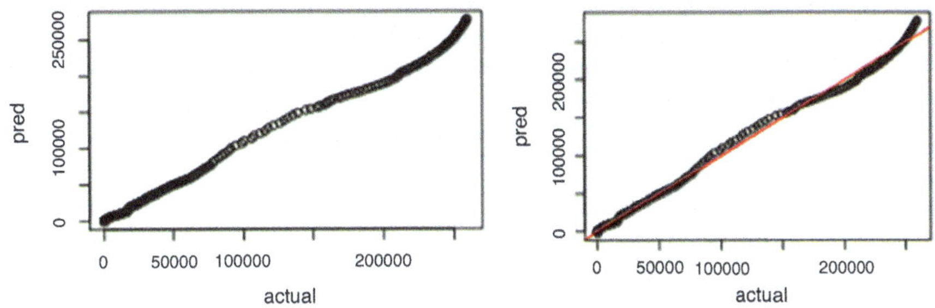

Fig. 6.4 Plot of model performance between actual and predicted

significantly increase, and the number of confirmed cases on average decrease on Tuesday and sharply increase from Wednesday to Saturday. This fluctuation on the plot depicts the process of reporting and checking for COVID-19. It does not mean people have a higher chance of getting the coronavirus on Wednesday to Saturday (Fig. 6.3).

The prediction model performance is plotted in Fig. 6.4 between actual and predict, which shows a linear pattern with upward trending. Using linear regression models tries to find the relationship between actual and predict value, which depicts no under or overestimation.

The summary of the linear regression model:

Linear Regression Model:

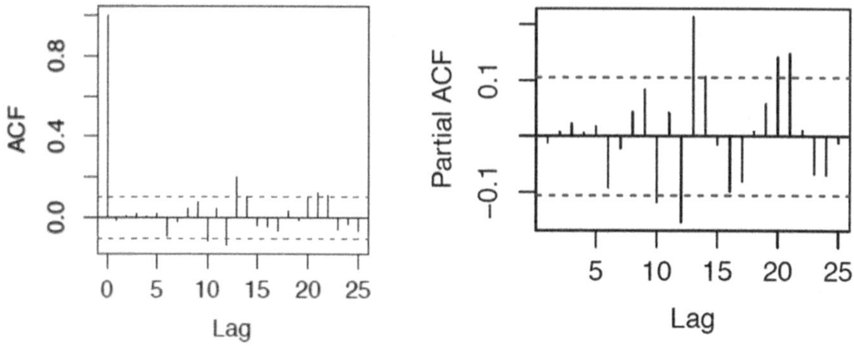

Fig. 6.5 Plot of data using autocorrelation function (ACF) and partial autocorrelation function (PACF)

Residual standard error: 4366 on 341 degrees of freedom
Multiple R-squared: 0.9974, Adjusted R-squared: 0.9974
F-statistic: 1.307e+05 on 1 and 341 DF, p-value: < 2.2e-16

Here, the adjusted R-squared value is 0.9974, which is very high, and the p-value is very low, which means the model is statistically significant.

For the Auto-Regressive Integrated Moving Average (ARIMA) model, check data autocorrelation using ACF, PACF, and augmented dicky-fuller test (ADF). It indicates highly autocorrelated with each other, and a high p-value signifies data are non-stationary. Using auto Arima with Akaike information criteria (AIC) test, the data becomes stationary or not by ACF, PACF, as shown in Fig. 6.5. It depicts that most of the line is below the blue dotted line in Fig. 6.5 and provides the best-fitted model.

Arima Model

Now refitting the best model(s) without approximations.

ARIMA(2,2,3): 5061.549

Best model: ARIMA(2,2,3)

From this best fitted ARIMA (2,2,3) model, Coronavirus confirmed cases are forecasted more than 300,000 to 1,500,000 from 2020 to 2022, as shown in Fig. 6.6. This prediction is relevant when applying the Ljung-Box Q test to check statistically.

Table 6.2 shows the ten number of predictions with a lower and higher confidence interval of 95%. Like 2020 lower confidence value will be 260788.770, and a higher 2662247.8 number of COVID-19 confirmed cases

After forecasting by the ARIMA model, its prediction is checked using the Ljung-Box Q test, which shows a p-value lower than 0.05, which means it is statistically significant.

Box-Ljung test

data: myforecast$resid

X-squared = 37.884, df = 15, p-value = 0.0009384

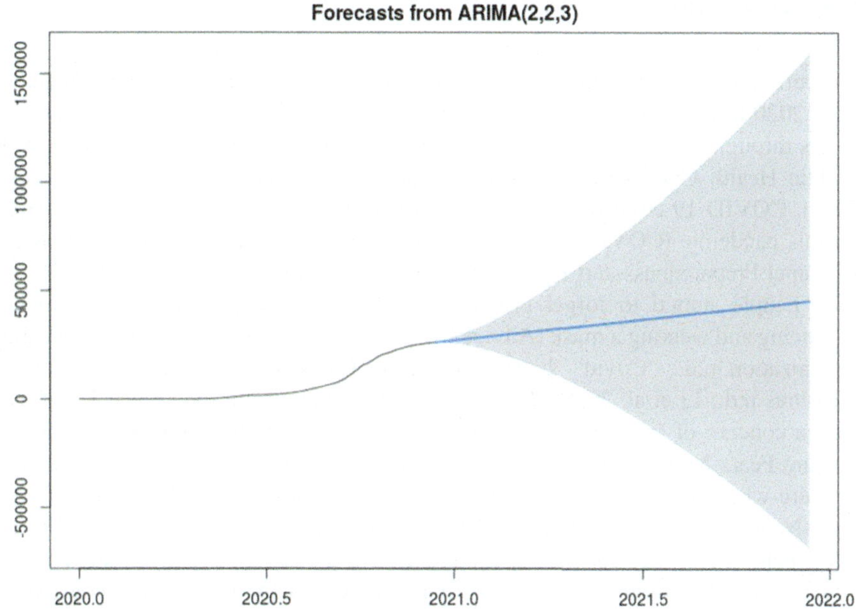

Fig. 6.6 Predication of COVID-19 confirmed cases using ARIMA model

Table 6.2 Forecast of COVID-19 confirmed cases

Point forecast		Lo 95 Hi 95	
2020.9479	261513.3	260778.770	262247.8
2020.9507	262082.3	260875.595	263289.1
2020.9534	262612.0	260743.488	264480.5
2020.9562	263117.9	260607.945	265627.8
2020.9589	263650.7	260517.259	266784.2
2020.9616	264186.5	260358.054	268014.9
2020.9644	264708.1	260131.715	269284.4
2020.9671	265232.8	259888.866	270576.8
2020.9699	265763.5	259616.512	271910.5
2020.9726	266290.7	259297.517	273283.9

Nepal's potential forecast was created using the Rstudio cloud from Johns Hopkins University (JHU) data repository in COVID-19 analytics package (Fig. 6.6). It shows an increasing trend of COVID-19 confirmed cases more than 300,000 to 1,500,000 from 2020 to 2022. Therefore, the result provides valuable information to the decision maker, government, health workers, hospitals, and other stakeholders to plan to reduce this pandemic's risk.

6.4 Discussion

The current models show that the cumulative number of confirmed cases plummeted from 2020 to 2022. The reason for significantly increasing cases is that the testing places throughout the country and people need test reports for their work and travel (Public Health Considerations While Resuming International Travel n.d.; Thapa 2021). COVID-19 confirmed cases highest with increasing risk and least prepared for this pandemic (COVID-19—Active Cases, Deaths and Recovered n.d.; Covid-19-Nepal-Preparedness-and-Response-Plan-(Nprp)-Draft-April-9.Pdf n.d.). However, people started to forget primary guidelines of coronavirus such as social distancing and wearing a mask (Advice for the Public on COVID-19—World Health Organization n.d.; "Covid" 2020; Do not Be Careless, Keep Following Safety Measures n.d.; Li et al. 2020). Five positive passengers from the United Kingdom raise a concern of spreading in the country (Five Arrivals from UK Test Positive, Raising Fears New Coronavirus Variant Is Here n.d.-b). Government officials are not sure whether it is new or old, stated proper assessment is necessary to find its type (Nepal Total Cases of COVID-19 - Google Search n.d.; Record 2020). Therefore, such a study will forecast upcoming days and cases to decide on quarantine, travel, hospital equipment preparation, and demand for the vaccine. Several rising confirmed cases in the future tell the people, hospitals, and government to prepare facilities, plans, strict rules of a mask, social distancing, and mass gathering. In the past few months, people have forgotten the minimum preventative measures and stopped testing and hiding their disease until it worsens, increasing the number of deaths and pandemic transmission in the community.

Few recent papers analyze infection of coronavirus spreading patterns in different countries. Similarly, previous studies employ machine learning and statistical approaches to capture the COVID 19 patterns and trends. Time series analysis is popular forecasts different diseases such as SARS, Ebola, pandemic influenza, and dengue. A study using the ARIMA (2,2,2) model predicts the future number of COVID-19 cases in Nepal. It has been concluded that patients in Nepal will increase exponentially, indicated that the instances rise of a new variant virus. The main motive is to help the government and medical workers plan ICU beds, ventilators, and isolation beds. In this study, the forecasts of coronavirus confirmed cases using time series prediction models.

6.5 Conclusions

These models determine the pattern and prediction of COVID's-19 confirmed cases in Nepal. It uses linear regression and the ARIMA model for time series data to show trends and forecasts from 2020 to 2022. The result shows the increasing trend in these three years of more than 300,000 to 1,500,000 confirmed coronavirus cases. This estimation is relevant if the current situation continues and forecasts for upcoming years. However, people take it lightly and perform daily activities without a mask and social distancing; then, the condition will be critical. Thus, people must follow the World Health Organization (WHO) guidelines and take necessary prevention and measures to limit physical contact and slow down coronavirus cases.

The future modifications to improve the predictive accuracy of the models will adopt multivariate time series modeling that considers other factors that are either directly or indirectly related to the spread of the pandemic. Another future ambition would be to use some form of transfer learning to bring learning from one country to another to know the majority parameters for the actual cause of the spread.

Declaration

Consent for Publication

I agreed to submit the final manuscript for this book and approved the submission.

Ethics Approval and Consent to Participant

Not applicable.

Availability of Data and Materials

The data used are cited with their sources; if data used in the manuscript are not precise, the author is agreed to clarify and send a dataset on request.

Competing Interests

There are no competing interests.

Funding

No funding.

Author's Contributions

The author performed analysis, evaluation, writing, editing of paper, and result.

Acknowledgments

I want to acknowledge the people who directly and indirectly contributed to this study. First, I want to thank Professor Dr. Bharatendra Rai for his YouTube video tutorial related to a similar topic for the USA and India that inspired me to study the same here in Nepal.

Author's Information (Optional)

Pawan Thapa, Lecturer, Department of Geomatics Engineering, Kathmandu University, Dhulikhel, Nepal.

References

20201005-weekly-epi-update-8.pdf (n.d.) Retrieved 2 January 2021, from https://www.who.int/docs/default-source/coronaviruse/situation-reports/20201005-weekly-epi-update-8.pdf

Advice for the Public on COVID-19 – World Health Organization (n.d.). Retrieved 3 January 2021, from https://www.who.int/emergencies/diseases/novel-coronavirus-2019/advice-for-public

Basu D, Salvatore M, Ray D, Kleinsasser M, Purkayastha S, Bhattacharyya R, Mukherjee B (2020) A comprehensive public health evaluation of lockdown as a non-pharmaceutical intervention on COVID-19 spread in India: national trends masking state level variations. MedRxiv, 2020.05.25.20113043. doi:https://doi.org/10.1101/2020.05.25.20113043

Coronavirus Spreads to Antarctic Research Station (2020, December 22) BBC News. https://www.bbc.com/news/world-latin-america-55410065

Coronavirus Update (Live): 84,460,368 Cases and 1,837,166 Deaths from COVID-19 Virus Pandemic - Worldometer (n.d.) Retrieved 2 January 2021, from https://www.worldometers.info/coronavirus/

COVID 19—Active Cases, Deaths and Recovered (n.d.) Retrieved 3 January 2021, from https://kathmandupost.com/covid19

Covid: How to Protest During a Global Pandemic (2020, October 20) BBC News. https://www.bbc.com/news/world-54477523

COVID-19 (n.d.) Retrieved 3 January 2021, from https://www.unicef.org/nepal/stories/novel-coronavirus

Covid-19-Nepal-Preparedness-and-Response-Plan-(Nprp)-Draft-April-9.pdf (n.d.) Retrieved 3 January 2021, from https://www.who.int/docs/default-source/nepal-documents/novel-coronavirus/covid-19-nepal-preparedness-and-response-plan-(nprp)-draft-april-9.pdf?sfvrsn=808a970a_2

Don't Be Careless, Keep Following Safety Measures (n.d.) Retrieved 3 January 2021, from https://thehimalayantimes.com/lifestyle/dont-be-careless-keep-following-safety-measures/

Five Arrivals from UK Test Positive, Raising Fears New Coronavirus Variant Is Here (n.d.-a) Retrieved 2 January 2021, from https://kathmandupost.com/health/2020/12/29/five-arrivals-from-uk-test-positive-raising-fears-new-coronavirus-variant-is-here

Five Arrivals from UK Test Positive, Raising Fears New Coronavirus Variant Is Here (n.d.-b) Retrieved 3 January 2021, from https://kathmandupost.com/health/2020/12/29/five-arrivals-from-uk-test-positive-raising-fears-new-coronavirus-variant-is-here

Gutierrez J, Wolgelenter M, Kannapell A (2021, January 1) 32 More countries have found the new covid-19 variant first seen in Britain. The New York Times. https://www.nytimes.com/live/2021/01/01/world/covid-19-coronavirus-updates

Hauck G (n.d.) Colorado confirms first known US case of new COVID-19 strain believed to be more contagious. USA Today. Retrieved 2 January 2021, from https://www.usatoday.com/story/news/health/2020/12/29/colorado-reports-first-known-us-case-new-covid-strain-uk/4079697001/

Li T, Liu Y, Li M, Qian X, Dai SY (2020) Mask or no mask for COVID-19: a public health and market study. PLoS One 15(8):e0237691. https://doi.org/10.1371/journal.pone.0237691

Nepal Not Sure If It Has New Coronavirus Variant with 5 Arriving from UK Testing Positive (n.d.) OnlineKhabar English News. Retrieved 2 January 2021, from https://english.onlinekhabar.com/nepal-not-sure-if-it-has-new-coronavirus-variant-with-5-arriving-from-uk-testing-positive.html

Nepal Total Cases of Covid 19—Google Search (n.d.) Retrieved 2 January 2021, from https://www.google.com/search?q=nepal+total+cases+of+covid+19&rlz=1C5CHFA_enNP911NP911&oq=Nepal+total+cas&aqs=chrome.1.0i20i263i457j0i20i263j0j69i57j0i22i30l2.7306j0j7&sourceid=chrome&ie=UTF-8

Petropoulos F, Makridakis S (2020) Forecasting the novel coronavirus COVID-19. PLoS One 15(3):e0231236. https://doi.org/10.1371/journal.pone.0231236

Ponce M, Sandhel A (2020) covid19.analytics: load and analyze live data from the CoViD-19 pandemic (2.0) [Computer software]. https://CRAN.R-project.org/package=covid19.analytics

Public Health Considerations While Resuming International Travel (n.d.) Retrieved 3 January 2021, from https://www.who.int/news-room/articles-detail/public-health-considerations-while-resuming-international-travel

Record T (2020, December 28) Nepal needs to be stricter about incoming flights. The Record. http://www.recordnepal.com/covid19/nepal-needs-to-be-stricter-about-incoming-flights/

Scudellari M (2020) How the pandemic might play out in 2021 and beyond. Nature 584(7819):22–25. https://doi.org/10.1038/d41586-020-02278-5

Thapa P (2021) Predicating COVID19 epidemic in Nepal using the SIR model. In: Oliva D, Hassan SA, Mohamed A (eds) Artificial intelligence for COVID-19. Studies in systems, decision and control, vol 358. Springer, Cham. https://doi.org/10.1007/978-3-030-69744-0_14

Wise J (2020) Covid-19: new coronavirus variant is identified in UK. BMJ 371:m4857. https://doi.org/10.1136/bmj.m4857

Chapter 7
A Multi-Step Predictive Model for COVID-19 Cases in Nigeria Using Machine Learning

Sakinat Oluwabukonla Folorunso, Ezekiel Adebayo Ogundepo, Joseph Bamidele Awotunde, Femi Emmanuel Ayo, Oluwatobi Oluwaseyi Banjo, and Abass Ishola Taiwo

Abstract COronaVIrus Disease 2019 (COVID-19) is a vastly communicable disease that has the whole world in a panic state with a huge negative impact. Forecasting for future occurrences for timely intervention requires a good model. The aim of this chapter is to model a multistep forecast for 14-days (2 weeks) incidence of COVID-19 daily cases in Nigeria with Machine Learning (ML) models.

The study dataset contains 241 instances (days) of the daily incidence of COVID-19 for confirmed, recovered, and patients who eventually died from February 27th to October 24th, 2020 in Nigeria. The proposed methodology jointly model multiple targets field (cases) simultaneously in order to capture dependencies between them. This 3-daily time-series study dataset was transformed to a supervised learning format for analysis by ML models. Linear Regression (LR), MultiLayerPerceptron Regressor (MLPR), Support Vector Regressor (SVR), *k*-Nearest Neighbor Regressor (*k*-NNR), and Random Forest Regressor (RFR) models were employed and their prediction errors were compared based on average values of Directional Accuracy (DAC), Mean Absolute Error (MAE), Mean Absolute Percentage Error (MAPE), and Root Mean Squared Error (RMSE).

The result obtained showed that SVR outperformed all other models and can accurately forecast a 14-day occurrence of COVID-19 cases in Nigeria based on all

S. O. Folorunso (✉) · O. O. Banjo · A. I. Taiwo
Department of Mathematical Sciences, Olabisi Onabanjo University, Ago Iwoye, Nigeria
e-mail: sakinat.folorunso@oouagoiwoye.edu.ng; banjo.tobi@oouagoiwoye.edu.ng; taiwo.abass@oouagoiwoye.edu.ng

E. A. Ogundepo
Data Science Nigeria, Yaba, Lagos, Nigeria

J. B. Awotunde
Department of Computer Science, University of Ilorin, Ilorin, Nigeria
e-mail: awotunde.jb@unilorin.edu.ng

F. E. Ayo
Department of Computer Science, McPherson University, Seriki Sotayo, Ogun, Nigeria

S. A. Hassan et al. (eds.), *Decision Sciences for COVID-19*, International Series in Operations Research & Management Science 320,
https://doi.org/10.1007/978-3-030-87019-5_7

the metrics used in this study. This study will aid the Government and health practitioners to plan ahead for the next 14 days (2 weeks). This 14-ahead step forecasting model could also be extended for longer terms.

ML model-based forecasting can effectively model a small sample size and is useful for decision-making and planning for future pandemics by the government and health service workers.

Keywords COVID-19 · Regression · Machine learning · Time-series · Pandemic decision science

7.1 Introduction

The transmission features of the COVID-19 epidemic have still not been properly recognized in the sense of global environmental changes. Likewise, the pace of global development, increased population size, nuanced encounters, and lack of medical security in developing countries all contribute to the complexities of COVID-19 prevention and regulation. The COVID-19 seems to have a greater impact on the global economy when compared to severe acute respiratory syndrome (SARS) that occurred in 2013. The bone of contention is how to fight the spread of these infectious diseases and how to contain its spread from human to human. Most countries depended on the observation of classic steps to monitor diseases and public health to contain the COVID-19 pandemic, similar to those used with SARS in 2003.

The coronavirus disease 2019 (COVID-19) epidemic poses a growing risk to the world. We are more connected than ever before, our cities are increasingly compactly populated and in recent years the world has observed successive waves of the new severe acute respiratory syndrome (SARS), Ebola virus, Zika virus and recently COVID-19 and influenza infectious diseases threats global pandemics. The coronavirus (COVID-19) pandemic is playing a thoughtful havoc in socio-communal systems, humanity, and creates economic crises worldwide (Guraya 2020). Many strategies have been used to manage and curtail COVID-19 outbreak, but many countries are still help-less in fighting and containing the outbreak.

The epidemiology complexity and the rapidly evolving disease patterns of the COVID-19 demand a couturier approach by the government, researchers, and medical experts in dealing with this overwhelming pandemic. Similarly, other methods such as social distancing, the use of facemask, isolation of confirmed patients among others have developed to curtail and manage the COVID-19 pandemic. The pandemic originated from China in a town called Wuhan in December 2019 and has advanced to 213 countries and territories. As of Friday, August 28, 2020, the World Health Organization (WHO) has reported a total of 24,926,312 COVID-19 cases with 840,662 deaths, and 17,799,210 patients recovered successfully from this deadly disease.

The occurrence of coronavirus (COVID-19) is greater than that of 2003 representing respiratory infections syndrome (SARS). As of Wednesday, August

12, 2020, the reported cases are more than the 73,435 confirmed and more than 2000 deaths cases worldwide, and both COVID-19 and SARS are distributed across regions, infecting living beings. By contrast, in 2003, SARS claimed 774 lives but in the shortest time, COVID-19 claimed more than that. But the major difference among them is that after 17 years of SARS other powerful tools have emerged, which could be used as an instrument in fighting this virus and keeping it within reasonable limits and Machine Learning (ML) is one of such tools. Recently, ML is causing a paradigm shift in the healthcare sector and its applicability in the COVID-19 outbreak might yield profit results especially in predicting and monitoring the outbreak. The application of ML in COVID-19 can be expediting the diagnoses and monitoring of COVID-19 and minimizes the burden of these processes (Awotunde et al. 2021).

COVID-19 is extremely infectious, and can spread complications before and after the onset. Monitoring and lockdown have to encompass anyone with symptoms and properly isolate persons who have been infected from those who are not, to allow good containment. Patients carrying the virus could either be minor symptomless (like fever, sore throat, and sneezing) or have serious clinical signs (such as pneumonia, respiratory failure, and eventually death) (Perrella et al. 2020). The transmittable SARS-CoV-2 condition is called "coronavirus disease" (COVID-19) (Kannan et al. 2020). Gratitude to the recent developments in analytical methods and information and communication technologies (ICTs), ML and big data will aid or manage the immense, unparalleled volume of data generated from patient monitoring, real-time tracking of disease outbreaks, now-casting/predicting patterns, daily situation briefings and public updates (Wong et al. 2019).

The use of machine learning and statistical methods to forecast uncertain or potential effects is referred to as predictive analysis (Brown et al. 2015; Jayanthi and Valluvan 2017). It responds to the question like what is the next step? It also uses past and present data to predict actions, patterns, and activities in the future. The prediction is rendered using quantitative questions, automated machine learning, and statistical analysis (Jayanthi and Valluvan 2017). Experts need to construct predictive models that are used for forecasting in predictive analytics (Jayanthi and Valluvan 2017).

Machine Learning (ML) is an umbrella word that depicts various tools for knowledge and control. The data science analysis using ML is newly evolving, with the goal of empowering health care systems and organizations to connect to harness information and convert it to usable knowledge and preferably personalized clinical decision-making. Utilizing deep learning, the implementation of ML in the field of infectious diseases has implemented a range of improvements in the modeling of knowledge generation (Folorunso et al. 2022). The hope of using ML in COVID-19 will have a great impact on the quality of outbreak diagnosis, prediction, and treatment and can be delivered quality care to patients across socioeconomic and geographic boundaries. The incidences of the COVID-19 on a daily basis could be forecasted for future plans by government agencies and health care services. These daily cases could be represented as a time series problem that can be modelled by ML.

A time series data is a chain of time-ordered data $\{T_t, t = 1, \cdots, n\}$, where t is the time, n is the count of instances (days) for the time period, and T_t is the value measured at time instant t (Folorunso et al. 2019). A multivariate time-series T_t consist of multiple time-series that can usefully contribute to forecasting T_{t+1}. Recently, machine learning models have been applied to make a future forecast on time series data (Folorunso et al. 2019; Ahmed et al. 2010). It can tell the underlying structure and unveiled patterns usually characterized by time series events.

This study proposes a robust ML model that can accurately and effectively forecast COVID-19 cases in Nigeria for the next 14 days (14-steps ahead). The advantage of using ML model is that it can be used as an alternative to epidemiological models, has ability to learn nonlinear patterns from the data, and can jointly model multiple targets field (cases) simultaneously in order to capture dependencies between multiple time series data. This study presented the Nigerian COVID-19 cases as a time series problem hence, model and analyze its incidences in Nigeria based on confirmed, recovered, and death cases from February 27th to October 24th 2020. Five (5) ML models such as LR, MLPR, SVR, *k*-NNR, and RFRR were modeled and their prediction errors were compared based on MAE, MAPE, RMSE, and DAC.

The major contributions of this research work are:

1. Transformed three times series data to a supervised learning regression problem with the lagging method
2. Jointly model multiple targets field simultaneously to capture dependencies
3. Build a 14-step ahead forecast model for 241 raw daily COVID-19 cases for Nigeria
4. SVR gave a superior performance than other Machine Learning models with a small sample size data
5. To the best of our knowledge, this is the first ML model to forecast raw daily COVID-19 cases (confirmed, recovered, and death) for Nigeria. Most authors modelled cumulative daily cases
6. The ML model-based forecasting can use to make decisions and plan for future pandemics by government and health service workers

The rest of the chapter is ordered as follows: Sect. 7.2 describes the review of related work while Sect. 7.3 present the materials and methods used for this study. Section 7.4 presents the results while Sect. 7.5 concludes the chapter alongside further work as a follow-up.

7.2 Related Work

The number of preliminary studies on analyzing the time series data associated with COVID-19 epidemic is growing daily. Researchers analyzed and forecasted the confirmed, recovered, and death cases of the OCIVD-19 pandemic for some countries or the whole world at large to better understand the true nature of the outbreak. Table 7.1 shows some related works with ML models.

Table 7.1 Related work

Authors	Aim	Data source	ML model	Optimal result
Parbat and Chakraborty (2020)	Predict the total, cumulative, and daily count of confirmed, deaths and recovered COVID-19 cases	Kaggle[a]	SVR with rbf, linear, and poly kernel	MAE, RMSE, R^2_{Score}, and accuracy. Accuracy = 97% for deaths, confirmed deaths, confirmed and recovered cases and 87% for new daily cases
Peng and Nagata (2020)	Predict the count of confirmed COVID-19 cases for the 12 most affected countries	JHU CSSE[b]	SVR with Linear, polynomial, Gaussian Kernel	MAE for Gaussian Kernel
Rustam et al. (2020)	Propose to use ML models to forecast the next 10 days for COVID-19	JHU CSSE[c]	LR, Least Absolute Shrinkage and Selection Operator (LASSO), Exponential Smoothing (ES), and SVM	ES model surpasses all other models based on R^2_{Score}, R_{Adjusted}, RMSE, MSE, MAE metrics for death and confirmed cases
Al-Qaness et al. (2020)	Proposed Flower Pollination Algorithm (FPA) + Salp Swarm Algorithm (SSA) + ANFIS	WHO[d] DS1[e] DS2-WHO[f]	ANN, *k*-NN, SVR, ANFIS, PSO, GA, ABC, FPA, FPASSA-ANFIS	FPASSA-ANFIS model outperform all other models based on RMSE = 5779, MAE = 4271, MAPE = 4.79, RMSRE = 0.07 R2 = 0.9645, Time = 23.30
Fong et al. (2020)	Proposes a methodology that augments data and selects the best forecast model using parameter tuning of each model	Chinese health authorities	Polynomial Neural Network with corrective feedback (PNN + cf), LR, SVM, Fast decision tree learner, MP5, Holt-Winters', ARIMA, Exponential	PNN + cf. performed best based on RMSE = 136.547
Gupta and Gharehgozli (2020)	Proposes a structure that builds analytical model to forecast the growth of COVID-19	Weather[g], pollution[h], GDP[i] for each state in USA	LR, SVM (Linear), SVM (radial), SVM (Poly), and Decision Tree (DT)	SVM (radial) Time-dependent performed best based on RMSE = 1.009 R2 = 0.405, MAE = 0.857 With exponential growth:

(continued)

Table 7.1 (continued)

Authors	Aim	Data source	ML model	Optimal result
				RMSE = 1.063, R2 = 0.465 MAE = 0.859 With Maximum Likelihood: RMSE = 0.795, R2 = 0.436 MAE = 0.642
Hu et al. (2020)	To develop an AI-inspired method for real-time 5-step ahead forecasting and evaluation of the impact of COVID-19	WHO[j]	Modified Auto Encoder	Average accuracy = 2.5%
Yadav et al. (2020)	Predicting the spread, analyzing the growth rates of COVID-19 across China, USA, Italy, South Korea, and India	JHU CSSE[k]	SVR, LR, and Polynomial Regression	SVR performed best
Tiwari et al. (2020)	Predict COVID-19 epidemic in India using China pattern of with ML	JHU CSSE[l]		MAE and RMSE
Sujatha et al. (2020)	Presented a model to predict the spread of COVID-2019	COVID-19 Kaggle data[m]	LR, MLP, and Vector Auto-regression (VAR)	MLP performed best
Folorunso et al. (2019)	Data mining techniques to forecast 1-step ahead of Nigeria electric power consumption from 2001 to 2017	Electricity[n]	RFR, LR, SVR, and Artificial Neural Network (ANN)	SVR model with MAE = 1.1523 MAPE = 5.9950 RMSE = 1.6497
Car et al. (2020)	Modelling of daily infected, recovered, and deceased patients in 406 locations over 51 days in South Korea	Johns Hopkins CSSE[o]	MLP with 4-hidden layers 16 total hidden neurons with 4-layers each	R^2 = 0.98599 for confirmed R^2 = 0.97941 for recovered R^2 = 0.99429 for deceased models
This research	Model a multistep forecast for 14 days (2 weeks) incidence of COVID-19 daily cases in Nigeria with ML models	Ogundepo et al. (2020)	LR, MLPR, SVR, *k*-NNR, and RFR	SVR based MAE, DAC, RMSE, and MAPE

(continued)

Table 7.1 (continued)

Authors	Aim	Data source	ML model	Optimal result

[a]https://www.kaggle.com/sudalairajkumar/novel-coronavirus-2019-dataset
[b]https://github.com/CSSEGISandData
[c]https://github.com/CSSEGISandData
[d]https://www.who.int/emergencies/diseases/novel-coronavirus-2019/situation-reports/
[e]https://www.cdc.gov/flu/weekly/
[f]https://www.who.int/influenza
[g]https://airnow.gov/
[h]https://mesonet.agron.iastate.edu/
[i]https://www.bea.gov/
[j]https://www.who.int/emergencies/diseases/novelcoronavirus-2019/situation-reports
[k]https://www.kaggle.com/sudalairajkumar/novel-corona-virus-2019-dataset
[l]https://www.kaggle.com/sudalairajkumar/novel-corona-virus-2019-dataset
[m]https://www.kaggle.com/imdevskp/corona-virus-report/data
[n]https://www.indexmundi.com/g/g.aspx?c=ni&v=81
[o]https://github.com/CSSEGISandData/COVID-19

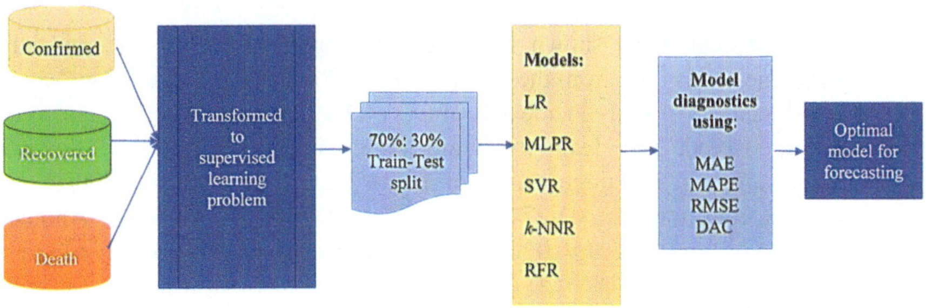

Fig. 7.1 Concept map workflow

7.3 Materials and Methods

This section presents the workflow, the dataset, and the models used in this study. It includes the original dataset, the transforming of the dataset to supervised learning format based on lagging instances, modelling using LR, MLPR, SVR, *k*-NNR, and RFR and evaluation metrics. The outline of the workflow is presented in Fig. 7.1.

7.3.1 Dataset

Dataset used for this study is obtained from (Ogundepo et al. 2020). The study dataset consists of 241 daily confirmed, recovered and death cases of COVID-19 in all the states in Nigeria from February 27th to October 24th 2020. The dataset is riven into three categories: Confirmed, Recovered, and Death. The participating

Table 7.2 Nigeria COVID-19 dataset

S/N	Features
1	Date
2	Confirmed
3	Recovered
4	Death

Table 7.3 A glimpse of Nigeria daily COVID-19 dataset

Date	Confirmed	Recovered	Death
2020-02-27	1	0	0
2020-02-28	0	0	0
2020-02-29	0	0	0
2020-03-01	0	0	0
⋮	⋮	⋮	⋮
2020-10-21	37	83	0
2020-10-22	138	105	2
2020-10-23	77	205	2
2020-10-24	48	95	0

variables in the dataset are shown in Table 7.2. A glimpse of the Nigeria daily COVID-19 time series data from the onset to 10/24/2020 is shown in Table 7.3.

The trend analysis of daily incidences of COVID-19 cases in Nigeria in terms of confirmed, recovered, and death cases are shown in Fig. 7.2, which shows a significant changes in the three cases till 10/24/2020. The onset of COVID-19 incidence was confirmed on 02/27/2020 in Nigeria with one (1) case. Another confirmed case occurred on 3/9/2020 and 3/17/2020, thus cause a sudden rise to 5 confirmed incidences on 3/18/2020. The highest number of confirmed cases (790) occurred on 07/01/2020. All the while, there were no recovery or death cases reported. The first 2 recovered cases occurred on 3/21/2020 while the first death occurred on 3/24/2020. A strong variation in all cases was observed from 2/27/2020 to 8/21/2020, especially for death cases. It is worth to note the astronomical increase in recovered cases on 8/4/2020 with a value of 11,188. The date of occurrence of COVID-19 is represented on the x-axis while the value of occurrence is represented on the y-axis.

The statistical description of the three (3) time series in the datasets is displayed in Table 7.4. The sum of daily cases and confirmed cases have the highest occurrences while the death cases have the lowest index, and all three occurrences are positively skewed. Their minimum values were 0 while the maximum value for recovered cases is 11,188.

7.3.2 Data Preprocessing

Lagging is the method of establishing a relationship between previous and recent values of a series. This is achieved by creating a "window" over a time period. The lag value used in this study is three (3). Table 7.5 displays the transformed data based

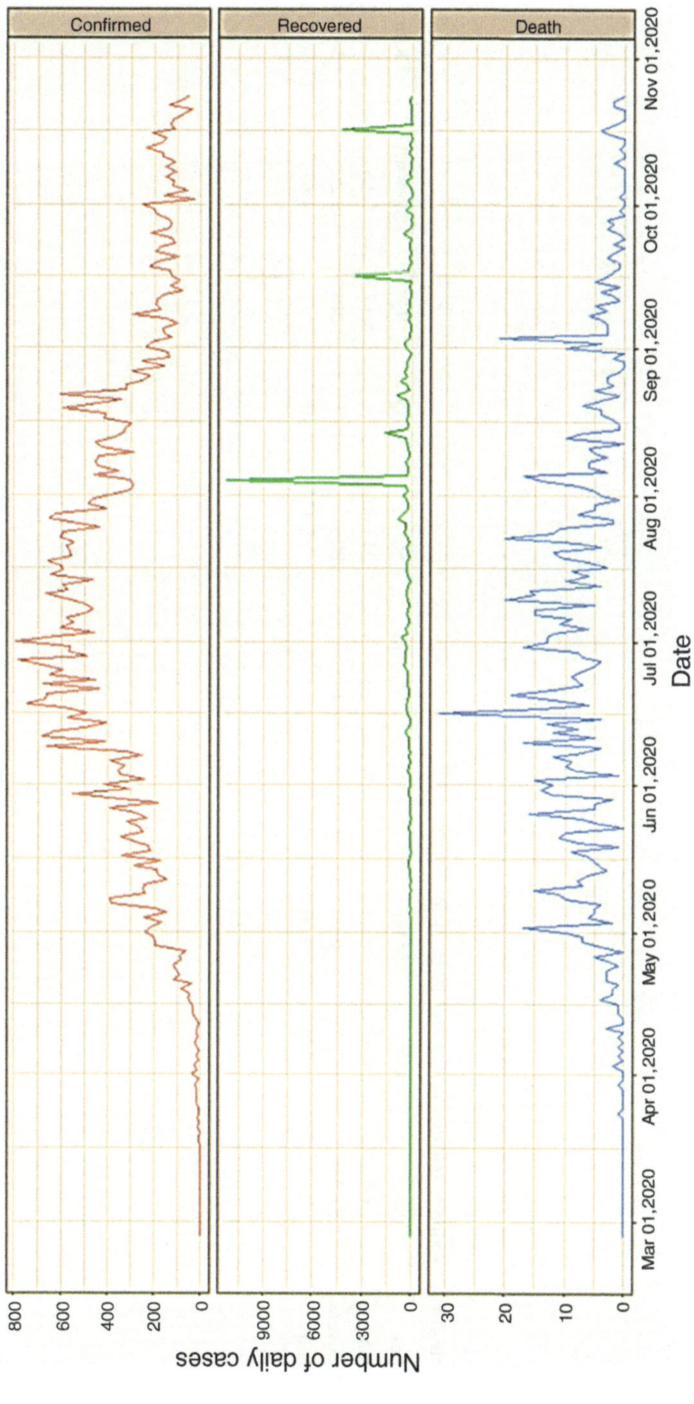

Fig. 7.2 Daily cases of COVID-19 in Nigeria

Table 7.4 Descriptive statistics of COVID-19 cases in Nigeria

Cases	Sum	Avg	Min	Max	Median	Std	1Q	3Q	Skewness	Kurtosis
Confirmed	61,930	256.97	0.00	790	204	213.60	79	417	0.56	−0.82
Recovered	57,285	237.70	0.00	11,188	121	806	17	249	11.29	143.56
Death	1129	4.68	0.00	31	3	5.25	0	7	1.47	2.52

Table 7.5 Transformed supervised learning of Nigeria COVID-19 dataset

S/N	Features	Data type
1	Confirmed	Numeric
2	Recovered	Numeric
3	Death	Numeric
4	Lag_Confirmed-1	Numeric
5	Lag_Confirmed-2	Numeric
6	Lag_Confirmed-3	Numeric
7	Lag_Recovered-1	Numeric
8	Lag_Recovered-2	Numeric
9	Lag_Recovered-3	Numeric
10	Lag_Death-1	Numeric
11	Lag_Death-2	Numeric
12	Lag_Death-3	Numeric

on three (3) lag values from time series to supervised learning (4 features to 12 features) based on lag values. The 3 transformed times series data with 12 features is the input to the ML model while forecasted values are the output. The 241 instances (days) are then split into 70–30 train_test split ratio. So, 30% of 241 which is 72 instances is now used for testing the model.

7.3.3 Time Series

The time series problem forecasts forthcoming values based on past and recent values of the time series. These past and recent values serve as input features to the ML model. One-step ahead forecast is regarded as short-term while multistep ahead forecast is regarded as a long-term forecast problem. A time series may be formally presented as a succession of scalar random events as shown in (7.1)

$$y_t = y_0, \cdots, y_{t-1}, y_t, y_{t+1} \tag{7.1}$$

The lag of the series is given by the delay used to transform the data (Sergio and Ludermir 2015). Then, time series forecasting means to predict a future value of the sequence, given by (7.2)

$$\widehat{y_t} = \widehat{f}\left(y_{t-d}, y_{t-d-1}, \cdots, y_{t-d-k+1}\right) + w(t) \tag{7.2}$$

where t is the time of event, d is the lag time, k is the step lag, w is the error/noise term, and f are the machine learning models. When k is 1, then this is called 1-step ahead forecast. If otherwise, it is called multistep or k-step ahead forecast. The purpose of this study is $\widehat{f}$ is (LR, MLPR, SVR, k-NNR, and RFR). The time lag $d = 3$, $k = 14$, and $y_t = 3$ as there are 3 time series data.

7.3.3.1 Time Series to Machine Learning Format

Supervised learning technique of machine learning method models a finite set of -instances and map the relationship between a set of independent inputs to one or more dependent output variables. A one-step forecasting time series can be transformed into a generic supervised learning regression problem provided a historical record S or n previous values (lag) of the series is available as suggested by Eq. (7.3).

The multistep approach adopted for this study is the recursive strategy (Sorjamaa et al. 2007), which firstly builds a 1-step model f

$$y_{t+1} = f(y_t, \cdots, y_{t-n+1}) + w_{t+1} \tag{7.3}$$

With $t \in \{n, \cdots, N - 1\}$ and then uses it recursively for returning a multistep forecast. A weakness of this strategy is that it is sensitive to the estimation error (Bontempi et al. 2013).

7.3.4 *Regression Model*

Regression analysis method investigates and models the relationship between variables (dependent and independent) (Montgomery et al. 2012). Regression analysis can be applied to a host of diverse areas as management, finance, physical and chemical sciences, and medicine.

7.3.4.1 Linear Regression

Linear Regression (LR) models the linear relationship between dependent and independent variables as shown by (7.4)

$$y = \beta_0 + \beta_1 x + \varepsilon \tag{7.4}$$

where y is the dependent variable, x is the independent variable, β_0 is the intercept, β_1 is the slope, and ε is the error term.

For a multiple linear regression problem where there are more than one independent variable (k) as shown by (7.5)

$$y = \beta_0 + \beta_1 x_1 + \beta_2 x_2 + \cdots + \beta_k x_k + \varepsilon \tag{7.5}$$

7.3.4.2 Multilayer Perceptron

Artificial Neural Network (ANN) is a biologically inspired nervous system that treats information like the brain. It is made up of many overly associated processing elements called neurons working together to solve a problem (Amid and Gundoshmian 2017). A multilayer perceptron (MLP) is a type of feedforward artificial neural network (FANN). For this study, MLP with one hidden layer using WEKA's Optimization class by minimizing the given loss function plus a quadratic penalty with the BFGS method (Folorunso et al. 2019; Bouckaert et al. 2010) presented by (7.6)

$$y = \varphi\left(\sum_{i=1}^{n} w_i x_i + b\right) = \varphi\left(w^T x + b\right) \tag{7.6}$$

where w is for the vector of weights, x is for the vector of inputs, b is for bias, and φ are the nonlinear activation function.

7.3.4.3 Support Vector Regression

Support Vector Regression (SVR) (Drucker et al. 1997; Shevade et al. 2000) model aims to detect the optimum midway between bias and the variance of a dataset. Basically, the functions of the SVR model are designed to provide a nonlinear mapping function that maps the training data to a high-dimensional feature space (Liu et al. 2018). In this study, the polykernel is adopted. The core benefit of using SVR is its ability to capture nonlinearity in the data and use it to enhance the forecasting cases and be amenable to small sample data size (Ribeiro et al. 2020). The training dataset pair is presented by $\{(x_i, y_i); i = 1, 2, \cdots, N; x_i \in \mathbb{R}^n, y_i \in \mathbb{R}\}$ where x_i is the ith input in the nth dimension, y_i is the real output, and N is the number of instances in the dataset. The objective function for SVR is presented by (7.7)

$$y = f(x_i) = w^T \varphi(x_i) + b \tag{7.7}$$

where $f(x)$ is the forecast values, $\varphi(x)$ is the input attribute function, w and b are adaptable coefficients. The penalty function for estimating w and b is $R(C)$ as presented by (7.8) where C is the cost coefficient and the error term ε (Liu et al. 2018).

$$R(C) = \frac{1}{2}\|w\|^2 + C.\frac{1}{n}\sum_{i=1}^{n} |y_i - f(x)|_\varepsilon \tag{7.8}$$

7.3.4.4 *k*-Nearest Neighbor Regression

k-Nearest Neighbor Regression (k-NNR) is a nonparametric method, which makes a prediction by assigning value to continuous target in a training set. With a training sample $S = \{x_1 y_1, \cdots, x_N y_N\}$, the goal of the objective function is to generalize $f: x \rightarrow y$. Let a dataset consisting of feature pairs $(x_i, y_i) \in X \times Y$. Learning a pattern x', k-NNR calculates the mean of function f values of its k-NN as presents by (7.9)

$$f(x') = \frac{1}{k} \sum_{i \in N_k(x')} y_i \tag{7.9}$$

with set $N_k(x')$ containing the indices of the k-NN of x' (Kramer 2011). The k closest points in the examples are based on the similarity measure between the target and the particular point as presented by (7.10) (Riza et al. 2019).

$$d(x, y) = \sqrt{\sum_{i=1}^{k} (x_i - y_i)^2} \tag{7.10}$$

7.3.4.5 Random Forest Regressor

Random Forest Regressor (RFR) is an ensemble-based regression model (Breiman 2001), which involves the combination of predictions from multiple learning algorithms to make a more accurate prediction than a single model. The bootstrap aggregating method is used to train the random forest algorithm. The overall prediction is made by averaging the predictions of each regression tree. This algorithm uses a modified tree learning algorithm that selects and split each learning process by a subset of random features (Khanday et al. 2020; Katuwal and Suganthan 2018).

7.3.5 Metrics

The metrics employed in this study to compare forecast performances are RMSE, MAE, MAPE, and DAC. The metrics are shown in Eqs. (7.11), (7.12), (7.13) and (7.14) respectively.

$$RMSE = \sqrt{\frac{1}{n}\sum_{t=1}^{n}\left(f_t - \widehat{f}_t\right)} \tag{7.11}$$

n is the count of samples for each of the cases
$\widehat{f}_t$ is the forecasts value at time t
$\widehat{f_{t-1}}$ is the previous forecast
f_t actual value of a point at time t
Sign $(\cdot)$ is the sign function

$$\text{MAPE} = \frac{100}{n}\sum_{t=1}^{n}\left|\frac{\left(f_t - \widehat{f}_t\right)}{f_t}\right| \tag{7.12}$$

$$\text{MAE} = \frac{1}{n}\sum_{t=1}^{n}\left|\left(f_t - \widehat{f}_t\right)\right| \tag{7.13}$$

$$\text{DAC} = \text{sign}\left(f_t - \widehat{f_{t-1}}\right) \tag{7.14}$$

7.4 Experimental Result and Discussion

This section presents the discussion and the results obtained in detail on the three times series data: confirmed, recovered, and death cases for COVID-19 in Nigeria for 14-steps ahead (2 weeks) forecast with 95% confidence intervals. Five ML models were evaluated and compared using MAE, MAPE, RMSE, and DAC metrics based on test data containing 72 instances. Usually, a multiple-step ahead forecasting involves a multiple one-step ahead modeling where the prediction of a current step is the input to the next step. The chain goes on until the desired step is achieved. This type of strategy is called recursive strategy (Bontempi et al. 2013). The study dataset for the three cases were from 2/27/2020 (Day 1) to 10/24/2020 (Day 241) (see Fig. 7.2). The train data start from 2/27/2020 to 08/12/2020. Test data starts from 08/13/2020 to 10/24/2020. The forecasted cases start from the end of test data from 10/25/2020 to 11/07/2020 (14 days) presented by Tables 7.10, 7.11, and 7.12. The overall best results based on metric used are presented in bold. All experimentations were performed on MS Windows 10 running 64-bit OS on an Intel® core™ i5–7200 CPU @ 2.50GHz to 2.70 GHz Pentium Windows computer with 8GB RAM on Weka (Bouckaert et al. 2010), an open-source machine learning platform.

7.4.1 Model Evaluations

This section presents the results obtained to compare all models deployed in this study based on MAE, MAPE, RMSE, and DAC. The results show that the lower the value of MAE, MAPE, and RMSE the better the model and vice versa for DAC. Hence, models with lesser errors are more efficient for forecasting. The model parameters used in the study are WEKA default values.

Figures 7.3, 7.4, 7.5, and 7.6 show the error distribution trends of all models based on MAE, DAC, MAPE, and RMSE metrics from 1 to 14 steps ahead for confirmed, recovered, and death cases. These three cases were forecasted from steps 1 to 14 (2 weeks), respectively. The error generated by a model for multiple step-ahead forecast accumulates from a previous step and this increases as the step ahead increases. This is the case as it is iterating the prediction errors obtained in the previous steps. This is a weakness common to recursive strategy (RS) for multiple steps ahead prediction. RS is sensitive to the estimation error and since estimated values are used rather than the actual values, these errors are propagated further in the future (Bontempi et al. 2013). As a result, the larger step ahead, the lesser is the accuracy of the data in the past event. Hence, the errors of the past prediction are accrued and MAPE, MAE, and RMSE errors become larger (Wei et al. 2019).

Figure 7.3a–c shows the superiority of the SVR model when compared to all models based on MAE error distribution for all models. There are very mild variations in its values over the 14-step ahead period. Poor performance of MLPR was noted across the three cases. The same pattern is noticed for MAPE and RMSE as SVR still performed best as shown in Figs. 7.5 and 7.6, respectively.

Figure 7.4 shows the DAC distribution trend of all models to be similar for the steps ahead forecast. SVR still achieved a relatively high value.

Tables 7.6, 7.7, and 7.8 present the averaged error results for all models in this study over the 14-steps ahead forecast for confirmed, recovered, and death cases in Nigeria. The bolded values indicate the best models for each of the models. Based on MAE, MAPE, and RMSE evaluation, SVR outperformed all other models for confirmed (65.9993, 51.4828, 87.6290, respectively and death (2.1796, 84.9544, 3.1620, respectively) cases. For recovered cases, SVR model performed best based on DAC, MAPE, and RMSE with 52.9039, 55.8743, and 688.0478 values, respectively.

7.4.2 Future Forecasts for 14-Step Ahead

This section presents the forecasted values for 14-step ahead (2 weeks) from the end of the test data at 95% confidence interval. Tables 7.9 and 7.10 present the forecasted values for 2 weeks for COVID-19 using the real 3 times series data of confirmed, recovered, and death cases with LR MLPR, SVR, *k*-NNR, and RFR models, respectively, at the end of the test data.

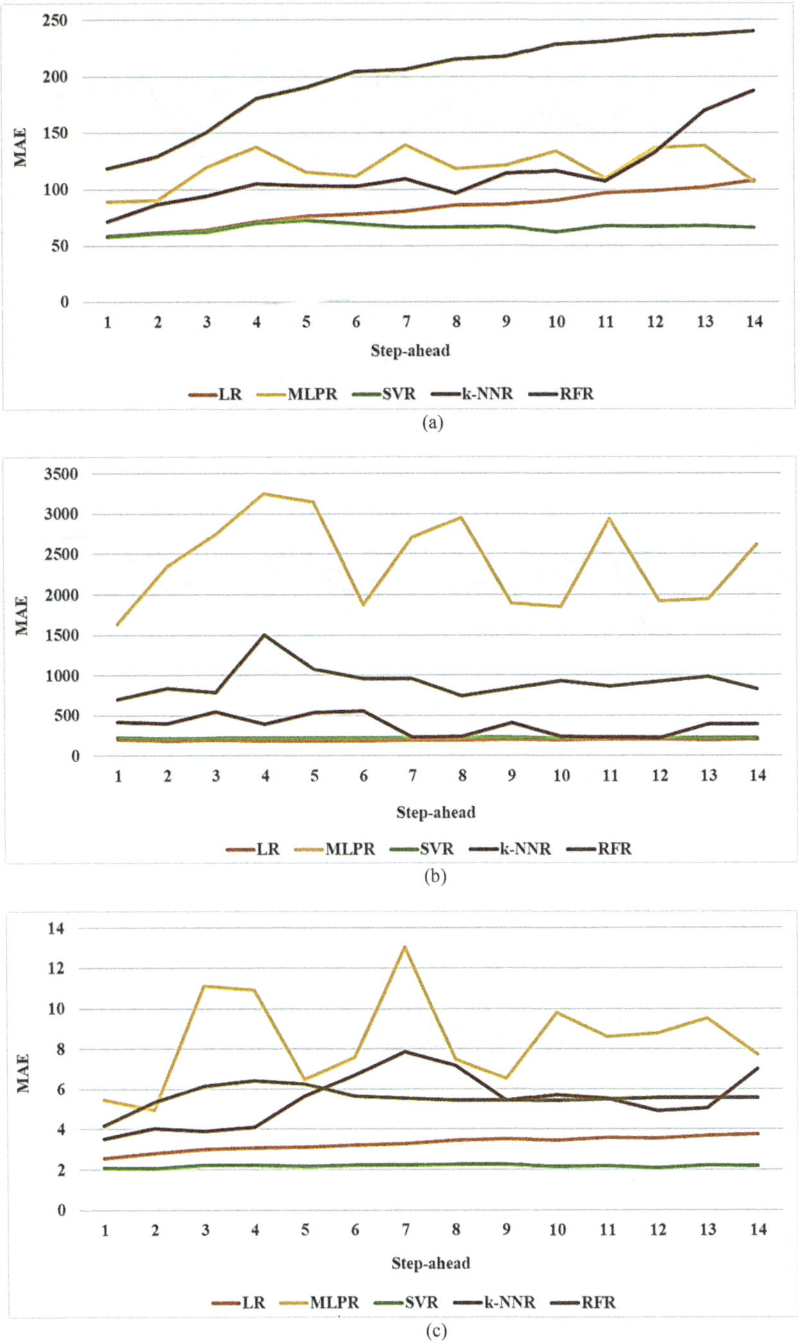

Fig. 7.3 The error distribution of MAE from 1 to 14-step ahead. (**a**) Confirmed, (**b**) Recovered, (**c**) Death

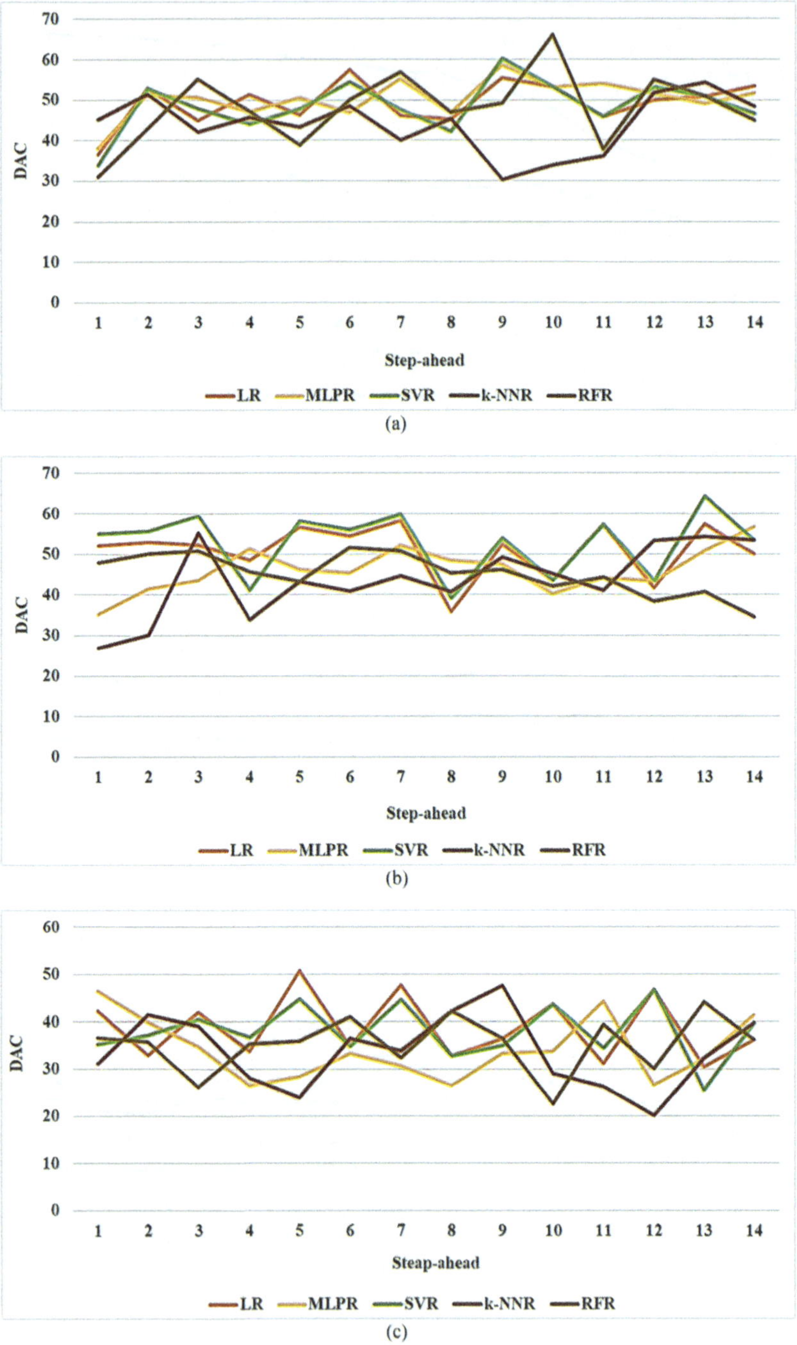

Fig. 7.4 The distribution of DAC from 1 to 14 step ahead. (**a**) Confirmed, (**b**) Recovered, (**c**) Death

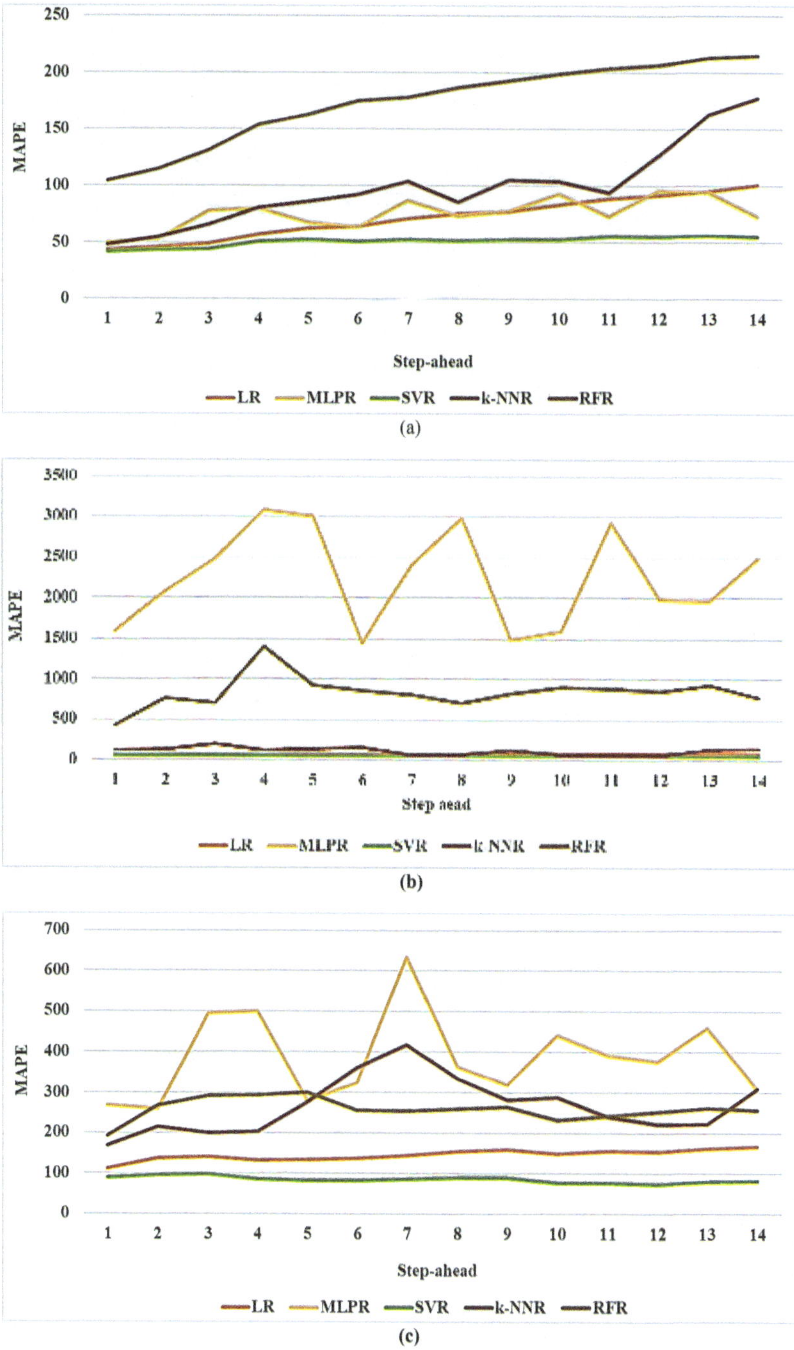

Fig. 7.5 The error distribution of MAPE from 1 to 14 step ahead. (**a**) Confirmed, (**b**) Recovered, (**c**) Death

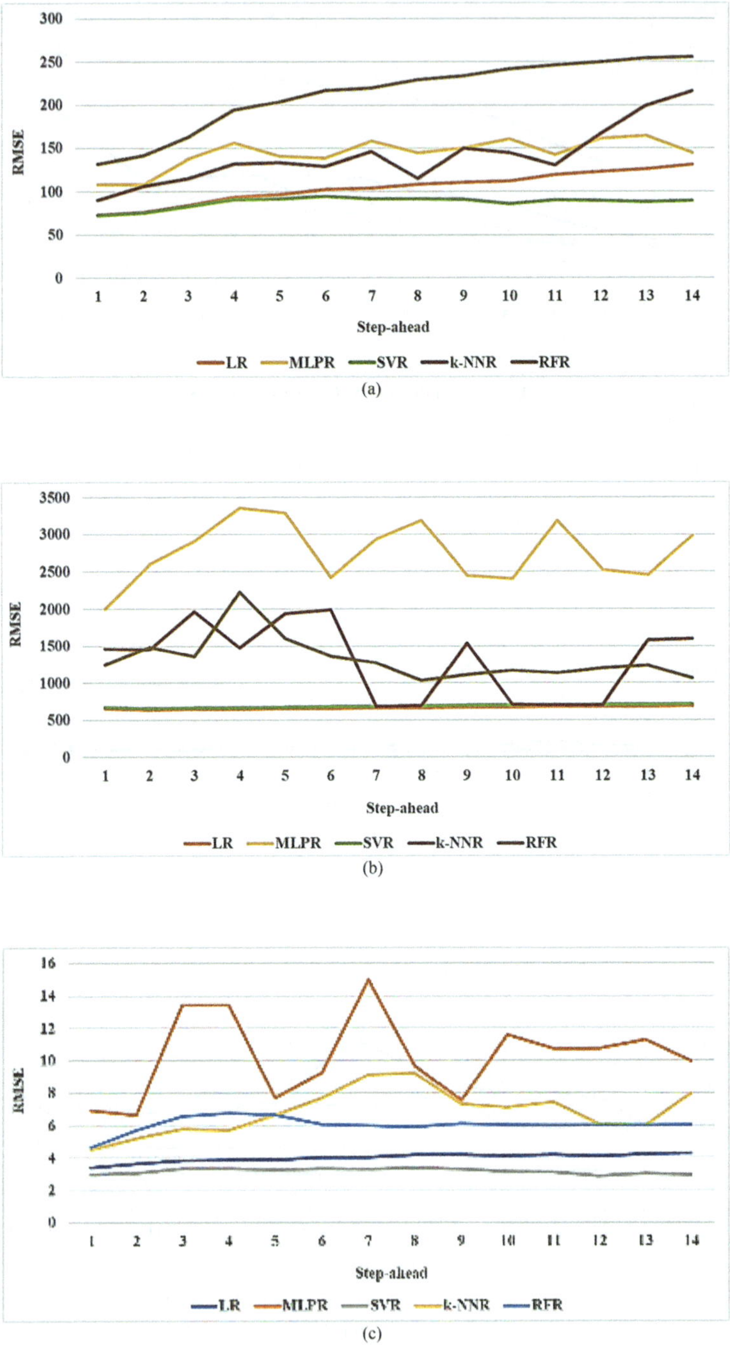

Fig. 7.6 The error distribution of RMSE from 1 to 14 step ahead. (**a**) Confirmed, (**b**) Recovered, (**c**) Death

Table 7.6 Averaged error values for confirmed cases

Models	MAE	DAC	MAPE	RMSE
LR	82.8248	49.2975	72.1419	104.6402
MLPR	119.078	**50.4153**	76.0893	143.9776
SVR	**65.9993**	48.6405	**51.4828**	**87.6290**
k-NN	113.9135	43.9623	98.9396	141.007
RFR	198.8781	48.0210	173.9336	212.842

Table 7.7 Averaged error values for recovered cases

Models	MAE	DAC	MAPE	RMSE
LR	**197.3562**	50.9953	72.1762	666.2736
MLPR	2415.247	46.23846	2256.307	2766.297
SVR	224.8166	**52.9039**	**55.8743**	**688.0478**
k-NN	370.877	43.6757	112.1235	1318.166
RFR	921.3331	45.0574	835.5158	1320.74

Table 7.8 Averaged error values for death cases

Models	MAE	DAC	MAPE	RMSE
LR	3.2860	**38.6892**	145.8573	3.9939
MLPR	8.4267	34.1765	387.888	10.2785
SVR	**2.1796**	37.9566	**84.9544**	**3.1620**
k-NN	5.4687	33.6074	266.2888	6.8417
RFR	5.5703	35.2606	258.0642	6.0234

Table 7.9 14 step ahead future forecast for confirmed cases at 95% confidence interval for all models

Date	LR	MLPR	SVR	*k*-NNR	RFR
2020-10-25	90.0817	36.8171	86.0772	117	240.6432
2020-10-26	84.3642	−9.2463	69.1536	91	241.2202
2020-10-27	82.716	−12.4956	64.7718	108	303.0755
2020-10-28	94.2927	−22.728	67.8692	114	296.5983
2020-10-29	100.6159	−32.503	65.729	87	310.1092
2020-10-30	105.6879	−8.0775	65.2893	91	364.3182
2020-10-31	112.2908	−31.0043	65.0656	64	362.795
2020-11-01	118.5572	65.0541	64.6711	195	372.9628
2020-11-02	124.3246	−20.8522	64.3208	196	354.1483
2020-11-03	130.1623	−25.9036	64.0322	204	352.8528
2020-11-04	135.9091	150.8703	63.7307	238	362.2033
2020-11-05	141.4743	60.3314	63.436	218	353.1285
2020-11-06	146.9366	−5.8387	63.1523	170	353.4485
2020-11-07	152.2956	8.2997	62.8708	244	353.5345

Tables 7.9 and 7.10 present the forecast values for all models for confirmed cases. For SVR model, there is an initial increase from 48 on 10/24/2020 to 86. 0772 on the next day. The model then maintained a consistent decrease to 62.8708 at the end of

Table 7.10 14 step ahead future forecast for recovered cases at 95% confidence interval for all models

Date	LR	MLPR	SVR	*k*-NNR	RFR
2020-10-25	117.3208	1723.2764	38.8299	9	213.1616
2020-10-26	79.5538	3472.483	29.9695	0	155.4343
2020-10-27	61.599	953.0008	18.8655	0	343.7031
2020-10-28	87.653	2605.6651	25.0308	11	193.9728
2020-10-29	84.1132	3720.9484	19.0308	14	297.932
2020-10-30	83.0927	−170.3725	16.9772	17	223.1383
2020-10-31	90.2602	32.0339	17.4836	16	342.3468
2020-11-01	94.1751	3720.9574	16.5257	0	244.821
2020-11-02	97.3154	−171.9659	16.2691	52	240.4892
2020-11-03	101.4034	−406.0696	16.0826	12	240.99
2020-11-04	105.2831	3720.7021	15.8754	32	238.0106
2020-11-05	108.8539	1104.2377	15.7021	34	229.7678
2020-11-06	112.4682	−180.1456	15.5509	15	227.1864
2020-11-07	116.0262	3720.9592	15.4011	17	220.036

the two weeks on 11/07/2020. This shows a gradual decrease in confirmed daily cases until it flattens out. This shows a gradual decrease in confirmed daily cases with mild fluctuations as against the purported second wave that the government is using as the basis for hoarding palliatives in warehouses.

For recovered cases there was an initial decrease from 95 on 10/24/2020 to 38.8299 by the next day 10/25/2020 for the SVR model. There is a steady but low rate of recovered cases by the end of the stipulated 2 weeks 11/07/2020 to 15.4011 cases. This shows a gradual decrease in recovered daily cases with mild fluctuations.

For death cases, there is an initial increase in death cases from 0 on 10/24/2020 to 0.1637 by the next day 10/25/2020 for the SVR model. Tables 7.9 and 7.10 show that the death cases will have risen to 1.1579 by the end of two weeks 11/07/2020. This shows a gradual increase in death daily cases with mild fluctuations (Table 7.11).

7.4.3 Analysis on Lockdown Dates

This section presents the rate of spread of COVID-19 cases in Nigeria during the lockdown phase. The lockdown phase is riven into three periods: pre-lockdown, lockdown, and easing periods as shown in Fig. 7.7 and Table 7.12.

The pre-lockdown period which ranges from 2020-02-27 to 2020-03-29 shows a very mild pattern. During pre-lockdown, 111 patients were confirmed to have COVID-19, 3 were recovered while only 1 patient died. The lockdown period which ranges from 2020-3-30 to 2020-05-4 observes a sizable change in COVID-19 cases as likened to the previous period. The number of deaths recorded had risen

Table 7.11 14 step ahead future forecast for death cases at 95% confidence interval for all models

Date	LR	MLPR	SVR	*k*-NNR	RFR
2020-10-25	1.1965	−3.7303	0.1637	3	3.96
2020-10-26	2.5908	4.3166	0.9343	3	7.39
2020-10-27	2.4926	2.0293	1.1657	3	8.15
2020-10-28	2.495	−16.2248	1.0415	1	10.39
2020-10-29	2.7598	8.0458	1.1853	3	8.62
2020-10-30	2.8416	7.3108	1.176	5	8.06
2020-10-31	2.9173	−19.4655	1.1772	0	9.41
2020-11-01	3.0326	6.1559	1.1826	4	7.64
2020-11-02	3.1316	7.8389	1.1795	7	8.62
2020-11-03	3.2217	−8.1901	1.1756	7	8.01
2020-11-04	3.3157	6.1591	1.1716	10	7.82
2020-11-05	3.4074	14.124	1.1672	17	8.08
2020-11-06	3.4957	−17.2205	1.1625	2	8.02
2020-11-07	3.5828	5.7487	1.1579	6	8.03

to 92, a number of recovered cases had also risen to 414 out of 2691 cases confirmed for lockdown period only. The easing period ranging from 2020-05-05 upwards witnessed an astronomical increase in the spread of the pandemic. Figure 7.3 also shows a high rise in the spread where 790 confirmed cases were reported on the 2020-07-01, 11, 188 COVID-19 cases were recovered on 08/04/2020 and 31 death cases recorded on 06-16-2020. A moderate downward trend was noticed toward October. The aggregate of the COVID-19 occurrences surpassed all the previous period.

7.4.4 *Weekday and Monthly Analysis of COVID-19 Cases*

The progressive spread of COVID–19 cases from February 27 to October 24 is presented in Fig. 7.2. From the onset of the pandemic till the end of March which was the first quarter of the year, the number of confirmed, recovered, and death cases were 139, 9, and 2, respectively. However, by the end of the second quarter which ended in June, the number of cases had steeply risen to 25,555, 9737, and 588 for confirmed, recovered, and death cases, respectively. By third quarter, yet the astronomical increase of confirmed and recovered cases to 33,154 and 40,612, respectively, while death cases reduced mildly to 522 cases.

Figure 7.8a shows that the highest number of confirmed cases were recorded on Saturdays with a value of 9431 followed by Thursday with a value of 9373. Sunday recorded the least number of confirmed cases with a value of 7660. Figure 7.8b shows the corresponding monthly analysis of the spread of confirmed cases. The month of July recorded the highest count of confirmed cases with the value of 17,457 closely followed by June with a value of 15,532 confirmed cases while February

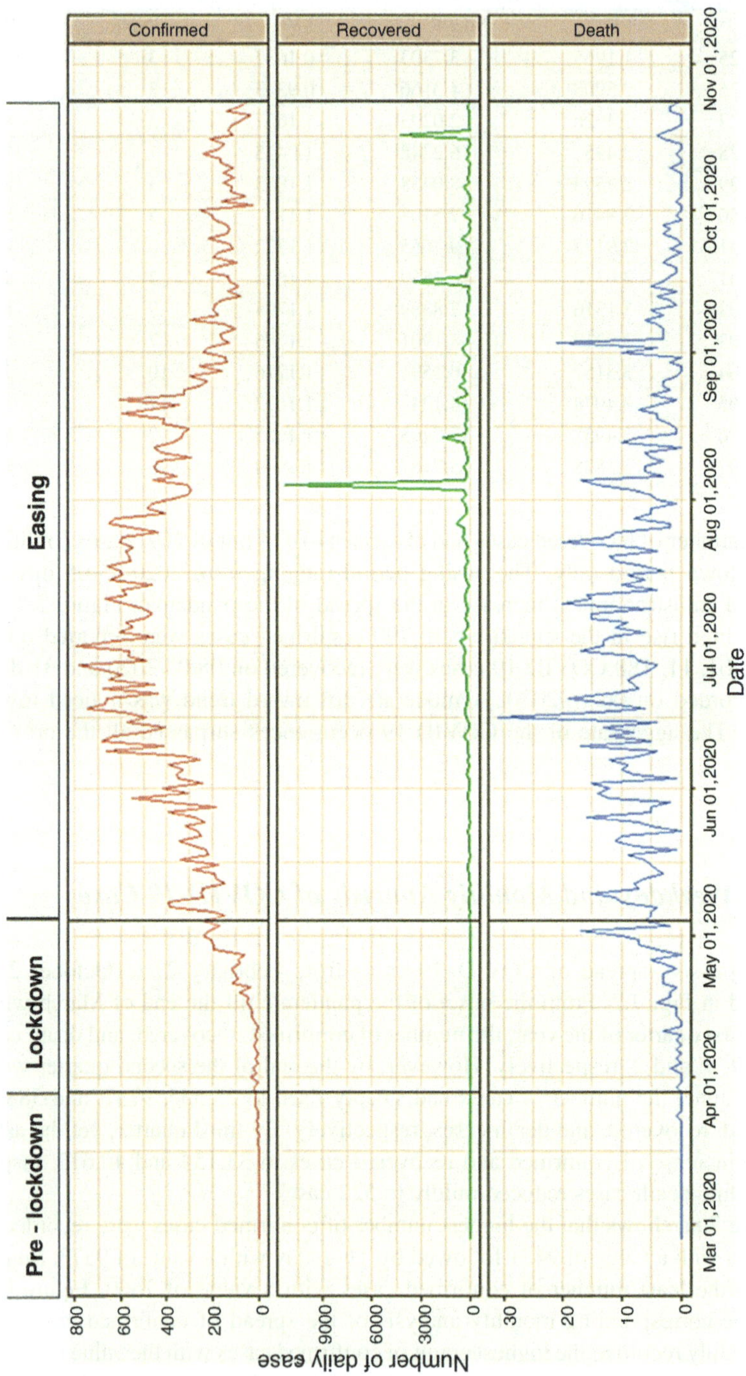

Fig. 7.7 Lockdown analysis of COVID-19 cases

Table 7.12 COVID-19 phases in Nigeria

COVID-19 phases	Period	Confirmed	Recovered	Death
Pre-lockdown	2020-02-27 to 2020-3-29	111	3	1
Lockdown	2020-3-30 to 2020-05-4	2691	414	92
Easing	2020-05-05 to 2020-10-24	59,128	56,868	1036

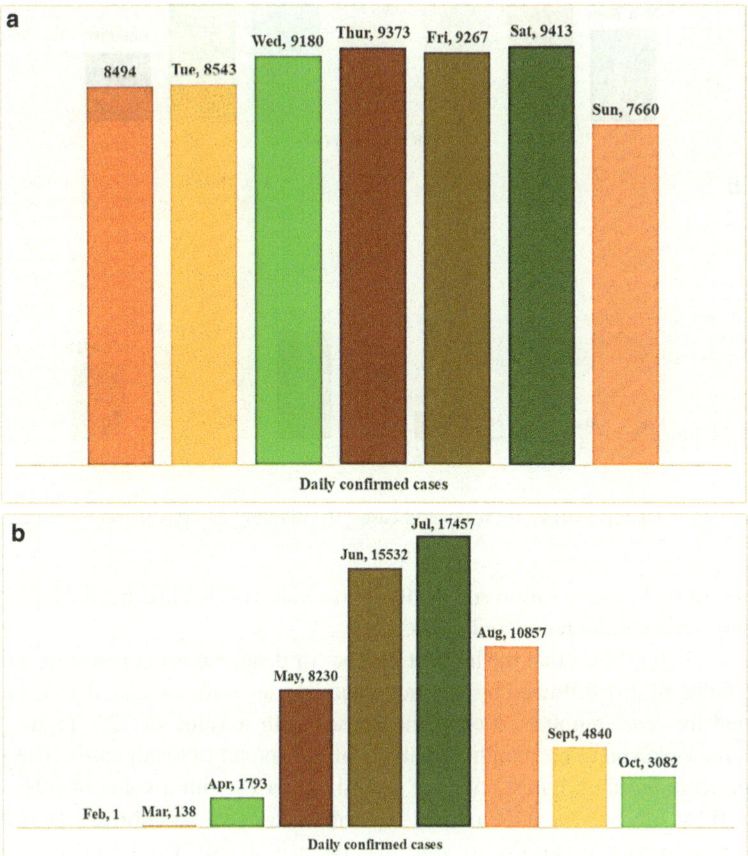

Fig. 7.8 (**a**) Weekday analysis for confirmed cases, (**b**) monthly analysis for confirmed cases

recorded only one (1) incident. However, it is worthy to note that there is a vast decline in the count of confirmed cases in the month of October with 3082 values.

Figure 7.9a shows that the highest number of recovered cases happened on Tuesday with a value of 16,240 followed by Saturday with a value of 10,215. Sunday recorded the least count of recovered cases with a value of 3780. Figure 7.9b shows the corresponding monthly analysis of the state of recovered cases. The month of August recorded the highest count of recovered cases with the

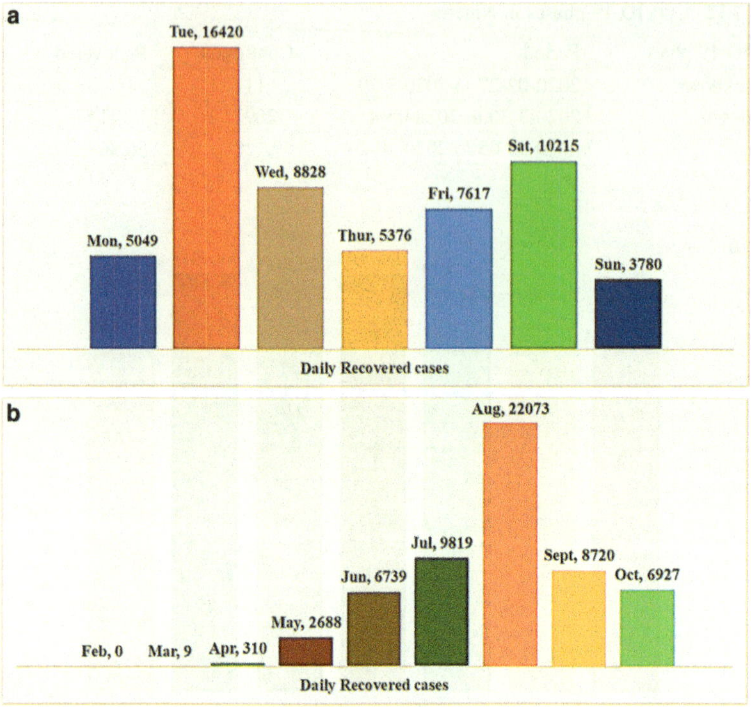

Fig. 7.9 (**a**) Weekday analysis for recovered cases, (**b**) monthly analysis for recovered cases

value of 22,073 closely followed by July with a value of 9819 recovered cases while February recorded no recovered cases.

Figure 7.10a shows that the highest number of death cases happened on Tuesday with a value of 201 followed by Thursday and Sunday with a value of 164. Monday recorded the least number of confirmed cases with a value of 123. Figure 7.10b shows the corresponding monthly analysis of the spread of death cases. The month of June recorded the highest count of confirmed cases with a value of 303 closely followed by July with a value of 289 confirmed cases while February recorded no death. However, it is worthy to note that there is a vast decline in the count of confirmed cases in the month of October with 17 values.

The average weekly analysis for the period of the study shows that Saturday has a high range of incidence cases while Sunday also recorded a low value for COVID-19 cases.

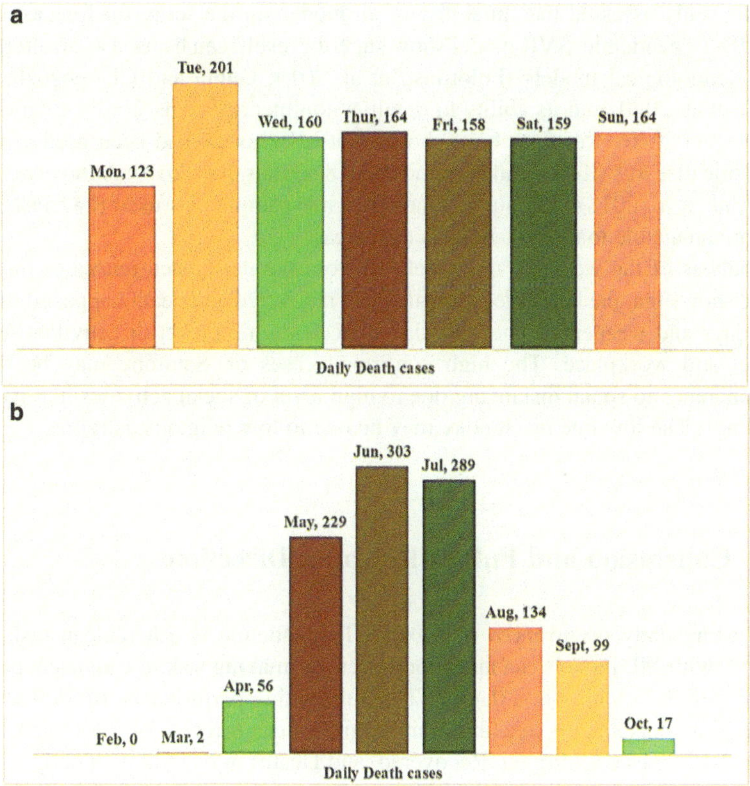

Fig. 7.10 (**a**) Weekday analysis for death cases, (**b**) monthly analysis for death cases

7.4.5 Discussion

Machine learning models for forecasting time series COVID-19 daily cases (Confirmed, Recovered, and Death) for Nigeria were considered. The data were preprocessed using a lagging scheme with 3 days to transform the time series data into a supervised learning format suitable for ML models. The test results for the five ML models under consideration (LR, MLPR, SVR, *k*-NN, and RFR) were used. WEKA default values for all the models. The performances of these models vary for the three cases based on MAE, MAPE, RMSE, and DAC. From the result obtained in Tables 7.7, 7.8, and 7.9, SVR is most suitable for confirmed, recovered, and death cases. Consequently, ML can model the three daily cases of COVID-19 incidence in Nigeria. Inferring from the multistep ahead forecast was tested. The progression of the outbreak by all models was shown by the forecasted values. ML can model the three time series together and can capture the dependencies between the three time series data. SVR is a good choice to model the three time series together. A holdout method of 70:30 ratios was used to avoid overfitting of the data.

This study assessed the suitability of all models in the study for forecasting the COVID-19 pandemic. SVR model show superior result, can be used as an alternative to epidemiological models (Folorunso et al. 2019; Gupta and Gharehgozli 2020; Yadav et al. 2020) and its ability to obtain a nonlinear patterns from the data (Peng and Nagata 2020). Research have shown that ML models had been used to model pandemic diseases (Koike and Morimoto 2018; Dallas et al. 2019; Anno et al. 2019; Muurlink et al. 2018). Although, small data size sample was used (241 instances), they are amenable to the ML models to forecast.

Analysis of the weekday and weekend from the daily occurrences of this pandemic shows that the high rate for confirmed, recovered and death happened on both weekdays and weekends. The weekday indicates a high level of activities such as market and workplace. The high confirmed cases on Saturday may be due to nonadherence to social distancing due to high level of social activities like partying weddings. The low rate on Sunday may be due to low religious activities.

7.5 Conclusion and Future Research Directions

Forecasting daily occurrences of COVID-19 pandemic is a herculean task. Data analysis with ML models can aid a good decision-making task. It can detect patterns and hidden information in a dataset. The outstanding performance of SVR showed achieved models that the machine learning model can forecast 14-days incidence of COVID-19 cases (Confirmed, Recovered, and Death) in Nigeria with daily cases as inputs. This result also established that ML can model the spread and rate of occurrence of infectious diseases attempting to forecast and also model the dependencies among the three variables. Automating the forecasting of the pandemic incidence can aid the government and health care providers to plan, train, and supply medical services to the community. For further work, this study may be extended for forecasting the COVID-19 pandemic for all states in Nigeria with more ML models and additional related data like economic, population, and weather data. This research quantitatively and qualitatively helps the government to ascertain the possibility of a second wave of virus outbreak and possibly the period of its occurrence as against what the government is saying, which is not backed up with data and does not reflect in Fig. 7.3.

References

Ahmed NK, Atiya AF, El Gayar N, El-Shishiny H (2010) An empirical comparison of machine learning models for time series forecasting. Econ Rev 29:5–6

Al-Qaness MAA, Ewees AA, Fan A, Abd El Aziz M (2020) Optimization method for forecasting confirmed cases of COVID-19 in China. J Clin Med 9(674):1–15

Amid S, Gundoshmian TM (2017) Prediction of output energies for broiler production using linear regression, ANN (MLP, RBF), and ANFIS models. Environ Progress Sustain Energy 36:577–585

Anno S, Hara T, Kai H, Lee MA, Chang Y, Oyoshi K, Mizukami Y, Tadono T (2019) Spatiotemporal dengue fever hotspots associated with climatic factors in taiwan including outbreak predictions based on machine-learning. Geospat Health 14:183–194. https://doi.org/10.4081/gh.2019.771

Awotunde JB, Folorunso SO, Jimoh RG, Adeniyi EA, Abiodun KM, Ajamu GJ (2021) Application of Artificial Intelligence for COVID-19 epidemic: an exploratory study, opportunities, challenges, and future prospects. Stud Sys Decis control 358:47–61

Bontempi G, Taieb SB, Le Borgne T (2013) Machine learning strategies for time series forecasting. In: Aufaure MA, Zimanyi E (eds) Business intelligence. eBISS 2012. Lecture notes in business information processing, vol 138. Springer, Berlin, pp 62–77

Bouckaert RR, Frank E, Hall MA, Holmes B, Reutemann P, Witten IA (2010) WEKA – experiences with a Java Open-Source Project. J Mach Learn Res 11:2533–2541

Breiman L (2001) Random forests. Mach Learn 45(1):5–32

Brown DE, Abbasi A, Lau RY (2015) Predictive analytics: predictive modeling at the micro level. IEEE Intell Syst 30(3):6–8

Car Z, Šegota SB, Anđelić N, Lorencin I, Mrzljak V (2020) Modeling the spread of COVID-19 infection using a multilayer perceptron. Comput Math Methods Med 2020(5714714):1–10

Dallas TA, Carlson CJ, Poisot T (2019) Testing predictability of disease outbreaks with a simple model of pathogen biogeography. R Soc Open Sci 6:190883

Drucker H, Burges CJ, Kaufman L, Smola A, Vapnik VN (1997) Support vector regression machines. Adv Neural Inf Process Syst 9:155–161

Folorunso SO, Taiwo AI, Olatayo TO (2019) Empirical comparison of time series data mining algorithms for electrical power prediction. Ind J Comput 4(2):109–118

Folorunso SO, Awotunde JB, Adeboye NO, Matiluko OE (2022) Data classification model for COVID-19 pandemic. Stud Sys Decis control 378:3–118

Fong SJ, Li G, Dey N, Crespo RG, Herrera-Viedma E (2020) Finding an accurate early forecasting model from small dataset: a case of 2019-nCoV novel coronavirus outbreak. Int J Interact Multimed Artif Intell 6(1):132–140

Gupta A, Gharehgozli A (2020). Developing a machine learning framework to determine the spread of COVID-19. Available at SSRN: https://ssrn.com/abstract=3635211, pp 1–19

Guraya SY (2020) Transforming laparoendoscopic surgical protocols during COVID-19 pandemic; big data analytics, resource allocation and operational considerations. Int J Surg 80:21–25

Hu Z, Ge Q, Li S, Boerwinkle E, Jin L, Xiong M (2020) Forecasting and evaluating multiple interventions for COVID-19 worldwide. Front Artif Intell 3(41):1–11

Jayanthi N, Valluvan KR (2017) A review of performance metrics in designing of protocols for wireless sensor networks. Asian J Res Soc Sci Human 7(1):716–730

Kannan S, Ali PSS, Sheeza A, Hemalatha K (2020) COVID-19 (Novel Coronavirus 2019)-recent trends. Eur Rev Med Pharmacol Sci 24(4):2006–2011

Katuwal R, Suganthan PN (2018) Enhancing multi-class classification of random forest using random vector functional neural network and oblique decision surfaces. Arxiv:1802.01240v1

Khanday AMUD, Rabani ST, Khan RQ, Rouf N, Din M (2020) Machine learning based approaches for detecting COVID-19 using clinical text data. Int J Infect Technol 12(3):731–739

Koike F, Morimoto N (2018) Supervised forecasting of the range expansion of novel non-indigenous organisms: alien pest organisms and the 2009 H1N1 flu pandemic. Glob Ecol Biogeogr 27:991–1000

Kramer O (2011) Unsupervised K-nearest neighbor regression. arXiv:1107.3600v2 [stat.ML]

Liu HH, Chang LC, Li CW, Yang CH (2018) Particle swarm optimization-based support vector regression for tourist arrivals forecasting. Comput Intell Neurosci 2018:6076475, p. 13

Montgomery DC, Peck EA, Vinning GG (2012) In: Balding DJ, Cressie NAC, Fitzmaurice GM, Goldstein H, Johnstone LM, Molenberghs G, Scott DW, Smith AFM, Tsay RS, Weisberg S (eds) Introduction to linear regression analysis, 5th edn. Wiley, Hoboken, NJ, p 679

Muurlink OT, Stephenson P, Islam MZ, Taylor-Robinson AW (2018) Long-term predictors of dengue outbreaks in Bangladesh: a data mining approach. Infect Dis Model 3:322–330

Ogundepo E, Folorunso S, Adekanmbi O, Akinsande O, Banjo O, Ogbuju E, Oladipo F, Abimbola O, Oseghale E, Babajide O (2020) An exploratory assessment of a multidimensional healthcare and economic data on COVID-19 in Nigeria. Data Brief 33:106424

Parbat D, Chakraborty M (2020) A python-based support vector regression model for prediction of COVID19 cases in India. Chaos Solit Fractals 138:109942

Peng Y, Nagata MH (2020) An empirical overview of nonlinearity and overfitting in machine learning using COVID-19 data. Chaos Solit Fractals 139:110055

Perrella A, Carannante N, Berretta M, Rinaldi M, Maturo N, Rinaldi L (2020) Editorial–Novel Coronavirus 2019 (Sars-CoV2): a global emergency that needs new approaches? Eur Rev Med Pharmacol Sci 24:2162–2164

Ribeiro MHDM, da Silva RG, Mariani VC, Coelho LDS (2020) Short-term forecasting COVID-19 cumulative confirmed cases: perspectives for Brazil. Chaos Solit Fractals 135:109853

Riza LS, Pertiwi AD, Rahman EF, Munir, Abdullah CU (2019) Question generator system of sentence completion in TOEFL using NLP and K-nearest Neighbor. Indones J Sci Technol 4 (2):294–311

Rustam F, Reshi AA, Mehmood A, Ullah S, On B, Aslam W, Choi GS (2020) COVID-19 future forecasting using supervised machine learning models. IEEE Access 8:101489

Sergio AT, Ludermir TB (2015) Deep learning for wind speed forecasting in northeastern region of Brasil In Brazilian conference on intelligent systems (BRACIS), 2015

Shevade SK, Keerthi SS, Bhattacharyya C, Murthy KRK (2000) Improvements to the SMO algorithm for SVM regression. IEEE Trans Neural Netw 11(5):1188–1193

Sorjamaa A, Hao J, Reyhani N, Ji Y, Lendasse A (2007) Methodology for long-term prediction of time series. Neurocomputing 70(16–18):2861–2869

Sujatha R, Chatterjee JM, Hassanien AE (2020) A machine learning forecasting model for COVID-19 pandemic in India. Stoch Environ Res Risk Assess 34:959–972

Tiwari S, Kumar S, Guleria K (2020) Outbreak trends of coronavirus disease–2019 in India: a prediction. Disaster Med Public Health Prep:1–6

Wei W, Wu H, Ma H (2019) An autoencoder and LSTM-based traffic flow prediction method. Sensors 19:2946

Wong ZXY, Zhou J, Zhang Q (2019) Artificial intelligence for infectious disease big data analytics. Infect Dis Health 24(1):44–48

Yadav M, Perumal M, Srinivas M (2020) Analysis on novel coronavirus (COVID-19) using machine learning methods. Chaos Solit Fractals 139(110050):1–12

Chapter 8
Nigerian COVID-19 Incidence Modeling and Forecasting with Univariate Time Series Model

Abass Ishola Taiwo, Adedayo Funmi Adedotun, and Timothy Olabisi Olatayo

Abstract The occurrence of COVID-19 has given rise to dreadful medical difficulties due to its hyper-endemic effects on the human population. This made it fundamental to model and forecast COVID-19 pervasiveness and mortality to control the spread viably.

The COVID-19 data used was from February, 28, 2020 to March 1, 2021. ARIMA(1,2,0) was selected for modeling COVID-19 confirmed and ARIMA (1,1,0) for death cases. The model was shown to be adequate for modeling and forecasting Nigerian COVID-19 data based on the ARIMA model building results. The forecasted values from the two models indicated Nigerian COVID-19 cumulative confirmed and death case continues to rise and maybe in-between 189,019–327,426 and interval 406–3043, respectively in the next 3 months (May 30, 2021). The ARIMA models forecast indicated an alarming rise in Nigerian COVID-19 confirmed and death cases on a daily basis.

The findings indicated that effective treatment strategies must be put in place, the health sector should be monitored and properly funded. All the protocols and restrictions put in place by the NCDC, Nigeria should be clung to diminish the spread of the pandemic and possible mortality before immunizations that can forestall the infection is developed.

Keywords COVID-19 · ARIMA model · Confirmed cases · Death cases · Modeling · Forecasting

---

A. I. Taiwo (✉) · T. O. Olatayo
Department of Mathematical Sciences, Olabisi Onabanjo University, Ago-Iwoye, Ogun State, Nigeria
e-mail: taiwo.abass@oouagoiwoye.edu.ng

A. F. Adedotun
Covenant University, Ota, Ogun State, Nigeria
e-mail: bisi.olatayo@oouagoiwoye.edu.ng

S. A. Hassan et al. (eds.), *Decision Sciences for COVID-19*, International Series in Operations Research & Management Science 320,
https://doi.org/10.1007/978-3-030-87019-5_8

8.1 Introduction

COVID-19 is considered an emergency of international importance for public health. About 15% of cases are in patients with a chronic form of the disease (Zhang et al. 2020). Aging individuals and those with pre-existing medical problems are likely to become severely ill (Holingue et al. 2020). The 2019 Coronavirus disease (COVID-19), which first took place in Wuhan, China, in December 2019, has spread to almost all of the world (Adenomon et al. 2020; Igwe 2020). argued that the global economy faces the worst-ever economic recession due to the outbreak of the COVID-19. He also noted that the shock from this virus would raise instability that will adversely impact the economic and financial environment of every country (Adenomon et al. 2020). reported on the Zombie epidemic that could be close to COVID-19, the study concluded that a mild zombie outbreak would lead to a decrease in GDP and the stock market of 23.44% and 29.30%, respectively, leaving one million people dead in the main developed world.

On February 27, 2020, the COVID-19 case was officially identified in Nigeria, and many more cases were identified after that (Adegboye et al. 2020). The effects of COVID-19 on Nigeria's economy and financial markets as to do with the economic lockout of major cities (Abuja, Lagos, Ogun States) on March 30, 2020, resulting in economic losses, particularly for the everyday earners of small–medium-sized firms, the removal of capital from the market by buyers and the collapse in oil prices (Abulude and Abulude 2020). Furthermore, if the impact of COVID-19 on the economy and financial markets are not adequately recognized and controlled, this virus has the potential to completely crash any country, such as Nigeria's health sector or economy (Ajibo 2020). Especially in the absence of a vaccine, while studies have shown that it is not possible to reinfect a recovered population (Mumbu and Hugo 2020). Chukwuka and Mma (2020). recognized the effect of the COVID-19 epidemic on the Nigerian economy, their result revealed that Nigeria's economy based on GDP should rise by 2.5% in 2020; this was truncated by the pandemic and led to a high increase in the nation's debt services and income ratio at 60% amid declining oil prices (Akanni and Gabriel 2020). stated that the COVID-19 pandemic has led to the destruction of operations and economic turmoil as the United Trade and Development Agency cost the economic loss to the outbreak around $2 trillion. Factors such as social distancing, stay at home, spending restrictions and supply factors have been shown to harm economic development. The National Bureau of Statistics (NBS) survey 2020 ranked Nigeria 21 among 181 counties with a high unemployment rate of 23.1%, with an estimated 87 million living with less than $2 a day benchmark.

Olufemi and Bolanle (2018) Investigated the diversification of the Nigerian stock market's foreign portfolio, the analysis was concluded using the vector autoregressive granger causality test for relationship revealed that there is no relationship between the stock market of Nigeria and the other five developed countries. However, applying the Generalized Moment Regression approach, the result revealed that established capital markets have an effect on the Nigerian stock market after the crisis period and before the crisis period. This was also inferred that before the COVID-19 crisis, Nigeria's stock market was safe for investors, but the COVID-19 situation has made investment impossible.

John (2020) discussed the COVID-19 pandemic, a battle to be fought and its economic implications for Africa. The study concluded that the International Monetary Fund has called on all official bilateral creditors to postpone all loan payments. The International Development Organization Countries are asking for forbearance to savage economic uncertainty as the effect of COVID-19 on the African economy cannot be decided yet, but as the situation progresses, there will be more insight into it.

Based on the discussion above, the purpose of this research is to use an Autoregressive integrated moving average model, which is a Univariate time series model, to forecast Nigerian COVID-19 incidences as a way to determine potential future occurrences.

Since forecasting is a process of predicting or estimating the future based on past and present data. It is important because it provides relevant and reliable information about past, present, and future events. It gives room for vivid planning and enhances the process of making important decisions and policies. Types of forecasting are qualitative and quantitative methods, where the latter involves the use of statistical methods. In particular, forecasting with time series analysis models, which include Autoregression (AR), Moving Average (MA), Autoregressive Moving Average (ARMA), Autoregressive Integrated Moving Average (ARIMA) and Seasonal Autoregressive Integrated Moving Average (SARIMA). The data patterns can be a trend, cyclical, seasonal, and irregular in nature. But before the data can be model, stationarity is firstly attained by differencing the series and check for a possible seasonal component. Therefore, the appropriate model for forecasting stationary series without seasonal components is an autoregressive integrated moving average (Box et al. 2015). The basic steps in forecasting are problem definition, gathering information, preliminary exploratory analysis, choosing and fitting models and evaluating the forecasting model. The forecasting model is evaluated for accuracy using the forecast error, which is the difference between the actual and the forecast for a given period. The measures of forecast accuracy often used are mean absolute error (MAE), root mean square forecast error (RMSE) and mean absolute percentage error (MAPE) (Forecast Errors 2002).

8.2 Materials and Methods

8.2.1 Data Collection

The data used was obtained from Nigeria Center for disease control (NCDC) website. The COVID 19 confirmed and death cases dataset obtained span between February 28, 2020 to March 1, 2021.

8.2.2 *Univariate Time Series Model*

A Univariate time series model is a class of specifications used to model and forecast a particular time series using only information contained in their past values and possibly current and past values of an error term.

8.2.3 *Autoregressive Integrated Moving Average (ARIMA) Model*

ARIMA model is a Univariate time series model that consists of an autoregressive polynomial, an order of integration (d) and a moving average polynomial. The usual forms of AR(p) and MA(q) are written as

$$x_t = \varnothing_1 x_{t-1} + \varnothing_2 x_{t-2} + \ldots + \varnothing_p x_{t-p} + e_t \tag{8.1}$$

and

$$x_t = \varepsilon_t + \theta_1 \varepsilon_{t-1} + \theta_2 \varepsilon_{t-2} + \ldots + \theta_q \varepsilon_{t-q} \tag{8.2}$$

where $\varnothing$ and θ are the autoregressive and moving average parameters, respectively. x_t is the observed value at time t and ε_t is the value of the random shock at time t. It is assumed to be independently and identically distributed with a mean of zero and a constant variance (σ^2). ARMA (p, q) model is comprised of AR and MA models, in which the current value of the time series is defined linearly in terms of its previous values as well as current and previous error series.

The ARMA (p, q) model is given in (8.3) as

$$x_t = \varnothing_1 x_{t-1} + \varnothing_2 x_{t-2} + \ldots + \varnothing_p x_{t-p} + \varepsilon_t + \theta_1 \varepsilon_{t-1} + \theta_2 \varepsilon_{t-2} + \ldots + \theta_q \varepsilon_{t-q} \tag{8.3}$$

This can be simplified by a backward shift operator B to obtain

$$(B)\nabla^d x_t = \theta(B) w_t \tag{8.4}$$

Therefore (8.4) can be expressed as ARMA(p, d, q) where $\nabla^d = (1 - B)^d$ with $\nabla^d y_t$ and dth consecutive differencing. The basic steps used for ARIMA model buildings involve model identification, estimation, diagnostic, and forecasting.

8.2.3.1 Model Identification

A correlogram is a chart of correlation statistics, and it includes an autocorrelations function (ACF) and partial autocorrelation functions (PACF). The ACF is used to measure the amount of linear dependence between observations in the time series, and the partial autocorrelation function is used to determine the possible order of the ARIMA model. The Autocorrelations function (ACF) is denoted by

$$\rho_k = \frac{E[(x_t - \overline{x})(x_{t-k} - \overline{x})]}{E[x_t - \overline{x}]^2} \tag{8.5}$$

and partial autocorrelation functions (PACF) is given as

$$x_t = \rho_0 + \sum_{k=1}^{K} \rho_{kk} x_{t-k} \tag{8.6}$$

where ρ_{kk} is the kth autoregressive coefficient, $k = 1, 2, \ldots, K$. After the identification stage, the smallest values of Akaike Information criteria (AIC), Schwartz Bayesian Information criteria (SBC) and Hannan Quinn Information criteria (HQC) will be used to choose the appropriate model.

8.2.3.2 Parameter Estimation

The coefficient of the ARIMA model will be obtained using the ordinary least squares estimation method. The coefficient will be obtained using

$$\widehat{\varnothing} = \sum\nolimits_{t=2}^{n} (x_{t-1})(x_t) / \sum\nolimits_{t=2}^{n} x_{t-1}^2 \tag{8.7}$$

8.2.3.3 Diagnostic Checking

The stability of the estimated model will be diagnosed using Modified Box-Pierce (Ljung-Box) Chi-Square Statistic. This is defined as

$$Q(m) = n(n+2) \sum_{j=1}^{m} \frac{r_j^2}{n-j} \tag{8.8}$$

where n is the number of usable data points after any differencing operations.

8.2.3.4 Forecasting

There are two kinds of forecasts, and these are sample period forecasts and post-sample period forecasts. The former will be used to develop confidence in the model, and the latter will be used to generate genuinely desired forecasts. In forecasting, the goal is to predict future values of a time series, x_{t+m}, $m = 1, 2, \ldots$ based on the data collected to the present, $x = \{x_t, x_{t-1}, \ldots, x_1\}$.

8.3 Result and Discussion

A daily cumulative dataset for confirmed and death cases of COVID-19 occurrences from February 28, 2020 to March 1, 2021 was considered. Time plots of the confirmed and death cases of COVID-19 data are presented in Figs. 8.1 and 8.2 to

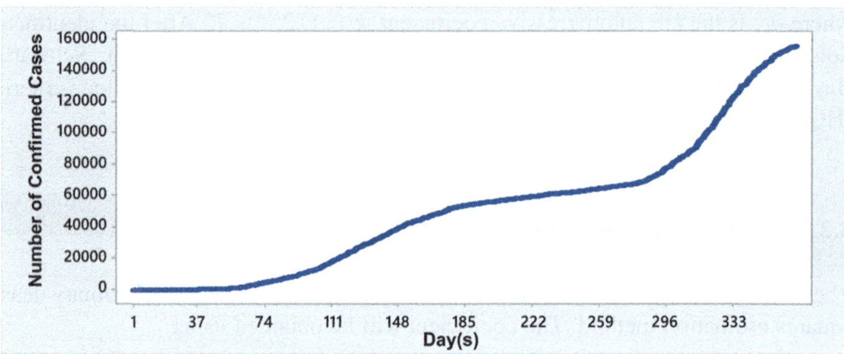

Fig. 8.1 Time plot of confirmed cases from February 28, 2020 to March 1, 2021

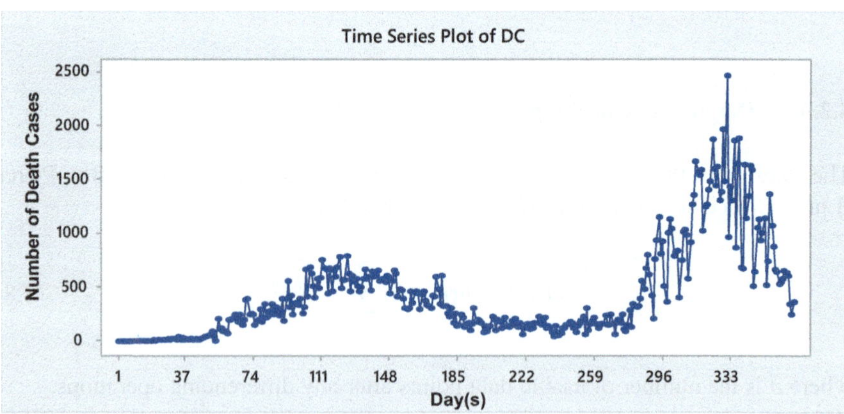

Fig. 8.2 Time plot of death cases from February 28, 2020 to March 1, 2021

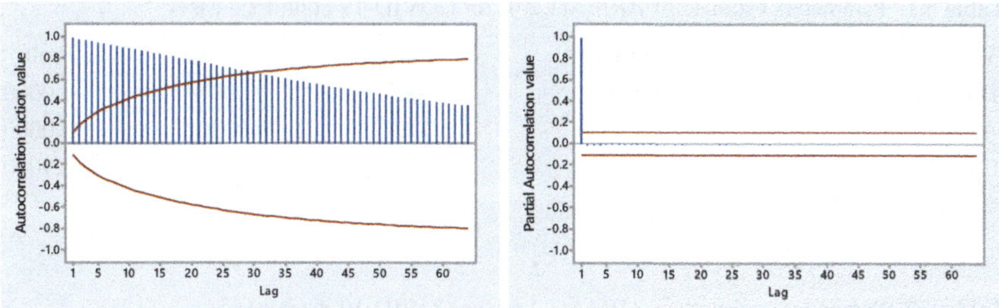

Fig. 8.3 (a) Autocorrelation function for confirmed cases. (b) Partial autocorrelation function for confirmed case

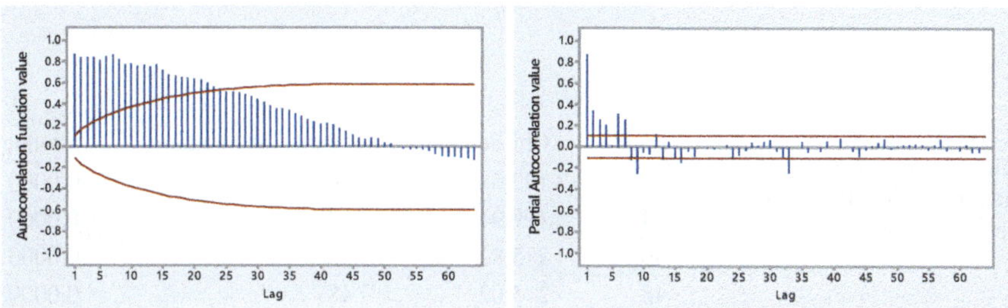

Fig. 8.4 (a) Autocorrelation function for death cases. (b) Partial autocorrelation function for death cases

investigate the trend and behavior of data concerning over a while. Figures 8.1 and 8.2 showed that confirmed and death cases of COVID-19 data exhibited a trend or secular movement. The Autocorrelation function plots in Figs. 8.3a and 8.4a and partial autocorrelation function plots in Figs. 8.3b and 8.4b were used to show that the COVID-19 confirmed and death cases are not stationary since autocorrelations diminish marginally. Consequently, the first difference was taken to balance out the mean of the COVID-19 information. To accomplish stationarity, the Dickey-Fuller unit root test was established, and from the outcome, the COVID-19 confirmed cases were stationary at the second difference ($d = 2$), and death cases were stationary at the first difference ($d = 1$).

The autocorrelation function (ACF) plots in Figs. 8.3a and 8.4a and partial autocorrelation function (PACF) plots in Figs. 8.3b and 8.4b were used to examine the randomness of the COVID-19 data. This is used to quantify the relation between lagged time series values. A further critical look showed that the autocorrelation function (ACF), and partial autocorrelation function (PACF) showed the ACF tailed off at lag 1 and PACF cut-off after lag 0. In essence, $p = 1$ and $q = 0$ or $p = 0$ and $q = 1$. Based on the values of p and q, the following models ARIMA(1,2,1), ARIMA

Table 8.1 Parameters estimate of ARMA(1,2,0) for COVID-19 confirmed cases

Variable	Coefficient	Std. Error	*t*-Statistic	Prob.
D(Confirmed Case,2)	0.0005	0.0007	0.0007	1.0000
AR(1)	−0.3962	0.0002	0.0030	0.0000
Constant	1.25	10.32	0.12	0.9040

Table 8.2 Parameters estimate of ARMA(1,1,0) for COVID-19 death cases

Variable	Coefficient	Std. Error	*t*-Statistic	Prob.
D(Death Case,1)	0.0003	0.0007	4.5374	1.0000
AR(1)	−0.3960	3.3728	0.0024	0.0000
Constant	1.24	1029	0.12	0.9040

Table 8.3 Modified Box-Pierce Chi-square statistic for COVID-19 confirmed cases

Lag	Chi-Square	Degree of freedom	*p*-value
12	143.68	10	0.0000
24	199.94	22	0.0000
36	215.82	34	0.0000
48	245.02	48	0.0000

Table 8.4 Modified Box-Pierce Chi-square statistic for COVID-19 death cases

Lag	Chi-Square	Degree of freedom	*p*-value
12	165.35	10	0.0000
24	274.09	22	0.0000
36	336.39	34	0.0000
48	361.90	48	0.0000

(0,2,1), ARIMA(1,2,0) were created for COVID-19 confirmed cases and ARIMA (1,1,1), ARIMA(0,1,1) and ARIMA(1,1,0) were created for COVID death cases.

ARIMA(1,2,0) was chosen for COVID-19 confirmed cases, and ARIMA(1,1,0) was chosen for COVID-19 death cases from the created models based on the least values the information criteria after the coefficients of the models were estimated using the least squares method for both confirmed and death cases of COVID-19 data. The fitted model results for confirmed and death cases of COVID-19 data are given in Tables 8.1 and 8.2. The adequacy of both models for forecasting Nigeria was confirmed, and death cases were determined based on Modified Box-Pierce (Ljung-Box) Chi-Square Statistic values given in Tables 8.3 and 8.4. In essence, the models can now be used to forecast Nigerian confirmed and death cases.

Figure 8.5 was used to present the fitted and predicted values for Nigerian confirmed and death cases for the next 90-days. Table 8.5 further showed Nigerian

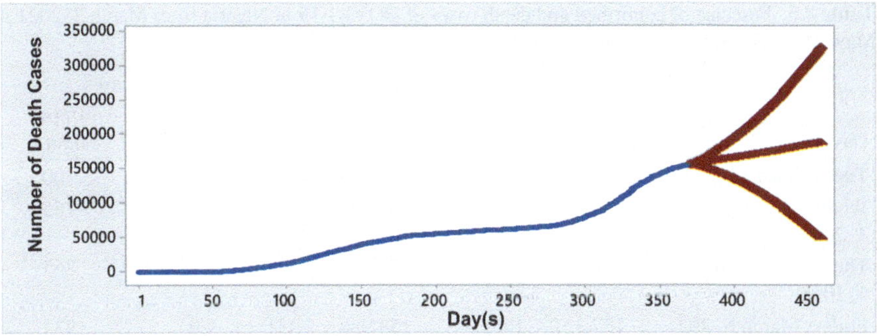

Fig. 8.5 Time plot of confirmed cases forecast from February 28, 2020 to May 30, 2021

cumulative confirmed and death cases forecast, upper and lower limit values. Based on Table 8.5 and Fig. 8.5, the Nigerian COVID-19 cumulative confirmed case continues to rise and maybe in between 189,019–327,426 confirmed cases in the next 3 months (May 30, 2021). While Fig. 8.6 depicts the fitted and predicted values for Nigerian cumulative death cases for the next 90-days. Table 8.5 as well is used to present the Nigerian cumulative death case forecast, upper and lower limit values, and this indicated that Nigerian COVID-19 death cases continue to increase, and this may rise between 406–3043 deaths in the next 3 months (May 30, 2021).

The ARIMA models forecast showed that Nigerian COVID-19 confirmed and death cases are increasing at an alarming rate on a daily basis. Accordingly, the ARIMA models forecast showed a disturbing ascent in Nigerian COVID-19 confirmed and death cases. These results benefit the Nigerian government at all level, local and international organizations and individuals that COVID-19 is a serious and deadly disease that can easily spread and harper human growth and existence. This as well gives a general awareness to Nigerian citizens that everyone has a role to play to put a stop to COVID-19 spread by adhering to all safety protocols recommended by World Health Organization (WHO) and National Center for Disease Control (NCDC), Nigeria. Therefore, based on the COVID-19 confirmed and death cases forecast, the Nigerian government must decide to put in place pro-activity and efficient treatment strategies. The health sector should be monitored and properly funded. All the protocols and restrictions put in place by the NCDC, Nigeria should be enforced in other to combat the spread of the pandemic, and possible mortality before immunizations that can forestall or fix the infection is created and readily available.

Table 8.5 Forecast of confirmed and death cases of COVID-19 in Nigeria from March 2, 2021 to May 30, 2021 with 95% confidence interval

	Forecast for confirmed cases			Forecast for death cases		
Day(s)	Forecast	Lower limit	Upper limit	Forecast	Lower limit	Upper limit
Tuesday, March 2, 2021	156,331	155,944	156,718	314	0	700
Wednesday, March 3, 2021	156,664	155,933	157,395	333	0	785
Thursday, March 4, 2021	156,991	155,820	158,162	327	0	865
Friday, March 5, 2021	157,321	155,655	158,988	331	0	933
Saturday, March 6, 2021	157,652	155,433	159,871	330	0	995
Sunday, March 7, 2021	157,983	155,163	160,804	332	0	1051
Monday, March 8, 2021	158,316	154,847	161,784	332	0	1103
Tuesday, March 9, 2021	158,649	154,490	162,809	333	0	1152
Wednesday, March 10, 2021	158,983	154,093	163,874	334	0	1199
Thursday, March 11, 2021	159,319	153,658	164,979	335	0	1243
Friday, March 12, 2021	159,655	153,187	166,122	336	0	1285
Saturday, March 13, 2021	159,992	152,683	167,300	337	0	1325
Sunday, March 14, 2021	160,329	152,146	168,513	338	0	1364
Monday, March 15, 2021	160,668	151,577	169,759	339	0	1402
Tuesday, March 16, 2021	161,008	150,979	171,037	340	0	1438
Wednesday, March 17, 2021	161,348	150,350	172,346	340	0	1473
Thursday, March 18, 2021	161,690	149,694	173,685	341	0	1507
Friday, March 19, 2021	162,032	149,010	175,054	342	0	1541
Saturday, March 20, 2021	162,375	148,299	176,451	343	0	1573
Sunday, March 21, 2021	162,719	147,562	177,876	344	0	1605
Monday, March 22, 2021	163,064	146,799	179,328	345	0	1636
Tuesday, March 23, 2021	163,410	146,012	180,807	346	0	1666
Wednesday, March 24, 2021	163,756	145,201	182,312	347	0	1696
Thursday, March 25, 2021	164,104	144,366	183,843	348	0	1725
Friday, March 26, 2021	164,453	143,507	185,398	348	0	1753
Saturday, March 27, 2021	164,802	142,626	186,978	349	0	1781
Sunday, March 28, 2021	165,152	141,722	188,582	350	0	1808

(continued)

Table 8.5 (continued)

	Forecast for confirmed cases			Forecast for death cases		
Day(s)	Forecast	Lower limit	Upper limit	Forecast	Lower limit	Upper limit
Monday, March 29, 2021	165,504	140,797	190,210	351	0	1835
Tuesday, March 30, 2021	165,856	139,850	191,861	352	0	1862
Wednesday, March 31, 2021	166,209	138,882	193,535	353	0	1888
Thursday, April 1, 2021	166,562	137,894	195,231	354	0	1913
Friday, April 2, 2021	166,917	136,885	196,950	355	0	1939
Saturday, April 3, 2021	167,273	135,856	198,690	356	0	1963
Sunday, April 4, 2021	167,629	134,807	200,452	356	0	1988
Monday, April 5, 2021	167,987	133,738	202,235	357	0	2012
Tuesday, April 6, 2021	168,345	132,651	204,039	358	0	2036
Wednesday, April 7, 2021	168,704	131,545	205,864	359	0	2060
Thursday, April 8, 2021	169,064	130,420	207,709	360	0	2083
Friday, April 9, 2021	169,425	129,276	209,575	361	0	2106
Saturday, April 10, 2021	169,787	128,115	211,460	362	0	2129
Sunday, April 11, 2021	170,150	126,935	213,365	363	0	2151
Monday, April 12, 2021	170,514	125,738	215,289	363	0	2173
Tuesday, April 13, 2021	170,878	124,524	217,233	364	0	2195
Wednesday, April 14, 2021	171,244	123,292	219,196	365	0	2217
Thursday, April 15, 2021	171,610	122,043	221,177	366	0	2238
Friday, April 16, 2021	171,977	120,777	223,177	367	0	2259
Saturday, April 17, 2021	172,346	119,495	225,196	368	0	2281
Sunday, April 18, 2021	172,715	118,196	227,233	369	0	2301
Monday, April 19, 2021	173,085	116,881	229,288	370	0	2322
Tuesday, April 20, 2021	173,455	115,550	231,361	371	0	2342
Wednesday, April 21, 2021	173,827	114,203	233,451	371	0	2363
Thursday, April 22, 2021	174,200	112,840	235,559	372	0	2383
Friday, April 23, 2021	174,573	111,461	237,685	373	0	2402
Saturday, April 24, 2021	174,947	110,067	239,828	374	0	2422
Sunday, April 25, 2021	175,323	108,658	241,988	375	0	2442
Monday, April 26, 2021	175,699	107,233	244,164	376	0	2461
Tuesday, April 27, 2021	176,076	105,794	246,358	377	0	2480
Wednesday, April 28, 2021	176,454	104,339	248,568	378	0	2499
Thursday, April 29, 2021	176,833	102,870	250,795	379	0	2518

(continued)

Table 8.5 (continued)

	Forecast for confirmed cases			Forecast for death cases		
Day(s)	Forecast	Lower limit	Upper limit	Forecast	Lower limit	Upper limit
Friday, April 30, 2021	177,212	101,386	253,039	379	0	2537
Saturday, May 1, 2021	177,593	99,888	255,298	380	0	2555
Sunday, May 2, 2021	177,975	98,375	257,574	381	0	2574
Monday, May 3, 2021	178,357	96,848	259,866	382	0	2592
Tuesday, May 4, 2021	178,740	95,307	262,173	383	0	2610
Wednesday, May 5, 2021	179,125	93,752	264,497	384	0	2628
Thursday, May 6, 2021	179,510	92,183	266,836	385	0	2646
Friday, May 7, 2021	179,896	90,600	269,191	386	0	2664
Saturday, May 8, 2021	180,282	89,004	271,561	387	0	2681
Sunday, May 9, 2021	180,670	87,394	273,947	387	0	2699
Monday, May 10, 2021	181,059	85,770	276,348	388	0	2716
Tuesday, May 11, 2021	181,448	84,133	278,764	389	0	2734
Wednesday, May 12, 2021	181,839	82,483	281,195	390	0	2751
Thursday, May 13, 2021	182,230	80,820	283,641	391	0	2768
Friday, May 14, 2021	182,622	79,143	286,102	392	0	2785
Saturday, May 15, 2021	183,016	77,454	288,577	393	0	2802
Sunday, May 16, 2021	183,410	75,751	291,068	394	0	2818
Monday, May 17, 2021	183,804	74,036	293,573	395	0	2835
Tuesday, May 18, 2021	184,200	72,308	296,092	395	0	2852
Wednesday, May 19, 2021	184,597	70,568	298,626	396	0	2868
Thursday, May 20, 2021	184,994	68,815	301,174	397	0	2884
Friday, May 21, 2021	185,393	67,049	303,737	398	0	2901
Saturday, May 22, 2021	185,792	65,271	306,313	399	0	2917
Sunday, May 23, 2021	186,193	63,481	308,904	400	0	2933
Monday, May 24, 2021	186,594	61,679	311,508	401	0	2949
Tuesday, May 25, 2021	186,996	59,864	314,127	402	0	2965
Wednesday, May 26, 2021	187,399	58,038	316,759	403	0	2980
Thursday, May 27, 2021	187,802	56,199	319,406	403	0	2996
Friday, May 28, 2021	188,207	54,349	322,066	404	0	3012
Saturday, May 29, 2021	188,613	52,486	324,739	405	0	3027
Sunday, May 30, 2021	189,019	50,612	327,426	406	0	3043

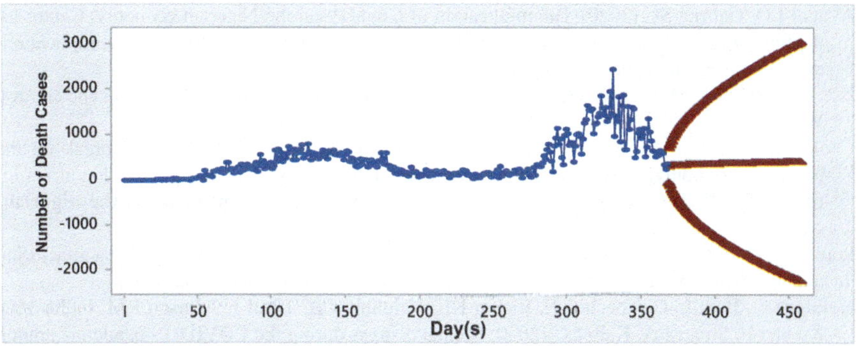

Fig. 8.6 Time plot of death cases forecast from February 28, 2020 to May 30, 2021

8.4 Conclusion

In this study, ARIMA models were used to model and forecast Nigerian confirmed and death cases as a result of the COVID-19 pandemic. This COVID-19 data considered was from February 28, 2020 to March 1, 2021. The ARIMA(1,2,0) and ARIMA(1,1,0) models were obtained as the appropriate models based on the steps of ARIMA model building. The forecasted values from the two models indicated Nigerian COVID-19 cumulative confirmed and death cases continues to rise and maybe in-between 189,019–327,426 confirmed cases and 406–3043 death cases respectively in the next 3 months (May 30, 2021). The ARIMA models forecast indicated an alarming rise in Nigerian COVID-19 confirmed and death cases on daily basis. These findings indicated that effective treatment strategies must be put in place, the health sector should be monitored and properly funded. All the protocols and restrictions put in place by the NCDC, Nigeria should be clung to diminish the spread of the pandemic and possible mortality before immunizations that can forestall the infection is developed and readily available.

References

Abulude FO, Abulude IA (2020). Impact of the COVID-19 pandemic: lesson from Nigeria. Available at SSRN 3629598. doi:https://doi.org/10.2139/ssrn.3629598. PPR: PPR242362

Adegboye OA, Adekunle AI, Gayawan E (2020) Early transmission dynamics of novel coronavirus (COVID-19) in Nigeria. Int J Environ Res Public Health 17(9):30–54

Adenomon MO, Maijamaa B, John DO (2020) On the effects of COVID-19 outbreak on the Nigerian stock exchange performance: evidence from GARCH models. Available at 2020040444. doi:https://doi.org/10.20944/preprints202004.0444.v1

Ajibo H (2020) Effect of Covid-19 on Nigerian socio-economic well-being, health sector pandemic preparedness and the role of nigerian social workers in the war against Covid-19. Soc Work Public Health 35(7):511–522

Akanni LO, Gabriel SC (2020) The implication of Covid19 on the Nigerian economy. Centre for the Study of the Economies of Africa (CSEA). Available at http://cseaafrica.org/theimplication-of-covid19-on-the-nigerian-economy/

Box GE, Jenkins GM, Reinsel GC, Ljung GM (2015) Time Series analysis: forecasting and control. Wiley, 709p

Chukwuka O, Mma AE (2020) Understanding the impact of the COVID-19 outbreak on the Nigerian Economy. Africa in focus. Available at https://www.brookings.edu/blog/africa-in-focus/2020/04/08/understanding-the-impact-of-the-covid-19-outbreak-on-the-nigerian-economy/

Forecast Errors (2002) In: Swamidass PM (eds) Encyclopedia of production and manufacturing management. Springer, Boston, MA. https://doi.org/10.1007/1-4020-0612-8_358

Holingue C, Badillo-Goicoechea E, Riehm KE, Veldhuis CB, Thrul J, Johnson RM, Fallin MD, Kreuter F, Stuart EA, Kalb LG (2020) Mental distress during the COVID-19 pandemic among US adults without a pre-existing mental health condition: Findings from American trend panel Survey. Prev Med 139(106231):1–21

Igwe PA (2020) Coronavirus with looming global health and economic doom. Afr Dev Inst Res Methodol 1(1):1–6

John EA (2020) COVID-19 pandemic, a war to be won: understanding its economic implications for Africa. Appl Health Econ Health Policy 5:1–4

Mumbu ARJ, Hugo AK (2020) Mathematical modelling on COVID-19 transmission impacts with preventive measures: a case study of Tanzania. J Biol Dyn 14(1):748–766

Olufemi AA, Bolanle A (2018) International portfolio diversification in the Nigerian stock market: a global financial crisis perspective. Future Bus J 4(2):189–194

Zhang L, Zhu F, Xie L, Wang C, Wang J, Chen R, Jia P, Guan HQ, Peng L, Chen Y, Peng P, Zhang P, Chu Q, Shen Q, Wang Y, Xu SY, Zhao JP, Zhou M (2020) Clinical characteristics of COVID-19-infected cancer patients: a retrospective case study in three hospitals within Wuhan, China. Ann Oncol 31(7):894–901

Chapter 9
Predicting the Spread of COVID-19 in Africa Using Facebook Prophet and Polynomial Regression

Cecilia Ajowho Adenusi, Olufunke Rebecca Vincent, Abiodun Folurera Ajayi, and Bukola Taibat Adebiyi

Abstract The COVID-19 pandemic is a noisy disease and a deadly one that has got the whole world's attention. This deadly disease led to the whole world's total lockdown for months before necessary measures were put in place for those who could not go out. Measures like regular hand washing, sanitizer, nose or face covering, social distances, and the like. This pandemic was first discovered in China and later in other parts of the world too. This study looked into the spread of COVID-19 in Africa using the US COVID-19 dataset, where data was extracted for analysis and prediction using Polynomial Regression. The results were further compared using a Facebook prophet. But at the end of the prediction, polynomial regression has the lowest Relative Mean Absolute Error (RMSE), which is now the model used for predicting the spread of COVID-19 in Africa.

Keywords COVID-19 · Africa · WHO · Facebook Prophet · Polynomial Regression · SIR · Prediction · Diseases

9.1 Introduction

The COVID-19 pandemic was discovered on 31 December 2019. This unknown pandemic was reported by the World Health Organization (WHO) in a city called Wuhan, in China, with a population of about 10 million people. As of March 16, 2021, COV confirmed more than 119,960,700 cases. 2,656,822 deaths were registered, which included 223 countries and territories worldwide. The Africa census statistics as of March 16, 2021, also around 4,041,835 confirmed cases,

C. A. Adenusi (✉)
Linux Professional Institute, Nigeria Master Affiliate, UI, Ibadan, Nigeria

O. R. Vincent · A. F. Ajayi · B. T. Adebiyi
Department of Computer Science, Federal University of Agriculture, Abeokuta, Nigeria

S. A. Hassan et al. (eds.), *Decision Sciences for COVID-19*, International Series in Operations Research & Management Science 320,
https://doi.org/10.1007/978-3-030-87019-5_9

108,064 death cases registered and 3,630,060 recovered cases. This disease has affected both the economy and the human race and it has led to a mass number of people destabilized by this virus. According to Bouktif et al., the stages of the pandemic affected countries and all over the world have a lot of, which they have been able to come up with this year 2021. Some people have already taken the vaccine, while others are still waiting for theirs. The COVID-19 test is also readily available everywhere. All these measures in place have drastically reduced the spread of this disease. So many dashboards that entail COVID-19 information were consulted, like the US COVID-19 dashboard and some researchers have made use of diverse models for analyzing and predicting the spread of COVID-19, such models as polynomial regression, Facebook prophet, SIR, SVM, CNN, CNN-LSTM, ANN, XGBoast Regression, Linear regression, and the likes. This study looks into how to predict the spread of COVID-19 in Africa for frontline health workers to make use of it in further decision-making. This chapter entails: Section 9.1 for the introduction, Sect. 9.2 for the literature review, Sect. 9.3 for the methodology, and Sect. 9.4 for Result and Implementation followed by the conclusion and references.

9.2 Related Works

According to the novel, Coronavirus (COVID-19) has been established to be extremely transmissible across the earth. In this research, a novel approach was suggested for estimating the density of the virus for each country for three different dates between April and May 2020. Sequentially, to explain this alarming spread, socio-economic indicators in each country may be studied to examine this alarming evolution further. In the literature, most machine learning applications built for spread prediction have tried to predict national and international statistics relating to total cases, total deaths, and total recoveries (Bandyopadhyay et al. 2020).

This research displays a relative analysis of machine learning and soft computing models to foretell the COVID-19 outbreak as a substitute for the susceptible–infected–recovered (SIR) and Susceptible-Exposed-Infectious-removed (SEIR) models. Access to precise outbreak prediction models is important for understanding the possible spread and consequences of infectious diseases. Because of the difficulty and large-scale nature of developing epidemiological models, machine learning (ML) has recently gained attention for developing outbreak prediction models.

The Novel COVID-19 (coronavirus disease) was in peril in Wuhan, China on December 31, 2019. It endangers the lives of everyone in this world. The World Health Organization (WHO) risk assessment in China for COVID-19 is very high and the global level is high, as well as at the regional level. A few researchers have begun to concentrate on the virus by using artificial intelligence methods to forecast, predict, and adapt learning to the new coronavirus, and others have indicated the benefit of artificial intelligence in pandemics and epidemic predictions. The chapter makes use of artificial intelligence and deep learning for the prediction of coronavirus through time series using the Non-Linear Regressive Network (NAR).

The network forecasted coronavirus cases and deaths in nine countries, namely, Egypt, Saudi Arabia, Jordan, the USA, Spain, Italy, France, Iran, the Russian Federation, during the period from March 23, 2020 to July 30, 2020. In this chapter, the Non-linear Auto-Regressive Network (NAR) network was used to solve the Coronavirus outbreak problem with the data available from WHO reports about the virus outbreak cases and deaths. The NAR is used to predict the future from past information. It is a type of artificial neural network that performs the same function as a traditional neural network, but it relies on prior knowledge. Its performance depends on network regression ability.

Coronavirus (COVID-19) is an infectious disease caused by various pathogens that can be transmitted from one person to another at a fast speed of transmission. There is presently no treatment. In earlier research, prediction methods for the occurrence, circulation, and change of infectious diseases mainly included regression prediction models, Markov chain models, Bayesian networks, and other machine learning methods for COVID-19, but there are precautions, including social distancing, and regular washing of hands. For this research, the development trend analysis of the cumulative confirmed cases, deaths, and recovered cases was carried out on the data from Wuhan, China from January 23, 2020, to April 6, 2020, using an Elman neural network, Long short-term memory (LSTM) and Support Vector Machine (SVM).

According to Ian Cooper et al., the effectiveness of the SIR model approach was studied regarding the spread of the novel COVID-19 disease within a community. The model (Susceptible-infected removed) investigated the time evolution of diverse populations and also monitored the significant parameters of the spread of COVID-19 to diverse communities, especially in Texas in the USA. The model also predicted the spread of the pandemic, using data from January to June 2020, and concluded that the pandemic is under control in Texas after strict policies and restrictions were implemented.

Lakshmanarao et al., efficient medication or vaccines is the real essence of the seriousness of this pandemic. The World Health Organization issued a warning to prevent the spread of this disease. From the start of the study until November 15, 2020, the Kaggle COVID-19 dataset was analyzed using linear regression, polynomial regression, and Holt's method. The results of the analysis are very useful for frontline workers to aid in the reduction of disease contamination. In November 2020, there were 58,900,547 COVID-19 cases worldwide, with India having 9,177,840 at the time. The survival of the virus varies from place to place and it takes a minimum of 14 days for the recovery period of the affected people.

Mohamed Lounis is a French actor (2021). In December 2019, the coronavirus disease pandemic 2019 (COVID-19) emerged in Wuhan Province, China, and has since spread to all countries. Using the FB prophet model, the researchers were able to estimate the active, death and healed rates of COVID-19 in Algeria for a future period of 35 days. The active rate and death rate decreased during the next several days, whereas the cured rate increased. The rates of active, cured, and death are projected to be 19.7%, 78.85%, and 2.55%, respectively. These findings underline the significance of the FB prophet model in COVID-19 prediction, which may aid national authorities in implementing the most effective preventive measures.

9.3 Research Methodology

This section entails the mode of data collection, how the analysis is being done, manipulation of data, the dataset used, model, and the algorithms used.

9.3.1 Data Collection

The collection of data for this chapter was done via the use of secondary data on COVID-19 for the USA, which was downloaded from Kaggle and Africa's data was extracted from it. The dataset used has three datasets, each on confirmed case, recovery and death, collected and carried out on.

9.3.2 Dataset Used

Three major data tables were used. Each of the datasets contains the same number of variables for the collation of information, and the already collated results will be used in this chapter. The datasets are in CSV format. The datasets are all from the Kaggle repository and are as follows:

/kaggle/input/novel-corona-virus-2019- dataset/time_series_covid_19_deaths_US.cs v

/kaggle/input/novel-corona-virus-2019-dataset/time_series_covid_19_recovered._US.csv

/kaggle/input/novel-corona-virus-2019-dataset/time_series_covid_19_confirmed_US.csv

The above tables denote the needed feature selection during the EDA process. Countries or regions in Africa with their respective confirmed cases, deaths, recovered, and current confirmed cases are required feature parameters. After this, the overall total cases in Africa were computed using pandas and NumPy libraries, and country-wise reports were generated in line with the features selected. Table 9.1 denoted the beginning of the year while Table 9.2 denoted the end of the year data exploratory analysis (Fig. 9.1).

This model is being used to predict the spread of COVID-19 in Africa, using three machine learning methods (Polynomial Regression and Facebook Prophet) for better accurate results.

Table 9.1 Exploratory data analysis of COVID-19 in Africa

	S. no.	Observation date	Country/ Region	Last update	Confirmed	Deaths	Recovered
0	265	01/27/2020	Côte d'lvoire	1/27/2023:59	1.0	0.0	0.0
1	1478	02/14/2020	Egypt	2020-02-14T23:53:02	1.0	0.0	0.0
2	1553	02/15/2020	Egypt	2020-02-14T23:53:02	1.0	0.0	0.0
3	1628	02/16/2020	Egypt	2020-02-14T23:53:02	1.0	0.0	0.0
4	1703	02/17/2020	Egypt	2020-02-14T23:53:02	1.0	0.0	0.0

Table 9.2 End of year 2020 Covid cases in Africa

	Observation date	Country/ Region	Confirmed	Deaths	Recovered	Current confirmed
14223	12/31/2020	South Africa	1.05716e+06	28469	879671	149021
14214	12/31/2020	Morocco	439193	7388	407504	24301
14228	12/31/2020	Tunisia	139140	4676	105364	29100
14195	12/31/2020	Egypt	138062	7631	112105	18326
14199	12/31/2020	Ethiopia	124264	1923	112096	10245
14209	12/31/2020	Libya	100277	1478	72107	26692

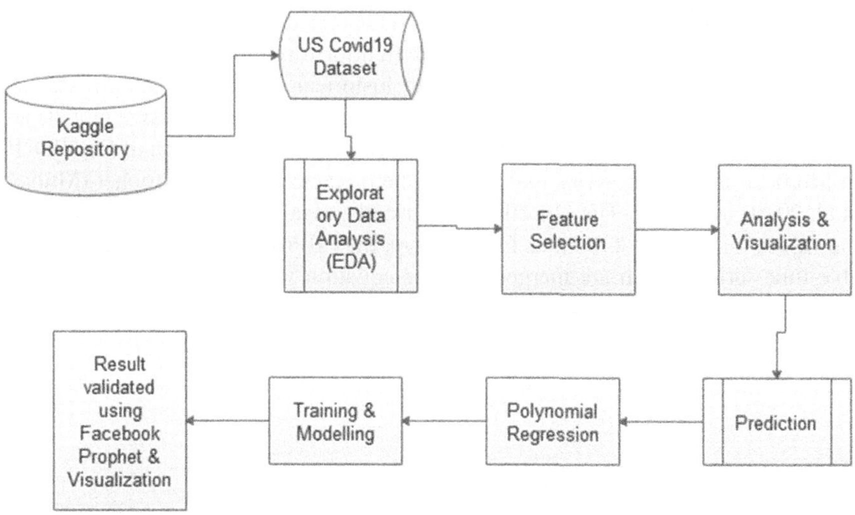

Fig. 9.1 COVID-19 predicted model

Machine learning (ML) is one of the very best branches of Artificial Intelligence that can be used as a predictive tool, because of its ability to be able to learn, train, model, fit, and predict. And there are three categories or types of ML, which are Supervised, Unsupervised, and reinforcement learning. The three techniques being used here are all machine learning methods.

9.3.3 Polynomial Regressions Analysis

In this study, a polynomial of degree two or more have been used for data analysis and the residuals in the model can be reduced by choosing a high degree of polynomial and is defined as Adjusted $R2$ for polynomial regression as follows:

$$R2\text{Adjusted} = 1 - (\text{SSResidualSSTotal}) * ((c-1)(c-d-1)).$$

N stands for the number of observations in COVID-19 dataset training; d also stands for the degree of polynomial. Both simple and adjusted R2 were computed.

9.3.4 Facebook Prophet

Facebook's team of data scientists created FBProphet, an open-source forecasting system. It is utilized for forecasting time series data using an additive model, which makes forecasting more accessible and straightforward. It works best with time series with substantial seasonal effects and historical data from multiple seasons (Mubayi et al. 2010). ARIMA, exponential models, and other regressive models are used by FBProphet. The number of accumulated confirmed cases (C) of COVID-19 in Morocco is the time series that needs to be predicted using FBProphet (Mubayi et al. 2010) in Python. The objective is to use historical time series data to forecast the values for the next 4 months. Furthermore, the FBProphet is built on decomposable time-series, which are merged into the equation $y(t) = G(t) + s(t) + h(t) + t$, which has three major components: • $G(t)$: represents the trend function, which uses piecewise linear or logical growth to account for non-periodic variations in the time series value.

9.4 Results and Discussion

This section includes the predicted outcome of COVID-19 in Africa, as well as a visualization of the topmost affected countries and a comparison of the most affected countries using python programming via Jupiter notebook to carry out the analysis and prediction outcomes (Fig. 9.2).

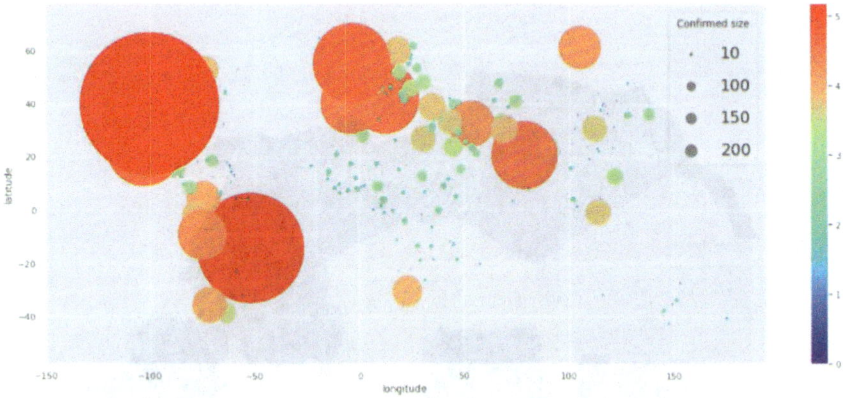

Fig. 9.2 The spread of COVID-19 mortality in the world longitude attributes

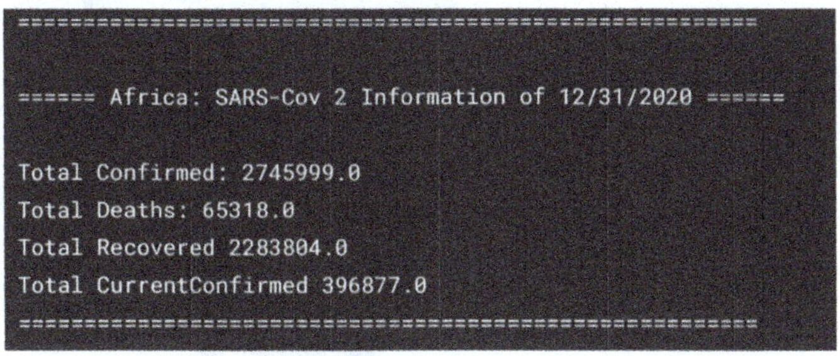

```
==========================================================

====== Africa: SARS-Cov 2 Information of 12/31/2020 ======

Total Confirmed: 2745999.0
Total Deaths: 65318.0
Total Recovered 2283804.0
Total CurrentConfirmed 396877.0
==========================================================
```

Fig. 9.3 Africa COVID-19 information for the year 2020

This demostrated that COVID-19 moves along the latitude and countries with cool climates were much more affected, while continents such as Africa were less affected due to hot climate conditions was very high in Europe before shifting to South and North America at the time of writing this chapter. Each large circle in the graph above represents a country with a large population and a high rate of COVID-19. Until the time of writing this chapter, the American continent had been the most affected.

The graph in Fig. 9.3 depicts a summary of all confirmed deaths, recovered and confirmed on the African continent by the end of 2020. The result was extracted from the World US dataset used for this research.

Figure 9.4 shows the confirmed cases in Africa from January 2020 till February 2021. The attributes, latitude, and longitude in the dataset were used to plot the map graph and the result is displayed in Fig. 9.4.

Figure 9.5 shows the time evolution of COVID-19 from 2020 to 2021. This shows the trend of the confirmed, recovered, death and current confirmed cases against the occurrence time.

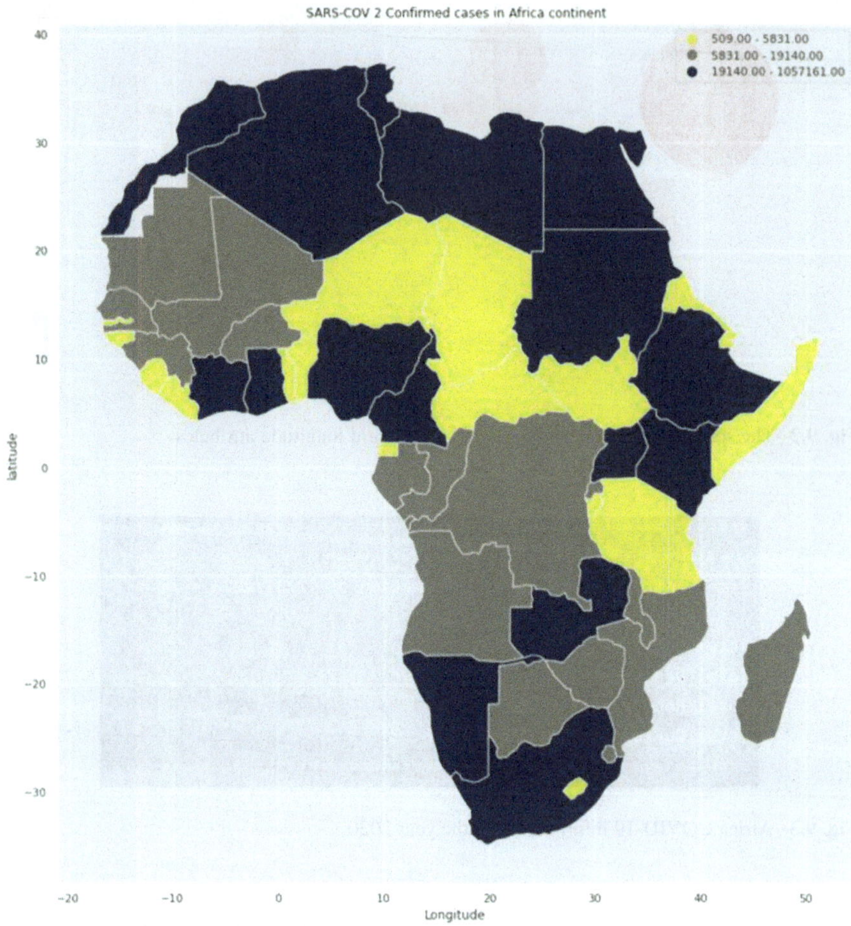

Fig. 9.4 Visualization map of confirmed cases in Africa till the year 2021

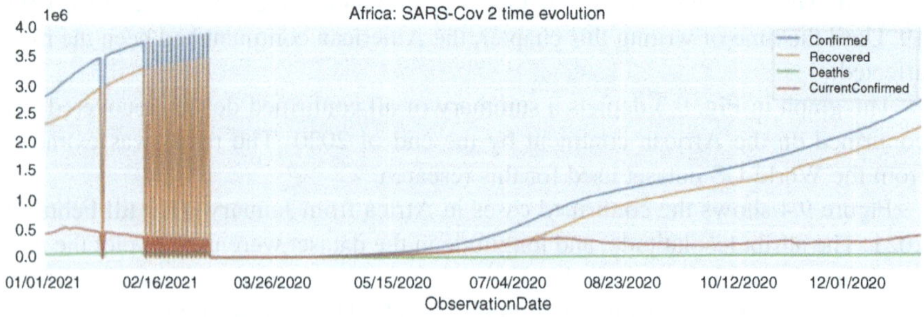

Fig. 9.5 The time evolution of COVID-19

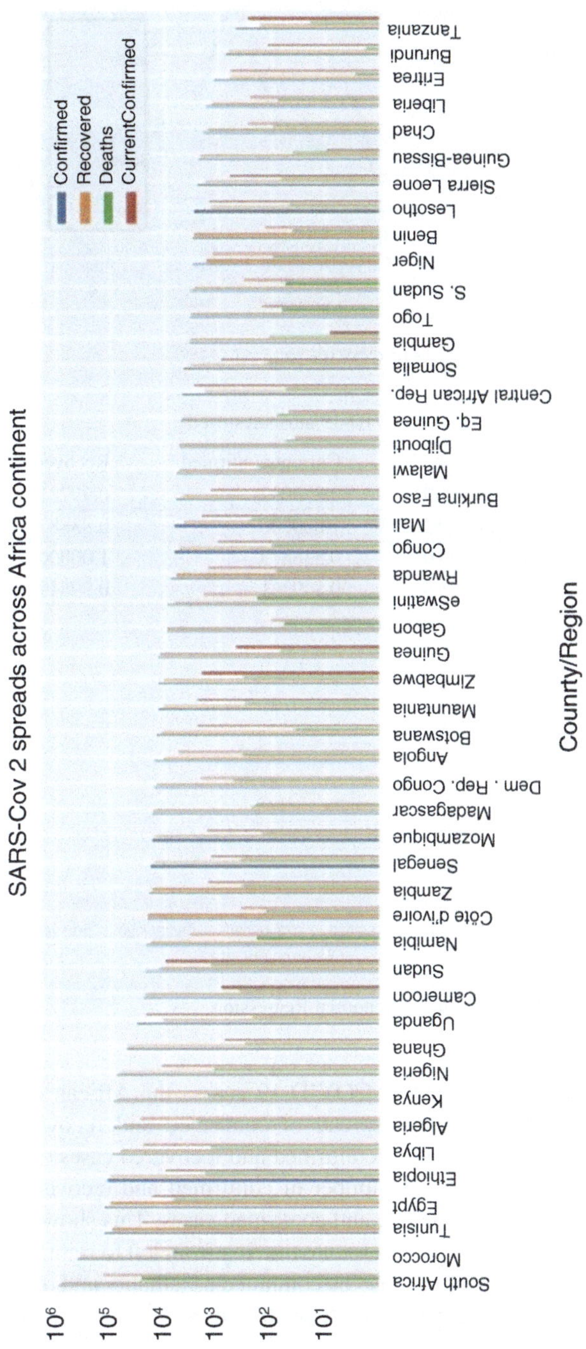

Fig. 9.6 The visual spread of Sars-Cov in Africa

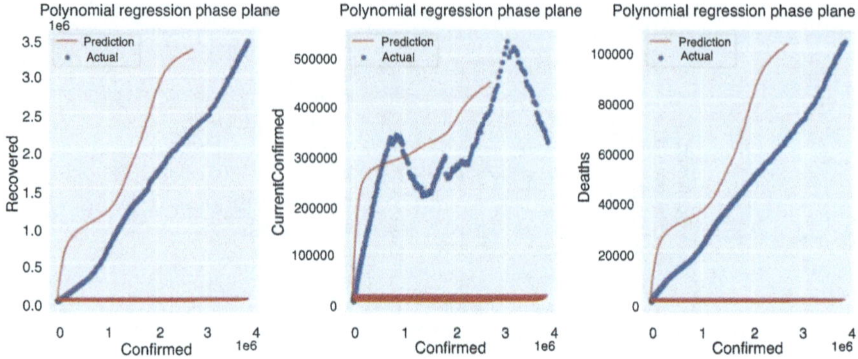

Fig. 9.7 Phase plane graph of spread of COVID-19 in Africa

Table 9.3 Correlation results of the COVID-19 attributes used

	Confirmed	Current confirmed	Recovered	Death
Confirmed	1.000000	0.873032	0.997810	0.997
Current confirmed	0.873032	1.000000	0.838873	0.849
Recovered	0.997810	0.838873	1.000000	0.998
Deaths	0.997798	0.849361	0.998492	1.000

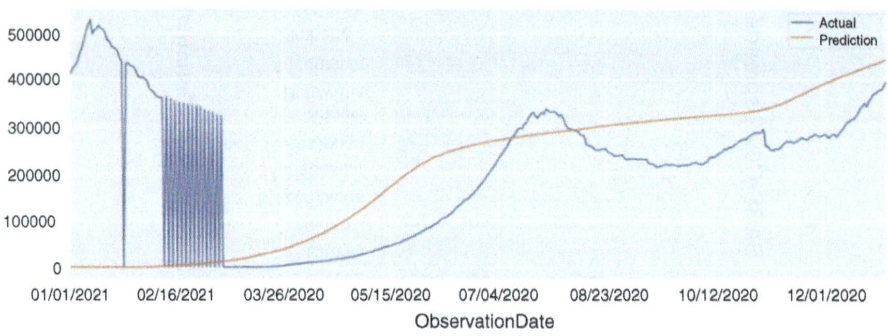

Fig. 9.8 Predicted COVID-19 using Polynomial Regression

Figure 9.6 shows the spread of COVID-19 across the African continent, with South Africa having the highest spread of confirmed and recovered cases and Tanzania with the lowest spread of confirmed and recovered cases.

Figure 9.7 depicts the current number of confirmed and recovered cases, confirmed and confirmed cases, deaths, and confirmed cases. This shows that there is a decrease in the phase plane graph, as we are moving from left to right. To confirm the death of confirmed cases, this can also be computed as lethality and recovered rate in the graph slope plane of recovery. The lethality rate is 3%, and the recovery rate is 97%.

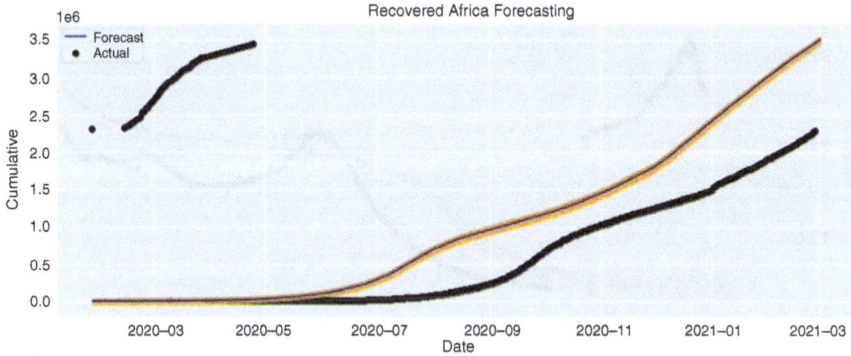

Fig. 9.9 Predicted COVID-19 using Prophet Method

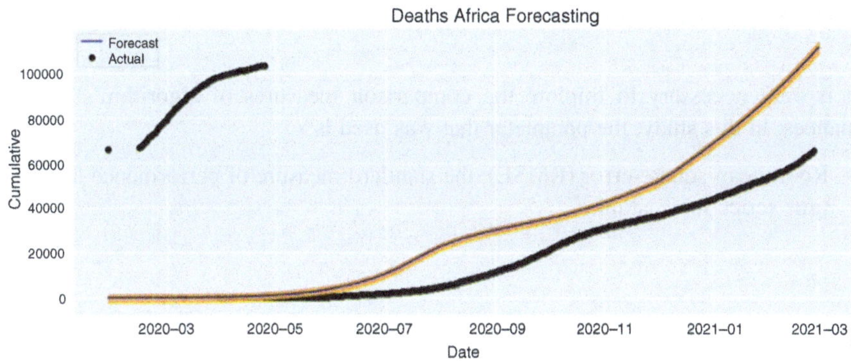

Fig. 9.10 Africa COVID-19 death forecast using Prophet

According to the results in Table 9.3, there are measures of the degree of association between the variables used.

Figure 9.8 shows the actual values of the COVID-19 attributes and the predicted values generated. The blue line color signifies the actual values and the red line signifies the predicted values of the probability of the spread of COVID-19.

Figure 9.9 depicts the rate at which people with confirmed case got recovery from COVID-19 and the graph above denote the increase rate and the graph also shows the actual growth rate and predicted growth rate of recovery cases.

Figure 9.10 shows the forecast record of death cases using the Prophet's words. The black line represents the actual value of the death rate, while the blue line represents the forecasted value.

Figure 9.11 shows the forecast record of confirmed cases in Africa.

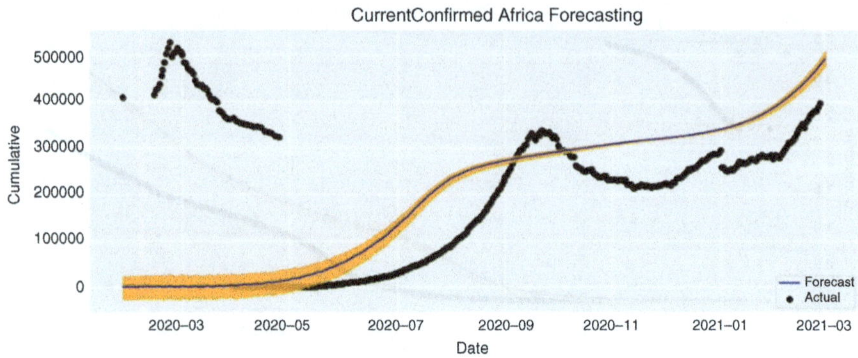

Fig. 9.11 Africa COVID-19 confirmed cases forecast using Prophet

9.5 Performance Metrics Measure

It is very necessary to implore the comparison measures of algorithm's performances. In this study, the parameter that was used is:

- Root-mean-square-error (RMSE): the standard measure of performance used for time series forecasting.

9.6 Conclusion

This research has been able to look into the genesis of COVID-19, which is from China, how it spread across the world and finally to Africa land. Many reviews on COVID-19 were consulted with their methods. The dataset used was US COVID-19 data, from which Africa Data were extracted.

The entire purpose of this study was to investigate this disease, known as COVID-19, analyze and visualize it, and then predict what would happen next. So, based on the results thus far, it appears that the spread of this COVID-19 is decreasing in Africa, and polynomial regression is a more appropriate method of prediction than prophets, as shown in Table 9.4. However, this does not imply that Africa should relax and disregard all of the COVID-19 guidance and regulations; rather, Africa should continue to follow all of the necessary protocols.

Table 9.4 Performance metrics of models used

Cases	RMSE
Using Polynomial Regression	
Confirmed cases	18071.32
Recovered cases	12941.58
Death cases	38541.89
Using Facebook Prophet	
Confirmed cases	54518.66
Recovered cases	50498.79
Death cases	58660.11

References

Bandyopadhyay SK, Dutta S (2020) Machine learning approach for confirmation of covid-19 cases: positive, negative, death and release. MedRxiv

Mubayi A, Zaleta CK, Martcheva M, Castillo-Chávez C (2010) A cost- based comparison of quarantine strategies for new emerging diseases. Math Biosci Eng 7(3):687

Chapter 10
Comparing Predictive Accuracy of COVID-19 Prediction Models: A Case Study

Dmitriy Klyushin

Abstract Decision-making in the face of COVID-19 outbreaks is highly dependent on the accuracy of forecasting the development of the epidemic. The success of the fight against coronavirus depends largely on the soundness of the decisions made, which often entail serious economic consequences. Having received the forecast of the epidemic curve, the decision-making center can work out an independent decision or use the experience of other regions. In the first case, it is necessary to correctly assess the accuracy of the obtained forecasts, and in the second, it is required to find out how the observed conditions are similar to the prototype, in particular, whether the epidemic data have the same distribution. Traditional models for forecasting epidemics, based on differential equations and a number of factors characterized by high uncertainty, often give inaccurate predictions, forcing experts to analyze possible optimistic and pessimistic scenarios, rather than the most plausible course of events. These models are being replaced by other numerical alternatives and machine learning models that predict time series based on training samples. The training accuracy and generalization error of such algorithms usually boil down to cross-validation. In this case, it also becomes necessary to compare the distribution of errors on each of the training samples. Therefore, the comparison of predictive models can be reduced to checking the homogeneity of the samples of their errors. This chapter describes a simplified variant of the Klyushin-Petunin test for testing the homogeneity of two samples and the results of its application to compare two methods for predicting the epidemic curve of COVID-19 (cube spline and hybrid Euler method) using the cases of Germany, India, Italy, South Africa, and South Korea. We demonstrate the effectiveness and practicality of the predictive models proposed by the evaluation method.

Keywords COVID-19 · Machine learning · Time series · Forecasting · Nonparametric test · Hybrid Euler · Cube spline · Variational series · Samples homogeneity · Hill assumption

D. Klyushin (✉)
Taras Shevchenko National University of Kyiv, Kyiv, Ukraine

S. A. Hassan et al. (eds.), *Decision Sciences for COVID-19*, International Series in Operations Research & Management Science 320,
https://doi.org/10.1007/978-3-030-87019-5_10

10.1 Introduction

The coronavirus is spreading very quickly. This requires decision-making centers to take effective and scientifically sound measures to minimize the risk of contamination and economic losses. In this environment, the importance of accurately predicting the COVID-19 outbreak is skyrocketing. Traditional compartmental models (SIR, SEIR, etc.) adopted in epidemiology include many weakly measurable factors, for example, the degree of population mobility, and its homogeneity. The large uncertainty associated with these parameters makes the predictions of the compartment models approximate, short term, and scenario based. They do not so much predict a plausible course of events as they describe possible alternatives. Apparently, this situation requires the use of other methods, for example, machine learning and interpolation methods, which do not use a priori assumptions about the shape of the epidemic curve and the values of the parameters, relying only on the results of observations.

In addition, since the COVID-19 outbreaks do not occur simultaneously in different countries and regions, decision centers have time to analyze similar situations and choose the measures to be taken by analogy. In this case, it also becomes necessary to compare the distributions of epidemic data in order to find the most similar prototype.

In this chapter, we describe the case study of COVID-19 forecasts for different countries and use the novel nonparametric two-homogeneity tests to assess the forecasting precision of two numerical models for predicting the epidemic curves. Section 10.2 describes the general problems in forecasting time series in the COVID-19 context. Section 10.3 outlines the Klyushin–Petunin test and its simplification. In Sect. 10.4, we apply the nonparametric tests for samples homogeneity and versions of the Klyushin-Petunin test for estimation of the forecasting time series precision for the cases of Germany, India, Italy, South Africa, and South Korea. Section 10.5 includes conclusions and a description of open questions.

10.2 Prognostic Models of the Epidemic Curves

When forecasting the epidemic situation, the complexity and reliability of predictive models come to the fore (Eker 2020). The accuracy and reliability of the model for forecasting the COVID-19 epidemic are of great importance, since the decisions made have a great impact on the population of regions and countries. Consequently, the estimation of forecast accuracy and comparison of models becomes an important urgent task.

10.2.1 Prognostic Models for COVID-19 Outbreak

Traditionally, in epidemiology, the forecast of epidemics is based on the compartment SIR and SEIR models and their variants (Kermack and McKendrick 1927; Aron and Schwartz 1984; Hethcote 2000). The letters S, E, I, and R indicate that the model includes susceptible (S), infected (I), recovered (R), and exposed (E) compartments. The SIR (Kermack and McKendrick 1927) and SEIR (Aron and Schwartz 1984) models are systems of ordinary differential equations that depend on parameters describing the transitions between compartments. The SIR and SEIR family of models are used in many countries to predict outbreaks of COVID-19 (Anand et al. 2020; Babu et al. 2020; Guirao 2020; He et al. 2020; Ifguis et al. 2020; Li et al. 2020; Nesteruk 2020; Wang et al. 2020). In theory, they are mathematically rigorous and accurate, but their parameters (for example, transition rate) are very difficult to estimate correctly from real data. As a result, these models become a tool for modeling scenarios, rather than predicting the most plausible course of events. On the other hand, the applications of numerical interpolation algorithms and machine learning methods use real training data and often give results that are more accurate.

Compartmental models are widely represented in the scientific literature devoted to predicting the COVID-19 epidemic. In particular, in the paper (Li et al. 2020) the authors tried to clarify the epidemiological characteristics of SARS-CoV-2 based on observations of reported cases of infection in China. These data include data on mobility, latent infection rates, and infectiousness. In the paper (Anand et al. 2020) the SIR model taking into account the realities of India is refined. This approach allows avoiding unrealistic assumptions on population homogeneity and makes predictions that are more accurate. Ifghuis and others (Ifguis et al. 2020) developed a SIR model for the population data of the Kingdom of Morocco. They showed that the SIR model is more accurate than the SEIR model in predicting the ultimate extent of the COVID-19 epidemic. All of these papers emphasize that conclusions depend on the credibility of the initial assumptions. Violation of these assumptions leads to inaccurate predictions.

To refine the compartment models, the researchers tried to apply a statistical approach. In particular, in the paper (Nesteruk 2020) the optimal values of the parameters of the SIR model were determined and a short-term forecast of the spread of coronavirus in China, using the exact solution of the equations included in the model was formulated. At the same time, the author emphasizes that the accuracy of his model cannot be ensured if its parameters are calculated using unreliable data. Babu et al. (2020) performed short-term prediction of COVID-19 in India using logistic curve and SIR models. The authors noted that the logistic curve model was more accurate than the SIR, which overestimates the number of new infections. In the paper (He et al. 2020) the SEIR model was modified. They included not only epidemic, but also environmental indicators. The authors demonstrated the reliability of their forecasts by the example of some regions of China, Italy, South Korea, and Iran. In addition, the authors, like many other researchers, stressed the enormous importance of correct data and the fact that in many cases they have great uncertainty. In the paper (Guirao 2020) simplified SIR or SEIR models were used to estimate the final scale under possible scenarios for making management decisions. Wang et al. (2020) also used modified SIR and SEIR models to predict the COVID-19 outbreak. They report that mathematical modeling of the COVID-19 epidemic is a challenging task and, in most cases, it is based on compartment models rather than on statistical tests or machine learning algorithms.

Faced with the unrealistic assumptions of compartment models, many researchers turned their attention to machine learning algorithms. In particular, in the paper (Swapnarekha et al. 2020) the authors analyzed numerous statistical models, machine and deep learning algorithms, and described their applications for predicting the COVID-19 epidemic. The analysis showed that these approaches make it possible to predict the COVID-19 pandemic very accurately. In the paper (Sujath et al. 2020) a linear regression model, multilayer perceptron, and vector autoregressive method were applied to COVID-19 Kaggle data to predict the COVID-2019 epidemic in India. It turned out that the multilayer perceptron is more accurate than the linear regression model and vector autoregression methods. This suggests that using machine learning methods to predict the epidemic curve will yield more accurate results.

Appadu et al. (2020) applied the Euler's iterative method and cubic spline interpolation to forecast the total number of infected people and the number of COVID-19 incidents in Germany, India, South Africa, and South Korea. They compare the accuracy measures of the two criteria using relative error.

In the paper (Tuli et al. 2020), the authors developed a cloud computing platform for their own mathematical model and tried to accurately predict the epidemic in real time. Distante et al. (2020) simulated the COVID-19 outbreak in different regions of Italy using a deep convolution autoencoder and used the SEIR model to predict peaks in the epidemic curve using data collected in China for training. It turned out that these data allow for accurate predictions.

Artificial neural networks are one of the main machine learning tools for predicting the COVID-19 epidemic. Using recurrent neural networks (RNN), Kolozsvári et al. (2020) predicted the development of the COVID-19 epidemic and substantiated the accuracy of their predictions. In the paper (Ibrahim et al. 2020) the LSTM-Autoencoder variation model for predicting the COVID-19 epidemic was developed. It takes into account not only epidemic factors but also demographic data as well as measures taken by decision-making centers. Their model is highly accurate for both short- and long-term forecasting. Hu et al. (2020) have developed artificial intelligence techniques to simulate the dynamics of the spread of COVID-19. The authors declare that their forecasting methods are more accurate than standard epidemiological models. In the paper (Guo and He 2021) an artificial neural network was developed to simulate the rate of infection and death from COVID-19. The authors proved that it is reasonably accurate.

It is also worth noting the work (Balli 2021) on a time-series forecasting model using linear regression, multilayer perceptron, random forest, and SVM to estimate the COVID-19 epidemic curve. The SVM model turned out to be the most accurate in terms of standard indicators. In the paper (Kafieh et al. 2021) the authors predicted the development of COVID-19 epidemics in Iran, Germany, Italy, Japan, Korea, Switzerland, Spain, China, and the United States using multilayer perceptron, random forest, and recurrent neural network LSTM (long short-term memory). The best accuracy was provided by a modified version of the LSTM. The authors of the paper (Rustam et al. 2020) applied linear regression (LR), least absolute compression and selection operator (LASSO), support vector machine (SVM), and exponential smoothing (ES) to predict the COVID-19 outbreak. All of the proposed models have proven to be very effective.

So, many different models are used to predict the coronavirus pandemic. Effective tools are needed to compare their accuracy, because naive comparison of average accuracy rates ignores the statistical nature of the errors. Nonparametric tests are widely used for statistically valid comparisons of the accuracy of predictive models. They have proven to be very effective in econometrics (Flores 1989; Diebold and Mariano 1995), in climate forecasting (DelSole and Tippett 2014), and other areas. These tests do not use assumptions about distribution of errors. Therefore, they can be considered the most realistic.

To evaluate predictive models, we have developed novel nonparametric tests. To demonstrate the effectiveness of these methods, we consider a case study using the example of two interpolation methods and five countries: Germany, India, Italy, South Africa, and South Korea.

10.3 Nonparametric Tests for Prognostic Models Estimation

The most used in time series forecasting is the Diebold–Mariano test (Diebold and Mariano 1995). Let $T_j, j = 1, 2$ be prognostic models, $t_i^{(j)}$ be forecasted values of a time series t_i, $i = 1, \ldots, n$ produced by the model T_j and $\varepsilon_i^{(j)}$, $i = 1, \ldots, n; j = 1, 2$ be errors of the model T_j following distribution function F_j. Let $J\left(\varepsilon_i^{(j)}\right)$ a loss function of accuracy. If models T_1 and T_2 have the identical precision then $r_i = J\left(\varepsilon_i^{(1)}\right) - J\left(\varepsilon_i^{(2)}\right) = 0$. Therefore, $E\left(J\left(\varepsilon_i^{(1)}\right)\right) = E\left(J\left(\varepsilon_i^{(2)}\right)\right)$. If a loss function is the standard deviation $\varepsilon_i^{(2)} - \varepsilon_i^{(1)}$, then we obtain the hypothesis $E\left(\varepsilon_i^{(1)}\right) = E\left(\varepsilon_i^{(2)}\right)$. We propose to generalize this hypothesis and test the more general null hypothesis $F_1 \equiv F_2$..

10.3.1 Nonparametric Tests for Samples Homogeneity

Tests for homogeneity of two samples may be consistent with any pair of alternatives (as Kolmogorov–Smirnov test (Smirnov 1939a, b)), or with pairs of alternatives of a given class (Dixon 1940; Wald and Wolfowitz 1940; Mathisen 1943; Wilcoxon 1945; Mann and Whitney 1947; Wilks 1961, etc.). Also, they could be purely nonparametric and conditionally nonparametric. Nonparametric tests do not use any assumption on the distribution function. Conditionally, nonparametric tests (Pitman 1937; Lehmann 1947; Rosenblatt 1952; Dwass 1957; Fisz 1960; Barnard 1963; Birnbaum 1974; Jockel 1986; Efron and Tibshirani 1993; Allen 1997; Dufour and Farhat 2001) directly or indirectly use some assumptions on distributions.

For example, the two sample Kolmogorov-Smirnov test (Smirnov 1939a, b) is purely nonparametric one. It estimates the maximum difference between empirical distribution functions. The Kolmogorov-Smirnov test works perfectly when samples have different means, but it fails when samples include outliers. In opposite, the Wald–Wolfowitz test (Wald and Wolfowitz 1940) is more powerful than the Kolmogorov-Smirnov test when samples have the same location and different variances.

There are a group of tests for location shift, namely: the Mathisen test (Mathisen 1943), the Wilcoxon sign rank test (Wilcoxon 1945), the Mann–Whitney test (Mann and Whitney 1947), the Wilks test (Wilks 1961), etc. However, if the assumption on the same location is not satisfied their power decreases.

Another drawback of the Kolmogorov-Smirnov test is the dependence of its sensitivity on the location of the maximum of the test statistics. The test is the most sensitive when the discrepancy between two samples has the maximum near the center of the distribution, and vice versa, if the maximum is at a tail. To mitigate this drawback, the Cramér–von Mises test (Cramér 1928) was proposed. It estimates the sum of the squared deviations between empirical distribution functions, but also loss sensitivity at tails. To fix this problem the Anderson–Darling test (Darling 1957; Anderson 1962) is used. It is a weighted version of the Cramer–von Mises test. However, for small cases this test is quite complicated. The Allen test uses the distances between probability densities of functions in different spaces (Allen 1997). The Allen test is nonparametric only if the samples are infinite. For finite samples, this test is conventionally nonparametric.

Tests from the final group (Pitman 1937; Dwass 1957; Efron and Tibshirani 1993) estimate the difference between the sample average values. When samples are finite, these tests depend on distributions and are parametric. In the paper (Dufour and Farhat 2001) the authors proposed to use arbitrary random permutations of the elements of the combined sample. Since all these permutations are equiprobable, a test rejecting the null hypothesis can be considered nonparametric.

As we see, the key problem is that there is no universal nonparametric two-sample test of homogeneity, that would be equally powerful when samples have the same mean but different variances and vice versa. To fill this gap, we apply the Klyushin-Petunin test (Klyushin and Petunin 2003). Since, the original version is quite computationally cost (we do exhaust search of intervals formed by order statistics) we propose its simplified version and show that it is equally sensitive and specific.

10.3.2 Original Version of the Klyushin-Petunin Test

The Hill assumption $A_{(n)}$ (Hill 1968) states that for exchangeable random values x_1, $x_2, \ldots, x_n$ following an absolutely continuous distribution we have.

$$P\left(x_{n+1} \in \left(x_{(i)}, x_{(j)}\right)\right) = p_{ij} = \frac{j-i}{n+1}, \quad j > i,$$

where x_{n+1} follows a distribution of $x_1, x_2, \ldots, x_n$, and $x_{(i)}$ is the i-th order statistics.

Suppose that samples $x = (x_1, x_2, \ldots, x_n)$ and $y = (y_1, y_2, \ldots, y_m)$ follow absolutely continuous distribution functions F_1 and F_2. Let $A_{ij}^{(k)}$ is an event $\{x_{(i)} < y_k < x_{(j)}\}$ and h_{ij} is its relative frequency. The Wilson confidence interval for the probability of $A_{ij}^{(k)}$ has the form:

$$p_{ij}^{(1)} = \frac{h_{ij}m + z^2/2 - z\sqrt{h_{ij}\left(1-h_{ij}\right)m + z^2/4}}{m+z^2},$$

$$p_{ij}^{(2)} = \frac{h_{ij}m + z^2/2 + z\sqrt{h_{ij}\left(1-h_{ij}\right)m + z^2/4}}{m+z^2}.$$

The lower and upper bounds of the Wilson confidence interval $I_{ij}^{(n,m)} = \left(p_{ij}^{(1)}, p_{ij}^{(2)}\right)$ depends on the parameter z. If $z = 3$ than the significance level of $I_{ij}^{(n,m)}$ is less than 0.05 (Klyushin and Petunin 2003). Introduce $N = (n - 1)n/2$ and $L = \#\left\{p_{ij} = \frac{j-i}{n+1} \in I_{ij}^{(n,m)}\right\}$. Then, $\rho(x, y) = L/N$ is a measure of homogeneity of samples x and y (the probability that they are drawn from the same population). We shall call this measure p-statistics. Since the p-statistics is a Binomial proportion, constructing the Wilson confidence interval $I = (p_1, p_2)$ for the p-statistics, we can formulate the following decision rule: if $p_2 \geq 0.95$ then the null hypothesis is accepted, else it is rejected.

Remark As far the samples x and y in the test play different roles (the sample x is ordered and used for construction of a variational series and the sample y is sieved through intervals formed by ordered statistics $x_{(i)}$), the p-statistics is nonsymmetrical, i.e., $\rho(x, y) \neq \rho(y, x)$. It is easy to see, that we can construct a symmetrical p-statistics by averaging $\rho(x, y)$ and $\rho(y, x)$:

$$\rho^*(x, y) = \frac{1}{2}(\rho(x, y) + \rho(y, x)).$$

10.3.3 Simplified Version of the Klyushin-Petunin Test

It is easy to see, that in the original version of the Klyushin–Petunin test we must exhaust all the pairs $x_{(i)}$ and $x_{(j)}$. Therefore, the algorithmic complexity of this test is $O(n^2)$. Now, we propose to reduce the complexity to $O(n)$ using the well-known fact that the relative frequency in the Bernoulli schemes is stabilized rather quickly.

Let us select M times random numbers i and j such that $i < j \le n$. Find the relative frequency of the event $A_{ij}^{(k)} = \left\{x_{(i)} < y_k < x_{(j)}\right\}$. Then, construct the Wilson confidence interval $I_{ij}^{(n,m)} = \left(p_{ij}^{(1)}, p_{ij}^{(2)}\right)$ and compute $L = \#\left\{\frac{j-i}{n+1} \in I_{ij}^{(n,m)}\right\}$. Find $\rho(x, y) = L/M$. Constructing the Wilson confidence interval $I = (p_1, p_2)$ for the p-statistics, we can formulate the following decision rule: if $p_2 \ge 0.95$ then the null hypothesis is accepted, else it is rejected. According to practical recommendations, we used 100 samples.

10.4 Case Study

For experiments, we use data published in (Appadu et al. 2020). The sizes of samples used for two forecasting methods (cube interpolation and hybrid Euler method) are provided in Table 10.1. The data encompass the period from 15 February 2020 to 31 May 2020. The forecast was made for the period starting from 1 June 2020. In the case study, we applied the original and simplified Klyushin–Petunin tests for estimation homogeneity of the recorded incidents (Tables 10.2, 10.3, 10.4, and 10.5) and errors of forecasted time series (Tables 10.6, 10.7, 10.8, and 10.9) for Germany, India, Italy, South Africa, and South Korea (Figs. 10.1 and 10.2).

Tables 10.2–10.5 and Figures 10.1 and 10.2 show that the distributions of the number of COVID-19 cases during the considered period in Germany, India, Italy, South Africa, and South Korea are totally different. Therefore, the centers of

Table 10.1 Sample sizes

	Germany	India	Italy	South Africa	South Korea
Cube interpolation	24	8	16	8	11
Hybrid Euler Method	42	12	16	10	21

Table 10.2 Upper limit of the p-statistics for records (cube interpolation)

	Germany	India	Italy	South Africa	South Korea
Germany	1.000	0.863	0.702	0.863	0.790
India	0.791	1.000	0.799	0.930	0.930
Italy	0.585	0.934	1.000	0.905	0.808
South Africa	0.719	0.930	1.000	1.000	0.930
South Korea	0.680	0.925	0.785	0.928	1.000

Table 10.3 Upper limit for the p-statistics for records (hybrid Euler method)

	Germany	India	Italy	South Africa	South Korea
Germany	1.000	0.841	0.641	0.774	0.546
India	0.591	1.000	0.751	0.889	0.633
Italy	0.487	0.848	1.000	0.861	0.670
South Africa	0.421	0.833	0.732	1.000	0.732
South Korea	0.371	0.773	0.664	0.818	1.000

Table 10.4 Upper limit of the p-statistics for records (cube interpolation, simplified version)

	Germany	India	Italy	South Africa	South Korea
Germany	1.000	0.914	0.775	0.900	0.824
India	0.520	1.000	0.741	0.907	0.900
Italy	0.578	0.954	1.000	0.954	0.959
South Africa	0.625	0.878	1.000	1.000	0.886
South Korea	0.623	0.855	0.733	0.900	1.000

Table 10.5 Upper limit for the p-statistics for records (hybrid Euler method, simplified method)

	Germany	India	Italy	South Africa	South Korea
Germany	1.000	0.900	0.758	0.840	0.625
India	0.680	1.000	0.784	0.855	0.671
Italy	0.480	0.832	1.000	0.900	0.715
South Africa	0.311	0.855	0.643	1.000	0.698
South Korea	0.418	0.784	0.707	0.855	1.000

Table 10.6 Upper limit of the p-statistics of relative errors (cube interpolation)

	Germany	India	Italy	South Africa	South Korea
Germany	1.000	0.998	0.999	0.999	1.000
India	0.870	1.000	1.000	1.000	1.000
Italy	0.999	1.000	1.000	1.000	1.000
South Africa	0.823	1.000	1.000	1.000	1.000
South Korea	0.936	1.000	1.000	1.000	1.000

Table 10.7 Upper limit of the p-statistics of relative errors (hybrid Euler method)

	Germany	India	Italy	South Africa	South Korea
Germany	1.000	0.999	0.822	0.999	0.908
India	0.971	1.000	0.857	1.000	0.889
Italy	0.700	0.893	1.000	1.000	0.984
South Africa	0.934	0.946	1.000	1.000	0.992
South Korea	0.723	0.919	0.993	0.999	1.000

Table 10.8 Upper limit of the simplified p-statistics of relative errors (cube interpolation)

	Germany	India	Italy	South Africa	South Korea
Germany	1.000	0.999	1.000	0.997	1.000
India	0.750	1.000	1.000	1.000	1.000
Italy	0.999	1.000	1.000	1.000	1.000
South Africa	0.758	1.000	1.000	1.000	1.000
South Korea	0.928	1.000	1.000	1.000	1.000

Table 10.9 Upper limit of the simplified p-statistics of relative errors (hybrid Euler method)

	Germany	India	Italy	South Africa	South Korea
Germany	1.000	1.000	0.840	1.000	0.965
India	0.928	1.000	0.871	1.000	0.871
Italy	0.698	0.863	1.000	1.000	0.981
South Africa	0.947	0.893	1.000	1.000	0.976
South Korea	0.871	0.941	0.981	1.000	1.000

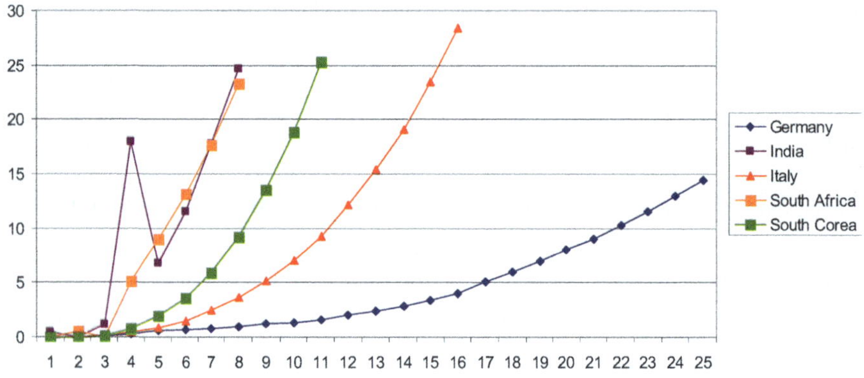

Fig. 10.1 Graphs of prediction errors (%) for cubic interpolation per day

decision-making in these countries cannot use an analogy between these countries to take measures.

As we see, the cube interpolation provides identical distributions of relative errors for most considered countries. In opposite, the hybrid Euler method generates different distributions in 9 of 25 cases. Therefore, the hybrid Euler method is not as robust as the cube interpolation.

To compare the accuracy of two proposed forecasting methods (cube interpolation and hybrid Euler method) we applied the simplified Klyushin–Petunin method to the relative errors (%) provided in (Appadu et al. 2020). The results are shown in Tables 10.6–10.9.

To test whether the cube interpolation and the hybrid Euler method produce identical distributions of relative errors, we applied the simplified Klyushin–Petunin

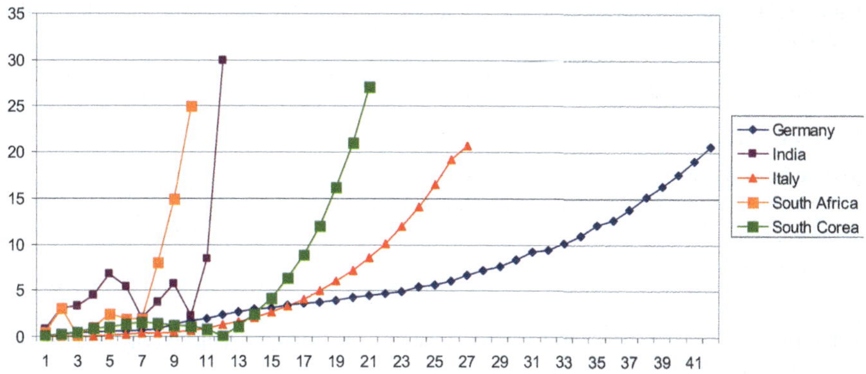

Fig. 10.2 Graphs of prediction errors (%) for hybrid Euler method per day

Table 10.10 Upper limit of the *p*-statistics of relative errors (cube interpolation and hybrid Euler method) for Germany

	Cube interpolation	Hybrid Euler method
	Original version	
Cube interpolation	1.000	0.999
Hybrid Euler method	0.998	1.000
	Simplified version	
Cube interpolation	1.000	1.000
Hybrid Euler method	0.997	1.000

test for data on the forecasting for Germany, India, Italy, South Africa, and South Korea (Tables 10.10, 10.11, 10.12, 10.13, and 10.14).

Tables 10.10–10.14 show that according to the original and simplified tests the relative errors of the cube interpolation and the hybrid Euler method follow identical distributions for almost all considered countries except India. Therefore, we can apply these tests effectively to obtain precise and robust forecasts.

10.5 Conclusion and Scope for the Future Work

The original and simplified versions of the Klyushin–Petunin test are effective tools for estimation the homogeneity of errors of different forecasting models. They may be successfully used both for comparing the distributions of epidemic data and the distributions of precision measures.

The future work on these tests is connected with the study of their power for unimodal distributions in general in specific cases.

Table 10.11 Upper limit of the *p*-statistics of relative errors (cube interpolation and hybrid Euler method) for India

	Cube interpolation	Hybrid Euler method
	Original version	
Cube interpolation	1.000	0.891
Hybrid Euler method	0.955	1.000
	Simplified version	
Cube interpolation	1.000	0.878
Hybrid Euler method	0.934	1.000

Table 10.12 Upper limit of the *p*-statistics of relative errors (cube interpolation and hybrid Euler method) for Italy

	Cube interpolation	Hybrid Euler method
	Original version	
Cube interpolation	1.000	0.951
Hybrid Euler method	0.980	1.000
	Simplified version	
Cube interpolation	1.000	0.971
Hybrid Euler method	0.981	1.000

Table 10.13 Upper limit of the *p*-statistics of relative errors (cube interpolation and hybrid Euler method) for South Africa

	Cube interpolation	Hybrid Euler method
	Original version	
Cube interpolation	1.000	0.977
Hybrid Euler method	1.000	1.000
	Simplified version	
Cube interpolation	1.000	0.934
Hybrid Euler method	1.000	1.000

Table 10.14 Upper limit of the *p*-statistics of relative errors (cube interpolation and hybrid Euler method) for South Korea

	Cube interpolation	Hybrid Euler method
	Original version	
Cube interpolation	1.000	0.974
Hybrid Euler method	1.000	1.000
	Simplified version	
Cube interpolation	1.000	0.965
Hybrid Euler method	1.000	1.000

References

Allen DL (1997) Hypothesis testing using L1-distance bootstrap. Am Stat 51:145–150. https://doi.org/10.1080/00031305.1997.10473949

Anand N, Sabarinath A, Geetha S, Somanath S (2020) Predicting the spread of COVID-19 using SIR model augmented to incorporate quarantine and testing. Trans Indian Natl Acad Eng 5:141–148. https://doi.org/10.1007/s41403-020-00151-5

Anderson TW (1962) On the distribution of the two-sample Cramér–von Mises criterion. Ann Math Stat 33:1148–1159. https://doi.org/10.1214/aoms/1177704477

Appadu AR, Kelil AS, Tijani YO (2020) Comparison of some forecasting methods for COVID-19. Alex Eng J 60:1565–1589. https://doi.org/10.1016/j.aej.2020.11.011

Aron J, Schwartz I (1984) Seasonality and period-doubling bifurcations in an epidemic model. J Theor Biol 110:665–679. https://doi.org/10.1016/S0022-5193(84)80150-2

Babu M, Marimuthu M, Joy M, Nadaraj M, Asirvatham E, Jeyaseelan L (2020) Forecasting COVID-19 epidemic in India and high incidence states using SIR and logistic growth models. Clin Epidemiol Glob Health 9:26–33. https://doi.org/10.1016/j.cegh.2020.06.006

Balli S (2021) Data analysis of Covid-19 pandemic and short-term cumulative case forecasting using machine learning time series methods. Chaos Solitons Fractals 142:110512. https://doi.org/10.1016/j.chaos.2020.110512

Barnard GA (1963) Comment on "The spectral analysis of point processes" by M.S. Bartlett. J R Stat Soc Ser B 25:294

Birnbaum ZW (1974) Computers and unconventional test-statistics. In: Prochan F, Serfling RJ (eds) Reliability and Biometry. SIAM, Philadelphia, PA, pp 441–458

Cramér H (1928) On the composition of elementary errors. Scand Actuar J 1:13–74. https://doi.org/10.1080/03461238.1928.10416862

Darling DA (1957) The Kolmogorov–Smirnov, Cramér–von Mises tests. Ann Math Stat 28:223–238. https://doi.org/10.1214/aoms/1177728589

DelSole T, Tippett MK (2014) Comparing forecast skill. Mon Wea Rev 142:4658–4678. https://doi.org/10.1175/MWR-D-14-00045.1

Diebold FX, Mariano RS (1995) Comparing predictive accuracy. J Bus Econ Stat 13:253–263. https://doi.org/10.1080/07350015.1995.10524599

Distante C, Pereira I, Gonçalves L, Piscitelli P, Miani A (2020) Forecasting Covid-19 outbreak progression in Italian regions: a model based on neural network training from Chinese data. MedRxiv. https://doi.org/10.1101/2020.04.09.20059055

Dixon WG (1940) A criterion for testing the hypothesis that two samples are from the same population. Ann Math Stat 11:199–204. https://doi.org/10.1214/AOMS/1177731914

Dufour JM, Farhat A (2001) Exact nonparametric two-sample homogeneity tests for possibly discrete distributions. Center for Interuniversity research in Quantitative Economics (CIREQ). Preprint 2001-23. California Press, pp 707–717

Dwass M (1957) Modified randomization tests for nonparametric hypotheses. Ann Math Stat 28:181–187. https://doi.org/10.1214/aoms/1177707045

Efron B, Tibshirani RJ (1993) An introduction to the bootstrap. Vol. 57 of Monographs on statistics and applied probability. New York: Chapman-Hall

Eker S (2020) Validity and usefulness of COVID-19 models. Humanit Soc Sci Commun 7(54). https://doi.org/10.1057/s41599-020-00553-4

Fisz M (1960) On a result be M. Rosenblatt concerning the Mises–Smirnov test. Ann Math Stat 31:427–429. https://doi.org/10.1214/aoms/1177705905

Flores BE (1989) The utilization of the Wilcoxon test to compare forecasting methods: a note. Int J Forecast 5:529–535. https://doi.org/10.1016/0169-2070(89)90008-3

Guirao A (2020) The Covid-19 outbreak in Spain. A simple dynamics model, some lessons, and a theoretical framework for control response. Infect Dis Model 5:652–669. https://doi.org/10.1016/j.idm.2020.08.010

Guo Q, He Z (2021) Prediction of the confirmed cases and deaths of global COVID-19 using artificial intelligence. Environ Sci Pollut Res 28:11672–11682. https://doi.org/10.1007/s11356-020-11930-6

He J, Chena G, Jiang Y, Jin R, Shortridge R, Agusti S, He M, Wu J, Duarte D, Christakos G (2020) Comparative infection modeling and control of COVID-19 transmission patterns in China, South Korea, Italy and Iran. Sci Total Environ 74:141447. https://doi.org/10.1016/j.scitotenv.2020.141447

Hethcote HW (2000) The mathematics of infectious diseases. SIAM Rev 42(4):599–653. https://doi.org/10.1137/S0036144500371907

Hill BM (1968) Posterior distribution of percentiles: Bayes' theorem for sampling from a population. J Am Stat Assoc 63:677–691. https://doi.org/10.1080/01621459.1968.11009286

Hu Z, Ge Q, Li S, Boerwincle E, Jin L, Xiong M (2020) Forecasting and evaluating multiple interventions of Covid-19 worldwide. Front Artif Intell 2020:00041. https://doi.org/10.3389/frai.2020.00041

Ibrahim MR, Haworth J, Lipani L, Aslam A, Cheng T, Christie N (2020) Variational-LSTM autoencoder to forecast the spread of coronavirus across the globe. MedRxiv. https://doi.org/10.1101/2020.04.20.20070938

Ifguis O, Ghozlani ME, Ammou F, Moutcine A, Abdellah A (2020) Simulation of the final size of the evolution curve of coronavirus epidemic in Morocco using the SIR Model. J Environ Pub Health 2020:9769267. https://doi.org/10.1155/2020/9769267

Jockel KH (1986) Finite sample properties and asymptotic efficiency of Monte Carlo tests. Ann Stat 14:336–347. https://doi.org/10.1214/aos/1176349860

Kafieh R, Arian R, Saeedizadeh N, Amini Z, Serej ND, Minaee S, Yadav SK, Vaezi A, Rezaei N, Javanmar SH (2021) COVID-19 in Iran: forecasting pandemic using deep learning. Comput Math Methods Med 2021:6927985. https://doi.org/10.1155/2021/692798

Kermack WO, McKendrick AG (1927) A contribution to the mathematical theory of epidemics. Proc R Soc Lond A 115:700–721. https://doi.org/10.1098/rspa.1927.0118

Klyushin DA, Petunin YI (2003) A nonparametric test for the equivalence of populations based on a measure of proximity of samples. Ukr Math J 55(2):181–198. https://doi.org/10.1023/A:1025495727612

Kolozsvári LR, Bérczes T, Hajdu A, Gesztelyi R, Tiba A, Varga I, Szőllősi GJ, Harsányi S, Garbóczy S, Zsuga J (2020) Predicting the epidemic curve of the coronavirus (SARS-CoV-2) disease (COVID-19) using artificial intelligence. MedRxiv. https://doi.org/10.1101/2020.04.17.20069666

Lehmann EL (1947) Consistency and unbiasedness of certain nonparametric tests. Ann Math Stat 22:165–179. https://doi.org/10.1214/aoms/1177729639

Li R, Pei S, Chen B, Song Y, Zhang T, Yang W, Wei B, Xin L, Wei X (2020) Substantial undocumented infection facilitates the rapid dissemination of novel coronavirus (sars-cov-2). Science 368:489–493. https://doi.org/10.1111/jebm.12376

Mann HB, Whitney DR (1947) On a test of whether one of the random variables is stochastically larger than other. Ann Math Stat 18:50–60. https://doi.org/10.1214/aoms/1177730491

Mathisen HC (1943) A method of testing the hypothesis that two samples are from the same population. Ann Math Stat 14:188–194. https://doi.org/10.1214/aoms/1177731460

Nesteruk I (2020) Statistics-based predictions of coronavirus epidemic spreading in Mainland China. Innov Biosyst Bioeng 4(1):13–18. https://doi.org/10.20535/ibb.2020.4.1.195074

Pitman EJG (1937) Significance tests which may be applied to samples from any populations. J R Stat Soc Ser A 4:119–130. https://doi.org/10.2307/2983647

Rosenblatt M (1952) Limit theorems associated with variants of the von Mises statistic. Ann Math Stat 23:617–623. https://doi.org/10.1214/aoms/1177729341

Rustam F et al (2020) COVID-19 future forecasting using supervised machine learning models. IEEE Access 8:101489–101499. https://doi.org/10.1109/ACCESS.2020.2997311

Smirnov NV (1939a) Estimate of difference between empirical distribution curves in two independent samples. Bull Mosk Gos Univ 2(2):3–14

Smirnov NV (1939b) On the deviations of an empirical distribution curve. Mat Sb 6(1):3–26

Sujath R, Chatterjee JM, Hassanien AE (2020) A machine learning forecasting model for COVID-19 pandemic in India. Stoch Environ Res Risk A. 34:959–972. https://doi.org/10.1007/s00477-020-01827-8

Swapnarekha H, Behera HS, Nayak J, Naik B (2020) Role of intelligent computing in COVID-19 prognosis: a state-of-the-art review. Chaos Solitons Fract 109947. https://doi.org/10.1016/j.chaos.2020.109947

Tuli S, Tuli S, Tuli R, Gill SS (2020) Predicting the growth and trend of COVID-19 pandemic using machine learning and cloud computing. Internet Things 11:100222. https://doi.org/10.1016/j.iot.2020.100222

Wald A, Wolfowitz J (1940) On a test whether two samples ate from the same population. Ann Math Stat 11:147–162. https://doi.org/10.1214/AOMS/1177731909

Wang N, Fu Y, Zhang H, Shi H (2020) An evaluation of mathematical models for the outbreak of COVID-19. Precis Clin Med 3(2):85–93. https://doi.org/10.1093/pcmedi/pbaa016

Wilcoxon F (1945) Individual comparisons by ranking methods. Biometrika 1:80–83. https://doi.org/10.2307/3001968

Wilks SS (1961) A combinatorial test for the problem of two samples from continuous distributions. Proc Fourth Berkeley Symp Math Stat Prob 1:707–717

Part III
Social Sciences

Chapter 11
AIC Algorithm for Entrepreneurial Intention in Covid19 Pandemic

Nguyen Thi Ngan and Bui Huy Khoi

Abstract This paper applied the AIC algorithm pointed out that the factors impacted the entrepreneurial intention in the Covid19 Pandemic. The COVID-19 epidemic had caused great harm to the start-up community when up to 50% of start-ups confirmed that they were operating in moderation and generating negligible income; while 23% of start-ups think that they are losing capital raising opportunities and expanding their market, 20% of start-ups choose to freeze their activities. We selected 178 students living in Ho Chi Minh City, Vietnam, to survey. The results suggest that the research results show that the factors affecting the Entrepreneurship intention of students from strong impact to weak impact are as follows: personality characteristics, subjective norm, feasibility perception, attitude towards entrepreneurship, financial approach impact entrepreneurship intention.

Keywords Covid19 pandemic · AIC algorithm · Entrepreneurial intention · R software · Vietnam

11.1 Introduction

After the Great Depression (1930) and the Great Depression (2008–2009), the world experienced a major blockade caused by the Covid-19 epidemic, with more than 98.2 million reported cases and over 2.1 million deaths globally since the start of the pandemic (WHO 2021) and caused shocks on both supply and demand angles. From there, the likelihood of system crashes increased, recession spread, and economies

Two authors contribute to the paper. Bui Huy Khoi contributed to the study of data and the gathering of research-related references. Nguyen Thi Ngan contributed to the compilation of data, and the manuscript was revised and conducted a data survey.

N. T. Ngan · B. H. Khoi (✉)
Industrial University of Ho Chi Minh City, Ho Chi Minh City, Vietnam
e-mail: buihuykhoi@iuh.edu.vn

S. A. Hassan et al. (eds.), *Decision Sciences for COVID-19*, International Series in Operations Research & Management Science 320,
https://doi.org/10.1007/978-3-030-87019-5_11

were isolated. According to the enterprise survey (PBS), 50% of businesses surveyed had to close in April 2020. Eighty percent of businesses resumed operations in June 2020, and 81% of surveyed businesses experienced a decrease in revenue in June 2020; accordingly, revenue decreased by over 40% over the same period. In addition, 15% of businesses surveyed have fired their employees, and the rest of the businesses choose to reduce working hours, wages, and leave. Besides, during the blockade period, about 50% of the surveyed enterprises partially operated, and about 35% of the enterprises temporarily stopped operating according to regulations (UEH 2021). After taking control of Covid-19, the economy resumed operations, with new challenges facing businesses. In that context, start-up students also have to change their thinking, management, and business ways to adapt and grow (Bao Da Nang 2021).

Entrepreneurial Intention has been regarded as motive power for economic development in many countries globally. As Vietnam has integrated more deeply into the global context, the interest in Entrepreneurial Intention, the establishment of a strong enterprise association, creation of a global nation's trademark is the way for Vietnam to catch up with international economies stably and sustainably. Recently, Vietnam has been focused on becoming a nation of start-ups. In order to achieve the target of 1 million enterprises operating in 2020, 1.5 million in 2025, and 2 million in 2030, facilitating Entrepreneurial Intention in the youths is a crucial orientation.

According to the "Start-up Student and Student Support Project" of the Ministry of Education and Training approved by the Prime Minister, by 2025, 70% of colleges and secondary schools have at least five bright ideas to start a business. Specifically, by 2020, 100% of universities, institutes, colleges, and secondary schools have plans to support start-up students and students; At least 90% of students and students are equipped with knowledge and skills in entrepreneurship before graduation. 66.6% of Vietnamese students currently do not know startup activities. These are data from a survey. The number of students who know about start-up programs is only 33.4%, and the fact that the number of students who annually participate in start-up programs initiated by the Vietnam Chamber of Commerce and Industry (VCCI) is only 0.016. %. Many experts assess that the start-up rate of students is low; most students, after graduating from colleges and universities, tend to apply to active businesses, only a few want to start a business by self-employment. Explaining the situation of only like to work as hired labor, do not like to be the owner of students, many opinions said that the current general education and university programs had not met the needs of knowledge about entrepreneurship in Vietnam; The curriculum focuses on theory, does not promote practicality and practical knowledge. There is also a shortage of entrepreneurship training units for college students and start-up tools services. As a result, many students today lack the knowledge, confidence, and vision necessary to start a business. These reasons lead to the fact that about 400,000 students graduate each year, but up to 225,000 students fail to find jobs. According to the National Startup Association's 1500 start-up businesses surveyed, most recent domestic graduates have not met the start-up's expectations.

11.2 Literature Review

11.2.1 Entrepreneurial Intention (EI)

Entrepreneurial Intention (EI) is a key concept in the early stages of the business process, a line of research has focused on understanding its precursor (Schlaegel and Koenig 2014). Entrepreneurship intent refers to a cognitive state that directs a person to start a business (Krueger 2017). Study on business purpose programs whose personality traits such as risk-taking tendencies, creativity, and openness to new experiences are positively associated to become an entrepreneur (Brandstätter 2011; Frese and Gielnik 2014; Rauch and Frese 2007; Zhao et al. 2010). Scholars have shown that intentions play an important role in the decision to start a new business (Barba-Sánchez and Atienza-Sahuquillo 2018) and have emphasized the importance of attitudes and perspectives to social and psychological aspects of the development of business intentions (Ajzen 1991; Ghani et al. 2013; Liñán and Santos 2007). However, up to this point, we have a limited understanding of the prerequisites needed to develop successful entrepreneurship intentions among students. In particular, little is known about the hierarchical structures of premise factors, including attitudes, social and psychological factors that work together to advance students' entrepreneurial intentions in Universities in Ho Chi Minh City, Vietnam.

11.2.2 Factors Affecting Entrepreneurial Intention

From the theoretical basis and related studies, based on inheriting and selecting the factors that affect the students' entrepreneurial intentions, the author has used the factors in the experimental studies of Eyel and Durmaz (2019), Wongnaa and Seyram (2014), Lüthje and Franke (2003), Miranda et al. (2017), Ambad and Damit (2016), and Thelken and de Jong (2020). Thereby, the research model proposed by the author includes 06 independent variables, including (1) personality characteristics, (2) subjective norm, (3) feasibility perception, (4) educational environment, (5) attitude towards entrepreneurship, (6) financial approach impact entrepreneurship intention of students at universities in Ho Chi Minh City.

According to research Wongnaa and Seyram (2014), Lüthje and Franke (2003), and Ambad and Damit (2016) have shown that character trait factor influences students' intention to start a business. From there, we have hypothesis H_1 as follows:

H_1: Personality characteristics have a positive impact on entrepreneurship intentions.

Applying this paper with previous studies of Miranda et al. (2017), Ambad and Damit (2016), Eyel and Durmaz (2019), and Thelken and de Jong (2020) showed that subjectivity affects the students' intention to start a business. From there, we have hypothesis H_2 as follows:

H_2: Subjective norm has a positive impact on entrepreneurship intentions.

According to research results of Eyel and Durmaz (2019), Ambad and Damit (2016), and Lüthje and Franke (2003) showed that the factor of Feasibility perception influences the idea of starting a business. From there, we have hypothesis H_3 as follows:

H_3: Feasibility perception has a positive impact on Entrepreneurship intentions.

The entrepreneurship education environment is a means to inspire students with entrepreneurship intentions, increasing students' ability to take risks in business. According to the research results of Wongnaa and Seyram (2014), Lüthje and Franke (2003), Ambad and Damit (2016), and Thelken and de Jong (2020) showed that educational environment factors influence Italy start-up of students. From there, we have hypothesis H_4 as follows:

H_4: The educational environment has a positive impact on entrepreneurial intention.

Applying this study, the author believes that the Attitude towards entrepreneurship should be measured on the personal means based on the scale of Eyel and Durmaz (2019), Ambad and Damit (2016), Miranda et al. (2017), and Thelken and de Jong (2020). Therefore, we have hypothesis H_5:

H_5: Attitude towards entrepreneurship has a positive impact on entrepreneurship intention.

According to the previous research results of Lüthje and Franke (2003) and Wongnaa and Seyram (2014), we find that the Financial Approach factor affects students' intention to start a business. Therefore, we have hypothesis H_6 as follows:

H_6: Financial Approach has a positive impact on entrepreneurship intentions.

11.3 Methods

11.3.1 Research Approach

In this study, collecting information for our quantitative analysis, questionnaires were used. The respondents' survey responses were the primary method for gathering information. The research was conducted in the year 2020. The research included questions about the status of the determinants that affected the Entrepreneurship Intention of students. According to Hair et al. (2006), the number of samples should be four or five times higher than the number of observed variables. In this study, the scale of the above model includes 30 observed variables; the expected number of samples is 150 (the number of samples is five times higher than the observed variable). Because this is an exploratory study, the sampling method is a non-probability sampling design with a convenient form of sampling, the subject of the survey is 3rd (41%) and 4th (59%) students. Respondents were selected using

Table 11.1 Statistics of sample

University Code	Number	Percentages (%)
IUH	65	36.5
IDT	19	10.7
SGU	26	14.6
HUTECH	32	18.0
HUFI	17	9.6
HUFLIT	19	10.7
Total	178	100.0

conventional methods, and a total of 178 students living in Ho Chi Minh City, Vietnam. Our study population consisted of 96 (53.9%) men and 82 (46.1%) women. Their amount is shown in Table 11.1.

IUH is accounting for 36.5% (65 students), followed by schools. HUTECH accounts for 18% (32 students), followed by TDT University accounting for 10.7% (19 students), followed by SGU accounting for 14.6% (26 students), and the last is HUFI accounting for 9.6% (17 students). Our study is composed of six universities in Vietnam. The IUH student accounts for the highest proportion.

We use the 5-point Likert scale to evaluate the level of consent for the related factors. Therefore, this paper also uses the 5-point Likert scale to evaluate the level of consent for all observed variables, with 1: Disagree and 5: Agree in Table 11.2. The mean of factors from 3.8 to 4.0 is good.

11.3.2 Blinding

For the duration of the study, all study staff and respondents were blinded. No one from the outside world had any contact with the study participants.

11.3.3 AIC in Model Selection

When it comes to finding the best model for predicting future observations, the Akaike Information Criterion (AIC) is more fitting. AIC considers a few canonical statistical problems and reports statistical optimality results for each of them. Its relationship with other model selection parameters, as well as some of its generalizations, is also discussed. Although the optimality is linked to Akaike's original motivation, as stated in, it does not proceed as a natural consequence. The problem of multivariate regression is one of the most important problems for which AIC can be used as a model selection procedure (Chakrabarti and Ghosh 2011).

Table 11.2 Factor and item

Unit	Factor and item	Mean
PC	*Personality characteristics*	4
PC1	You like to choose industries that require creativity and exploration	
PC2	You want to challenge your abilities in business in Covid19 pandemic	
PC3	You dare to take risks in business in Covid19 pandemic	
PC4	You are confident in your ability to start a business in Covid19 pandemic	
PC5	You can manage businesses in Covid19 pandemic	
SN	*Subjective norm*	3.8
SN1	Your family supports you when you run your business in Covid19 pandemic	
SN2	Your friends support your entrepreneurial spirit in Covid19 pandemic	
SN3	The university encourages you to pursue entrepreneurial intentions	
SN4	Good infrastructure in Vietnam is the premise to support start-ups in Covid19 pandemic	
FP	*Feasibility perception*	3.8
FP1	You know how to develop a project in Covid19 pandemic	
FP2	The probability of success is high if you start a business	
FP3	A high degree of autonomy if you start a business	
FP4	High satisfaction if you become an entrepreneur	
FP5	Entrepreneurship helps you take advantage of the educational background	
EE	*Educational environment*	4.0
EE1	University instructors equip you with full knowledge of entrepreneurship in Covid19 pandemic	
EE2	The program you get during your studies is adequate for entrepreneurship in Covid19 pandemic	
EE3	Social skills and knowledge are enough to get you started in Covid19 pandemic	
EE4	University always organizes programs to develop business potential	
ATE	*Attitude towards entrepreneurship*	4
ATE1	You find the intention to start a business very attractive in Covid19 pandemic	
ATE2	You want to create a side business if you have the opportunity and resources	
ATE3	You will feel satisfied if you become an entrepreneur	
ATE4	You think it will be successful if you decide on the business	
FA	*Financial approach*	3.9
FA1	Your family and friends can provide you with loans to start a business in Covid19 pandemic	
FA2	You can raise capital (banks, credit funds, ...) or start-up programs in Covid19 pandemic	
FA3	You can accumulate capital (save, work overtime, ...) in Covid19 pandemic	
EI	*Entrepreneurship intention*	4.0
EI1	You are determined to create a business in the future	
EI2	You want to be an entrepreneur	
EI3	You have a solid intention to start a company	
EI4	You will try to start and run your own company	
EI5	You have serious thought when starting your own company	

11.4 Results

11.4.1 Akaike Information Criterion Selection

AIC (Akaike's Information Criteria) was utilized to choose the best model by R software. AIC has been used in the theoretical context for model selection. And when multicollinearity occurs, the AIC approach can handle multiple independent variables. As a regression model, AIC can be applied, estimating one or more dependent variables from one or more independent variables. An essential and useful measurement for deciding a complete and straightforward model is the AIC. Based on the AIC information standard, a model with a lower AIC is selected. The best model will stop when the minimum AIC value is in Table 11.3 (Burnham and Anderson 2004; Khoi 2021).

In the above results, R reports show every step of searching for the optimal model. The first step is to start with all 06 independent variables with AIC = −401.17. The second step is to find a model, R stops with a model of 05 independent variables (PC, SN, FB, AIE, FA) in Table 11.3 with AIC = −402.7.

All variables have a *p*-value lower than 0.05 (Hill et al. 2018), so they are correlated with EI, which are in Table 11.4. (PC) Personality characteristics, (SN) Subjective norm, (FP) Feasibility perception, (ATE) Attitude towards entrepreneurship, (FA) Financial Approach impact Entrepreneurship Intention in Table 11.4.

11.4.2 Variance Inflation Factor

The multicollinearity phenomenon occurs when there is a high degree of correlation between the independent variables in the regression models. Gujarati and Porter (2009) showed some signs of multicollinearity in the model when the VIF coefficient is lower than 10. According to Table 11.5, VIF (Variance inflation factor) for the

Table 11.3 Akaike information criterion selection

Model	AIC
EI = f (PC, SN, FB, EE, AIE, FA)	−401.17
EI = f (PC, SN, FB, AIE, FA)	−402.7

Table 11.4 The coefficients

EI	Estimate	Std. Error	*t*-value	*p*-value	Decision
Cons	0.04934	0.23771	0.208	0.000	Accepted
PC	0.26133	0.04161	6.280	0.000	Accepted
SN	0.23176	0.04064	5.703	0.000	Accepted
FB	0.14271	0.03499	4.079	0.000	Accepted
ATE	0.17682	0.04870	3.631	0.000	Accepted
FA	0.20410	0.04223	4.833	0.000	Accepted

Table 11.5 Model test

	PC	SN	FB	ATE	FA
VIF	1.415037	1.035846	1.139268	1.559737	1.432551
Heteroskesdaticity	Goldfeld-Quandt test		chi2	df	p
			1.5446	83	0.0246
Autocorrelation	Durbin-Watson		Test for autocorrelation		
	2.0552		p-value = 0.6338		
Model evaluation	Adjusted R-squared 0.6345		F-statistic 62.46		p-value: 0.00000

independent variables is smaller than 10 (Miles 2014), so there is no collinearity between the independent variables.

11.4.3 Heteroskesdaticity

One of the fundamental assumptions of the classical linear regression model is that the random error must have a constant (Homoskedasticity). Obviously, in practice, this assumption is unlikely to exist. On the contrary, if the variance of the random error for each observation is different, we meet heteroscedasticity.

Goldfeld-Quandt test shows that p-value = 0.0246 and lower than 0.05 (Godfrey 1978), so it can be concluded that there is no heteroskedasticity in Table 11.5.

11.4.4 Autocorrelation

Durbin-Watson Test shows that there is no autocorrelation from the model in Table 11.5 because p-value = 0.6338 is greater than 0.05 (Durbin and Watson 1971).

11.4.5 Model Evaluation

According to the results from Table 11.5, Personality characteristics, Subjective norm, Feasibility perception, Attitude towards entrepreneurship, Financial Approach impact Entrepreneurship Intention is 63.45%. The above analysis shows the regression equation below is statistically significant (Greene 2003).

$$EI = 0.04934 + 0.26133\ PC + 0.23176\ SN + 0.14271\ FB + 0.17682\ ATE + 0.20410\ FA$$

11.4.6 Discussion

The results of the AIC Algorithm for the Entrepreneurial Intention showed that five independent variables Personality characteristics, Subjective norm, Feasibility perception, Attitude towards entrepreneurship, Financial Approach have a positive impact on the Entrepreneurship Intention of students because their p-value is greater than 0.05. Compare the impact level of these five variables on the dependent variable Entrepreneurial intentions in descending order as follows: Personality characteristics (0.26133), Subjective norm (0.23176), Financial Approach (0.20410), Attitude towards entrepreneurship (0.17682), Feasibility perception (0.14271). The educational environment of entrepreneurship (EE) has no impact (p-value lower than 0.05). Thus, hypotheses H_1, H_2, H_3, H_5, and H_6 are all accepted at the 95% confidence level.

11.5 Conclusion

This article aims to concentrate on the quantitative model and the results of the application for the Entrepreneurship Intention of students at universities in Ho Chi Minh City, Vietnam, in the Covid19 Pandemic Environment. The paper examines that the research results show that the factors affecting the Entrepreneurship intention of students from strong impact to weak impact are as follows: (1) Personality characteristics, (2) Subjective norm, (3) Feasibility perception, (4) Attitude towards entrepreneurship, (5) Financial Approach impact Entrepreneurship Intention of students at universities in Ho Chi Minh City, Vietnam.

The COVID-19 epidemic had caused great harm to the start-up community when up to 50% of start-ups confirmed that they were operating in moderation and generating negligible income; while 23% of start-ups think that they are losing capital raising opportunities and expanding their market, 20% of start-ups choose to freeze their activities, which means stopping all production and business activities; 4% of start-ups have to stop all advertising activities on all platforms including online and offline to save costs, and only 3% will be affected in a limited, negligible way. The total damage estimated in the first 4 months of 2020 of the total number of start-ups participating in the survey is 152.6 billion VND, with the amount of common damage falling between 200 and 500 million VND per start-up. Therefore, the need for loans to maintain and restore the Start-up production and business activities is currently very large, notably, 100% of the start-ups participating in the survey need medium and long-term loans with a total amount of 274.8 billion. The popular loan needs of each Start-up range from 500 million to 2 billion. We need a financial solution to revive the Start-up right now, otherwise, start-ups will die “from starvation” before dying “from the virus” (Bao Anh 2020).

First, Vietnamese students need to show their personality characteristics in the Covid19 Pandemic, such as creativity and exploration, abilities in business, confidence, and business management. Second, Vietnamese students create a good trust for their family and friends in the COVID-19. Third, they need to be supported Financial Approach of family, bank, and friends. Fourth, they need to attitude towards entrepreneurship confidently. Last, Feasibility perception is good projects, successful probability, and educational background to start-up.

Acknowledgments This research is funded by the Industrial University of Ho Chi Minh City, Vietnam.

References

Ajzen I (1991) The theory of planned behavior. Organ Behav Hum Decis Process 50(2):179–211

Ambad SNA, Damit DHDA (2016) Determinants of entrepreneurial intention among undergraduate students in Malaysia. Proc Econ Finance 37:108–114

Bao A (2020) The reality of Vietnam innovative entrepreneurship under the impact of COVID translation - 19 (Vietnamese). Accessed from http://doanthanhnien.vn/tin-tuc/khoi-nghiep-lap-nghiep/thuc-trang-khoi-nghiep-doi-moi-sang-tao-viet-nam-duoi-tac-dong-cua-dich-covid%2D%2D-19

Bao DN (2021) Start-up students after covid-19 (Vietnamese). Accessed from https://baodanang.vn/channel/5404/202006/sinh-vien-khoi-nghiep-sau-covid-19-3462939/

Barba-Sánchez V, Atienza-Sahuquillo C (2018) Entrepreneurial intention among engineering students: the role of entrepreneurship education. Eur Res Manag Bus Econ 24(1):53–61

Brandstätter H (2011) Personality aspects of entrepreneurship: a look at five meta-analyses. Personal Individ Differ 51(3):222–230

Burnham KP, Anderson DR (2004) Multimodel inference: understanding AIC and BIC in model selection. Sociol Methods Res 33(2):261–304

Chakrabarti A, Ghosh JK (2011, 2011) AIC, BIC and recent advances in model selection. Philos Stat:583–605

Durbin J, Watson GS (1971) Testing for serial correlation in least squares regression. III. Biometrika 58(1):1–19

Eyel CŞ, Durmaz İBV (2019) Entrepreneurial intentions of generation-Z: compare of social sciences and natural sciences undergraduate students at Bahçeşehir university. Proc Comput Sci 158:861–868

Frese M, Gielnik MM (2014) The psychology of entrepreneurship. Annu Rev Organ Psychol Organ Behav 1(1):413–438

Ghani M, Salanghouch MH, Mobaraki MH, Mirzaei E (2013) The effect of psychological capital on university of Tehran students' entrepreneurial intention. Asian J Res Bus Econ Manag 3 (10):313–321

Godfrey LG (1978) Testing for multiplicative heteroskedasticity. J Econ 8(2):227–236

Greene WH (2003) Econometric analysis. Pearson, New York

Gujarati DN, Porter D (2009) Basic econometrics. McGraw-Hill, New York

Hair JF, Black WC, Babin BJ, Anderson RE, Tatham RL (2006) Multivariate data analysis, vol 6. Pearson Prentice Hall, Upper Saddle River, NJ

Hill RC, Griffiths WE, Lim GC (2018) Principles of econometrics. Wiley, Boca Raton, FL

Khoi BH (2021) Factors influencing on university reputation: model selection. In: AIC data science for financial econometrics. Springer, New York, pp 177–188

Krueger NF (2017) Entrepreneurial intentions are dead: long live entrepreneurial intentions. In: Revisiting the entrepreneurial mind. Springer, New York, pp 13–34

Liñán F, Santos FJ (2007) Does social capital affect entrepreneurial intentions? Int Adv Econ Res 13(4):443–453

Lüthje C, Franke N (2003) The 'making'of an entrepreneur: testing a model of entrepreneurial intent among engineering students at MIT. R&D Manag 33(2):135–147

Miles J (2014) Tolerance and variance inflation factor. Wiley StatsRef: Statistics Reference Online

Miranda FJ, Chamorro-Mera A, Rubio S (2017) Academic entrepreneurship in Spanish universities: an analysis of the determinants of entrepreneurial intention. Eur Res Manag Bus Econ 23 (2):113–122

Rauch A, Frese M (2007) Let's put the person back into entrepreneurship research: a meta-analysis on the relationship between business owners' personality traits, business creation, and success. Eur J Work Organ Psy 16(4):353–385

Schlaegel C, Koenig M (2014) Determinants of entrepreneurial intent: a meta-analytic test and integration of competing models. Entrep Theory Pract 38(2):291–332

Thelken HN, de Jong G (2020) The impact of values and future orientation on intention formation within sustainable entrepreneurship. J Clean Prod 266:122052

UEH (2021) Digital platform - way for Vietnamese businesses to adapt and recover from the covid-19 pandemic (Vietnamese). Accessed from https://www.ueh.edu.vn/tin-tuc/nen-tang-so-cach-thuc-de-cac-doanh-nghiep-viet-nam-thich-ung-va-phuc-hoi-sau-dai-dich-covid-19_56145

WHO (2021). Accessed from https://www.who.int/publications/m/item/weekly-epidemiological-update%2D%2D-27-january-2021

Wongnaa CA, Seyram AZK (2014) Factors influencing polytechnic students' decision to graduate as entrepreneurs. J Glob Entrep Res 4(1):1–13

Zhao H, Seibert SE, Lumpkin GT (2010) The relationship of personality to entrepreneurial intentions and performance: a meta-analytic review. J Manag 36(2):381–404

Chapter 12
Sentiment Analysis for COVID Vaccinations Using Twitter: Text Clustering of Positive and Negative Sentiments

Dwijendra Nath Dwivedi and Shailendra Pathak

Abstract The objective of the chapter is to analyze Twitter data to extract sentiments and opinions in unstructured text. The team attempted to use contextual text analytics to categorize Twitter data to understand the positive or negative sentiments for COVID vaccinations and wish to highlight key concerns. Text clustering has been performed on positive and negative sentiments to understand the key themes behind them. We followed a two-step process. In the first step, we identified positive and negative sentiments from Twitter feeds. In the second step, we aggregated all sentiments into categories to deduce what the Twitterati is thinking about COVID-19 vaccinations.

The whole analysis was performed using Python, including TextBlob and Vader libraries. TextBlob library uses the Naïve-Bayes (probabilistic algorithms using Bayes's Theorem to predict the category of a text) classifier to assess the polarity of a sentence and generates a score ranging between −1 (strongly negative) and +1 (strongly positive). The Naïve Bayes classifier categorizes based on probabilities of events. Although it is a simple algorithm, it performs well in many text classification problems. On the other hand, the Vader library uses a lexical approach that uses preassigned scores labeled positive and negative for different words found in a text. These scores are based on pre-trained models classified as positive/negative by actual human beings.

We then performed the topic extraction that discovers the keywords in sentiments that capture the recurring theme of a text and is widely used to analyze large sets of sentiments to identify the most common topics easily and efficiently. We found a large segment as neutral (53%) followed by a positive sentiment segment that contributed 36% of tweets. However, at the same time, many people (10%+) remain on the fence regarding the potential repercussions of COVID vaccines as they are relatively new and yet untested over longer periods of time. It is reasonable to expect

---

D. N. Dwivedi (✉)
Krakow University of Economics, Kraków, Poland

S. Pathak
IGIDR, Mumbai, Maharashtra, India

S. A. Hassan et al. (eds.), *Decision Sciences for COVID-19*, International Series in Operations Research & Management Science 320,
https://doi.org/10.1007/978-3-030-87019-5_12

that people are a bit skeptical about vaccinations. Text clustering of negative sentiments identified late vaccinations and side effects being the key concerns. Positive sentiments mainly were driven by the readiness of other vaccines and weak reactions following vaccinations. The study contributes to text mining literature by providing a framework for analyzing public sentiments. This can help to understand the key themes in negative sentiments related to COVID vaccinations and can help in adjusting policies.

Keywords Sentiment analytics · Twitter · COVID-19 · Text clustering · Topic modeling · Opinion extraction · Topic extraction · Post COVID · Intention mining

12.1 Introduction

As the COVID-19 pandemic has spread throughout the world, it has impacted economic activity and health systems of all countries. The pandemic has severe consequences for health, economics, geopolitics, and other aspects of society. Most of the countries have started COVID vaccinations. This study wants to understand and analyze the sentiments after quick tweets post COVID vaccinations. Also, the intent is to understand if the majority of the tweets are having positive or negative sentiments. At the same time, it would be essential to understand the key clusters and segments of positive and negative tweets.

The research aims to address the following primary questions:

- What is the distribution of positive and negative sentiments emerging from public tweets in the context of COVID-19 pandemic vaccinations?
- What are the key segmentations and clusters in the positive and negative sentiments?

12.2 Literature Study

The following is the referenced literature for the two steps and stages in our work. We initially performed the sentiment analysis followed by text clustering of positive and negative sentiment tweets.

Sentiment Analytics Using Twitter Asghar et al. (2019) analyzed consumers' views for major automobile brands using Twitter data. He found that Audi has 87% of the positive tweets compared to 74% for BMW, 84% for Honda, 70% for Toyota, and 81% for Mercedes. Godea et al. (2015) explored the general sentiments and information dissemination concerning electronic cigarettes or e-cigs using Twitter and found that Twitter users are mainly concerned with sharing information (33%) and promoting e-cigs (22%). Olorunnimbe and Viktor (2015) were able to capture the sentiments of individuals as they evolve over time while exploring political sentiments on Twitter for opinion mining. Dinkić et al. (2018) used Twitter data to determining the popularity of city locations of interest and public spaces in general. Esiyok and Albayrak (2015) compared performance of the Naïve Bayes and Maximum Entropy classification methods for predicting marketing trends. Choi and Kim (2013) performed sentiment analysis for tracking breaking events and found that his study offers diverse evidence to prove that Twitter has valuable information for tracking breaking news throughout the world. Rao and Srivastava (2014) evaluated the use of public opinion tweets in driving investment decisions. Steede et al. (2018) analyzed the sentiment and content analysis of Twitter content regarding the use of antibiotics in livestock.

Text Clustering Methods To manage the explosion of electronic document archives, new techniques or tools are required that deal with organizing, searching, indexing, and reviewing large collections of data in a time-efficient manner (Alghamdi and Alfalqi 2015). As a generalization, there are two broad approaches to process text—natural language processing (NLP) and statistical-based programs like topic modeling (Hofmann 2001). Unlike NLP methods that tag parts-of-speech and grammatical structure, statistical-based models like topic models are largely based on the "bag-of-words" (BoW) assumption. In BoW models, the collection of text documents is quantified into a document-term matrix (DTM) that counts the occurrence of each word (columns) for each document (rows). In the case of most topic models like LDA, the DTM is one of two model inputs along with the number of topics (Wesslen 2018). Deerwester et al. (1990) presented one of the first topic models using latent semantic analysis (LSA) and singular value decomposition (SVD), in which a large DTM is decomposed into a set of about 100 orthogonal factors from which the original matrix can be approximated by linear combination. They assumed the presence of some underlying latent semantic structure and used statistical techniques to estimate this latent structure. Asmussen and Møller (2019) presented a framework to leverage the topic modeling technique for performing an exploratory literature review on a large collection of papers. The framework proposed by them enables a large volume of documents to be reviewed in a transparent, efficient, and reproducible way using LDA method. In general, there are two methods for automatically processing documents—Supervised learning and unsupervised learning. Supervised learning includes manually coding a collection of documents before conducting an analysis, which involves a high amount of time to achieve the result. On the other hand, unsupervised learning methods, such as topic modeling, do not have the pre-requisite to manually code the documents, which results in saving a lot of time for an exploratory review of a large collection of papers. Gottipati et al. (2018) leveraged topic modeling and data visualization methods to analyze student feedback comments from seven undergraduate courses taught at Singapore Management University. They assessed rule-based methods and statistical classifiers for extracting the topics. Al-Obeidat et al. (2018) further proposed opinions sandbox for topic extraction, sentiment analysis for extracting topics and their associated sentiments from a database. They used LDA for topic extraction, and the "bag-of-words" sentiment analysis algorithm where polarity is determined based on the frequency of occurrence of positive/negative words in a document. Benedetto and Tedeschi (2016) highlight the common approaches of sentiment analysis in social media streams and the related issues with cloud computing. Big data is divided into four features, namely four V's of big data—volume, velocity, variety, and veracity. Volume is the largest amount of data that should be stored and processed. Velocity is frequency of the incoming data. Variety describes different types of data whereas Veracity refers to trustworthiness and accuracy of the data available. Gupta et al. (2021) attempted to use contextual text analytics to identify product or service features that drives the sentiment of the users.

12.3 Data Source

This dataset has been extracted from Twitter and includes 4047 tweets from different parts of the world, all related to the COVID-19 pandemic. We analyzed around 4047 tweets regarding COVID-19 vaccinations between Dec-2019 to Jan-2020 provided by Kaggle. Kaggle allows users to find and publish data sets, explore and build models in a web-based data-science environment, work with other data scientists and machine learning engineers, and enter competitions to solve data science challenges and encourages work for research. We used data only for these 2 months to capture early opinions about vaccinations while they had just started in the world.

12.4 Results

12.4.1 Data Cleanup Steps

The data has been extensively cleaned before the sentiment polarity calculations. The steps included:

- Removal of stop words like ("also," "other," etc.) which do not have much impact on the meaning of the sentence. In most text mining projects, stopwords are a big challenge because stop words can bring a lot of noise to the data, which can throw off the insights hidden in the raw text.
- Striping the HTMLs—Since this was Twitter data, it contained many HTML links and mentions to other websites which needed to be removed. Sometimes the HTML links have words that may inadvertently reflect human emotions completely unrelated to what the person writing the tweets meant.
- Removal of noise and special characters—Another step intended to increase the accuracy of the sentiment polarity estimation.
- Removal of non-ASCII entries.
- Removal of punctuations—In text mining, punctuations do not impart valuable information as far as the sentiment of a statement is concerned. Hence, all

punctuations were removed from the texts by extensive usage of the Regex library of Python.
- Replacement of numbers—Numbers do not add any value in sentiment calculation.
- Stemming of words—Using the simplest form of each word in a sentence. It removes or stems the first or last few characters of a word. Stemming suffers from some limitations in that it can create nonsensical words from the perfectly legitimate text by removing a few letters from the beginning or the end of the world. However, it is a very useful technique to bring a lot of extended words into their base forms.
- Lemmatization—resolves words to their dictionary form. It considers every word in each context, and to do so, it looks up into detailed dictionaries of words that are meaningful and, in our case, are very useful to extract the sentiments of the text. One very important thing about lemmatization is that a lemma reduces. All the inflectional additions to a word and thereby return the base form of the word irrespective of how the word has been used in a sentence. This means that the Stemming can return the same output for a lot of different words and may impart more noise into text mining. However, lemmatization can give the same concise version of multiple words and can be more meaningful.

Sentiment analysis is a natural language processing technique used to determine whether data is positive, negative, or neutral. Sentiment analysis for free-flowing text like Twitter data can effectively combine natural language processing (NLP) and machine learning algorithms to deduce sentiment scores to a sentence or phrase. Sentiment analysis can give an idea about public opinion, brand reputation, key concerns, customer experience, and customer perception and an overall index of how optimistic the common public is about a topic of interest, which in our case is the attitude of the public toward COVID vaccinations.

12.4.2 Sentiment Analysis

Primarily there are four layers of a sentiment analysis engine:

- The first step is to break down raw text into a list of different components, sometimes referred to as tokens. The tokenization process breaks down the raw text into keywords meaningful keywords that reflect the emotions of the person.
- The second step is to isolate the part of the tokens that are sentiment bearing and ignore the rest.
- The third step is to assign polarity scores to each component of the tokens. This is done by referring to different sentiment libraries depending on the program employed. For example, we utilized the TextBlob and Vader library in Python to calculate the sentiment polarity of each tweet in our data. We transformed the text data and created word vectors. We then used TextBlob library to calculate the sentiment scores for the tweets. TextBlog uses naïve Bayes (probabilistic

Table 12.1 Sentiment segmentations

Sentiment score category	Count of tweets
Highly negative	189
Negative	224
Neutral	2159
Positive	950
Highly positive	525
Total	4047

algorithms that use Bayes's Theorem to predict the category of a text) classifier to assess the polarity of a sentence and generates a score ranging between −1 (strongly negative) and +1 (strongly positive). This is the key aspect of sentiment analysis as it tries to understand the opinion, or the orientation of the sentiments expressed in a text. This aspect is quantified with a positive or negative value, called polarity. Identifying the range of polarity scores to be classified as Positive, Neutral, or Negative Sentiment: Read it extensive analysis of the polarity scores and looked at the distribution of these scores along with man Willie eyeballing the sentiment expressed in a text.

- Lastly, we created score ranges that define whether the sentiment of a text was very positive, positive, neutral, very negative, or negative.

Following is the data distribution of the sentiments in the data as per below (Table 12.1):

Reasoning for Using Only 4047 Tweets We have only used 4047 tweets since we wanted to capture early sentiments from relevant tweets only and that also strictly related to vaccination and not get overshadowed by COVID-19 sentiments in general. The data also represents that most people were not explicitly talking about vaccinations yet as possibly they were also overwhelmed by COVID-19 itself.

We believe that 4000 tweets are too low for a definitive answer, however, is representative enough for inferring what the common public is feeling about vaccinations.

12.4.3 Text Clustering

The fundamental idea of applying classical data mining techniques to topic modeling relies on transforming text data (unstructured) to numbers (structured).

12.4.3.1 Text Clustering of Negative Sentiments

As can be seen from Table 12.2, the largest cluster for Negative sentiments is the late vaccine containing 27% of total negative sentiments, which is primarily related to late vaccinations. People are concerned about them getting late vaccinated, and this has to do with the availability of vaccines. The second biggest cluster is the "side

Table 12.2 Cluster: negative sentiments segmentations

Cluster	Cluster description	Frequency
1	Side work + effect + day today + hard post feel arm care amp first + approve vulnerable +dose	86
2	Pfizer Norway 'dead Norway' + die + hour people pfizervaccine + receive dead soreness + death + vaccination biontech + update arm	65
3	+base matthancock + country science borisjohnson government covidvaccine + late +number + question + minute + year covid19vaccine moderna + little	104
4	Virus + bad life Uk vaccine + worry due careful effectiveness news second + feel + wait + wear bit	67
5	+reaction allergic + suffer alaskan covid19 severe + worker serious + 'covid19 vaccine' + vaccine + 'pfizerbiontech vaccine' pfizerbiontech health + risk pfizercovidvaccine	59

Table 12.3 Cluster: positive sentiments segmentations

Cluster	Cluster description	Frequency
Other vaccine readiness	World data astrazeneca + know + number mrna covid_19 + develop far + team covidvaccination + work able morning + great	378
Effective	+vaccine pfizer covid-19 moderna effective biontech + 'covid-19 vaccine' + 'covid19 vaccine' + variant pfizer-biontech + shipment covid19 coronavirus + test	370
First Phizer dose	+dose first + receive president-elect second 'dose vaccine' joebiden grateful today covid19vaccine + delight + 'vaccine dose' + 'pfizerbiontech vaccine' + feel pfizerbiontech	187
Week reaction	Amp UK + year + jab end + old 'end tunnel' tunnel + vaccinate + week + reaction health people + approve science	273
Post vaccination	Covidvaccine + day feeling good + hour great + good +happy + great + vaccination nhs + vaccinate today + hope grateful	266

effect" that has 22% of the negative tweets. So, the side effects of vaccinations are the second largest reason for negative sentiments regarding vaccinations. Seventeen percent of negative sentiments were concerned with the second wave of COVID19.

12.4.3.2 Text Clustering of Positive Sentiments

As can be seen from Table 12.3, the largest cluster for positive sentiments is the readiness of other vaccines containing 26% of total positive sentiments, which is primarily related to the readiness of other vaccines like AstraZeneca. People are happy about the effectiveness of vaccines, and this cluster consists of 25% of tweets. Almost 19% of tweets are happy about the weak reaction, and 18% of the tweets were related to post vaccinations.

12.5 Discussion

The present study tries to analyze opinions and sentiments expressed on a free online medium of Twitter. As such, there are too many variations and shades in people's opinions expressed over the free media. However, by applying statistical methods on raw text, it can be inferred fairly logically that COVID vaccinations are a complicated subject, and it also reflects that people are very indecisive when it comes to expressing what they really feel about COVID vaccinations. This study demonstrates that early in the days when COVID vaccinations started, most people were unsure or positive about the impact of vaccinations. The results of our analysis lend support to continuous development and research towards further vaccine development and acceptance by all. To further add, if the governments and authorities are stonewalled by negative perceptions and propaganda about vaccines and any side effect—this study shows that people in general are not so negative about vaccines, and more information and focussed educative efforts from governments may actually persuade people to vaccinate.

12.6 Conclusion

The sentiment scores clearly highlight that overall, there is a high proportion of positive sentiment toward COVID vaccinations. However, most tweets indicated neutral sentiments. This is expected because a lot of tweets simply do not show an opinion but may indicate a fact. At the same time, a lot of people remain on the fence about the potential repercussions of COVID vaccines as these vaccines are relatively new and yet untested over longer periods of time. It is reasonable to expect that people are a bit skeptical about vaccinations. There are few limitations to this study. First, it does not cover all tweets on the topic, and as a next step, we would extract more data as recent months will have more feedback on COVID vaccinations. Second, we use one method of topic modeling—LSA and do not compare how it fares with other methods.

References

Alghamdi R, Alfalqi K (2015) A survey of topic modeling in text mining. Int J Adv Comput Sci Appl 6(1):147–153. https://doi.org/10.14569/ijacsa.2015.060121

Al-Obeidat F, Kafeza E, Spencer B (2018) Opinions sandbox: turning emotions on topics into actionable analytics. In: Lecture Notes of the Institute for Computer Sciences, Social-Informatics and Telecommunications Engineering, LNICST, vol 206. Springer, Cham, pp 110–119. https://doi.org/10.1007/978-3-319-67837-5_11

Asghar Z, Ali T, Ahmad I, Tharanidharan S, Nazar SKA, Kamal S (2019) Sentiment analysis on automobile brands using Twitter data. In: Communications in computer and information science, vol 932. Springer, Singapore. https://doi.org/10.1007/978-981-13-6052-7_7

Asmussen CB, Møller C (2019) Smart literature review: a practical topic modelling approach to exploratory literature review. J Big Data 6(1). https://doi.org/10.1186/s40537-019-0255-7

Benedetto F, Tedeschi A (2016) Big data sentiment analysis for brand monitoring in social media streams by cloud computing. Stud Comput Intell 639. https://doi.org/10.1007/978-3-319-30319-2_14

Choi D, Kim P (2013) Sentiment analysis for tracking breaking events: a case study on Twitter. In: Lecture notes in computer science (including subseries lecture notes in artificial intelligence and lecture notes in bioinformatics), vol 7803. Springer, Cham, pp 285–294. https://doi.org/10.1007/978-3-642-36543-0_30

Deerwester S, Dumais ST, Furnas GW, Landauer TK, Harshman R (1990) Indexing by latent semantic analysis. J Am Soc Inf Sci 41(6):391–407

Dinkić N, Džaković N, Joković J, Stoimenov L, Đukić A (2018) Using sentiment analysis of Twitter data for determining popularity of city locations. Adv Intell Syst Comput 665:156–164. https://doi.org/10.1007/978-3-319-68855-8_15

Esiyok C, Albayrak S (2015) Twitter sentiment tracking for predicting marketing trends. Adv Comput Vis Pattern Recogn 66:47–74. https://doi.org/10.1007/978-3-319-14178-7_2

Godea AK, Caragea C, Bulgarov FA, Ramisetty-Mikler S (2015) An analysis of twitter data on e-cigarette sentiments and promotion. In: Lecture Notes in Computer Science (Including Subseries Lecture Notes in Artificial Intelligence and Lecture Notes in Bioinformatics), vol 9105. Springer, Cham, pp 205–215. https://doi.org/10.1007/978-3-319-19551-3_27

Gottipati S, Shankararaman V, Lin JR (2018) Text analytics approach to extract course improvement suggestions from students' feedback. Res Pract Technol Enhanc Learn 13(1). https://doi.org/10.1186/s41039-018-0073-0

Gupta A et al (2021) Understanding consumer product sentiments through supervised models on cloud: pre and post COVID. Webology 18(1):406–415. https://doi.org/10.14704/web/v18i1/web18097

Hofmann T (2001) Unsupervised learning by probabilistic latent semantic analysis. Mach Learn 42 (1–2):177–196. https://doi.org/10.1023/A:1007617005950

Olorunnimbe MK, Viktor HL (2015) Tweets as a vote: exploring political sentiments on twitter for opinion mining. In: Lecture Notes in Computer Science (Including Subseries Lecture Notes in Artificial Intelligence and Lecture Notes in Bioinformatics), vol 9384. Springer, Cham, pp 180–185. https://doi.org/10.1007/978-3-319-25252-0_19

Rao T, Srivastava S (2014) Twitter sentiment analysis: how to hedge your bets in the stock markets. Springer, Cham, pp 227–247. https://doi.org/10.1007/978-3-319-05912-9_11

Steede GM, Meyers C, Li N, Irlbeck E (2018) A sentiment and content analysis of Twitter content regarding the use of antibiotics in livestock. J Appl Commun *102*(4):1B

Wesslen R (2018) Computer-assisted text analysis for social science: topic models and beyond. ArXiv

Chapter 13
Participation and Active Contribution of Private Universities in the Prevention of the Covid-19 Pandemic Transmission

Mahyudin Ritonga, Ahmad Lahmi, Sandra Ayu, Yufi Latmini Lasari, Sri Wahyuni, Desminar, Armalena, and Fitria Sartika

Abstract The study aims to reveal how active participation forms and real contributions of higher education institutions in preventing the transmission of the Covid-19 pandemic. To find data related to the research focus, this study used qualitative methods, the data sources were university leaders, students, and several community members who were selected by random sampling, data collection was carried out by interview, observation, and documentation study. The results of the study can be concluded that Muhammadiyah universities in Indonesia contribute significantly in handling the transmission of Covid-19 through the production and distribution of Hand Sanitizer, production and distribution of medical face shields, distribution of groceries, online transfer of lectures, the provision of distance learning scholarships, work from home instruction, and socialization through operational cars. This research is limited in the scope of participatory in helping campus communities and communities living around the campus. The practical implication of this research is that every institution and community can take action to reduce the impact of Covid-19 in accordance with their respective abilities and positions. University leaders everywhere can mobilize all potentials in their institutions to handle and reduce the impact of Covid-19, because such policies have a real contribution to the existing community. People get the attention of every institution around them.

M. Ritonga (✉) · A. Lahmi · S. Wahyuni · Desminar · Armalena · F. Sartika
Muhammadiyah University of West Sumatera, Padang, Indonesia

S. Ayu
Universitas Islam Negeri Imam Bonjol Padang, Padang, Indonesia

Y. L. Lasari
Institute Agama Islam Negeri Batusangkar, Batusangkar, Indonesia

S. A. Hassan et al. (eds.), *Decision Sciences for COVID-19*, International Series in Operations Research & Management Science 320,
https://doi.org/10.1007/978-3-030-87019-5_13

Keywords Contribution · Prevention · Covid-19

13.1 Introduction

The corona virus that has swept the world has caused various new problems for human life. In the health aspect, Covid-19 has caused many victims to die in the world (Zhang et al. 2020), in the economic field the corona virus has made it difficult for many people to make ends meet (Ozili and Arun 2020; Roy 2020; Susilawati et al. 2020), while in the educational aspect, each level of education transfers the online learning process (Ritonga et al. 2020a, b; Abidah et al. 2020; Suparman et al. 2020). Likewise, in the field of work, many companies have terminated employment relations with employees due to difficulties in paying their salaries (Inayah and Surisman 2020; Rahma and Arvianti 2020).

The extent of the impact of Covid-19 on human life requires the involvement of all parties to be actively involved in finding solutions according to their respective abilities and positions, because if you only hope for the government, handling the impact of Covid-19 will be slower. Muhammadiyah as one of the largest community organizations in Indonesia has long shown its active participation in finding solutions to all problems that arise in the Indonesia country (Rusydi 2016; Syahputra 2014). This participation can be seen from the involvement of various business charities in seeking solutions to every problem faced by the nation whether problems in the fields of religion, social, economy, health, and so on (Suyadi et al. 2020).

One of the Muhammadiyah organization's charities is Muhammadiyah and Aisyiyah colleges. The existence of Muhammadiyah universities in various provinces and cities in Indonesia also has a function to provide enlightenment in finding solutions faced by the nation as befits other charities owned by Muhammadiyah. Therefore, this research is intended to reveal the form of participation and real contribution of Muhammadiyah higher education in preventing the transmission of Covid-19.

Therefore, Muhammadiyah and Muhammadiyah Higher Education played an important role in overcoming the impact of this Covid-19. Muhammadiyah University of West Sumatra as one of the Muhammadiyah Universities has taken concrete steps in order to deal with the pandemic problem. In this scientific article, the author will describe several policies and concrete actions that have been taken by this university as a form of concern and efforts to deal with the impact of Covid-19, both on campus and on the wider community. With the aim of achieving the goals of the Tri Dharma of higher university, and the existence of Muhammadiyah Higher Education can be felt by the community, especially in difficult times such as the conditions during the Covid-19 pandemic.

13.2 Methodology

13.2.1 Research Approach

The research was carried out using a multidisciplinary scientific and interdisciplinary approach, this is based on a typical research that requires an approach in various scientific fields, namely the Covid-19 problem that affects various sectors of life must also be disclosed according to related fields. While methodologically the approach used in this research is qualitative, because what will be revealed is related to the meaning contained in actions and facts in the research location.

13.2.2 Research Setting

This research was conducted in four cities in the West Sumatra Region, Indonesia, to be precise Padang City, Padang Panjang City, Bukittinggi City, and Payakumbuh City, these four cities were used as research sites because the West Sumatra Muhammadiyah University Campus is located in these four cities, while the time needed for the completion of this research lasted for 6 months, from April to September 2020.

13.2.3 Research Participant

Participants who were involved in this research were university leaders and students, the leaders in question were starting from the level of the head of the institution with 3 people, 12 heads of administration for each faculty, and 3 people from the PR department. Other participants who were included in this study were students who were selected purposively, according to the research that had been carried out by purposive selection of informants, the number of students who became informants was 48 people, and this number was based on the taking of each faculty as many as four people.

13.2.4 Data Collection

Research data obtained from informants described above, is related to the active participation of Muhammadiyah universities in preventing the transmission of Covid-19, for that data collection techniques used in this study are interviews, documentation, and observation studies, the form of interview used is structured, this is because get information focused on the steps taken by the leadership in preventing Covid-19. The documentation in question is to collect data from documents available at the Muhammadiyah University of West Sumatra related to the prevention of Covid-19. Observation is used in this study to collect data that occurs in the field by being actively involved.

13.2.5 Data Analysis

All data collected were then analyzed through four stages, namely data display, data reduction, interpretation, and conclusion. To ensure the validity of the data, this study used triangulation of sources, techniques, time. Source triangulation is meant here is that data obtained from certain sources is verified with other sources, as well as data obtained through certain techniques, verification is carried out using other techniques, and time triangulation means that the data obtained at a certain time is verified at a certain time different.

13.3 Results and Discussion

In accordance with the results obtained, it is known that as a form of concern for Muhammadiyah Higher Education in overcoming the impact of the Covid-19 pandemic, there are seven policies and concrete actions that have been taken by Muhammadiyah University of West Sumatra for the internal campus community and the wider community in West Sumatra, participation and contribution can be seen in the following table:

The data contained in Table 13.1 can be explained that the Muhammadiyah University of West Sumatra shows its participation in anticipating the transmission of Covid-19 by producing hand sanitizers, this is in accordance with the results of the documentation that the researchers did as in Fig. 13.1:

The success in producing this hand sanitizer cannot be separated from the encouragement of both the morale and material of the university leadership to the pharmacy faculty to contribute in anticipating the transmission of Covid-19. The

Table 13.1 Muhammadiyah University contribution in overcoming Covid-19 transmission

No.	Form participant	Parties who get
1	Hand sanitizer production	Campus internal and external
2	Produce and share medical faces	Internal and external
3	Groceries	External
4	During	Internal
5	During college scholarships	Internal
6	Work from home	Internal
7	Socialization	Internal and external

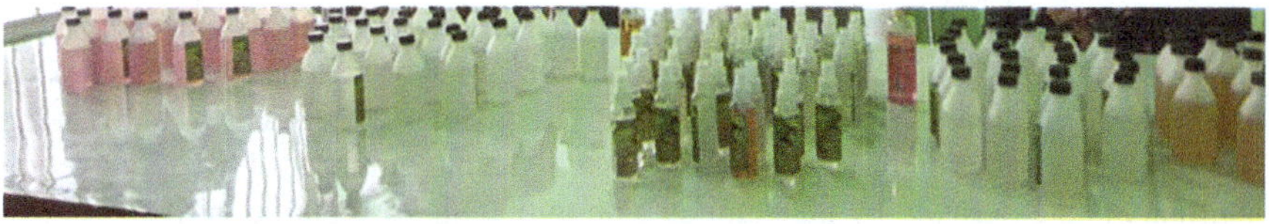

Fig. 13.1 Production results of hand sanitizer, Faculty of Pharmacy, UM Sumatra Barat

hand sanitizers that have been produced are then distributed to internal and external parties for free, especially places that are public facilities such as mosques, according to the results of interviews conducted by researchers with mosque administrators, they expressed their gratitude for the distribution of this hand sanitizer, because in Covid-19 conditions besides significantly increased prices are also difficult to obtain.

As reported by merdeka.com, hand sanitizer is an antiseptic hand sanitizer that comes from a mixture of chemicals and can be used to sterilize the human body which is infested by microorganisms such as germs, bacteria, and bad viruses that reproduce in the cells of living things, including Corona virus that is currently endemic. The importance of washing hands regularly during this pandemic has made some people hunt for hand sanitizers even though the price has soared. As recommended by the World Health Organization (WHO) to produce their own hand sanitizers, lecturers and students of the UM Sumatra Barat Faculty of Science and Technology, Pharmacy Studies study program collaborated in producing hand sanitizers to be distributed to the community around the campus and public facilities.

The UM Sumatra Barat Public Relations writer interviewed by cellular telephone on Friday, May 8, 2020, explained that the hand sanitizers produced by lecturers and students of Pharmacy Studies have been distributed to the community around the campus, students living around the campus, and distributed to facilities public such as hospitals and other public facilities in the region or region through the Muhammadiyah Covid-19 Command Center (MCCC). This contribution can be said to be a form and effort to deal with the spread of Covid-19, because it is in accordance with several research results that washing hands is part of real action to anticipate the transmission of Covid-19 (Jing et al. 2020; Berardi et al. 2020). Apart from that, Islam also reminds us of the importance of maintaining cleanliness (Amin et al. 2020) one of which can be done by washing hands.

UM Sumatra Barat also produces mass production of medical face shields, as reported by haluanriau.co, the leadership of the University revealed that UM Sumatra Barat has taken various forms of concern as an effort to prevent Covid-19 transmission, both for internal and external campus. After the production of hand sanitizers by lecturers and students of Pharmacy Studies, it was continued with the production of Medical Face Shield by the Faculty of Engineering. This was done as a form of UM Sumatra Barat concern for medical personnel who are fighting at the forefront of fighting the spread of Covid-19.

Medical Face Shield this has been distributed by the UM Sumatra Barat Covid-19 Care Team to several cities in West Sumatra, especially hospitals that handle Covid-19 patients directly. This contribution is expressed by the researchers that MFS is needed in the Covid-19 conditions (Rizki and Kurniawan 2020; Matuschek et al. 2020). Because by using MFS, transmission of virus-19 through its attachment to the face will be minimized (Chua et al. 2020; Dwirusman 2020; Worby and Chang 2020). The contribution of UM Sumatra Barat in the procurement of MFS can be felt internally and externally by the campus without having to spend money.

In addition, the Muhammadiyah University of West Sumatra also participates in distributing basic necessities. Facing this pandemic period, the community seems like a dilemma, on the one hand, they have to work to make a living to meet their daily needs, while, on the other hand, they have to obey the rules to stay at home (Chee 2020; Chen et al. 2020). Moreover, people generally work outside the home, such as traders, laborers, drivers, and the like. The need for foodstuffs is very urgent to be in stock during this pandemic, while finance is no longer adequate, including students who have to survive overseas for the smooth running of online lectures. Therefore, UM Sumatra Barat took action to distribute groceries to the community and also compensation to students affected by Covid-19.

As explained by the Public Relations Division of UM Sumatra Barat a few days ago, these basic necessities and compensation have been distributed to students who live around the campus or stay in their boarding houses. Staple foods were also distributed to local communities affected by this pandemic. UM Sumatra Barat in collaboration with MCCC also distributes basic necessities to the people of the West Sumatra region, and the distribution is submitted to MCCC.

To anticipate the transmission of Covid-19, UM Sumatra Barat also takes a policy so that lectures are carried out during the course. In accordance with the Ministry of Education and Culture Circular No. 4 of 2020 concerning the Implementation of Education Policies in the Covid-19 Emergency Period, that since March 16, 2020 all forms of learning activities in all educational institutions in Indonesia are closed or transferred to homes, including universities. UM Sumatra Barat issued UM Sumatra Barat Chancellor's Circular No.0253/II.3.AU/F/2020 concerning the Policy of the UM Sumatra Barat regarding the Prevention of COVID-19, that all forms of lecture activities are transferred to online or online lectures until the next circulation is issued. This Circular Letter was followed by Circular of the Chancellor of UM Sumatra Barat No.0323/II.3.AU/F/2020, that online lecture activities will be effective until 02 June 2020. This was done in response to the West Sumatra government's protocol, in an effort to avoid direct contact and associations that caused crowds. So that social distancing, physical distancing, and stay at home must be applied, and it has even been upgraded to Large-Scale Social Restrictions (PSBB) in the administrative area of West Sumatra.

This learning distance is conducted through various platforms or the best application features that support the implementation of distance learning (Suparman et al. 2020). Interestingly, UM Sumatra Barat also collaborates with Telkomsel to provide free Internet quota for students, as an effort to reduce student expenses in getting Internet quota to carry out lectures online.

Since the online lecture period has to be extended because Large-Scale Social Restrictions (PSBB) have been enforced in the West Sumatra region, UM Sumatra Barat provides Distance Learning scholarships to assist students in implementing online lectures. This scholarship is given in order to help students buy Internet package quotas during online lectures. This is intended so that students are not overwhelmed by the use of quite a lot of Internet quota during this online college period.

This learning distance scholarship is prioritized for students who come from economically vulnerable or underprivileged families, the requirements for getting the scholarship are accompanied by a Certificate of Disability (SKTM) from the sub-district where the student is domiciled, then all data is inputted online via the link provided by UM Sumatra Barat.

Based on UM Sumatra Barat Chancellor's Circular No.0323/II.3.AU/F/2020 regarding the Policy of the University of Muhammadiyah Sumatra West regarding the Prevention of COVID-19, that all activities of the UM Sumatra Barat academic community at all levels and faculties must be carried out online from their respective homes or WFH, both students, lecturers, employees and other employees. The entire UM Sumatra Barat academic community is also prohibited from traveling abroad in any way, and it is recommended not to travel outside the region. In this Circular, it is also explained that all forms of campus administration services are carried out online, and within a certain time administrative staff shifts are applied. The instructions in this Circular Letter are valid until 02 June 2020.

The UM Sumatra Barat Public Relations Division said that this socialization was carried out by the UM Sumatra Barat in collaboration with Muhammadiyah Peduli using ambulance. Officers who have been appointed to travel around the city of Padang use the car to urge the public to adhere to the rules for staying at home, washing their hands regularly, maintaining a clean and healthy lifestyle, avoiding gatherings or crowds, checking themselves into health facilities when experiencing illness or have a fever, spray disinfectant around each house, and when leaving the house it is expected to comply with the physical distancing appeal, namely maintaining a safe distance from other people in order to protect oneself, family, and others from spreading the corona virus. Through this, a contact number that the public can contact to update information related to Covid-19 is also informed. This socialization is considered very beneficial for the general public, because not all people get the same information from each village they live in.

From Table 13.1, it can also be known that UM Sumatra Barat also provides groceries for people outside the campus. In accordance with the information obtained from the head of public relations he said that the community at the time of Covid-19 is experiencing economic difficulties, so that all parties without exception educational institutions must have attention to them (PR, Interviews 2020). Another informant mentioned that during Covid-19 they have received three times the assistance in the form of food from UM Sumatra Barat (Informant, Interview 2020).

The Covid-19 pandemic has had a wide impact on various aspects of people's lives, especially on the economic aspect. Varona and Gonzales warned of uncertainty when the end of Covid-19 has impacted the economic recession and reduced levels of economic activity (Varona and Gonzales 2021). Reduced economic activity automatically results in a decrease in the level of the community economy.

Table 13.2 Form of scholarship during Covid-19

No	Jenis Beasiswa	Periode
1	Pemotongan SPP	Semester
2	Free semester exam money	Semester
3	Free registration	Registration
4	Internet quota	Covid-19 period

UM West Sumatra also provides scholarships to its students during Covid-19. In accordance with the information obtained, both through documentation studies and interviews, the types of scholarships provided are as in Table 13.2:

From the results of interviews with several informants it is known that UM Sumatra Barat provides tuition waivers to them, as expressed by students of the Faculty of Economics who stated that in normal times they must pay the full 2.500.000,-IDR/semester, but during Covid-19 they are only obliged to pay 1,750,000,-IDR/Semester (Informant, Interview 2020). When confirmed with the vice rector of finance he stated that UM Sumatra Barat realizes that the rights of students during the Covid-19 period are not entirely obtained, therefore the policy that can be made is to reduce the SPP that they must pay (Informant, Interview 2020).

From the existing documents it is also known that UM Sumatra Barat provides tuition waiver for all students of 750,000,-IDR/Semester during Covid-19 (Documentation Study 2020). The results of interviews and documentation studies prove that UM Sumatra Barat makes a real and beneficial contribution to students.

In addition, other scholarships are also given in the form of free exam money, as a private campus that does not get financing from the State, UM Sumatra Barat in normal times requires students to pay the final exam fee of 10,000,-IDR/SKS, but during the Covid-19 period the fee is not required to students. According to the information obtained, this policy was established because UM Sumatra Barat is aware of the economic difficulties experienced by parents of students.

UM Sumatra Barat also provides fee waivers for all prospective students, namely free registration money. In normal conditions, UM Sumatra Barat does not provide access to prospective students before they pay to get a PIN, but in the Covid-19 situation everyone who wants to enroll in UM Sumatra Barat can register on https://pmb.umsb.ac.id/form without having to pay off for free. This data shows the high concern of UM Sumatra Barat to the economic condition of the community due to the impact of Covid-19.

Another participatory in Covid-19 response conducted by UM Sumatra Barat is in the form of instruction to learn and work from home. In accordance with the rector's instructions, it is stated that the learning process is carried out on a during basis. Students and lecturers required to utilize various platforms for learning sustainability, such as google meetings, zoom meetings, http://siak.umsb.ac.id/ and others. This policy is all to avoid the spread of the Covid-19 virus.

In line with the implementation of learning and working from home, UM Sumatra Barat also provides Internet quota to all students, lecturers, and employees. Because without the provision of Internet quota, it is impossible for the policy to run as expected. Therefore, according to the data obtained Internet quota is valid for all human resources at UM Sumatra Barat who are registered handphone number on the UM Sumatra Barat database.

The implementation of the policy of studying and working from home during Covid-19 is also widely carried out by other institutions and universities. In his research Vyas and Butakhio explained that working from home provides new experiences to employees and employers to be able to get the job done well (Vyas and Butakhieo 2020). Completion of work and learning will not be possible to get good results without being supported by sufficient Internet quota. Mursal and friends reminded not a few parents who have to ask neighbors for help for the continuity of their child's education in the time of Covid-19 (Mursal et al. 2021).

13.4 Conclusion

From the explanation above, it can be concluded that there were several policies taken by the UM Sumatra Barat in the face of this Covid-19 pandemic. There are three policies for campus internals such as WFH instruction, online lectures, and PJJ Scholarships. Then four other actions apply to the internal and external environment of the campus or the wider community, namely the production and distribution of hand sanitizers, production and distribution of medical face shields for medical personnel, distribution of compensation and basic necessities, and socialization through operational cars to urge the public to make efforts, i.e. efforts to prevent the transmission of Covid-19. The caring action and real contributions that have been taken by the UM Sumatera Barat together with Muhammadiyah Cares and the Muhammadiyah Command Center can overcome the impact of the Covid-19 pandemic, so that UM Sumatera Barat and Muhammadiyah West Sumatra deserve to be role models in society in general.

References

Abidah A, Hidaayatullaah HN, Simamora RM, Fehabutar D, Mutakinati L (2020) The impact of Covid-19 to Indonesian education and its relation to the philosophy of "Merdeka Belajar.". Stud Philos Sci Educ (SiPoSE) 1(1):38–49. http://scie-journal.com/index.php/SiPoSE

Amin J, Siddiqui AA, Ilyas M, Alshammary F, Alam MK, Rathore HA (2020) Quarantine ad hygienic practices about combating contagious disease like COVID-19 and Islamic perspective. J Crit Rev 7(13):3369–3705. https://doi.org/10.31838/jcr.07.13.564

Berardi A, Perinelli DR, Merchant HA, Bisharat L, Basheti IA, Bonacucina G, Cespi M, Palmieri GF (2020) Hand sanitisers amid CoViD-19: a critical review of alcohol-based products on the market and formulation approaches to respond to increasing demand. Int J Pharm 584 (119431):1–14. https://doi.org/10.1016/j.ijpharm.2020.119431

Chee SY (2020) COVID-19 pandemic: the lived experiences of older adults in aged. Milleninial Asia 11:299–317. https://doi.org/10.1177/0976399620958326

Chen P, Mao L, Nassis GP, Harmer P, Ainsworth BE, Li F (2020) Coronavirus disease (COVID-19): the need to maintain regular physical activity while taking precautions. J Sport Health Sci 9 (2):103–104. https://doi.org/10.1016/j.jshs.2020.02.001

Chua MH, Cheng W, Goh SS, Kong J, Li B, Lim JYC, Mao L, Wang S, Xue K, Yang L, Ye E, Zhang K, Cheong WCD, Tan BH, Li Z, Tan BH, Loh XJ (2020) Review article face masks in the new COVID-19 normal: materials, testing, and perspectives. AAAS Res 2020:1–40. https://doi.org/10.34133/2020/7286735

Dwirusman CG (2020) Peran Dan Efektivitas Masker Dalam Pencegahan Penularan Corona Virus Desease 2019 (Covid-19). JMH Jurnal Medika Hutama 02(01):412–420

Inayah I, Surisman S (2020) Work termination during the Covid-19 pandemic in the perspective of positive law in Indonesia. Legal Standing Jurnal Ilmu Hukum 4(1):247–254

Jing JLJ, Yi TP, Bose RJC, McCarthy JR, Tharmalingam N, Madheswaran T (2020) Hand sanitizers: a review on formulation aspects, adverse effects, and regulations. Int J Environ Res Public Health 17(3326):1–17. https://doi.org/10.3390/ijerph17093326

Matuschek C, Moll F, Fangerau H, Fischer JC, Zänker K, van Griensven M, Schneider M, Kindgen-Milles D, Knoefel WT, Lichtenberg A, Tamaskovics B, Djiepmo-Njanang FJ, Budach W, Corradini S, Häussinger D, Feldt T, Jensen B, Pelka R, Orth K et al (2020) Face masks: benefits and risks during the COVID-19 crisis. Eur J Med Res 25(32):1–8. https://doi.org/10.1186/s40001-020-00430-5

Mursal M, Ritonga M, Sartika F, Lahmi A, Nurdianto T, Alam L (2021) The contribution of Amil Zakat, Infaq and Shadaqah Muhammadiyah (LAZISMU) institutions in handling the impact of Covid-19. J Sustain Finance Invest 0(0):1–7. https://doi.org/10.1080/20430795.2021.1886550

Ozili P, Arun T (2020) Spillover of COVID-19: impact on the global economy. JEL Classification 21(March):1–27. https://doi.org/10.2139/ssrn.3562570

Rahma VS, Arvianti GF (2020) The impacts of covid-19 pandemic in Indonesia and china's hotel industry: how to overcome it? Jelajah J Tour Hosp 2(1):55–64

Ritonga AW, Ritonga M, Nurdianto T, Kustati M, Lahmi A (2020a) E-Learning process of Maharah Qira'ah in higher education during the Covid-19 pandemic. Int J High Educ 9 (6):227–235. https://doi.org/10.5430/ijhe.v9n6p227

Ritonga M, Lahmi A, Bambang A, Firdaus S, Asmaret D, Afdhal S (2020b) Curriculum development strategy management for student mental health in Covid-19 pandemic. Int J Pharm Res 12 (2):4489–4494. https://doi.org/10.31838/ijpr/2020.SP2.562

Rizki SA, Kurniawan A (2020) Efficacy of cloth face mask in reducing COVID-19 transmission: a literature review. Natl Public Health J 15(1):43–48. https://doi.org/10.21109/kesmas.v15i2.3893

Roy S (2020) Economic impact of Covid-19 pandemic. A Preprint, July, 0–19

Rusydi R (2016) Peran Muhammadiyah (Konsep Pendidikan, Usaha-usaha di Bidang Pendidikan, dan Tokoh). Jurnal Tarbawi 1(2):139–148. https://doi.org/10.26618/jtw.v1i2.367

Suparman UIHT, Suwadi PA, Andriyani HES, Hairun Y, Ritonga M (2020) The implementation of community partnership program to improve the quality of online learning during the Covid-19 pandemic. Univers J Educ Res 8(11B):6134–6138. https://doi.org/10.13189/ujer.2020.082249

Susilawati S, Falefi R, Purwoko A (2020) Impact of Covid-19's pandemic on the economy of Indonesia. BIRCI J 3(2):1147–1156. https://doi.org/10.33258/birci.v3i2.954

Suyadi S, Nuryana Z, Fauzi NAF (2020) The Fiqh of disaster: the mitigation of covid-19 in the perspective of Islamic education-neuroscience. Int J Disaster Risk Reduc 51:101848. https://doi.org/10.1016/j.ijdrr.2020.101848

Syahputra H (2014) Peran Politik Muhammadiyah Tahun 2010-2014. Jurnal Ilmu Pemerintahan Universitas Brawijaya Malang, September 1–30

Varona L, Gonzales JR (2021) Dynamics of the impact of COVID-19 on the economic activity of Peru. PLoS One 16(1 January):1–30. https://doi.org/10.1371/journal.pone.0244920

Vyas L, Butakhieo N (2020) The impact of working from home during COVID-19 on work and life domains: an exploratory study on Hong Kong. Policy Design Practice 4(1):1–18. https://doi.org/10.1080/25741292.2020.1863560

Worby CJ, Chang H-H (2020) Face mask use in the general population and optimal resource allocation during the COVID-19 pandemic. Nat Commun 11(101):1–9. https://doi.org/10.1038/s41467-020-17922-x

Zhang B, Zhou X, Qiu Y, Song Y, Feng F, Feng J, Song Q, Jia Q, Wang J (2020) Clinical characteristics of 82 cases of death from COVID-19. PLoS One 15(7):1–13. https://doi.org/10.1371/journal.pone.0235458

Chapter 14
Effects on Mental Health by the Coronavirus Disease 2019 (COVID-19) Pandemic Outbreak

Sanjay Kumar Rout and Duryodhan Sahu

Abstract Coronavirus disease 2019 (COVID-19) contagion contributes to a substantial amount of psychological concerns and fitness complications, which has been affecting the eco-systems globally. The erratic outbreak of COVID-19 has the prospective of harmfully upsetting mental health on distinct and public altitudes. Presently all exertions are concentrated on the knowledge of medical landscapes, sanitation, approach of diffusion, neutralizing the banquet of the virus, and confrontation of universal health sustainability, whereas critically important mental health has been ignored in this effort. COVID-19 promotes severe psychosocial concerns and covering mental health shapes a subordinate health anxiety all over the domain. There is growing anxiety around the cerebral health contests of the health specialists, elderly people, students, overall inhabitants, COVID-19-infested survivors, and near associates. This chapter is to assess previous occurrences to apprehend the level of hostile properties on mental fitness, mental disaster interference, and psychological health supervision tactics.

Keywords COVID-19 · Contagion · Resilience · Stigma · Coping

14.1 Introduction

The present-day outburst of coronavirus disease 2019, generally termed as COVID-19, is posturing a critical warning to the health system of the worldwide community and causing an environment of fear, societal disorder, establishment downfall,

S. K. Rout (✉)
Department of Chemistry, Konark Institute of Science and Technology, Bhubaneswar, Odisha, India

D. Sahu
Department of Chemistry, National Institute of Science and Technology, Berhampur, Odisha, India
e-mail: duryosahu@nist.edu

S. A. Hassan et al. (eds.), *Decision Sciences for COVID-19*, International Series in Operations Research & Management Science 320,
https://doi.org/10.1007/978-3-030-87019-5_14

technical doubt, and consequently, these continuously have produced an unadorned tragedy to the society and ecosystem. World Health Organization (WHO) reported that in excess of 220 countries, areas, or territories with an overall established diseased numbers of 135,057,587 and deceased 2,919,932 numbers owed to coronavirus as on April 11, 2021. In the present scenario, COVID-19 is characterized as both an epidemic and pandemic. However, people have generally named the phenomenon as an outbreak. The word "endemic" denotes the continuous existence along with the traditional occurrence of a disease or infective agent within residents in the inner of a topographical space. Pandemic indicates to the epidemic that has blown out across nations, states, and the mainland—the worldwide spread of the new infection. It was conventional that COVID-19 was prompted by SARS-CoV-2 known as Severe Acute Respiratory Syndrome Corona Virus 2, which was originally acknowledged in the last part of the year 2019, in China at Wuhan, but it was declared as a pandemic by WHO in 2020 (Zhu et al. 2020). The pandemic is an epidemic and a contamination that has blown out over a massive range, that is, its "widespread over the whole country, landmass, or the whole flora and fauna, or the ecosystem." An epidemic "happening universally, or else in excess of a very extensive area, inters sectioning the intercontinental borders which usually disturb a huge number of individuals" but an outbreak is a "sudden contravention out or occurrence" or "outburst." As soon as denoting a contagious disease, an outbreak is obviously a quick increase in numbers, particularly when it is solitary or up to now disturbing, comparatively confined to a small area.

At present, there is a topmost health disaster happening all over the region of the flora and fauna as COVID-19 is upsetting the societies and universally accepting it as a pandemic. In excess of the previous few months, there is a substantial escalation in transience and ill health effects due to COVID-19 contagion. Along with this, it initiates vigilant measures for instigating protective besides adjusting processes, then nurturing deal with desirable flexibility are stimulating topographies; misrepresentation along with untruth of the beginning, measure, transmission, indications, signs, inhibition with management; universal social and economic disaster along with factors affecting it. Sports, workshop vulnerability controller; travel restrictions; a typical lifestyle due to self-isolation, lockdown time limit, quarantine and social distancing; conspiracy philosophies, suspension and invalidation of pious, social and show biz occasions; anxiety buying and advertisement; occurrences of prejudice, racial intolerance, perception, disgrace, emotional stress output, sidelining and forcefulness; overawed curative hubs and health establishments, and over-all effect on teaching, culture, policies, nature of economy in the society, surroundings and environment—remain as the number of the hazard features that can intensify and promote problems.

At the existing time, the most important points put by the WHO are inhibiting the enlargement of technical hitches in diseased individuals, regulating the blowout of contamination to near associates and curative authorities, hindrance of human-to-human transmission, loneliness and solitary confinement ability establishment, availing investigative and test site facility, exploration to yield precise cure besides injection and diminishing the social and economic influence on the civic. It was observed in excess of the previous few months that in the course of this outbreak of COVID-19 contamination, the total number of mental health issues has increased amongst the overall inhabitants, healthcare professionals, seniors, college students, refugees, manual workers, and patients with COVID-19 infection. Up to now, there are no precise endorsements from international bodies on the subject of talking about the mental fitness topics in due course of this COVID-19 infection. Investigation has revealed that rapid occurrence can deteriorate the state of affairs of the mental fitness in persons having previous health-related sickness (Ho et al. 2020). The stocks and movement of financial as well as physical resources are broken up. A lowering of overall gross domestic product with subsequent increase in health budget is assured to influence the economy of the whole world. Similarly, travel and tourism are also affected by COVID-19. Finally, the pandemic will have influence in all provinces of the existing flora and fauna, preliminary from economy, the general public, health and, finally influence the forthcoming rule constructing universal, local, and kingdom side by side.

14.1.1 History and Indications of Coronavirus

At present, the names of 40 different coronaviruses are approved according to their characteristics by the worldwide Committee, consisting of a large number of countries. Total numbers of known coronaviruses together with the current COVID-19 have touched seven, and all of these bugs appear to upset the wildlife. Out of seven bugs, four bugs are picked up by the public, and spread over the inhabitants persistently for a long era of period. Conversely, Middle East Respiratory Syndrome Coronavirus, known as MERS-CoV. Similarly, Severe Acute Respiratory Syndrome Coronavirus commonly known as SARS-CoV and together with SARS-CoV-2 are recent epidemics that are related through great transience rates. It is informed that, along with these three bugs, acute lung injury and acute respiratory distress syndrome also triggered by H5N1 influenza A and H1N1 2009 totally resulting in breathing failure and eventual fatalness (Shereen et al. 2020). It is a fact that each corona bug is generated in wildlife and pass on to humans through transformation, rejoining, and alteration. The virus can survive in snow as well as cold and can be communicated to zones of humid and hot environments. All groups of people are prone to get infected with COVID-19 regardless of their stage of development and time of life. Senior people with causal health illnesses such as chronic bronchitis, diabetes mellitus, and cardiovascular disorder were very susceptible; similarly, children are vulnerable to the infection. It was reported that bats are the primary cause of COVID-19, but the transitional cause before diffusion to humans is not yet recognized (Shereen et al. 2020). Researches demonstrate that public may gain the corona bug with the help of air and after coming in contact with the infected substances. Inventors' reported that the bug is identified for about 3 h in aerosol, remains for 1 day on cardboard, is able to remain for 4 h on copper, and is equipped with 2–3 days on stainless steel and plastic (Van Doremalen et al. 2020). The high contamination level, recombination level, and transformation level of these animal corona bugs increase the possibility of mutations aptitude to communicate into one more entertainer. It is anticipated that the harshness of infection considerably becomes fast between the new entertainer and the corona bug at the new round of alteration. COVID-19 causes breathing territory contaminations in human beings which may possibly result in general cold. Different persons are affected by COVID-19 viruses in altered techniques. Maximum sick persons grow minor to adequate signs due to respiratory disease arising from COVID-19 and recover without needing special treatment. The indications of this disease include dry cough, tiredness, respiratory illness, fever, and pneumonia. Less common symptoms include: sore throat, littleness of inhalation, dull pain and troubles, and very scarce persons describe diarrhea, vomiting, or an excess nasal drainage (World Health Organization n.d.).

14.1.2 COVID-19: Protection and Transferring Behavior

In order to protect our body from COVID-19 and considering public health perspective, we use different scientific masks; maintain social distancing, and self-quarantine. Bugs are often diffused through respiratory dewdrops created by sternutation and coughing. Different scientific fasciae are fabricated to eradicate maximum amounts (95%) of these smallest pieces of matter. Separate effectiveness of these technical fasciae shows that they are capable of separating almost all smallest pieces of matter. So, these fasciae are proficient in separating most free virions. Therefore, different fasciae likely protect against several modes of viral transferring.

Person, contaminated with COVID-19 infection was reported by the WHO, at last part of December 2019 and considers as the first cases of COVID-19. The first case was related to straight contact to the Huanan Seafood Wholesale Marketplace of Wuhan, China, so it was supposed that the animal-to-human diffusion was the key device of transmission but related successive cases were not observed with this contact device. In this way, it was predicted that the bug could also be diffused from human to human, and suggestive people are the most recurrent cause of COVID-19 blowout. The probability of diffusion before indications grow appears to be rare, even though it cannot be omitted. Additionally, there are proposals that persons who keep on asymptomatic could communicate the bug. This result recommends that the use of separation is the finest mode to get rid of this rampant (Features, Evaluation, and Treatment of Coronavirus n.d.). According to the available data from the research conducted by the CDC known as the Chinese Centre for Disease Control and Prevention, resident CDCs, and from the first COVID-19 personal belongings in Wuhan, the nurture time could be usually inside 3–7 days and equal to 2 weeks as the lengthiest time from contagion to indications was 12.5 days (Li et al. 2020). The time too indicated that at every 7 days, this epidemic doubled up, while the basic reproduction number (R_0) is 2.2. On the other hand, on average, every survivor transfers the contagion to an extra 2.2 persons. It is imperative that approximations of the R_0 of the SARS-CoV epidemic in 2002–2003 were in the region of three (Bauch et al. 2005). As reported earlier, the sources of the respiratory infection were SARS-CoV-2. In Johns Hopkins, a new analysis was conducted in the School of Public Health (SPH) and was supervised in the same organization by different scientists and investigators at Bloomberg. The analysis was based on the publicly obtainable statistics and information on contaminations from the SARS-CoV-2. Finally, it was established that SARS-CoV-2 was accountable for COVID-19 and was created in the median infection incubation time approximately of about 5.1 days (Lauer et al. 2020). Considering this augmented median time, 14 days should be the quarantine time for infected people. In the USA, CDCP popularly known as Centres for Disease Control and Prevention executed 14 days as the quarantine time for persons who will be expected to come in contact with the coronavirus. The exploration also endorses that approximately 97.5% of individuals having mature with the indications of SARS-CoV-2 contamination will organize subsequently within 11.5 days of contact. Social distancing strategies are required for community health point of view but, social distancing can have adverse effects on financial condition. In order to save human lives and to keep the environment justifiable medical treatments and hospitalization is also necessary as per the requirements to control the spread of respiratory illness.

Individuals who have causal medical conditions and senior persons have a higher jeopardy of emerging severe sickness and expiry. The ecosphere administrations are working at full peak with the survival of health and well-being of civilizations. Different health organizations are building all probable hard work to institute therapy to decrease the catastrophic belongings of COVID-19. Since there is no perfect treatment for the control of COVID-19, the researchers and scientists have been constructing drastic hard work to find procedures for remedy or for at least decreasing the sickness due to COVID-19 (Zhou et al. 2020). Therefore, appropriate medical management of COVID-19 survivors and proper supervision will reduce the shattering effects of COVID-19.

14.1.3 COVID-19 and Mental Health

Conferring to cure regulatory principle, COVID-19 infected individuals should be cured in isolated contagious clinics. In this way, due to societal separation, apparent jeopardy, ambiguity, bodily uneasiness, medicine sideways effects, terror of diffusion of the bug to others, and devastating demonstration of harmful news bulletin in mass broadcasting, persons with COVID-19 in these environments can experience aloneness, irritation, apprehension, misery, restlessness, and indications of trauma pressure that could adversely disturb the societal and professional working of persons and the worth of existence (Bo et al. 2021). Additionally, in this situation, mental/emotional stress, principally unintended upset instigated by the COVID-19 contagion, should not be overlooked, all of which, totally, powerfully affects widespread well-being and mental fitness (Li et al. 2020; Bauch et al. 2005). So, the COVID-19 widespread is a public fitness disaster of universal worry and postures an encounter to emotional or mental flexibility (Dong and Bouey 2020). With the growing figure of infested cases and demises, numerous survivors face corporeal sorrow and excess cerebral suffering (Bo et al. 2021). The recent fitness crisis circumstances and state of apprehension resulting from the contagion affirmed by the COVID-19 occurrence is considerably impelling the sensitive state of the whole inhabitants consequently leading to vigilant actions, terror, and anxiety of the entire societies. Even though awareness can help to raise individual and assembly managing mechanisms, however, if the fear is extreme, it creates adverse mental responses, not very active on an individual and societal extent, which can activate apprehension disorders and downheartedness, between others (SESCAM 2020).

At present, there is a lack of evidence on the mental otherwise emotional effect of the overall civic regarding suspected with confirmed COVID-19 numbers. A similar phenomenon also rises particularly in the framework of mystical fitness effect of curative workforce and rule implementation mediators throughout the outbreak of COVID-19 contagion. The phenomenon has turn out to be even more significant resulting in unpredictability with uncertainty turning round the occurrence of corona bug contagion through supreme extent with passion. Intrigue concepts, untrue entitlements, misrepresentation along with deception which arises due to unbreakable, unstoppable, and unbeatable nature of coronavirus, subsequently these are aggravating the mental calmness of civilization and creating intense environmental effect among citizens. COVID-19 occurrence based many enquiries emphasis on detecting the hygienic, scientific landscapes, quantifiable topographies, statistics on tactic of broadcast and its path, genomic characterization of the virus, nurture time, reservoirs, indications and clinical results, neutralizing the spread of the virus along with its stabilization; including endurance and mortality rates; and supervision of universal health control (Chen et al. 2020; Corman et al. 2020; Huang et al. 2020; Lu et al. 2020; Mukhtar 2020). The enthralling disaster encourages the wide-ranging depth of investigation on psychosomatic fitness and mental well-being of the civic in the fascia of COVID-19.

It was detected that there exists a neuropsychiatric connection among SARS (Severe Acute Respiratory Syndrome) and psychological fitness complications with acute psychotherapy comorbidities identical to anxiety attacks, hopelessness, fear occurrences, suicidal deaths, psychomotor enthusiasm, confusion, and psychotic indications (Xiang et al. 2020). Also, the survival and day-today activities influenced by COVID-19 are at further stick owing to the prolonged perspective of hostile belongings. For illustration, through tour constraints besides suspension and invalidation of game, spiritual activities, social and show biz events, persons in isolation can experience loneliness, irritation, dullness and nervousness, and signs of cough, myalgia, fever, and exhaustion can cause emotive grief with a mental state of fear in shrinking COVID-19 (Xiang et al. 2020). Despite the fact that clinicians, scientists, homegrown and worldwide health establishments, experts and community health experts are engaging to solve numerous unreciprocated queries of this unusual occurrence, overall community, universal mass broadcasting and influential personalities are answering this doubt created on a partial established/unverified awareness.

This has also risen up the significances of the survival of the people in the rouse of COVID-19 and calls for the new databank of inquiries on mental fitness. However, the influence of spread of disease-causing infection on mental or else has not been accepted in its wholeness, which contests the survivors, the overall inhabitants, and to the civilization.

Investigation leaflets and theoretical features are needed to improve tactics to decrease adverse mental or else psychological effects and psychiatric indications throughout the prevalent so the studies were further studied and partitioned for the applicability of literature appraisal. The present interpretation concentrates on the existing COVID-19 contagion mental well-being apprehensions or else psychological interferences for affected persons, elderly, students, unprotected inhabitants, and health maintenance workforce.

14.2 Effects on Mental Health of Civilizations by Coronavirus Pandemic

14.2.1 Annotation, Readiness, and Initiative for Controlling the Infection

Health care professionals of different health organizations were concentrating on regulating the COVID-19 contagion by endorsing public loneliness, separation in addition to self-isolation along with the minor prominence on the effect of intellectual fitness (Coronavirus disease (COVID-2019) situation reports 2020). The occurrence, dominance, and conduction of COVID-19 are away from intellectual well-being, and sensitive agony, fear, nervousness, civic dishonor, discrimination, racial intolerance, discernment, delayed-stress disorder signs, and insomnia are some of the concerns on psychosomatic well-being. The bodily corrective processes for handling COVID-19 comprise of prompt detection with a good-bye to the doubted persons, collection of biotic and medical information, consent for expert curative participations, foundation of isolation units, and firming of curative workforce in the simulate areas (Ford-Jones and Chaufan 2017; Report of the WHO-China joint mission on coronavirus disease 2019 (COVID-19) 2020; Severance et al. 2011). These infections along with inhibition and its regulator perform intrusions, which have been established to be active for struggling the contagion but have grave psychosomatic health influences on the curative crews and in the overall community (Rubin and Wessely 2020). The aim and importance of fitness maintenance establishments through the universe are to confirm readiness and to take initiative for regulating the infection through the fitness maintenance system and to deliver the essential support for life survival along with supply of some materials for the benefit of public health. Cooperative work along with pharmaceutical establishments and public health division; availability of PPE (personal protective equipment) with good medical facilities; and recognized health care units on inhibition, supervision and to regulate the contagion together with little care on intellectual fitness are beneficial for the society to escape from the pandemic. Availability of disaster money and economical supervision instruction for community indication about COVID-19 plus regulating measures along with the availability of resources determine and depend on the readiness to control the COVID-19 pandemic.

14.2.2 Mental Health Influence on Survivors, Fitness Maintenance Workforce, Rule Implementation Representatives, Children and Overall Civic

The mental/psychosomatic and instant anxiety consequences on how sufferer was evaluated in previous studies and states that the sufferer who was isolated all through MERS (Middle East Respiratory Syndrome) exhibited greater influence occasions mark on anxiety, lack of feeling, sleep, depression, and the subsequent effects of psychosomatic grief, which were steady with exaggerated persons in Nigeria on the impact of Ebola bug (Lee et al. 2018; Mohammed et al. 2015). Supplementary studies describing the influence of psychosomatic distress of remembrance in the case of MERS indicated that enduring entities were branded, freeze out and generally inaccessible even afterward fruitful cure (Shigemura et al. 2020; Sim 2016). Elongated time of nurture lengthier than normal owing to community ambiguity is the consequence of penetrated disinformation along with misrepresentation and consequently creates an impact on social broadcasting or mass newspapers. The confined nature of COVID-19 clasps resemblance to that of SARS besides MERS as analogous rights are flowing on communal mass broadcasting regarding the insecurity and changeability of COVID-19, prompting terror, fright, nervousness, and apprehensions in common civic. Further analysis (Batawi et al. 2019; Cheung et al. 2008) on psychologist influence fighters of SARS discovered a new type of PTSD (Posttraumatic stress disorder) known as an extremely delayed-stress disorder, a new indication that aggravated brutal signs. This interrelated to the recklessness demises of grown person and little worth of lifetime among afflicted persons and unspoilt general public, respectively (Mak et al. 2009). Earlier reports on the transferrable occurrences of MERS along with SARS and also Ebola publicized the harshness of sensitive regret not only in the overall civic but also between numerous curative general practitioner and law execution representatives who handled nervousness, PTSD, overtiredness, melancholy and wear out at the beginning, throughout and even afterward the occurrence of such epidemics (Lee et al. 2018).

The occasion of COVID-19 postures extra substantial emotional fitness deteriorating since remedial consultants and nurses are in the same way vulnerable to the contagious communicating infection owing to insufficient PPE or personal protective equipment, frustration, poor health, hopelessness, tiredness, isolation, discernment, sufferers with adverse sentiments and deficiency of interaction with their relatives (Kang et al. 2020). This universal communal well-being worries the character and accountability of curative personnel, universal influence of contagion, effect of fiscal accomplishments on movement and exchange limitations along with justifiable care of communal prosperity and discrete privileges throughout the escalation of contagions. Mental fitness effects can be reduced by evading extreme interaction with COVID-19 mass media exposure (a widespread countrywide activity, particularly binge-watching news networks), sustaining optimistic standard of living, and compassionately calming others as well. Flexibility in working out courses for law enforcement agents, curative team, and overall civic to manage the repercussions of the contagion of this harshness and strength have to be presented through the following points: (i) Enlightening and making group of people for contagions and epidemics in the forthcoming; (ii) Household effort and life span equilibrium; (iii) Dependable, reliable and well-timed merged evidence about the contagious infection and its significances on psychological well-being; and (iv) Authenticating with respecting frontline's personnel contribution.

It is a fact that during COVID-19 children come to be physically less dynamic and have got much time to see television, laptop, mobile, and other screen activities. During this period, children also take less favorable food resulting in irregular sleep patterns. All these activities lead to a gain of weight and a loss of cardiorespiratory appropriateness of the children. Childcare responsibilities play an important role in healthcare professions. But, it is a matter of regret that due to unplanned childcare responsibilities closures of the schools come into picture either directly or indirectly. Due to these in most of the countries of the universe child care affected and which is related to the current situation. Subsequently, the mental health of the children are also affected.

14.2.3 Mental Health Impact on Elderly

Amongst the maximum susceptible and hazard assemblies throughout contamination are the older people (Girdhar et al. 2020; Meng et al. 2020), because they frequently have related to many diseases. The occurrence of total overexcited tension in persons who completed 60 years of age rank among 45.5 and 63.1%, heart disaster books for 3.8%, and Chronic Obstructive Pulmonary Disease (COPD) is found in 23.7% (Danielsen et al. 2017; Shahid et al. 2020), and that of diabetes mellitus is about 16.8 and 26.8% (Danielsen et al. 2017; Shahid et al. 2020). These illnesses can possibly upset the prospects of sufferers with COVID-19, as they possibly will injury to muscular constructions, reduced immunity, and causes impaired lung function (Zheng et al. 2020). Correspondingly, the senior obviously have a comparatively less operative resistant organism than early persons and are extra vulnerable to emerging serious sicknesses (Meng et al. 2020; Zheng et al. 2020). So, the senior inhabitants can be measured at high jeopardy of sickness headway and demise in COVID-19. Thus, the precise idea of susceptibility, preceding comorbidities such as heart disaster, and COPD that rise the jeopardy of worry syndromes, depression, and practical restrictions instigated by these comorbidities can be noteworthy stressors for the psychological suffering of these citizens.

Lack of communal isolation and interaction activities throughout the COVID-19 epidemic aggravate psychological complaints which increase the jeopardy of misery and apprehension in the senior inhabitants (Reynolds et al. 2015; Armitage and Nellums 2020). It is reported that approximately 37.1% of the senior inhabitants had proficient misery and apprehension all through the contagion (Meng et al. 2020). Along with separation, stress and fear also donate to the commencement and intensification of preexistent mental health complaints. Persons having obsessive compulsion complaints have greater probabilities of undergoing preoccupied views owing to cautionary actions (Haider et al. 2020). From the data of the cross-references analysis piloted in China, it is established that around 20% show depressive symptoms and 33% of people display anxiety complaints. Still, it claimed that these statistics have to be lesser in more progressive age assemblies (Huang and Zhao 2020).

A forceful projecting feature for intrapersonal comorbidity is insanity, which is general in higher age bar (Brown et al. 2020a; Wang et al. 2020). Subjects with insanity and mental deficiency have inadequate entree to exact evidence and particulars about the contagion (Wang et al. 2020). Also, they could not properly monitor the agreements to decline the extent of COVID-19, because they cannot able to recall processes and vital evidence (Wang et al. 2020; Banerjee 2020). Distinct properties of communal separation are also simulated in individuals with insanity owing to removal from vital non-pharmaceutical treatments to cure comorbidity, like physical exercises, public events, and assembly healings (Brown et al. 2020a). Potential distress originating from these alterations can further quicken cerebral deterioration. As per the themes with insanity are extra possible to have cardiac ailment with diabetes (Brown et al. 2020a), it can be supposed that this assembly is at an even advanced jeopardy of ill health and even death from COVID-19.

Derangement desires exceptional care like that of dementia. Communal separation procedures can rise psychogenic sufferers' anxiety, just like safety measures linked to the blowout of disease and have been related with higher neurotic (Brown et al. 2020b). The surplus evidence can also exaggerate neurotic signs, producing uncertainties in concerning well-being (Banerjee 2020). In this case, sufferers with derangement are less inspired to fulfill the suggested trials (Brown et al. 2020b), overriding them to circumvent the communal separation and isolation processes (Banerjee 2020). Outcomes indicate that COVID-19 has been accompanying by a 25% rise in the occurrence of sick rashes circumstances (Brown et al. 2020b; Hu et al. 2020). In the senior, there has also been an upsurge in the jeopardy of hypomania, as the average age for sufferers freshly detected with hypomania altered from 39–50 years (Hu et al. 2020). The harshness of warning signs and steroid amalgamation seem to subsidize the beginning of sick indications (Turner et al. 2018). In the same way, there are information of topical derangement in infested entities, and SARS-CoV-2 may have a neuropathogenic device that would activate these warning sign (Brown et al. 2020b; Ferrando et al. 2020).

14.2.3.1 Suggested Proposals to Decrease the Jeopardy of Mental Health Influence and Thoughtful Illnesses in the Senior

Due to the crisis triggered by the COVID-19 pandemic, preventive measures and intervention must be executed to alleviate and decrease the jeopardy of mental or else emotional influence and thoughtful disarrays in the senior (Meng et al. 2020; Armitage and Nellums 2020; World Health Organization 2017; Druss 2020), which are discussed briefly as follow:

(a) Using a subsection of telemarketing "telehealth" as a variety of facilities comprising thoughtful evaluations, therapy, survivor instruction, and treatment supervision.
(b) Telehealth too involved in straight contact among a doctor and the survivor. Hence, elderly people with mild/moderate psychiatric disorders benefitted from telepsychiatry.
(c) The elderly and their family members take the help of expanded telehealth services to reply queries about different signs, instituting connection to observer entree/treatment supervision and recommend nonphysical complementary remedy (Girdhar et al. 2020; Armitage and Nellums 2020).
(d) Fitness specialists on the basis of their previous skills prepared the training materials to be suitable for them, to deliver safekeeping and act as multiple of noble cerebral fitness performing best in the contagion (Girdhar et al. 2020; Hu et al. 2020).
(e) Proposing educational materials and advertisement to make individuals conscious of the want to work together, safekeeping and respect their senior relations, the necessity to preserve steady interaction through online or over the mobile (Banerjee 2020) throughout the contagion, and fitness advancement actions to contest COVID-19 and cerebral fitness complaints (Meng et al. 2020).
(f) Familiarizing communal safety procedures to contest the fiscal eradication of these persons (Druss 2020; Van Hal 2015).

14.2.4 Effects of Quarantine and Social Distancing on Mental Health Impact

Persons who have come hypothetically in connection with the contagion asked them to remain as separate at house or in an isolation facility by many countries during the outbreak of COVID-19. Among older adults, the phenomena of social loneliness or isolation assists as a severe public health problem owing to their bigger jeopardy of neurocognitive, auto-igniting, cardiovascular, and cerebral health harms; communal loneliness creates a higher risk of depression and anxiety for older adults (Armitage and Nellums 2020). Quarantine and isolation are the two extreme forms of social estrangement and are the cause of depression besides anxiety respectively in current pandemics. On the basis of these types of effects people are separated from their dear ones, are lacking individual autonomies, and are aimless owing to changed means of support. This can donate toward the stumpy frame of mind, frustration, dullness, and potentially depression (Venkatesh and Edirappuli 2020). The execution of unprecedented stringent quarantine measures of the COVID-19 contagion has reserved a huge figure of persons in loneliness and exaggerated several features of people's existence. This condition has initiated widespread emotional difficulties, such as depression and anxiety complaints. New investigators have described adverse emotional outcomes, counting confusion, indications of complete exhaustion distress, and irritation. The aggravations which are generally incorporated in the enlarged isolation period are dissatisfaction, financial losses, dullness, insufficient goods, and fears of infection, inadequate information, and stigma. Several researchers have proposed long-term consequences of separation, and also after 3 years of loneliness, incidents of exhaustion strain have been stated (Vidal 2020; Brooks et al. 2020). In several circumstances, the returns to societal separation and confinement are very alarm, which causes fear and consequently proceeding through unsteadiness in the face of public leading to an undistinguishable future. In any circumstance, it appears that quarantine will yield a psychological peal on the whole occupants, irrespective of their previous cerebral health position (SESCAM 2020).

14.2.4.1 Suggested Proposals to Decrease the Effects of Quarantine and Social Distancing

In order to decrease the effects of quarantine and social distancing some suggested proposals should be implemented as given below.

(a) During the said phase, it is vital to stay in connection with the health care department, social media, etc., to grasp the support when we need it, even as we cut back on in-person socializing. So, priority should be given to stay in touch with family members, friends and to think about scheduling steady chat, phone, or Zoom dates to counteract the serious state.
(b) Social media can be also an influential device—not only for connecting us with relatives, and associates—but for our feelings connected in a greater sense to the society, nation, and the flora and fauna. It recapitulates us we are not alone.
(c) During the period of isolation, in-person visits are restricted, but, face-to-face communication through video chatting, works like a "vitamin" for mental health, reducing the jeopardy of depression and facilitating ease of stress and anxiety.

14.2.5 Mental Health Impact on Students

Enlarged intensities of somatization, anxiety, misery, and stress were considerably associated with academic complications and describe the capability to concentrate on academic work. It is similarly vital to report that, to the overrate extent, all these practices of cerebral fitness liability were considerably accompanying with the problems of electronically connected (online) education. But, as soon as the consequences of other clarities were taken into accounts, trouble with online/electronically connected learning was in no way a weighty forecaster of mental fitness problems. Reduced wages, reduced work hours, and loss of job also significantly associated with a higher level of mental depression. Even though problems in finding drug and sanitation deliveries were considerably connected with enlarged anxiety ranks. School students who rely on occupations to backing them and/or relatives may be principally susceptible to sadness and anxiety owing to loss of jobs and subsequent financial poverty. As doubts round the upcoming carry on this may result in deteriorating mental fitness position, predominantly between early entities (Ozamiz-Etxebarria et al. 2020). The present contagion has lifted undergraduates' precedence and several are bothered regarding the health of their families, fitness of their relatives, their own health, or struggling financially, possibly they become less attentive on education, and facing growing educational complications. A sudden movement of old-style confrontational teaching to the electronically connected (online) style can be mostly hard for pupils who are used to attain the offline courses and to persons from intellectually peripheral assemblies of undergraduate pupil (Xu and Jaggars 2014), or else the academic-based syllabus which should be suited not satisfactorily for online education (Jaggars 2014). Academic activities through electronically connected teaching may also worsen emotional fitness suffering between undergraduates. All the augments can have suggestions for scholars' retaining frequency, good-bye institution bureaucrats, and mentors along the assignment to build inventive elucidations to the unpredicted encounter through the contagion of the pandemic. Academy backing amenities can have a critical character in serving schoolchildren and circumnavigate the lifespan contests accompanying with the contagion may benefit to develop their mental fitness.

14.2.5.1 Suggested Proposals to Decrease the Effects of Mental Health Impact on Students

In order to decrease the effects of mental health impact on students some suggested proposals should be implemented as given below:

(a) Noble interactions and good social networks along with religious affiliation, appear to have a protective influence against mental health problems of students.
(b) Psychiatric disorder, is widely customary in the student inhabitants, and this may have an important influence on academic performance. Monetary pressures and academic concerns are consistently identified as important contributors to mental health symptoms. Counselors, working in education organizations, are well-known for their understanding of the connections between psychological, monetary, and academic difficulties. It is usual for them to create awareness of how to minimize psychiatric disorders, monetary pressures, and academic difficulties among students.

(c) It is habitual, for counselors, to offer consultation to staff apprehensive about students, workshops for students and staff, online information about how to help with study and mental health difficulties, printed materials to monitor students and staff in their response to students in agony, collaboration with others with responsibility for mental well-being within their own organizations, training (including suicide consciousness) for students and staff, a series of therapeutic work to students and to staff, contributions to institutional policy-making on mental health problems respectively.
(d) The low degrees of treatment acceptance, by the students with psychological health issues, are highlighted by numerous studies. There is a requirement to identify the social, cultural, and demographic correlations of treatment to access and to consider what steps could be taken to eliminate this problem.

14.2.6 Mental Pressure of Productivity and Stigma

The general civic have a permeating impression about the lockdown as per a huge assembly of individuals expected it as leaves or breaks and compelled others for ideal working in using their period and powerfully engaging in job-related or educational accomplishments. Quotation marks circulated in general mass media for instance 'In this isolation if you don't come out is a novel ability, your secondary occupation in progress, or extra facts enlarged . . . then you on no occasion want for period, you be deficient in discipline' are more cooperating the cerebral fitness of persons and civilization. Communications such as for example 'if you don't appear at this isolation with a new talent, your secondary occupation begin, or extra awareness added, then you are performing precisely fair' would be dispersed as not everybody can distinguish a distressing incidence as a chance of education. This defaming mental burden has more intensified approaches of sadness, fault, disgrace, overwhelmed feelings, anger, negative self-talk, regret, self-pity, internalized emotions, impracticable prospects, and apparent logic of disappointment. The mental burden of contending in gathering more jobs than supplementary aristocrats, coupling mob of supporters and donor, creating job-related and educational results in the compulsion of effort or household, and downplaying the shock in individual and remnant will have distressing effects on mental happiness (Mukhtar 2020).

14.2.6.1 Suggested Proposals to Decrease the Effects of Mental Pressure of Productivity and Stigma

In order to decrease the effects of mental pressure of productivity and stigma some important proposals should be implemented as given below:

(a) Efforts can be made within workplaces to improve the situation for staff persons, struggling with mental suffering or psychoses. Sensitization workshops help, in educating and informing staff about mental health problems and contribute to decreasing stigma, discrimination, and fear in the workplaces.

(b) Identification of signs and symptoms of mental distress should be made to be an integral part of leadership training so that managers can employ the necessary tools and measures to backing their employees. An exchange of thoughts and dissertation about mental well-being and distress has to be actively encouraged in workplaces to foster an environment of support, comfort, and optimal productivity.

(c) Nurturing a healthy work atmosphere also entails establishing a culture that is favorable to supporting employee's mental health by raising consciousness of workplace programs and policies that promote mental and physical health and wellness.

(d) Capitalization in mental healthcare programs, employee insurance that covers mental health, on-call mental health professionals, right of entry to professional counsellors, mandated time-off are some of the steps progressive organizations are taking to combat the issue.

14.2.7 Mental Health Impact on Vulnerable Population

COVID-19 is a single as well as a group of traumatic incidents which in a straight line or ramblingly has exaggerated each singular species in the environment. All attempts would be concentrating on diminishing the undesirable properties of this shocking COVID-19 contagion occasion on "patients." Lockdown brings about activities of social distancing, quarantine, and self-isolation distant beyond the spare period trips for value-added working—it is a combined shocking incident which postures grave risk to persons and has caused in huge destruction of lives and assets for every human being in society (Mukhtar 2020). Examples of susceptible inhabitants are elder grownups, children, expectant females, persons having surviving corporeal and mental sicknesses, sufferers of exploitation and forcefulness, existing with addicts and culprits, persons existing beneath the scarcity mark, and other persons are vulnerable of not just constricting the coronavirus but additionally, the emotional upset. Numerous individuals are going through interactive upsetting actions too in addition to the collective shocking COVID-19 such as abuse, loneliness, fear of losing family, emotional and behavioral problems, financial burden, mental health issues, fatalities, gender-based violence resulting in domestic violence, grief and bereavement, and physical injuries. Different scientific-technical innovative skills should be implemented and promoted for the survival of the vulnerable population and to escape from the combined interactive traumatic events and collective traumatic COVID-19 effect.

14.2.7.1 Suggested Proposals to Decrease the Effects of Mental Health Impact on Vulnerable Population

In order to decrease the effects of mental health impact on vulnerable populations some suggested proposals should be implemented as given below:

(a) The effects of mental health impact on frontline workers can be decreased by Confirming ergonomic environment through providing proper and adequate protective gear, Detecting their situation of moral injury and addressing them, Ensuring workplace respect and safety, Providing incentives to them and their families for endangering their lives to take care of others, Identifying their selfless efforts and rewarding them properly and Ensuring workplace respect and safety.
(b) The effects of mental health impact on children's can be decreased by creating ways to keep in touch with their friends, making a home learning routine, Handling children's anxieties by detecting their emotional requirements, and to avoid from negative news by decreasing screen-time but providing clear information regarding involvement in creative and mentally stimulating indoor activities.
(c) The effects of mental health impact on migrant workers can be decreased by treating every migrant worker with dignity, respect, fellow feeling and displaying sympathy individually without generalization, Confirming respectful fiscal support and assurance, Emphasizing the need to stay away from their families plus providing the guarantee of mental along with physical support and Constant systematic assurance, effective counseling and providing the basic needs.
(d) The effects of mental health impact on a geriatric population can be decreased by Giving out clear, concise, and necessary information in a respectful way, Assurance and assistance to the more vulnerable, Engage family members and support workers carefully to deal with mental health issues, Ensuring enough medications for those in need to address any insecurities and Connecting with loved ones living away engaging in entertaining activities.
(e) The effects of mental health impact on people with COVID-19, contacts, survivors, family members can be decreased by Helping them and cope with an emotional loss if they have lost a family or friend to the grave pandemic, Speaking the grief and trauma faced by people with COVID-19 and their family creating self-help platforms and recognizing the survivors and providing them the mental and physical comfort at their isolation sites or hospitals.
(f) The effects of mental health impact on people with existing mental illnesses can be decreased by effectively modifying their counseling periods, to help them, to cope with the pandemic along with their already existing illness, providing entree to treatment through telemedicine consultations and video consultations, and Involving family members with their concern and consideration.

14.3 Management of the COVID-19 Infodemic: Endorsing Healthy Performances and Moderating the Harm from Misinformation and Disinformation

14.3.1 Preserving Resilience, Coping, Mindfulness, and Welfare

Ecosystem created by the nature. "Ecosystem, Society and Men are disturbed by nature but not by the thing" is the logical source of the instruction, dynamic, organized and timespan interference tactic to mitigate, report, extravagance, evaluate and accomplish the overabundance of emotional, cerebral, developmental, societal and even the interaction of psychological, social and spiritual province at the outlay of COVID-19. Mental fitness amenities, and specific thoughtful cure groups together with psychiatrists, therapists, and psychogenic nurses must be recognized to speech mental health anxieties in the overall civic. Individuals and groups of people can consciously nurture flexibility, well-managing approaches, alertness, and welfare. The prospective for flexibility, managing, alertness, and welfare is neither single independences that individual own (or not) nor results of nonappearance of most traumatic anxiety and strain. The proficiency for flexibility is course of variation. Handling device is cultured design of performance which solitary grows above the stretch of time. Alertness is the emotional procedure of intentionally carrying one's devotion in the current instant, which one nurtures with repetition. Welfare indicates the skill of being relaxed in their state of affairs. All these are procedures and they can be assimilated with exercise. Resilience indicates flexibility, coping indicates managing ability, mindfulness, and welfare are not lone dichotomous outcomes deliberate—firming up these procedures intentionally strengthened repetition by facing and culture with passion.

The government administrators and healthcare organizations are concentrating on the regulator of COVID-19 contagion by implementing numerous precautionary approaches, but there is minute attention delivered to the cerebral well-being position of the inaccessible and fright people. Owing to the absence of regular day-to-day actions, consistent exercises, and enduring at home for a lengthier time will influence their cerebral health position. It was reported that unanticipated occurrences can deteriorate the mental fitness situations of persons with pre-existing cerebral health sickness (Ho et al. 2020). To avoid the stressful situation, individuals should start thinking positively and not get unprotected excessively to mass media coverage, to continue a healthy and hearty association, get in trace with groups and family memberships on a steady intermission. It is reported that, in the modern age of machinery, health maintenance amenities can familiarize providing online emotional provision facilities for those individuals who are missing their near intimate followers due to COVID-19 (Ho et al. 2020). Further, to backing the confidence and mental health of the combat zone fitness care authorities, fitness maintenance administrations and governments should announce reduced working times, regular disruptions, and rotational shifts duties (Ho et al. 2020). Persons can manage with

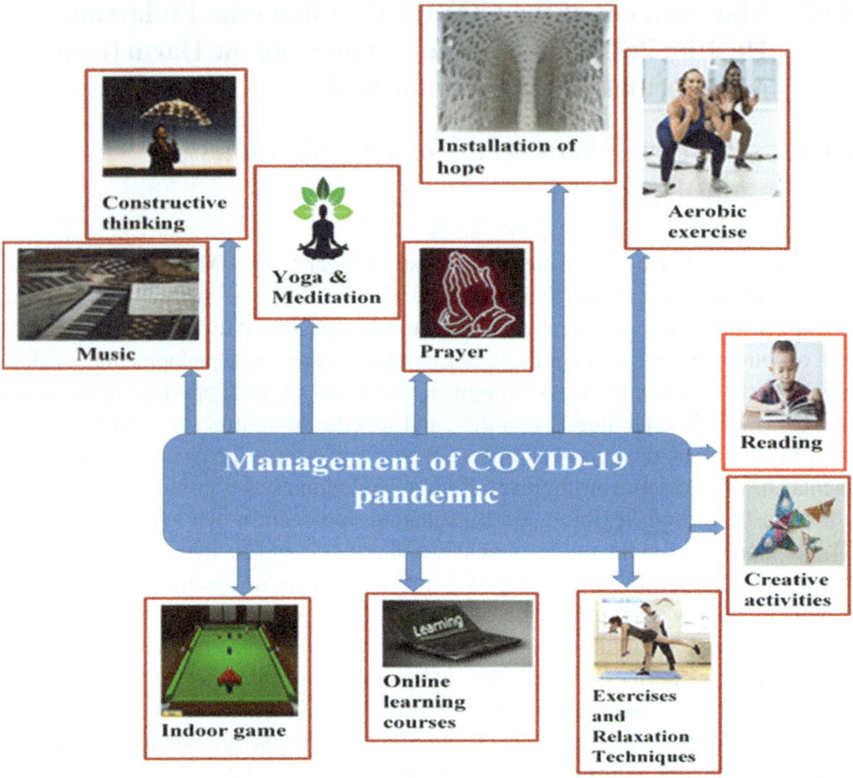

Fig. 14.1 Management of COVID-19 Pandemic (Kar et al. 2020)

cerebral fitness encounters by accommodating several lifespan activities. Relaxation techniques, indoor and creative activities, physical along with spiritual activities are valuable tools to help the human being to remain calm and continue to protect the health from the adverse effects of COVID-19. Kar et al. (2020) stated that different indoor activities (music, online learning, reading, indoor game), spiritual activities (prayer, yoga, and meditation), relaxation techniques (aerobic and other exercises), creative activities, constructive thinking, and installation of hope are important key activities responsible for the management of COVID-19 pandemic as shown in Fig. 14.1.

Kar et al. (2020) also stated different advice for progressive mental health during the COVID-19 pandemic as shown in Fig. 14.2. According to their report regular exercise, positive thinking, balanced diet, relaxation exercise, taking a break from routine work but maintaining regular daily routine, socialization and other healthy lifestyle measures are beneficial for progressive mental health and human being should ensure to do it to maintain the good mental health during the pandemic. On the other hand, use of substance, excessive watching of television (mostly news), excessive online activity, believing the fake news, partying, travel, and focusing on

Advices for Progressive Mental Health	
DO'S	Don'ts
◆ Regular exercise ◆ Positive thinking ◆ Balanced diet ◆ Relaxation exercise ◆ Taking a break from routine work of life ◆ Socialization (online) ◆ Regular daily routine ◆ Other healthy lifestyle measures	◆ Substance use ◆ Excessive watching television (mostly News) ◆ Excessive online activity ◆ Believing fake news ◆ Partying, Travel ◆ Focusing on the negative aspects of COVID-19 pandemic

Fig. 14.2 Advises for Progressive Mental Health (Kar et al. 2020)

the negative aspects of the COVID-19 pandemic were adversely affect the mental health and human being should ensure to avoid it during COVID-19 pandemic.

14.3.2 Prevention of (Mis) Infodemic, Disinformation, and Misinformation

It is very important that for a sustainable society, awareness should be generated between vulnerable populations, and in order to make it a great success, different initiatives should be taken. Considering it, observed then technical evidence about the COVID-19 incidence, inhibition, regulating in addition to cure proposal, development of data, and modernized prominence of fitness instructions (in inherent literatures) ought to be circulated to combat zone of the therapeutic squads and regulation implementation mediators, caregivers, survivors, families, and overall civic. Media accrues collusion philosophies, misrepresentation, and propaganda of the source, measure, signs, indications, communication, inhibition and cure; mass media publicity also gathers always developing fears, and recurrent contact to these occasions enhances the signs of suffering. Distortion, machination concepts, wrong privileges, and deception are solitary irritating the cerebral equanimity of the overall public. The accredited fitness administrations and government should implement then confirm dependable operational evidence over dependable distribution podiums to deliver and endorse tele-psychological recommend and psychoneurotic treatments to decrease the effect of mental or else emotional well-being in COVID-19.

14.3.3 PCI (Psychological Crisis Intervention) and PFA (Psychological First Aid)

Early interferences of PCI commonly known as Psychological Crisis Intervention and PFA (psychological first aid) centers on the emotional fitness of the exaggerated persons and offer an intended device by given that psychosocial provision to alleviate suffering through the occurrences such as COVID-19. Both PFA and PCI are the primary intrusions that emphasize on the psychological health of the exaggerated persons and recommend a planned device by providing that psychosocial backing to alleviate the suffering during COVID-19 occurrences. Presently PCI along with PFA are necessary for the crisis supervision to familiarize expressively incredulous stairs over interaction, charming, security, practical assistance, luxury, and by lecturing anxiety-based responses. PFA design (Everly Jr et al. 2012) comprises of developing relationship over dynamic and imagined thoughtful attending, calculation, and valuation of emotional necessities, ordering contingent on the harshness of growing cases, reasoning, and social interferences to alleviate suffering, and nature and continuation till equilibrium of the circumstances over continual backing and steady checking.

14.3.4 Prospective Policies

Even though the influence of COVID-19 contagion on worldwide mental fitness is not yet recorded and dignified, analogous evidence from preceding investigation working can give an elucidation and vision. Mental health practitioners should be delivered early and timely psychiatric intrusions to manage the occurrence of high-fatality communicable ailments (Mukhtar 2020; Srivatsa and Stewart 2020). The present COVID-19 pandemic is instigating shattering psychically health anxieties for example: anxiety, terror, depressing signs, distress, anger, sleep turbulences, frustration, rejection, and suspicion in the overall public (Mukhtar 2020; Rana et al. 2020). For the curative workforce, these emotional hitches are linked to devotion and executive capabilities which can obstruct the combat in contradiction of COVID-19. The occurrence of emotional difficulties in the overall inhabitants has been fluctuating from 4% to 41% of traumatic indications and 7% of gloomy indications (Kang et al. 2020). In the course of any public disaster, people strive for related occasions to attain and to radiate the fear of the unknown which points to greater apprehension, and in the case of confusing misrepresentation, distorted perception of risk, disinformation on social media, dangerous terror of unidentified/insecurity and community fright may lead to disgrace, criticize and victims (Mukhtar 2020; Mowbray 2020). Even though analysis on COVID-19 is rare, more than a few writers have projected the probable consequences on emotional and physical fitness not only vulnerable but also affecting the overall inhabitants (Kang et al. 2020). Psychological interferences and psychosocial backing would develop civic mental health throughout the contagion COVID-19 occurrence. Inhibition is the finest attitude to battle with the COVID-19 contagion. Till now, no sure cure is obtainable for the cure of the COVID-19 contamination. According to Kar et al. (2020) and their findings, generally two categories of preventive or safety measures chosen to be taken for the active inhibition of COVID-19, which are given below along with their descriptions:

(a) *Broad-spectrum of preventive measures:* This one is intended to aim at everyone in the public. This preventive measure includes sufficient hand and respiratory sanitization, social distancing, and avoidance of traveling during COVID-19 contagion (Fig. 14.3).

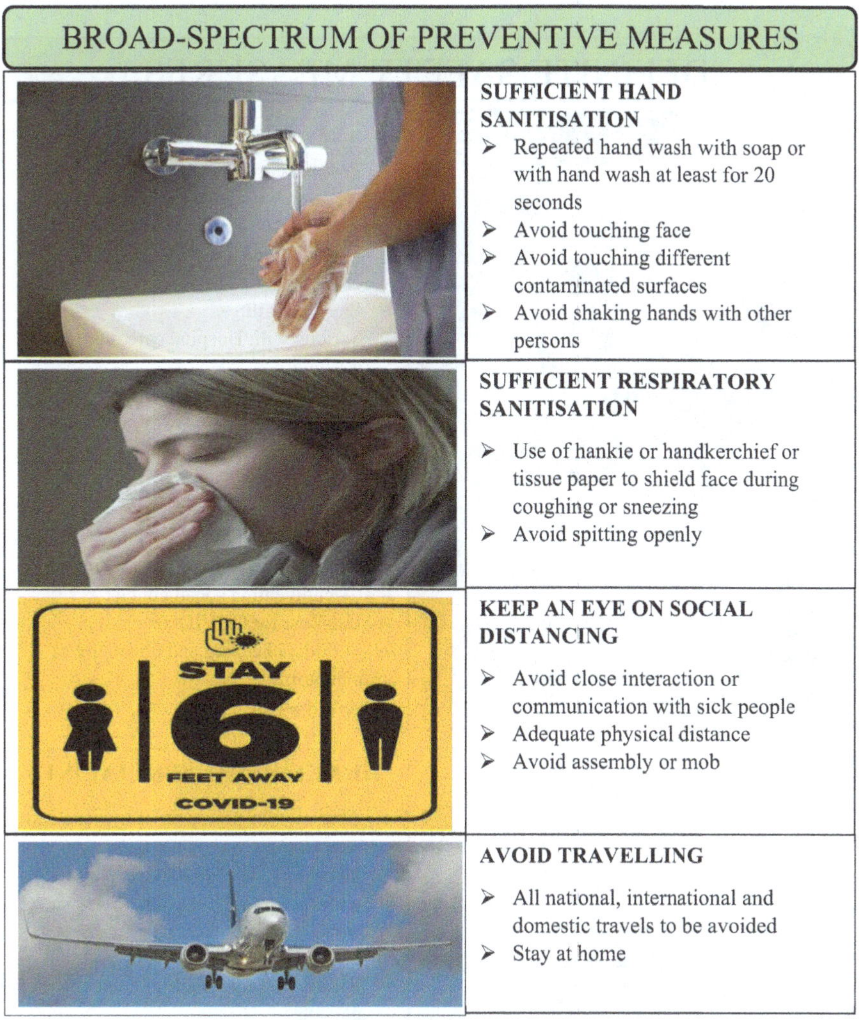

Fig. 14.3 Broad-spectrum of preventive measures during COVID-19 Contagion (Kar et al. 2020)

Sufficient hand sanitization should be followed by repeated handwashing with soap at least for 20 s, avoid touching face along with different contaminated surfaces, and avoid shaking hands with other persons should be followed. Sufficient respiratory sanitization should be followed by the use of a hankie or handkerchief or tissue paper to shield the face during coughing or sneezing. Social distancing should be followed by avoiding close interaction or communication with sick people, social distancing also maintained by adequate physical distance. All the national, international, and domestic travels should also be avoided.

(b) *Definite safety measures:* The definite safety measures pertinent for sick persons, close acquaintances, and healthcare specialists (Fig. 14.4).

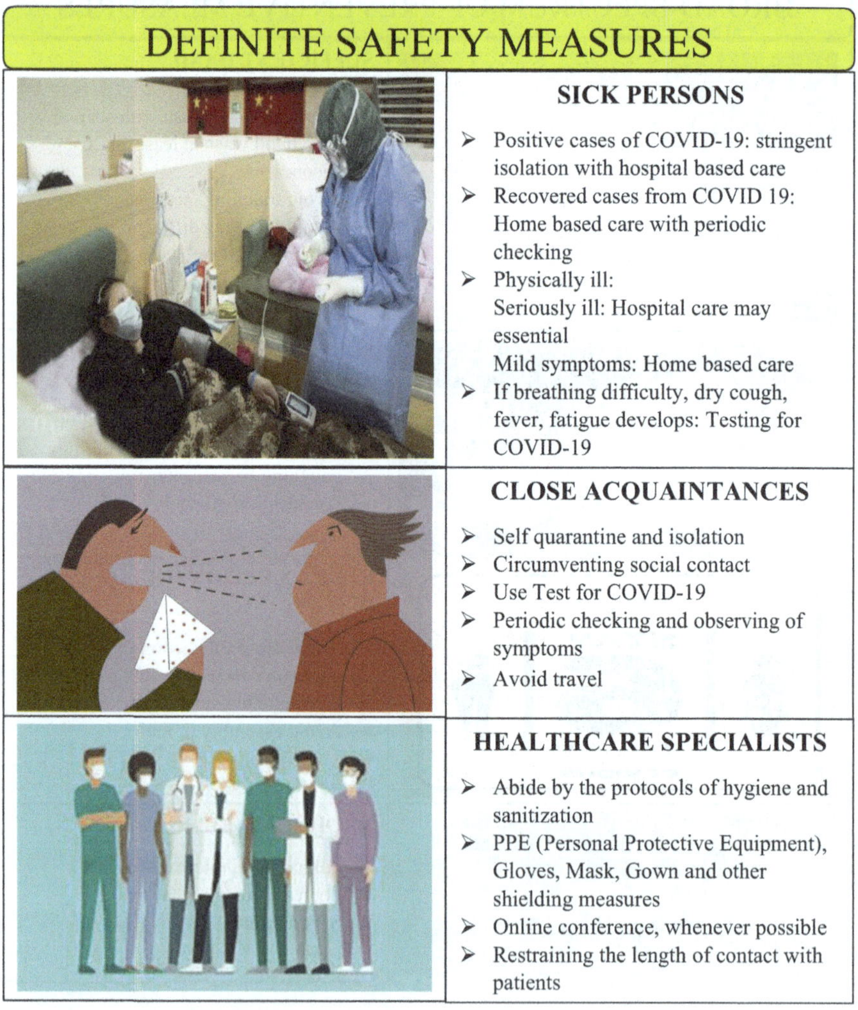

Fig. 14.4 Definite safety measures during COVID-19 Contagion (Kar et al. 2020)

Among the sick person's stringent isolation and hospital-based care adopted for positive cases of COVID-19. Home-based care intended for recovered COVID-19 cases. Hospital-based care was essential for seriously ill cases. On the other hand, breathing difficulty, dry cough, fever cases are intended for COVID-19 testing. Close acquaintances cases envisioned for COVID-19 testing along with self-quarantine and isolation. In these cases, travel should also be restricted. Healthcare specialists should abide by the protocols of hygiene, sanitization, PPE, and other shielding measures. Healthcare specialists also restrain the length of contact with patients.

Mostly, all are divided into three assemblies of inhabitants as mentioned below:

(a) Healthcare workers
(b) Common inhabitants
(c) COVID-19 individuals and nearby associates

The safety measures and endorsements are battered to speak the requirements of the overhead three sets of the inhabitants. Persons who show perseverance in suffering may look for assistance from the cerebral health authorities through accessible help or in clinics in the occurrence of crisis circumstances. Kar et al. (2020) reported the risk cruelty as shown in Fig. 14.5.

Patients and healthcare patrons have been categorized as high-risk severity and belong to severe symptoms-indicated strategy. However, family participants, close interactions, persons having more than one disease, and elderly inhabitants have been categorized as intermediate risk severity and belong to moderate to severe symptoms—a number of strategy. But, general inhabitants are categorized as low-risk severity and belong to mild to moderate symptoms—worldwide strategy.

Kar et al. (2020) described the supervision tactic to mental health difficulties during COVID-19 contagion (Fig. 14.6). According to the report suffering connected to COVID-19 may be mild and sporadic or moderate to severe and persistent. Reassurance, firming up social backing, relaxation, and lifestyle measures can control the mild and sporadic suffering. Taking the help of mental health specialists can control the moderate to severe and persistent suffering. Figures 14.5 and 14.6 describe the endorsements rendering jeopardy harshness and supervision tactic to cerebral fitness problems during COVID-19 contagion, respectively.

14.4 Problem Analysis

COVID-19 pandemic can be reduced by implementing some proposals through analysis of the related problems. The proposals were as given below:

(a) Enabling mental health research to understand the psychiatric characteristics of the pandemic and insertion of psychiatrists in the task force to combat the COVID-19 pandemic.

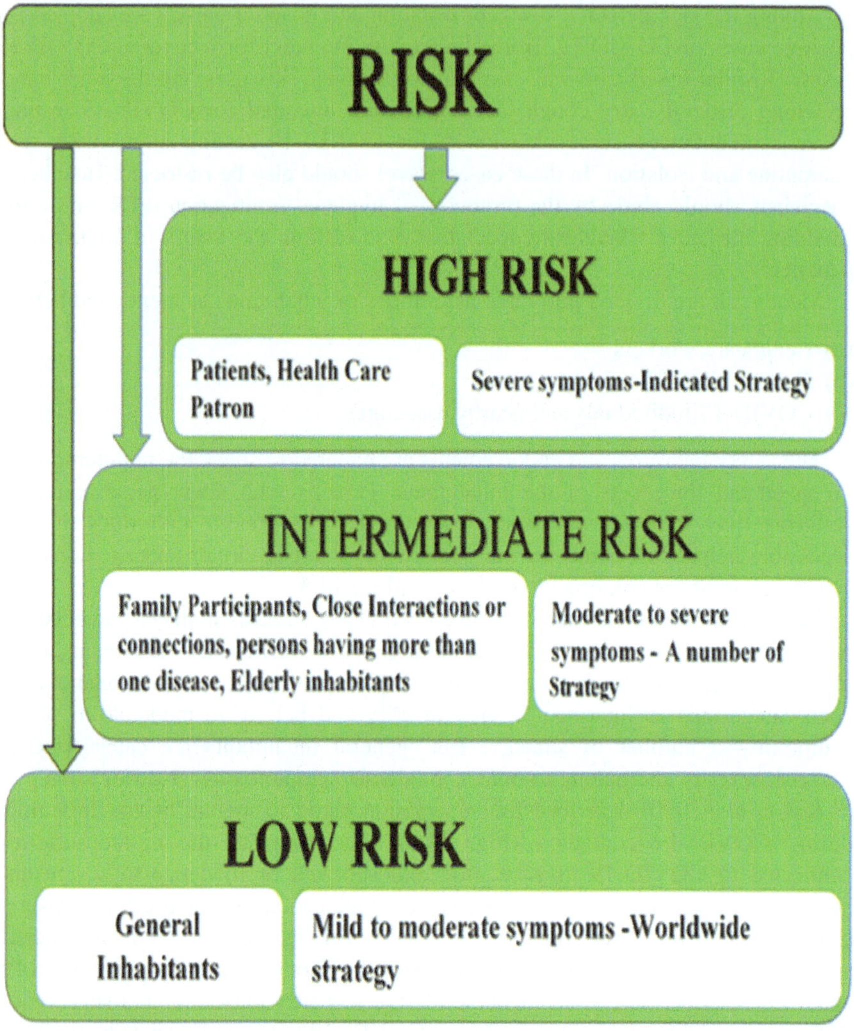

Fig. 14.5 Endorsements rendering the COVID-19 risk cruelty (Kar et al. 2020)

(b) Accessibility of psychotropic medications in hospital settings with the adequate amount and ensuring continuous supply of good-quality medicines.
(c) Presently, Telepsychiatry consultation has brought possible access to the health care system through networking coordination and by telephonic conversation. So, these amenities should be implemented all through the universe.
(d) There is a need to improve the existing infrastructure, to avail essential medications, addressing the needs of the marginalized population; assessing the needs of healthcare providers, removing stigma, and to keep a close watch on the after-effects of infection in the survivors of the pandemic.

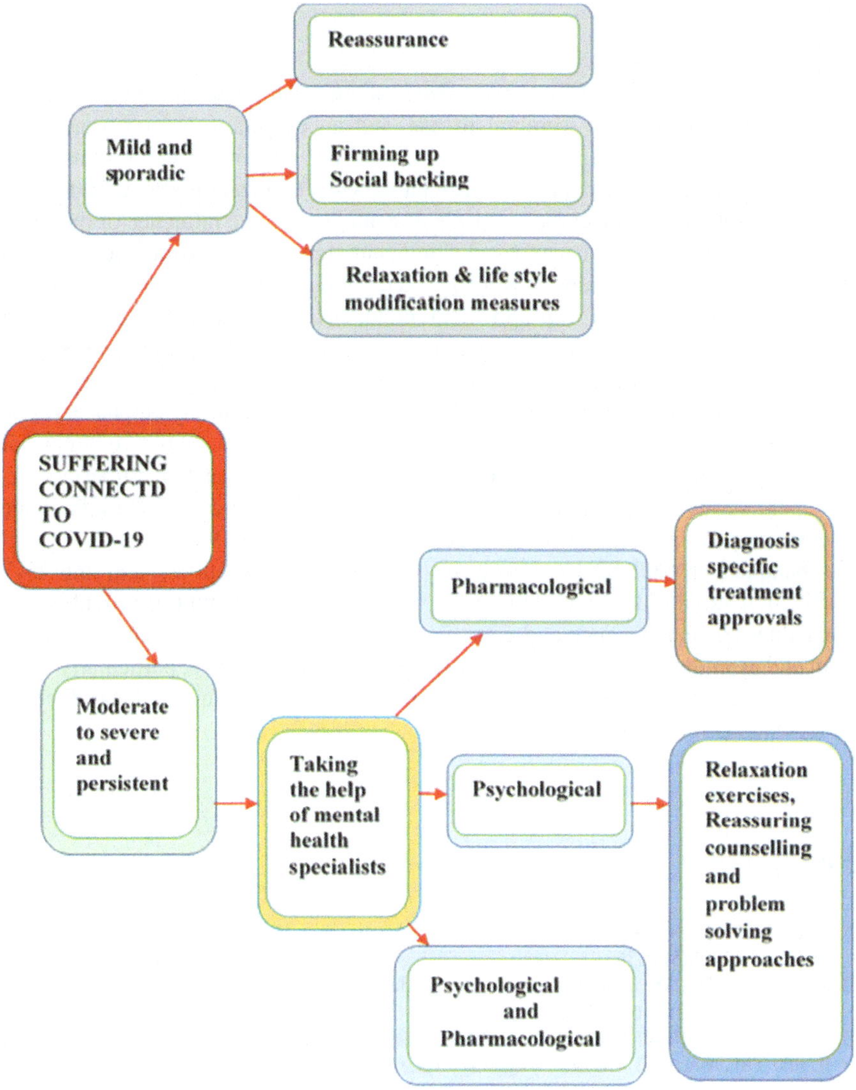

Fig. 14.6 Supervision tactic to mental health difficulties during COVID-19 contagion (Kar et al. 2020)

(e) Enhancing mental resilience will help to combat with the coronavirus pandemic effectively.
(f) Regular contact of the daily activities along with indoor recreational activities, Positive thinking, and installation of hope.
(g) Approaching the healthcare organization, if any symptoms develop and get vaccinated with a COVID-19 vaccine when available.

14.5 Conclusions

At present, the COVID-19 contagion creates an unusual, unanticipated, and incredible encounter to publics' cerebral fitness, and an escalation in occurrence with projected through the utmost critical stage of the epidemic, and a rise in the occurrence of psychological fitness difficulties when civilization give back to common normality. From the earlier analysis of contagion on the influence of mental or else emotional fitness, it has exposed significant signs of psychosocial concerns in the exaggerated entities along with the overall inhabitants. The long-term health problems communicated through quarantine, isolation, anxiety, psychosocial stressors, feelings of helplessness, mental distress, posttraumatic indications, stigma, fear, and xenophobia may develop the emerging universal cerebral fitness issues comparative to an epidemic like COVID-19. It is very vital to lay emphasis on the mental health and its well-being (societal, physical, mental, emotive, financial, development, mystical, psychological along with appealing action, superiority of life span, survival happiness, and domain-specific gratification) of the inhabitants over preemptive psychological interferences in the COVID-19 contagion. Through the banquet of COVID-19, health institutions, administrations, and the civic civilization should intensely replicate on the matter in an attempt to provide even-handedness of maintenance and construct crisis in civic rules to handle the small-term and extended-term psychiatric consequence of the contagion, particularly in the utmost susceptible groups. Further studies and researches are required to brighten the endorsements for upcoming interferences.

References

Armitage R, Nellums LB (2020) COVID-19 and the consequences of isolating the elderly. Lancet Public Health 5(5):e256

Banerjee D (2020) The impact of Covid-19 pandemic on elderly mental health. Int J Geriatr Psychiatry 35(12):1466–1468

Batawi S, Tarazan N, Al-Raddadi R, Al Qasim E, Sindi A, Johni SA, Al-Hameed FM, Arabi YM, Uyeki TM, Alraddadi BM (2019) Quality of life reported by survivors after hospitalization for Middle East respiratory syndrome (MERS). Health Qual Life Outcomes 17(1):1–7

Bauch CT, Lloyd-Smith JO, Coffee MP, Galvani AP (2005) Dynamically modeling SARS and other newly emerging respiratory illnesses: past, present, and future. Epidemiology 16 (6):791–801

Bo HX, Li W, Yang Y, Wang Y, Zhang Q, Cheung T, Wu X, Xiang YT (2021) Posttraumatic stress symptoms and attitude toward crisis mental health services among clinically stable patients with COVID-19 in China. Psychol Med 51(6):1052–1053

Brooks SK, Webster RK, Smith LE, Woodland L, Wessely S, Greenberg N, Rubin GJ (2020) The psychological impact of quarantine and how to reduce it: rapid review of the evidence. Lancet 395(10227):912–920

Brown EE, Kumar S, Rajji TK, Pollock BG, Mulsant BH (2020a) Anticipating and mitigating the impact of the COVID-19 pandemic on Alzheimer's disease and related dementias. Am J Geriatr Psychiatry 28(7):712–721

Brown E, Gray R, Monaco SL, O'Donoghue B, Nelson B, Thompson A, Francey S, McGorry P (2020b) The potential impact of COVID-19 on psychosis: a rapid review of contemporary epidemic and pandemic research. Schizophr Res 222:79–87

Chen N, Zhou M, Dong X, Qu J, Gong F, Han Y, Qiu Y, Wang J, Liu Y, Wei Y, Xia J, Yu T, Zhang X, Zhang L (2020) Epidemiological and clinical characteristics of 99 cases of 2019 novel coronavirus pneumonia in Wuhan, China: a descriptive study. Lancet 395(10223):507–513

Cheung YT, Chau PH, Yip PS (2008) A revisit on older adults suicides and Severe Acute Respiratory Syndrome (SARS) epidemic in Hong Kong. Int J Geriatric Psychiatry 23 (12):1231–1238

Corman VM, Landt O, Kaiser M, Molenkamp R, Meijer A, Chu DK, Bleicker T, Brünink S, Schneider J, Schmidt ML, Mulders DG, Haagmans BL, van der Veer B, van den Brink S, Wijsman L, Goderski G, Romette JL, Ellis J, Zambon M, Peiris M, Goossens H, Reusken C, Koopmans MP, Drosten C (2020) Detection of 2019 novel coronavirus (2019-nCoV) by real-time RT-PCR. Eurosurveillance 25(3):2000045

Coronavirus disease (COVID-2019) situation reports (2020). https://www.who.int/emergencies/diseases/novel-coronavirus-2019/situation-reports

Danielsen R, Thorgeirsson G, Einarsson H, Olafsson O, Aspelund T, Harris TB, Launer L, Gudnason V (2017) Prevalence of heart failure in the elderly and future projections: the AGES-Reykjavík study. Scand Cardiovasc J 51(4):183–189

Dong L, Bouey J (2020) Public mental health crisis during COVID-19 pandemic, China. Emerg Infect Dis 26(7):1616–1618

Druss BG (2020) Addressing the COVID-19 pandemic in populations with serious mental illness. JAMA Psychiat 77(9):891–892

Everly GS Jr, Barnett DJ, Links JM (2012) The Johns Hopkins model of psychological first aid (RAPID—PFA): curriculum development and content validation. Int J Emerg Ment Health 14 (2):95–103

Features, Evaluation, and Treatment of Coronavirus (n.d.). https://www.ncbi.nlm.nih.gov/books/NBK554776/

Ferrando SJ, Klepacz L, Lynch S, Tavakkoli M, Dornbush R, Baharani R, Smolin Y, Bartell A (2020) COVID-19 psychosis: a potential new neuropsychiatric condition triggered by novel coronavirus infection and the inflammatory response? Psychosomatics 61(5):551–555

Ford-Jones PC, Chaufan C (2017) A critical analysis of debates around mental health calls in the prehospital setting. Inquiry 54:0046958017704608

Girdhar R, Srivastava V, Sethi S (2020) Managing mental health issues among elderly during COVID-19 pandemic. J Geriatric Care Res 7(1):32–35

Haider II, Tiwana F, Tahir SM (2020) Impact of the COVID-19 pandemic on adult mental health. Pak J Med Sci 36:S90

Ho CS, Chee CY, Ho RC (2020) Mental health strategies to combat the psychological impact of COVID-19 beyond paranoia and panic. Ann Acad Med Singap 49(1):1–3

Hu W, Su L, Qiao J, Zhu J, Zhou Y (2020) COVID-19 outbreak increased risk of schizophrenia in aged adults. PsyChinaXiv

Huang Y, Zhao N (2020) Generalized anxiety disorder, depressive symptoms and sleep quality during COVID-19 outbreak in China: a web-based cross-sectional survey. Psychiatry Res 288: 112954

Huang C, Wang Y, Li X, Ren L, Zhao J, Hu Y, Zhang L, Fan G, Xu J, Gu X, Cheng Z, Yu T, Xia J, Wei Y, Wu W, Xie X, Yin W, Li H, Liu M, Xiao Y, Gao H, Guo L, Xie J, Wang G, Jiang R, Gao Z, Jin Q, Wang J, Cao B (2020) Clinical features of patients infected with 2019 novel coronavirus in Wuhan, China. Lancet 395(10223):497–506

Jaggars SS (2014) Choosing between online and face-to-face courses: community college student voices. Am J Dist Educ 28(1):27–38

Kang L, Li Y, Hu S, Chen M, Yang C, Yang BX, Wang Y, Hu J, Lai J, Ma X, Chen J, Guan L, Wang G, Ma H, Liu Z (2020) The mental health of medical workers in Wuhan, China dealing with the 2019 novel coronavirus. Lancet Psychiatry 7(3):e14

Kar SK, Arafat SY, Kabir R, Sharma P, Saxena SK (2020) Coping with mental health challenges during COVID-19. In: Coronavirus disease 2019 (COVID-19). Springer, Singapore, pp 199–213

Lauer SA, Grantz KH, Bi Q, Jones FK, Zheng Q, Meredith HR, Azman AS, Reich NG, Lessler J (2020) The incubation period of coronavirus disease 2019 (COVID-19) from publicly reported confirmed cases: estimation and application. Ann Intern Med 172(9):577–582

Lee SM, Kang WS, Cho AR, Kim T, Park JK (2018) Psychological impact of the 2015 MERS outbreak on hospital workers and quarantined hemodialysis patients. Compr Psychiatry 87:123–127

Li Q, Guan X, Wu P, Wang X, Zhou L, Tong Y, Ren R, Leung KS, Lau EH, Wong JY, Xing X, Xiang N, Wu Y, Li C, Chen Q, Li D, Liu T, Zhao J, Liu M, Tu W, Chen C, Jin L, Yang R, Wang Q, Zhou S, Wang R, Liu H, Luo Y, Liu Y, Shao G, Li H, Tao Z, Yang Y, Deng Z, Liu B, Ma Z, Zhang Y, Shi G, Lam TTY, Wu JT, Gao GF, Cowling BJ, Yang B, Leung GM, Feng Z (2020) Early transmission dynamics in Wuhan, China, of novel coronavirus–infected pneumonia. N Engl J Med 382(13):1199–1207

Lu R, Zhao X, Li J, Niu P, Yang B, Wu H, Wang W, Song H, Huang B, Zhu N, Bi Y, Ma X, Zhan F, Wang L, Hu T, Zhou H, Hu Z, Zhou W, Zhao L, Chen J, Meng Y, Wang J, Lin Y, Yuan J, Xie Z, Ma J, Liu WJ, Wang D, Xu W, Holmes EC, Gao GF, Wu G, Chen W, Shi W, Tan W (2020) Genomic characterisation and epidemiology of 2019 novel coronavirus: implications for virus origins and receptor binding. Lancet 395(10224):565–574

Mak IWC, Chu CM, Pan PC, Yiu MGC, Chan VL (2009) Long-term psychiatric morbidities among SARS survivors. Gen Hosp Psychiatry 31(4):318–326

Meng H, Xu Y, Dai J, Zhang Y, Liu B, Yang H (2020) Analyze the psychological impact of COVID-19 among the elderly population in China and make corresponding suggestions. Psychiatry Res 289:112983

Mohammed A, Sheikh TL, Gidado S, Poggensee G, Nguku P, Olayinka A, Ohuabunwo C, Waziri N, Shuaib F, Adeyemi J, Uzoma O, Ahmed A, Doherty F, Nyanti SB, Nzuki CK, Nasidi A, Oyemakinde A, Oguntimehin O, Abdus-Salam IA, Obiako RO (2015) An evaluation of psychological distress and social support of survivors and contacts of Ebola virus disease infection and their relatives in Lagos, Nigeria: a cross sectional study– 2014. BMC Public Health 15(1):1–8

Mowbray H (2020) In Beijing, coronavirus 2019-nCoV has created a siege mentality. BMJ 368

Mukhtar MS (2020) Mental health and psychosocial aspects of coronavirus outbreak in Pakistan: psychological intervention for public mental health crisis. Asian J Psychiatr 51:102069

Ozamiz-Etxebarria N, Idoiaga Mondragon N, Dosil Santamaría M, Picaza Gorrotxategi M (2020) Psychological symptoms during the two stages of lockdown in response to the COVID-19 outbreak: an investigation in a sample of citizens in Northern Spain. Front Psychol 11:1491

Rana W, Mukhtar S, Mukhtar S (2020) Mental health of medical workers in Pakistan during the pandemic COVID-19 outbreak. Asian J Psychiatr 51:102080

Report of the WHO-China joint mission on coronavirus disease 2019 (COVID-19) (2020). https://www.who.int/docs/default-source/coronaviruse/who-china-joint-mission-oncovid-19, final report.pdf

Reynolds K, Pietrzak RH, El-Gabalawy R, Mackenzie CS, Sareen J (2015) Prevalence of psychiatric disorders in US older adults: findings from a nationally representative survey. World Psychiatry 14(1):74–81

Rubin GJ, Wessely S (2020) The psychological effects of quarantining a city. BMJ 368

SESCAM [Regional integrated strategy: coping with emotional distress related to the health emergency-COVID-19 Regional Coordination of Mental Health]. Modelo Intersectorial de Trabajo en Red para la Intervención en Crisis en Castilla-La Mancha, 2020

Severance EG, Dickerson FB, Viscidi RP, Bossis I, Stallings CR, Origoni AE, Sullens A, Yolken RH (2011) Coronavirus immune reactivity in individuals with a recent onset of psychotic symptoms. Schizophr Bull 37(1):101–107

Shahid Z, Kalayanamitra R, McClafferty B, Kepko D, Ramgobin D, Patel R, Aggarwal CS, Vunnam R, Sahu N, Bhatt D, Pharm KJ, Golamari R, Jain R (2020) COVID-19 and older adults: what we know. J Am Geriatr Soc 68(5):926–929

Shereen MA, Khan S, Kazmi A, Bashir N, Siddique R (2020) COVID-19 infection: origin, transmission, and characteristics of human coronaviruses. J Adv Res 24:91–98

Shigemura J, Ursano RJ, Morganstein JC, Kurosawa M, Benedek DM (2020) Public responses to the novel 2019 coronavirus (2019-nCoV) in Japan: mental health consequences and target populations. Psychiatry Clin Neurosci 74(4):281–282

Sim M (2016) Psychological trauma of Middle East Respiratory Syndrome victims and bereaved families. Epidemiol Health 38

Srivatsa S, Stewart KA (2020) How should clinicians integrate mental health into epidemic responses? AMA J Ethics 22(1):10–15

Turner AM, Osterhage KP, Taylor JO, Hartzler AL, Demiris G (2018) A closer look at health information seeking by older adults and involved family and friends: design considerations for health information technologies. In: AMIA annual symposium proceedings. American Medical Informatics Association, New York, pp 1036–1045

Van Doremalen N, Bushmaker T, Morris DH, Holbrook MG, Gamble A, Williamson BN, Tamin A, Harcourt JL, Thornburg NJ, Gerber SI, Lloyd-Smith JO (2020) Aerosol and surface stability of SARS-CoV-2 as compared with SARS-CoV-1. N Engl J Med 382(16):1564–1567

Van Hal G (2015) The true cost of the economic crisis on psychological well-being: a review. Psychol Res Behav Manag 8:17–25

Venkatesh A, Edirappuli S (2020) Social distancing in covid-19: what are the mental health implications? BMJ 369

Vidal M "Everyone in Wuhan suffers from trauma." Specialists prepare for the psychological consequences of quarantine in the original focus of the pandemic. El País; 10 ABR, 2020. https://elpais.com/sociedad/2020-04-10/todo-el-mundo-en-wuhan-padece-un-trauma.html

Wang H, Li T, Barbarino P, Gauthier S, Brodaty H, Molinuevo JL, Xie H, Sun Y, Yu E, Tang Y, Weidner W, Yu X (2020) Dementia care during COVID-19. Lancet 395(10231):1190–1191

World Health Organization. Mental health of older adults (2017). Available from https://www.who.int/news-room/fact-sheets/detail/mental-health-of-olderadults

World Health Organization (n.d.). https://www.who.int/health-topics/coronavirus#tab=tab_3

Xiang YT, Yang Y, Li W, Zhang L, Zhang Q, Cheung T, Ng CH (2020) Timely mental health care for the 2019 novel coronavirus outbreak is urgently needed. Lancet Psychiatry 7(3):228–229

Xu D, Jaggars SS (2014) Performance gaps between online and face-to-face courses: differences across types of students and academic subject areas. J High Educ 85(5):633–659

Zheng Z, Peng F, Xu B, Zhao J, Liu H, Peng J, Li Q, Jiang C, Zhou Y, Liu S, Ye C, Zhang P, Xing Y, Guo H, Tang W (2020) Risk factors of critical & mortal COVID-19 cases: a systematic literature review and meta-analysis. J Infect 81(2):e16–e25

Zhou M, Zhang X, Qu J (2020) Coronavirus disease 2019 (COVID-19): a clinical update. Front Med 14(2):126–135

Zhu N, Zhang D, Wang W, Li X, Yang B, Song J, Zhao X, Huang B, Shi W, Lu R, Niu P (2020) A novel coronavirus from patients with pneumonia in China, 2019. N Engl J Med 382:727–733

Chapter 15
Multimodal Analysis of Cognitive and Social Psychology Effects of COVID-19 Victims

V. Kakulapati, S. Mahender Reddy, and Sriman Naini

Abstract COVID-19 is an expanding social, economic, and health epidemic. Present COVID-19, as it has led to tremendous increases in psychiatric problems, has a leading national influence on this secondary disease. COVID-19 caused widespread hysteria, socioeconomic injury, and a high infection rate and mortality to the psychosocial effect. The virus is predictable to pose a big global mental health problem that already has a huge impact on millions of people's physical health. Emotional and cognitive risks. Discussed various perceived risks when the perceived surveillance reduced health risk chance. The prediction model incorporates biological, psychological, and social variables in diagnosis, prognosis, and treatment of COVID-19 by logistic regression, decision tree, random forest, RNN (Recurrent Neural Network), and PNN (Probability Neural Network). In order to provide doctors and patients with information about their use, efficacy, and deficiencies, this research includes a design evaluation for the topic, diagnosis, and assessment of moderate or extreme neurocognitive impairments. The adverse effects of fear, cold, and depression increase the health risk; obsession raises the health risk, risks between individuals and mental health, and uncertainty; Finally, positive mental states enhance health risk perception. Further, positive survivor techniques can help ease emotional distress that causes tension, while pessimistic coping mechanisms can intensify emotional symptoms due to stress.

Keywords Psychosocial · COVID-19 · Coronavirus · SARS-CoV2 · Emotions · Pandemic · Machine learning

---

V. Kakulapati (✉)
Sreenidhi Institute of Science and Technology, Hyderabad, Telangana, India

S. Mahender Reddy
Otto-Friedrich University of Bamberg, IsoSySc, Bamberg, Germany
e-mail: mahender-reddy.sheri@stud.uni-bamberg.de

S. Naini
Rosenheim Technical University of Applied Science, Rosenheim, Germany

S. A. Hassan et al. (eds.), *Decision Sciences for COVID-19*, International Series in Operations Research & Management Science 320,
https://doi.org/10.1007/978-3-030-87019-5_15

15.1 Introduction

Novel influenza of coronavirus spread globally in 2020. Not only has this led to a public health crisis but also serious socio-psychological and economic implications. The findings after the COVID-19 epidemic show that people are susceptible to pandemic-related psychiatric illnesses such as depressive thoughts, fear and signs of depression, and post-traumatic (Kang et al. 2020). The pathophysiology of mental health conditions and long-term sequelae of mental health should also be understanding. Many researchers are suggesting neurological effects, including headaches, altered consciousness, and dyspnea in COVID-19 patients.

Furthermore, it is testifying that the virus could affect the nervous system. Together, these studies point to the potential involvement of the virus in developing acute COVID-19 clinical symptoms and long-term neuropsychiatric consequences. The neurological dysfunctions linked to the infection by COVID-19 are likely to influence neurological functions in the long term. Observations from other respiratory virus conditions, such as severe acute respiratory syndrome (SARS), indicate that psychological illnesses, extremely durable neuropsychiatric disorders, and cognitive can progress. This traumatic condition's psychological and social ramifications expect to impair emotional well-being and eventually cause detrimental impacts over the long term (Holmes et al. 2020).

India witnesses the second coronavirus wave. The new wave is spreading more quickly than last year, according to analysts, because it was mainly older people that got COVID-19 the previous year. But patients are mostly infants and pregnant women this time around. The second coronavirus wave has shown new infection symptoms. Fever, body ache, loss of smell and taste, chills, breathlessness are common symptoms of COVID-19. Some reports say red eyes, gastronomical conditions, and impaired listening may also be due to virus infection and should not be working light heartedly (https://www.dnaindia.com/health/report-second-wave-of-coronavirus-brings-these-new-symptoms-check-if-you-have-them-2885488).

The people are suffering from several aspects those are (a) Well-being uncertainty and interpretation of communications are the main causes of tension during the pandemic (Brooks et al. 2018), (b) subconscious biases, community isolation (Qiu et al. 2020), (c) financial difficulties and joblessness, and (d) uncertain infobesity (Gao et al. 2020). The intensity of fear increased with improved pandemic reports and the number of confirmed cases of infection following the outbreak of the COVID-19 (Lima et al. 2020). During the COVID-19 pandemic, there was a spike in psychiatric issues like anxiety and depression (Duan et al. 2020).

Emotional effects of outbreaks are disorder perceptions, mental stress, and aggressive reactions: terror, resentment, isolation, and rage. A rare condition is known as a "news article stress disorder" can occur through new influenza: it is obvious by a strong traumatic event, stress and anxiety, constant media coverage that can lead to physical symptoms, including palpitations and insomnia (Dong et al. 2020). A psychological difficulty is a recurrent deficiency in decision-making caused by inaccurate, motivating influences, and adaptations to nature (Mata 2012). It must be borne in mind that in addition to those at risk of serious illness, individuals who have previous medical problems such as chronic lung disorder, coronary heart disease, or Mellitus, are still concerned with COVID-19 (Menotti et al. 2016).

"The patients" cognitive function was unknown before the COVID, and the findings also do not indicate long-term rehabilitation, meaning that any cognitive consequences may be short term (https://timesofindia.indiatimes.com/life-style/health-fitness/health-news/covids-cognitive-costs-some-patients-brains-may-age-10-years/articleshow/78892984.cms). A unique manifestation of the form of anxiety. A specific type of anxiety perception. Cognitive aspects such as obsessive concerns about perceptions, extraction bias, and strain imaginations. Behavioral aspects include dysfunction, evasion, addiction to behavior, mental aspects of apprehension, agitation, and annoyance. Coronaviral anxiety physiological component (i.e., sleep disturbances, somatic distress, and tonic immobility).

The framework created mental well-being—anxiety, depression, and after-traumatic stress, which intends to estimate the probability of psychological issues for an individual patient 3 months after Posttreatment. Maturity, loss of social care, painful experiences, and psychotic disorders evaluated at Posttreatment were significant predictors in that model. Postoperative to identify patients who were high-risk and who might need post-ICU acceptance was anticipate for the model.

The prediction of two different kinds of dependent variables is complete using logistic regression. The first category is a dichotomous dependent variable in one of two mutually exclusive countries. For example, "pessimism against optimism," "right vs. wrong," and "depression vs. anxiety." An asymptotic outcome measure is the second category of predictor variables. In this case, a pessimism analysis will give the participants who fear significant depression pathologies a value of 0 and the participants who lack optimism a value of one.

Decision trees (DTs) are a process of classification and regression, nonparametrically training set. The strategy is to create a model that forecasts the model's performance, training from the data's effective decision laws. A tree is viewing as a bit by bit.

The RF (Random Forests) developed as a single method of data analysis are highly predictive, creating multiple decision trees by random sampling in the same data set, combining them, and eventually forecasting target variables (Shameem et al. 2015). In addition, RFs have an outstanding predictive performance. Several covariates are added to the model to identify relationships between explanatory variables and an illness and prevent over-fitness (Strobl et al. 2007).

RNN (Recurrent neural network) is a feed-forward ANN (artificial neural networks) used for observational clinical data as an efficient forecasting model. It enables thoughts and actions between dimensions by displaying sequential transient characteristics for time sequences. It can provide semi-cooperative analysis of data collected as a clinically proven tool (Ghazi et al. 2018).

The PNN (probabilistic neural network) describes a concurrent application of the Bayes model optimization method. The supervised learning techniques in Bayes are creating processes that mitigate the "anticipated chance." An evaluation of model parameters in respect of each category is the principal challenge utilizing Bayes cognitive functioning.

The remaining chapter is splitting into different parts. Section 15.1 describes the research work, and Sect. 15.2 addresses the literature. Section 15.3 discusses the chapter's topic, and Sect. 15.4 outlines the methods of the work suggested and the results of the experiment in question summarized in Sect. 15.5. Section 15.6 addresses the proposed approach and its importance, and in Sect. 15.8, the future directions are supplementing by the final remarks in Sect. 15.7.

15.2 Related Work

The advancement coronavirus (COVID-19) pandemic created a massive challenge to the general demographic's physical and mental health (Yan et al. 2021). The correlation between perceived tension and mental distress in the first wave of epidemic is investigating. Also, psychological discomfort, such as anxiety, terror, obsession, panic, anorexia, and social phobia, is related to psychological trauma. The interaction between perceived tension and psychological distress was highly and significantly influencing by boredom.

Numerous investigations in the past have practiced with diversified designs to evaluate the psychological effects of COVID-19 and other outbreaks' cognitive symptoms (Rubin 2020; Brooks et al. 2020; Taylor 2019). These works mainly focus on different pandemic situations such as lockdown, segregation, social distances, and emotional responses against an outbreak.

In one investigation, COVID-19 predicts to cause significant damage to the nervous system (Wu et al. 2020) and also demonstrated adverse psychiatric conditions (e.g., fear and depression) are associating with improvements in the respiratory system (Rajkumar 2020). Thus, the nervous system can be improving from neuroscience by eliminating psychological discomfort every day, preserving the good quality of sleep, consistency of the diet, and ensuring sustainable life and physical exercise. This will eliminate the amount of COVID-19 infection (Kang et al. 2020) if individuals have healthy immune responses. This study provides significant references to the patient-related evidence that may assist psychologists with systemic functional with psychological issues because of an incidence of COVID-19 (Zhou et al. 2015) effectively.

A significant factor is a psychological effect on physicians who practice through the disease outbreak (Greenberg et al. 2020; Dewey et al. 2020). Stressor includes physiological adaptation to meet expected growth (Charney 2004). If treatment involves no end to severe physiological reaction, harmful effects can occur on neurophysiological functions (Lapa et al. 2017). In contrast, most people who anticipated pandemic of influenza among those working but on the poverty level than those who were unemployed or had sufficient income were significantly higher (Choi et al. 2018). Income level affecting proportional risk (Barr et al. 2008). Middle-income users were more likely to evaluate less risk (Jacobs et al. 2010).

Many communities have been deprived massively due to the disappearance mortality of family members. The tragic deaths associated with the COVID-19 issue have intensified fear and tension. The devastation has led to elevated psychological disorders and anticipates boosting posttraumatic stress disorders (Horesh et al. 2020). The ordinary condition of people's quality of life has been compromising such that people face massive psychological falls in the entire disease outbreak. A complex and multifaceted depression recognition model investigates how social networking and lexicon evolution affect multiple forms of interaction. Cumulative by assessing emotional understanding of the social perceptions. Three specific applications are discussing in Kakulapati et al. (2020), and those applications are mathematical models and lexical psychological phenomenon data.

Several COVID-19 recoverees have psychological, cognitive, and behavioral disorders as well (Liotta et al. 2020). It is anticipating that certain symptoms occur in nearly three-fourths of COVID-19 inpatients. These indications also seem to persist even after the initial infection is healed and have serious consequences for patients' health and well-being (Serrano-Castro et al. 2020). This demonstrates that COVID-19 will influence the brain (Delorme et al. 2020) increasingly. COVID-19 cognitive effects are including ischemic stroke, pulmonary edema, leukemia, and psychiatric conditions in the peripheral regions. About 30% of the experienced exercise physiology were well assuming throughout compromised cognition, diagnosis, cognitive ability, short memory, and neurocognitive. A previous study has found that the risk of cognitive failure often exceeds in patients who have been admitting to ICU with respiratory agony (Pandharipande et al. 2013).

There have been several psychological complications and significant mental health impacts since the epidemic of the COVID-19, including fatigue, angst, sadness, anger, and uncertainty (Duan et al. 2020). Basic psychological reactions to the mass quarantine to mitigate the transmission of COVID-19 have been widespread panic and widespread community distress, usually correlated with illnesses, and escalated in new cases alongside insufficient, media-led details. The emotional effects of the COVID-19 outbreak may originate from fear or communal hysteria (Rubin et al. 2020) to general feelings of depression and anxiety, such as suicide. Pertinently, abnormally high stress will affect other health measures.

15.3 About Psychological and Cognitive Feelings of Patients

The mental health pandemic, which has been labelling the "third wave." According to the *Indian Journal on Psychology* by India on the possible effects of COVID-19, over two-fifths of patients suffer from fear and depression. In March 2020, social distancing began worldwide with the outpatient unit and "innovative" interactive consultations, and respondents experience that their emotional distress is increasing steadily or deteriorating. Interestingly, the pathological depression of unfamiliar realities of working house, temporary jobs, schooling, and decline in physical interaction between all friends and family and peers significantly contributed to these cognitive. Even then, roughly two-thirds of the ventilated patients' instances of COVID-19 happen to have true psychotic symptoms, which includes confusion, uncertainty, and dysexecutive mental disorder, and the impact of the coronavirus on cognitive functions.

Many patients with real cognitive implications do not report and reject such love. Family members or relatives always get patients affordable healthcare as individuals find significant inconsistencies in their everyday activities or repetitive tasks. Some patients are complaining that they have low alertness. However, only a handful of them struggles with psychological pathogenesis and a much more neurocognitive assessment.

Many patients with genuine neurological consequences do not disclose it and refuse to accept such an adoration. Patients are often certain psychiatric help from loved ones who find noticeable deficiencies in successfully implementing daily activities or routine tasks. However, only a good number have neurological epidemiology and will involve sufficient neuropsychological evaluation and testing to accomplish this identification.

The impact is starting with a contextual sequence of initial intellectual interpretation problems. Threshold psychological spheres with fever, disturbed consciousness, focal neuronal deficiencies, and epilepsy should lift the alarm of brain attack, brain swelter, and encephalopathy for a neurological reason. Both these are more frequent in people with infection with COVID-19. Need supplementary assessment of brain images by MRI (Magnetic resonance imaging) and CSF (cerebral spinal fluid) assessments following vertebrae valve. This analysis involves imagery for cerebrospinal fluid. In marginalized communities such as patients at the end of the lifespan, or with significant comorbidities, or before the mental or cognitional conditions, the analysis limit should be minimal. The risk assessment and effective control of all COVID-19 infections and recovery processes will timely diagnose such catastrophic abnormalities.

15.3.1 The Coronavirus Disease Psychological Effects

The emotional stress of depression with mild cognitive patients and the perception of wrenching about themselves or the society lead to the total effects of such a condition in such neurological disorders. A revolutionary change mostly challenged psychological neurorehabilitation with several vulnerable relatives, and these limitations influenced the consciousness. Coincidentally, including interactions with family, kids, families, colleagues, and communities, groups expressed beneficial effects of the scramble on the "interpersonal aspect." This was well-thought-out to be due to more leisure moments and much less atmospheric pressure.

Other aspects of "social awareness" are significantly related and affect patient social behavior. In addition, even after COVID-19 span vaccination, a progressive cooperative process will eliminate and strengthen negative emotional psychology and neurophysiology effects, which include antisocial behavior, depressive symptoms at an initial phase. For prevention of psychological and cognitive impact, WHO (World Health Organization) recommended some aspects such as emotion; interpersonal activity, such as community cohesion and generosity; participation with the utilization of healthcare facilities; adherence with others and commitment to medical regulations; rumor denial, racism, and gratification in drug or alcohol misuse; regulating dietary, taking a nap and exercising workouts; reducing screen time; preventive measures; interaction antiquity; and advancement of real and COVID-19 facts and anxieties. Everyone must particularly gratefully acknowledge the medical professionals of various governments and all associated with COVID-19 for everyone's self-sacrifice contributions and the efforts made by the communities throughout this even unpredictable disease outbreak. Emotional stability is also a societal responsibility to physicians (https://www.timesnownews.com/health/article/coronavirus-symptoms-impact-of-covid-19-on-psychological-and-cognitive-health/717878).

The disease outbreak is raising the need for the diagnosis of psychological disorders. Repentance, loneliness, economic impairment, and anxiety worsen or intensify severe psychological issues. Numerous patients will face high beverage and drug consumption, sleeplessness, and distress. Till then, COVID-19 will lead to neurological and mental health issues such as unconsciousness, restlessness, and a brain hemorrhage. Those with pre-existing psychological, cognitive, or alcohol utilization disorders can also have a higher probability of severe outcomes or even mortality due to complications with SARS-CoV-2 (severe acute respiratory syndrome coronavirus 2) (https://www.who.int/news/item/05-10-2020-covid-19-disrupting-mental-health-services-in-most-countries-who-survey#:~:text=Bereavement%2C%20isolation%2C%20loss,outcomes%20and%20even%20death).

One-third of patients demonstrate levels of cognitive dysfunction such that neurocognitive evaluation performance is approximately equal to medium brain damage. In normal experiences, neurological impacts on perception, concentration, and management can contribute to problems in preventing and controlling infections, money management, the interpretation of the content, and communication with loved ones. The psychiatric consequences of the emergency department tend to stay often identified, including anxiety, distress, and post-traumatic mental illnesses.

15.3.2 COVID's Susceptible Psychological Consequences

COVID can affect neurological damage by straightforward inflammation, peripheral arterial disease, and deficiency of ventilation. However, head trauma is extremely common among patients with acute illness or disability hospitalization, with its normally evident consequences.

15.3.3 COVID Virus Time Series Consequences

COVID contamination also progresses to cognitive impairment—particularly in patients, especially in the elderly. Though head trauma has become obvious and aggravates psychological dysfunction, trauma is often minor, and may lead to problems with continuous treatment.

Many nations acknowledged that a variety of psychological and cognitive effects:

- More than 60% of patient emotional stability complications have recognized too vulnerable adults.
- About 67% of patients have had rehabilitation and psychotherapeutic dysfunction.
- More than 35% of emergency dysfunction, such as long-term panic attacks and underlying medical conditions.
- 30% indicated interference with access to medicines for psychiatric, physiological, and substance usage conditions.

The two concepts frequently analyzed in psychiatric disorders are tension and anxiety. Tension is a psychological or mental tension caused by a certain circumstance that allows a person to be depressed, frustrated, or nervous. It seems to be a real or perceived agitation originating from any occurrence or perception that is stressful, angry, or nervous to a patient. It has adverse cognitive effects and perhaps even impacts the brain (MedlinePlus 2020).

Psychological first aid is describing as "an initial attempt to support and provide disaster relief to foster protection, stabilize disaster survivors and link people to aid and resources" (Psychological First Aid: American Psychological Association n.d.).

Aspects cognitive relating character traits, to be conscious is related to high sensitivity. However, other "expressive" persons could have—facilitating depression probably due to more probably fantasies of possible implications and consequences. Cognitive factors, such as behavior and self, tend to be associated with a risk evaluation. These aspects are often known as temporary protection tendencies; risk assessment is much more influential than the alternative.

Contagion and isolation have a neurological impact mostly related to the perception of the virus. For example, some disease outbreak aspects, the distance from relatives and friends, the deprivation of autonomy, uncertainty about the progression of this virus, and the perception of helplessness (Li et al. 2020), impact the growth further. Such considerations seemed to have substantial implications (Weir 2020), including increasing suicides (Kawohl et al. 2020). The emotions of frustration associate with the traumatic situation are commonly associated with aggressive tendencies in those able to live in the most affected areas (Suicide Awareness Voices of Education 2020). Considering these consequences, it is necessary to properly analyze the prospective of isolation, taking massive emotional problems into account (Mazza et al. 2020).

15.3.4 Patient Privacy Aspects

15.3.4.1 Resilientness

Patients with psychological disorders, anxiety, community disruptions, and neurotoxic effects with the disease outbreak can be notoriously problematic. Risk in practice, patient quality of life, and quarantine can worsen associated mental disorders such as depression, anxiousness, and chronic post-traumatic depression. In addition, the risk of disorder dissemination could exacerbate infection issues in patients with obsessive-compulsive syndrome and Hypochondria, perhaps experiencing suspect conceptualization. While contagious protection from coronavirus spreads, this involves depression and boredom, resulting in severe cognitive anxiety and potentially cause or aggravate psychiatric disorders (Vahia et al. 2020).

15.3.4.2 Community Support

The healthcare system is effective for cognitive well-being and quality of life, ensuring that enhanced health assistance will increase patients' emotional wellness efficiency (Lin et al. 1979). Community assistance is protective for psychological disorders; it performs an explicit and implicit function across social interactions and inhibits stress and worry (Dour et al. 2014). According to various evidence, there is evidence that social support may effectively reduce the risk of anxiety in patients and health professionals (Khan et al. 2010). In the same way, a significant number of investigations show the tension level to be reciprocal with behavior modification in terms of anxiety analysis. In many cases, welfare support protects from panic and can foresee the prevalence of distress from an inadequate perceived assistance level (Barbee et al. 1993). Consequently, this becomes particularly important to prioritize welfare to the normal community during the dissemination and influence of COVID-19.

15.3.5 Precautionary Approaches

Effective prevention strategies can alleviate the psychological and behavioral implications of the COVID-19 outbreak at the Societal level, such as the effective interaction and the delivery of adequate psychological services. Health must be intensifying electronically. Psychological anxiety in connection with COVID-19 is addressing correctly. At the same time, discrimination and inequality must be predictable as important impediments to strengthen the concerns in a world crisis. Clinical measures for effective and data improving response activities to implement appropriate safety measures must be providing to medical professionals (https://academic.oup.com/qjmed/article/113/8/531/5860841).

15.3.6 COVID-19 Patients' Significant Psychological Effects

One-third of chronic patients in an emergency department face so significantly psychological dysfunction that neurocognitive function is consistent among those with moderated brain damage when acute pulmonary hypertension or due to discomfort from certain symptoms. Certain cognitive effects on retention, perception, and strategic functioning can lead to risk factors in the development of addiction, risk management, interpretation of written information, and talking among friends and family members. Long-term mental consequences of ICU stay often include nausea, depression, and post-traumatic stress disorder (PTSD). It is forecast that the impact of COVID ICU remains will be close—a forecast proved in the UK, Canada, and Finland above (https://www.health.harvard.edu/blog/the-hidden-long-term-cognitive-effects-of-covid-2020100821133#:~:text=There%20is%20one%20inevitable%20conclusion,to%20difficulties%20with%20sustained%20attention).

15.3.7 COVID-19 Impacts on Neurological and Wellness

The contagion has catastrophic results for the early detection of psychological disorders on emotional stability and well-being. Many investigations reveal that about 95% of patients are negative impacts caused by social and health implications. The most commonly mentioned consequences were depression, loneliness, uncertainty, confusion, anguish, and frustration.

15.4 Methodology

The key objective is to predict anxiety and depression levels in patients. Involvement in cognitive impairment reduces the risk for acute mental health disorders and distress, prevents them, removes the severe and negative consequences, and supports patients in strengthening stressful situations or healing psychologically (Wang et al. 2002; Fan 2003). It became proposed that psychological intervention in a predicament would allow patients to recover and achieve emotional alignment quickly and efficiently. The emotional and behavioral aspects of the inhabitants after a catastrophe are constant. Their mental health should also be an individual's focus of the intervention in social and cognitive disorders.

1. **LR (Logistic regression):** The prediction of a definite consequence from multiple regressors is an appropriate analytical process. The correlation is estimating by the LR of the categorical factor's various or two relationships between the independent variables by evaluating the possible outcomes (Le Cessie et al. 1992).

2. **DT (Decision Trees):** Processes binary tree categorization or regression methods. Which employs conditional statements law, and that for labelling is mutually exclusive in an effective manner. By using training effectiveness, the guidelines are trained sequences. The entries affected by the policies are extracting when a rule is to observe. This procedure is iterating once the goal of the learning is accomplishing. The structure is creating in splitting and conquering sequential, top-down method. As its recipient recognizes the quantitative rules, small DTs have the added value of transparency. Nevertheless, the simpler DTs normally underperform complicated DTs as Random Forest (Breiman 2001) and XGBoost.
3. **RF (Random Forests)**: This type of classification technique obtains a randomized experience of several DTs and consists of the learning phase that incorporates multiple DTs and analysis phases that characterize and anticipate received data (Devi and Rajagopalan 2011). RFs have improved predictive capacity and intensity than DT to eliminate overfitting (Hussain and Hazarika 2014).

   Potentially, RF can be safe from overfitting but does not compromise much between noise and external factors (Chen et al. 2020). When trying to reduce classification biases, it could create a highly precise performance. RF is, moreover, often probable to provide an overhead line, indicating that many more forests proceed significantly downward. It is much more potential when an insignificant external stimulus be selected those trees were more complicated.
4. **RNN (Recurrent Neural Networks):** Deep neural networks that distinguish from feed-forward networks by providing slow access correlations to neurons with the same layer and therefore have enough knowledge. RNNs require quantitative vectors for the input. Key phrases probably translate into computational descriptions using modularizing strategies, using hidden layers to maximize interpretations under constraints.
5. **PNN (Probability Neural Network):** It is a sophisticated feed-forward network of 3 layers comprising a feeder layer, a radial basis, and an ability to compete for layer. The radial basis layer determines separation from the activation function to the feature vector of the processing. It produces a vector where the members indicate as connected each input to the training input is. The third layer combines all representations to a set of probabilities for every encoder class. Ultimately, a balanced input signal in the third layer performance measures the higher among these probabilities and provides 1 for such a category and 0 for other classes. There seems to be no incremental learning process necessary for PNNs to control model complexity better than neural network models in MLPs. The consequence of these PNNs is the behavior of the Bayesian methodology.

15.5 Experiment Analysis

The dataset (https://data.mendeley.com/datasets/8rjrmncrhw/2) collects from a Croatian source where the entire dataset is translated from Croatian to English and then preprocessed the dataset into a numerical dataset.

Elementary school 0, high school 1, undergraduate 2, graduate 3, postgraduate 4, abnormal 0, border 1, normal 2, informed 1, uninformed 0, male 1, female 0, pessimism 0, optimism 1 output2, Anxiety 0, depression 1. Output1, Shape life events, label covid sum, Red life events, depression, Blue pessimism, covid sum, Life events = Significant life events that happened in last week.

Da = Yes (Death of close person, Termination of partnership, Changes in the work environment, Change of residence, covid).

It contains certain factors such as sumcovid = indicating the knowledge on corona (transmission, symptoms, vaccine, etc.), KatA describing the anxiety category (normal, abnormal, border), KatD depression category (normal, abnormal, border) where both values are the sum of the values of anxiety and depression related questions, i.e., (fear, worries, anxiety, panic) and then Output1 (Anxiety 0, depression 1) and similarly optimism and pessimism with output2 (pessimism 0, optimism 1).

Coming to the diagrams, circular-free visualization diagram with more blue dots indicates the life events effect on corona awareness, depression, anxiety (output1), i.e., corresponding to the correlation using Pearson's correlation and with more red dots is of COVID sum with respect to the optimism and pessimism correlation and other factors.

All other factors are indicating in the visualization.

Bar graphs of sum COVID (COVID awareness) output2 (pessimism, optimism) are relatively plotting with respect to life events.

There is a violen plot with red and blue colors, depression, and anxiety pessimism. It shows the extent of red (depression), blue (anxiety), and pessimist thinking, leading to mental disorders.

The main focus of the questionnaire is on their emotional stability: how they perceive situations every day, what they expect, how they feel, how they like or dislike ordinary achievement, anxiety, optimistic thoughts, and negative thoughts.

The assessment recognizes the individual and psychological traits that make a patient more neurologically conscious than elevated of the reduced level of the various aspects of healthcare. Theoretical perspective provides no insight on the emotional health of an individual's model. The technique used is to implement logistic regression contributing to the individual patient's state, such as pessimism and optimism, through categorical dependent variables (i.e., the appropriate psychological wellness measure).

The machine learning approach for the decision tree is based on the tree data structures to make choices at various levels—for predictive problems since the construction is robust and simple to read. The classification tree models with a variable objective for a separate set of values and cover regression (Li and Zhang 2010).

Confusion matrices were generating by the implementation in all three groups of stress, anxiety, and depression in all five techniques, i.e., LR, DT, RF, RNN, and PNN.

According to a certain value, the RF method will evaluate the features such that there are fewer significant characteristics that can be minimized and eliminated. The

Table 15.1 Confusion matrix for Random Forest

		Predicted		
		0	1	Σ
Actual	0	1182	53	1235
	1	202	433	635
	Σ	1384	486	1870

Table 15.2 Confusion matrix for Recurrent Neural Network

		Predicted		
		0	1	Σ
Actual	0	1209	26	1235
	1	25	610	635
	Σ	1234	636	1870

Table 15.3 Confusion matrix for Recurrent Neural Network (%)

		Predicted		
		0	1	Σ
Actual	0	98.0 %	4.1 %	1235
	1	2.0 %	95.9 %	635
	Σ	1234	636	1870

core principle is to assess the extent to which the RF forecast reduces by incorporating noise to every feature (Tables 15.1, 15.2, 15.3, and 15.4).

The efficiency, error rate, specificity, recall, sensitivity, and f1 score of every selected feature classification are achieving. The analysis in research work presents that the RNN, LR, and RF were the most accurate for all three levels of anxiety, depression, and stress. Moreover, in Table 15.5, the results showed that strategic classes are imbalanced, such as anxiety, depression, and stress confusion, which

Table 15.4 Performance measurements of RT, RNN, and LR

Model	AUC	CA	F1	Precision	Recall
Random Forest	0.938	0.863	0.859	0.0.865	0.863
Neural network	0.97385	0.955	0.935	0.9448	0.9348
Logistic regression	0.964	0.954	0.943	0.952	0.942

have generated mild and typical effects. Additionally, infected individuals persisted, medium cases, and highly serious issues accordingly to the scales of anxiety, depression, and stress; thus, the classification was also adversely affected. Precision alone is not enough in such situations, and the f1 score is a key measure in deciding the best model. F1 is a harmonic mean of accuracy and recall, the value of which is greater if accuracy and reminder are higher. In this way, an f1-score higher, even though its precision is poorer, is the better model in cases of grade imbalance. For stress and depression, the f1 marking was highest, although it was poor for all anxiety algorithms. It was experiential that all algorithms' accuracy is around 90% or above in all three instances. This is also an important parameter in healthcare, as it means that adverse outcomes are also accurately categorized. In this analysis, all the algorithms were also extremely efficient in negative situations (Fig. 15.1).

Figure 15.2 demonstrates the results for DT predictions of depression psychological symptoms. The total precision, the pessimism and optimism ratio, that the model properly predicts (Figs. 15.3, 15.4, 15.5, and 15.6).

PNN: COVID-related awareness with a pessimistic way of thinking (factors) categorized into depression (normal, abnormal, border) (Table 15.6).

ROC Curve for LR and RNN Model

The ROC curve for the logistic regression and recurrent neural network is presenting in Fig. 15.7. The LR model showed good predictive accuracy.

15.6 Discussion

Contagion has significant adverse effects on patient wellness. There will also be substantial socioeconomic impacts that will harm the psychological and emotional well-being—all policymakers and decision makers, which will effectively make sensible preventive strategies to reduce epidemic disorder. The catastrophic impact on patient well-being should reflect the psychological impact of depression and panic due to its rapid spreading of the disease outbreak as the urgency for public sanitation activity. The results indicated the sense of emotional assistance for COVID-19 patients, which might irritate their psychological issues. Patients with psychopathology are much more influenced by the contagion COVID-19, leading to a worsening or degradation of the chronic psychological disorders as contrasted with

Table 15.5 Pearson correlational points

1	-0.108	Age	sumcovid
2	-0.107	ID	sumcovid
3	-0.092	raregood	sumcovid
4	+0.090	enjoyasusual	sumcovid
5	+0.078	Education	sumcovid
6	+0.075	relax	sumcovid
7	+0.070	enjoyfreetime	sumcovid
8	-0.065	enthusiasm	sumcovid
9	+0.061	sumcovid	tenseupset
10	-0.056	KatD	sumcovid
11	-0.053	futureoptimistic	sumcovid
12	+0.043	bestuncertain	sumcovid
13	-0.040	brighterside	sumcovid
14	-0.039	badend	sumcovid
15	+0.038	Depression	sumcovid
16	+0.033	fear	sumcovid
17	-0.032	notfavour	sumcovid
18	-0.029	comfort	sumcovid
19	-0.028	Pessimism	sumcovid
20	+0.028	negativethink	sumcovid
21	+0.028	sumcovid	upsetmoveo
22	-0.026	successexpect	sumcovid
23	-0.026	fate	sumcovid
24	+0.026	disapointed	sumcovid
25	-0.021	dontexpect	sumcovid
26	-0.021	sumcovid	worries
27	-0.018	Optimism	sumcovid
28	+0.017	panick	sumcovid
29	+0.016	goodmood	sumcovid
30	-0.016	positiveside	sumcovid
31	+0.015	Anxiety	sumcovid
32	-0.011	KatA	sumcovid
33	-0.008	postivethink	sumcovid
34	-0.006	rightway	sumcovid
35	+0.005	generalwell	sumcovid
36	+0.004	concerncorona	sumcovid
37	+0.001	sloweddown	sumcovid

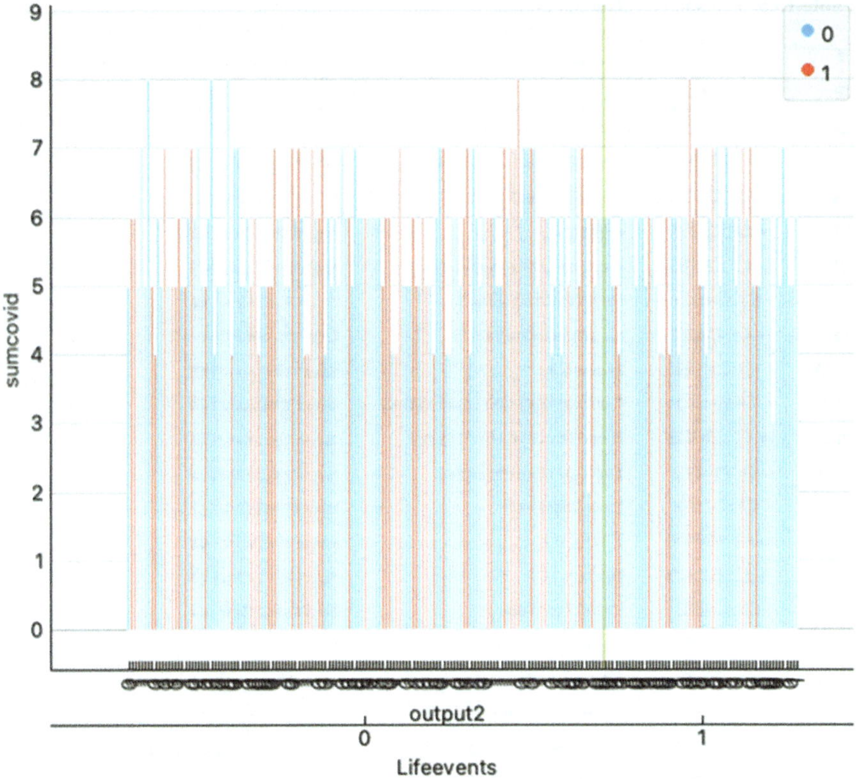

Fig. 15.1 COVID anxiety pessimism of life events

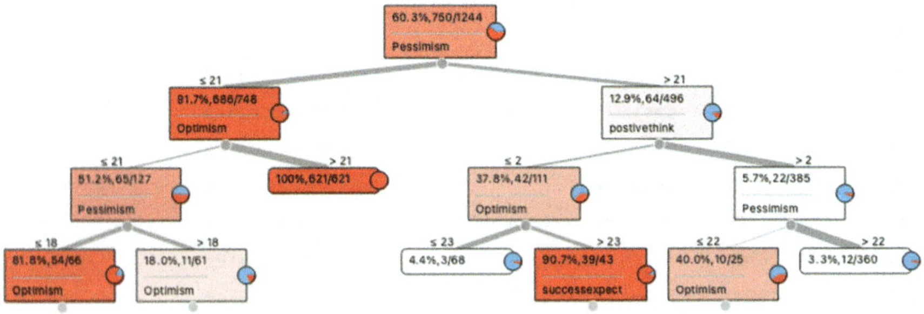

Fig. 15.2 Decision tree view of optimism pessimism

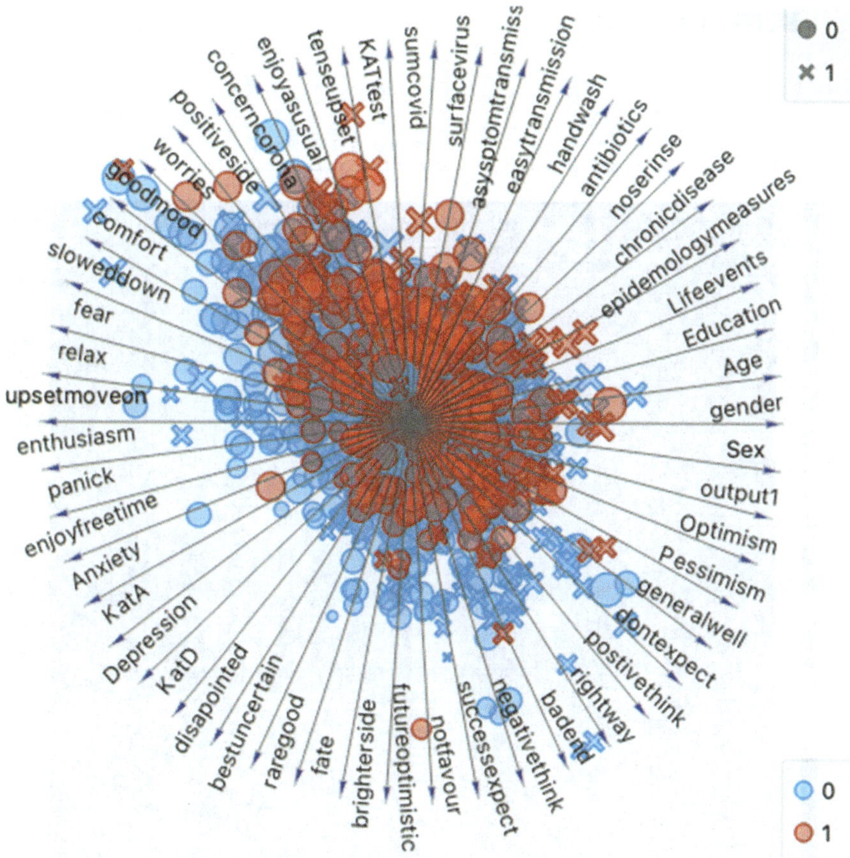

Fig. 15.3 Effect of life events on depression COVID-19

the common public due to the increased vulnerability to anxiety. Mentally ill patients experienced several complications and difficulties. The organized logistic methodology to the emotional well-being analysis adopted and discusses the impact of a specific number of significant factors on calculating the probabilities that psychological well-being is low or strong in a given facet.

In enabling patients to enhance the positive reinforcement of COVID-19 to overcome their psychological disorders, the appropriate detection can support them in their treatment from COVID-19. Patients who affect COVID-19 are depressed, and their psychological and cognitive consequences on their potential lives are discussing.

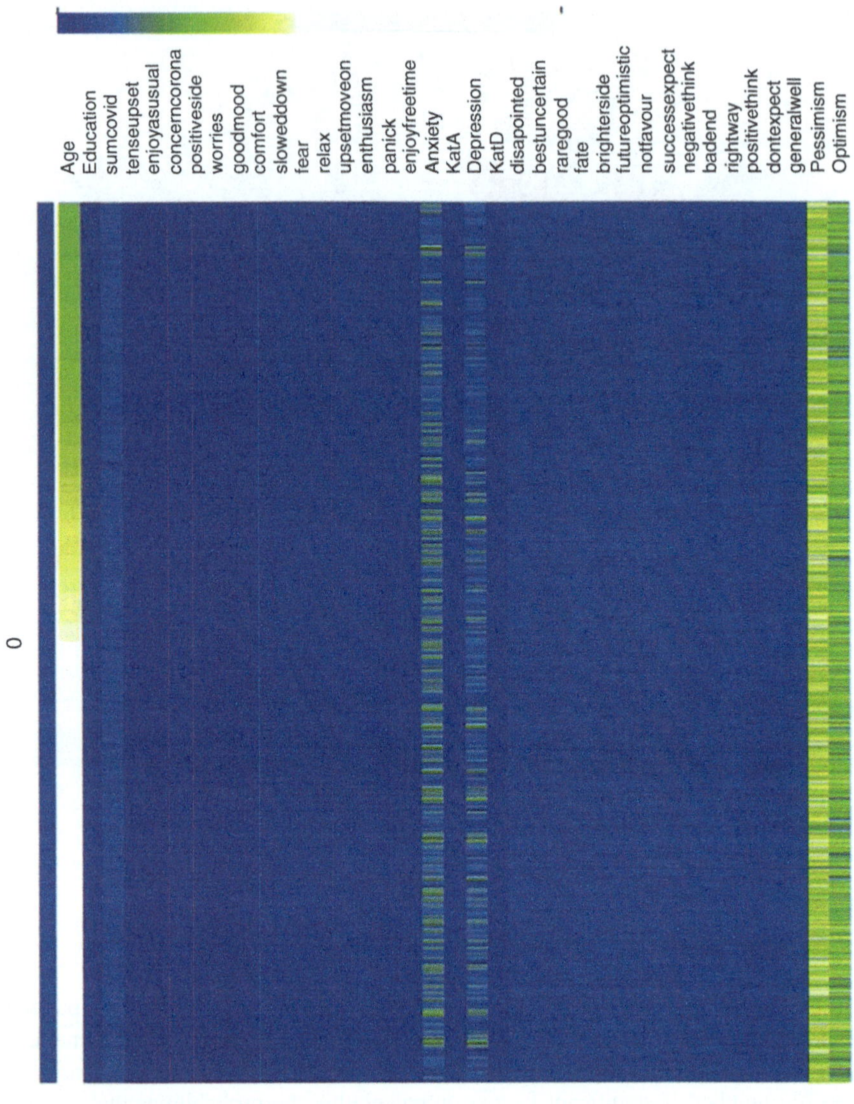

Fig. 15.4 Heatmap data

15.7 Conclusion

In the community of identified and potential outbreaks, an epidemic of COVID-19 is predicting to affect severe psychiatric disorders and neurological mortalities. Individuals infected by COVID-19 may experience neurobiological complications for numerous factors psychological complications, disease progression, adverse effects

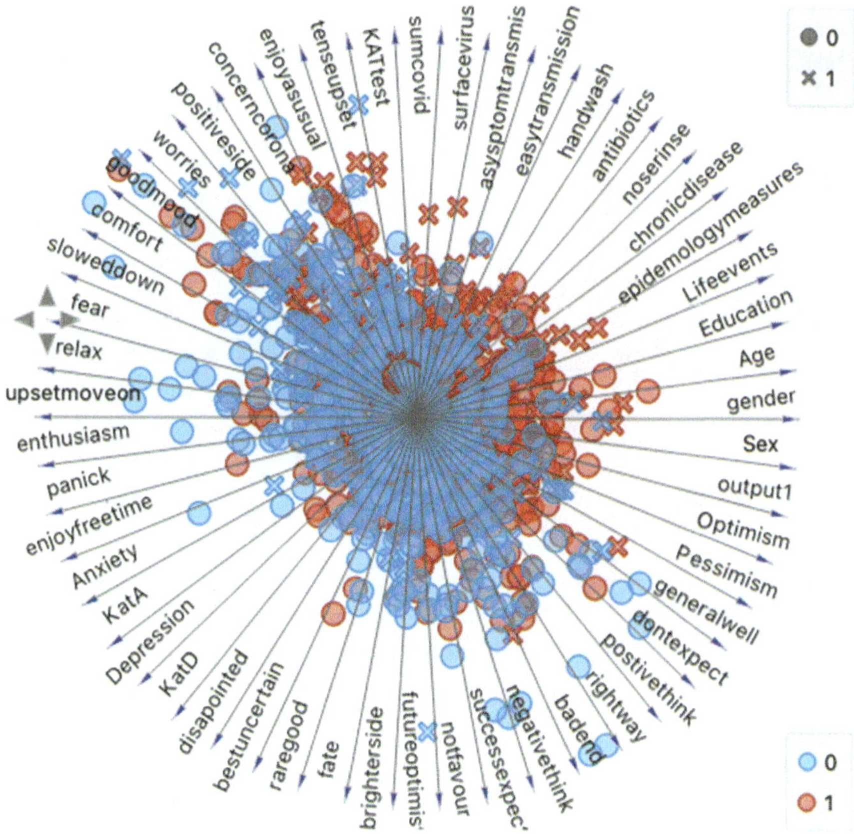

Fig. 15.5 Optimism and pessimism of COVID-19 life events

on medications, potential risks, the anxiety of virus transmission to everyone, and emotional instability.

Five machine learning methods were employed—the DT, LR, RF, RNN, and PNN. The LR's accuracy was the highest, while the PNN was the better model. Since it concerns imbalanced classes, the effective model selected to rely on the f1 score for imbalanced partitioning situations. The major factors for stress, anxiety, and psychological stress were "normal," "abnormal," and "border."

The psychological and cognitive analysis of the COVID-19 patients, RNN 98%, RF 87%, and more probabilities of panic, depressive symptoms, and uncertainty utilizing logistic regression is predicting. Pessimism and optimism result accurately from the DT. Cognitive and psychological implications of the COVID-19 depression level are probably analyzing in the recurrent neural network. Observations indicate that COVID-19 has a significant psychological effect. Before recovery, 96.2% of psychologically supportive patients found significant post-traumatic psychological stress, which seems to reduce their well-being and mental stability. The considerable

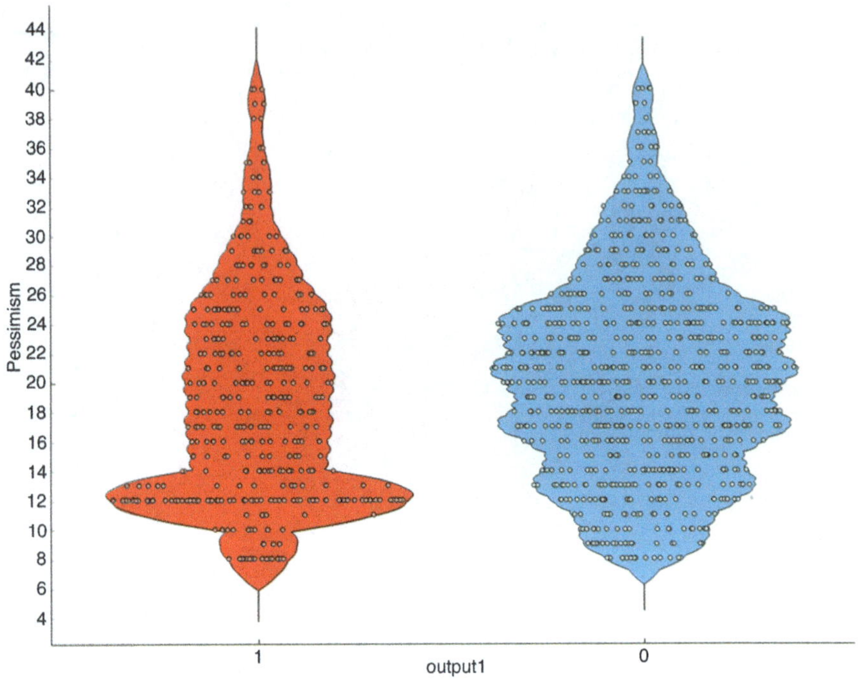

Fig. 15.6 Violen plot dep_anx pessimism

Table 15.6 Probability neural network for pessimism for Depression

S. no.	Sumcovid	Pessimism	Depression
0	6	21	Normal
1	6	37	Border
2	5	28	Normal
3	6	17	Normal
4	6	14	Normal

c_abnormal = 3.72
c_border = 2.66
conormal = 19.0
Depression: Normal. Best value: 19.0

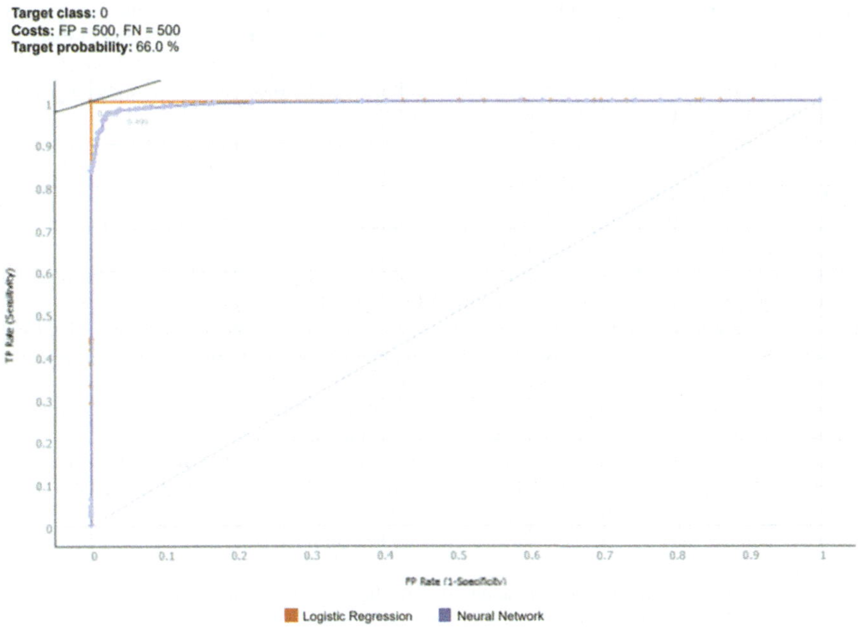

Fig. 15.7 The ROC curve for Logistic Regression and Recurrent Neural network

prevalence of anxiety may be due to the specific stages of the disease. A constructive survivor approach may help relieve stressful, emotional pain, while negative processes can exacerbate stress-related psychological disorders.

15.8 Future Enhancement

In the future, emotional and mental or influencing factors, patient's experience of fitness, family status, smoking, alcohol, etc. Expanding analysis into cognitive impairment by incorporating similarity methods and including the complementary aspects of depression are recommendations. An associated sadness could be essential to measure if various datasets are using to analyze forecasting performances and social network reviews.

References

Barbee AP et al (1993) Effects of gender role expectations on the social support process. J Soc Issues 49:175–190

Barr M et al (2008) Pandemic influenza in Australia: using telephone surveys to measure perceptions of threat and willingness to comply. BMC Infect Dis 8:1–14. https://doi.org/10.1186/1471-2334-8-117

Breiman L (2001) Statistical modeling: the two cultures (with comments and a rejoinder by the author). Stat Sci 16:199–231. https://doi.org/10.1214/ss/1009213726

Brooks SK et al (2018) A systematic, thematic review of social and occupational factors associated with psychological outcomes in healthcare employees during an infectious disease outbreak. J Occup Environ Med 60(3):248–257

Brooks SK et al (2020) The psychological impact of quarantine and how to reduce it: a rapid review of the evidence. Lancet 395:912–920

Charney DS (2004) Psychobiological mechanisms of resilience and vulnerability: implications for successful adaptation to extreme stress. Am J Psychiatr 161:195–216

Chen J et al (2020) A machine learning ensemble approach based on random forest and radial basis function neural network for risk evaluation of regional flood disaster: a case study of the Yangtze River Delta. China Int J Environ Res Public Health 17:49

Choi D et al (2018) Exploring risk perception and intention to engage in social and economic activities during the South Korean MERS outbreak. Int J Commun 12:3600–3620

Delorme C, On behalf of CoCo-Neurosciences study group and COVID SMIT PSL study group et al (2020 Aug 15) Covid-19-related encephalopathy: a case series with brain FDG-PET/CT findings. Eur J Neurol. https://doi.org/10.1111/ene.14478

Devi SN, Rajagopalan SP (2011) A study on feature selection techniques in bioinformatics. Int J Adv Comput Sci Appl 2(1):138–144

Dewey C et al (2020) Supporting clinicians during the COVID-19 pandemic. Ann Intern Med 172:752–753

Dong M et al (2020) Letter to the editor: headline stress disorder caused by Netnews during the outbreak of CoViD-19. Health Expect 23:259–260

Dour HJ et al (2014) Perceived social support mediates anxiety and depressive symptom changes following primary care intervention. Depress Anxiety 31:436–442

Duan L et al (2020) Psychological interventions for people affected by the COVID-19 epidemic. Lancet Psychiatry 7(4):300–302. https://doi.org/10.1016/S2215-0366(20)30073-0

Fan FM (2003) SARS crisis intervention and psychological counseling model. Chin Mental Health J 09:600–602

Gao J et al (2020) Mental health problems and social media exposure during COVID-19 outbreak. Available at SSRN 3541120

Ghazi MM et al (2018) Robust training of recurrent neural networks to handle missing data for disease progression modeling. arXiv preprint arXiv:1808.05500

Greenberg N et al (2020) Managing mental health challenges faced by healthcare workers during a covid-19 pandemic. Br Med J 368:m1211

Holmes EA et al (2020) Multidisciplinary research priorities for the COVID-19 pandemic: a call for action for mental health science. Lancet Psychiatry 7:547–560

Horesh D et al (2020) Traumatic stress in the age of COVID-19: a call to close critical gaps and adapt to new realities. Psychol Trauma Theory Res Pract Policy 12(4):331

Hussain S, Hazarika GC (2014) Educational data mining model using rattle. Int J Adv Comput Sci Appl 5(6):22–27

Jacobs J et al (2010) Factors associated with increased risk perception of pandemic influenza in Australia. Influenza Res Treatment 2010:1–7. https://doi.org/10.1155/2010/947906

Kakulapati V et al (2020) Lexical analysis and mathematical modelling for analysing depression detection of social media reviews. In: Gunjan V (ed) Advances in cybernetics, cognition, and machine learning for communication technologies, Lecture notes in electrical engineering, vol 643. Springer, Singapore

Kang L et al (2020) The mental health of medical workers in Wuhan, China dealing with the 2019 novel coronavirus. Lancet Psychiatry 7(3):e14. https://doi.org/10.1016/S2215-0366(20)30047-X

Kawohl W et al (2020) COVID-19, unemployment, and suicide. Lancet Psychiatry 7:389–390. https://doi.org/10.1016/s2215-0366(20)30141-3
Khan A et al (2010) Social support as a moderator of positive psychological strengths and subjective well-being. Psychol Rep 106:534–538
Lapa TA et al (2017) Burnout syndrome and well-being in anaesthesiologists: the importance of emotion regulation strategies. Minerva Anesthesiol 83:191–199
Le Cessie S et al (1992) Ridge estimators in logistic regression. Appl Stat 41:191–201
Li L, Zhang X (2010) Study of data mining algorithm based on a decision tree. In: 2010 International Conference on Computer Design and Applications. IEEE 1:155–158
Li S et al (2020) Analysis of influencing factors of anxiety and emotional disorders in children and adolescents during home isolation during the epidemic of novel coronavirus pneumonia. Chin J Child Health 2020:1–9
Lima CKT et al (2020) The emotional impact of coronavirus 2019-nCoV (new coronavirus disease). Psychiatry Res. https://doi.org/10.1016/j.psychres.2020.112915
Lin NSR et al (1979) Social support, stressful life events, illness a model and an empirical test. J Health Soc Behav 20:108–119
Liotta EM et al (2020) Frequent neurologic manifestations and encephalopathy-associated morbidity in Covid-19 patients. Ann Clin Transl Neurol 7:2221–2230. https://doi.org/10.1002/acn3.51210
Mata R (2012) Cognitive bias. In: Encyclopedia of human behaviour, vol 1. Academic Press, New York, pp 531–535. https://doi.org/10.1016/B978-0-12-375000-6.00094-X
Mazza C et al (2020) Nationwide survey of psychological distress among Italian people during the COVID-19 pandemic: immediate psychological responses and associated factors. Int J Environ Res Publ Health 17:3165. https://doi.org/10.3390/ijerph17093165
MedlinePlus (2020) Stress and your health. Retrieved from https://medlineplus.gov/ency/article/003211.htm
Menotti A et al (2016) Cardiovascular and other causes of death as a function of lifestyle habits in a quasi-extinct middle-aged male population. A 50-year follow-up study. Int J Cardiol 210:173–178. https://doi.org/10.1016/j.ijcard.2016.02.115
Pandharipande PP et al (2013) Term cognitive impairment after critical illness. N Engl J Med 369:1306–1316. https://doi.org/10.1056/NEJMoa1301372
Psychological First Aid: American Psychological Association (n.d.). https://www.apa.org/practice/programs/dmhi/psychological/first-aid
Qiu J et al (2020) A nationwide survey of psychological distress among Chinese people in the COVID-19 epidemic: implications and policy recommendations. Gen Psychiatr 33(2):e100213
Rajkumar RP (2020) Ayurveda and COVID-19: where psychoneuroimmunology and the meaning response meet. Brain Behav Immun 87:8–9. https://doi.org/10.1016/j.bbi.2020.04.056
Rubin JG (2020) The psychological effects of quarantining a city. BMJ 368:m313
Rubin GJ et al (2020) Coronavirus: the psychological effects of quarantining a city. BMJ Opinion 368:m313
Serrano-Castro PJ et al (2020) Impact of SARS-CoV-2 infection on neurodegenerative and neuropsychiatric diseases: a delayed pandemic? Neurologia 35(4):245–251
Shameem A et al (2015) Analysis of significant factors for dengue infection prognosis using the random forest classifier. Int J Adv Comput Sci Appl 6(2):240–245
Strobl C et al (2007) Bias in random forest variable importance measures: illustrations, sources, and solutions. BMC Bioinformatics 8(1):25
Suicide Awareness Voices of Education (2020) Preventing suicide during and after the COVID-19 pandemic. Available from https://save.org/blog/preventing-suicide-covid-19-pandemic. Accessed 5 June 2020
Taylor S (2019) The psychology of pandemics: preparing for the next global outbreak of infectious disease. Cambridge Scholars, Newcastle upon Tyne
Vahia IV et al (2020) COVID-19, mental health and aging: a need for new knowledge to bridge science and service. Am J Geriatr Psychiatry 28(7):695

Wang XD et al (2002) Epidemiological characteristics and risk factors of post-traumatic stress disorder. Chin J Epidemiol 05:11–14

Weir K (2020) Grief and COVID-19: mourning our bygone lives. American Psychological Association, Washington

Wu YS et al (2020) Nervous system involvement after infection with COVID-19 and other coronaviruses. Brain Behav Immun 87:18–22. https://doi.org/10.1016/j.bbi.2020.03.031

Yan L et al (2021) The relationship between perceived stress and emotional distress during the COVID-19 outbreak: effects of boredom proneness and coping style. J Anxiety Disord 77:102328. https://doi.org/10.1016/j.janxdis.2020.102328

Zhou XB et al (2015) Risk factors of severity of post-traumatic stress disorder among survivors with physical disabilities one year after the Wenchuan earthquake. Psychiatry Res 228(3):468–474

Chapter 16
Public Transport Passenger's Density Estimation Tool for Supporting Policy Responses for COVID-19

Nilton A. Henao-Calle, Mateo Arroyave-Quintero, Semaria Ruiz-Alvarez, and Danny A. J. Gómez-Ramírez

Abstract We propose a strategy for estimating the public transport system occupancy using open data. Specifically, we use the origin–destination matrix, the population density, and routes' data to determine the traveler's density in the public transportation vehicles. This density estimation is intended to serve as a data supply for the government and entities in charge of analyzing the contagions and the policies required to avoid the global propagation of COVID-19. We have taken as a study case, the Metropolitan area of the Aburrá Valley located in the department of Antioquia, Colombia.

16.1 Introduction

Since the beginning of COVID-19 pandemic, the World Health Organization (WHO) has reported 146.06 million COVID-19 confirmed cases by April 26th, 2021, and has affected all the economies around the world. Particularly, in Colombia the reports talk about 2.74 million confirmed cases (World Health Organization 2021).

Nowadays, the metropolitan area of Aburrá Valley faced a third contagion wave that brings new challenges in public transportation. Primary, one sees such a

N. A. Henao-Calle (✉) · M. Arroyave-Quintero
Department of System Information, Instituto Tecnolólogico Metropolitano, Medellín, Colombia
e-mail: nahenao@unal.edu.co

S. Ruiz-Alvarez
Department of Energy and Processes, Universidad Nacional de Colombia, Medellín, Colombia
e-mail: seruizal@unal.edu.co

D. A. J. Gómez-Ramírez
Parque Tech at the Institución Universitaria Pascual Bravo and Visi6n Real Cognitiva S.A.S, Itagüí, Colombia

S. A. Hassan et al. (eds.), *Decision Sciences for COVID-19*, International Series in Operations Research & Management Science 320,
https://doi.org/10.1007/978-3-030-87019-5_16

challenge in the Metro system, a massive transport system that supports a big part of travels through the metropolitan area.

Despite the fact that the use of mass transport systems is considered a mean of high spread of the virus, it is important to add that a decrease in the spread of the virus is evidenced with the implementation of biosecurity measures for users and staff, who are in contact with the system (Hsiao et al. 2020). However, the spread rate tends to increase considerably with the prolonged use of the public transportation system, due to the growth in the time of exposure to virus agents (travel time) (Hörcher et al. 2020) (for a more general discussion and general contextualization see Dzisi and Dei (2020), Gkiotsalitis and Cats (2020), Shen et al. (2020), World Health Organization (2020), Zhao et al. (2020), and Brooks et al. (2021). In this sense, the use of technologies to mitigate the impact of COVID-19 has been diverse, allowing a greater level of control and contingency in the face of high rates of spread. This goes from the application of monitoring systems of the specific body temperature of subjects and of groups of people, until the implementation of data collection systems that allow understanding propagation patterns and the identification of various key areas. Here, in such an area the level of exposure to the virus is higher, and therefore the probability of contracting it is larger, facilitating the observation of the phenomenon of virus dispersal from a broader panorama and with more solutions and mitigation aspects (Hu et al. 2021).

In this regard, we have implemented an algorithm intended to quantitatively estimate the contagion exposure of persons traveling in the massive public transportation system. This with the purpose of supply the required data, for governmental entities in charge of development and implement the policies to mitigate the COVID-19 contagion.

16.2 Understanding the Context

Medellín is the second-largest city in Colombia, located at 6.217 degrees of latitude and −75.567 degrees of longitude.

Medellín is the main city in the metropolitan area of Aburrá Valley which encloses an area of 1157.39 km^2, and includes 10 municipalities: Barbosa, Girardota, Copacabana, Bello, Medellín, Envigado, Itagüí, Sabaneta, La Estrella, and Caldas.

Each municipality is subdivided in small regions called *communes*. Particularly, considering the subdivision in communes for the biggest municipality, Medellín, the resultant map is the one shown in Fig. 16.1. This is the division from which we will consider making posterior calculations.

Furthermore, Fig. 16.1 illustrates the population density, which is approximately 4 million inhabitants, concentrated mainly in the northern part of the region.

However, the industrial zone is located at the south of the Valley, which generates the need for the working population to cross the city every day, mainly using public transportation modes. This fact can be observed in Fig. 16.2 depicting the travels in

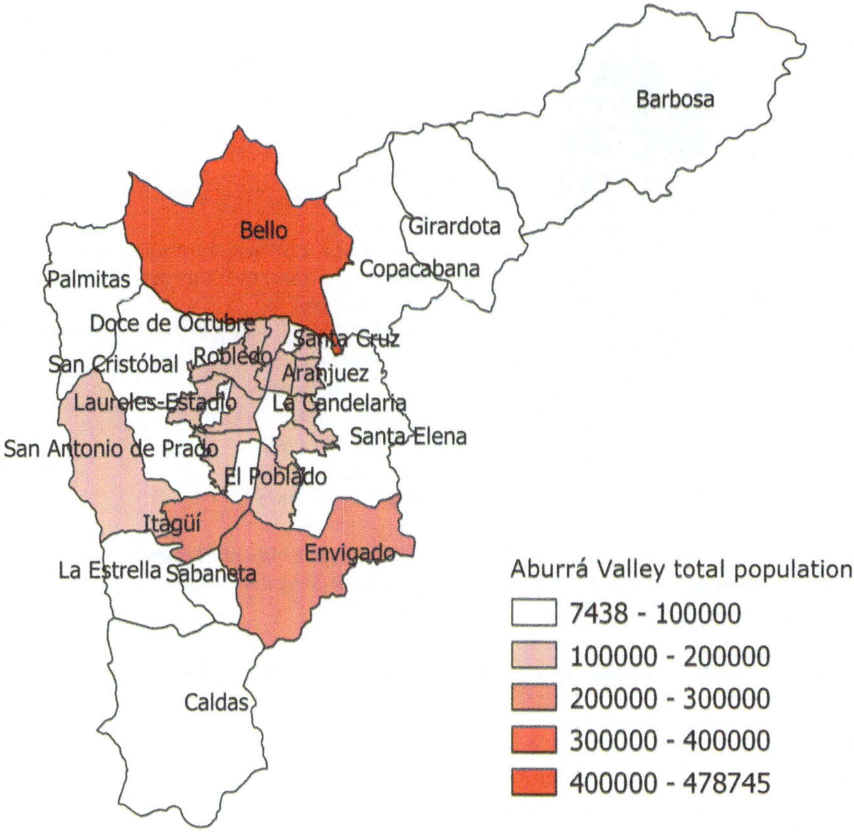

Fig. 16.1 Illustration of the Aburrá Valley's total population for 2018. Own elaborated from data provided by Area Metropolitana Valle de Aburrá (2019) and Alcaldía de Medellín (2019a)

public transport modes for a typical day, obtained from the origin–destination matrix data.

Specifically, the public transport systems in the Aburrá Valley region move around 1.5 million passengers per day, and, as we will explain in Sect. 16.2.2, our interest is to estimate the passenger density generated by the usual travelers. To perform this task, we need as input data, the estimated passenger demand of the public transport modes in the region, not only as an average amount, but geospatial and temporal information about travelers is also required. This information in the Aburrá Valley region can be obtained from different sources, as shown in the following section.

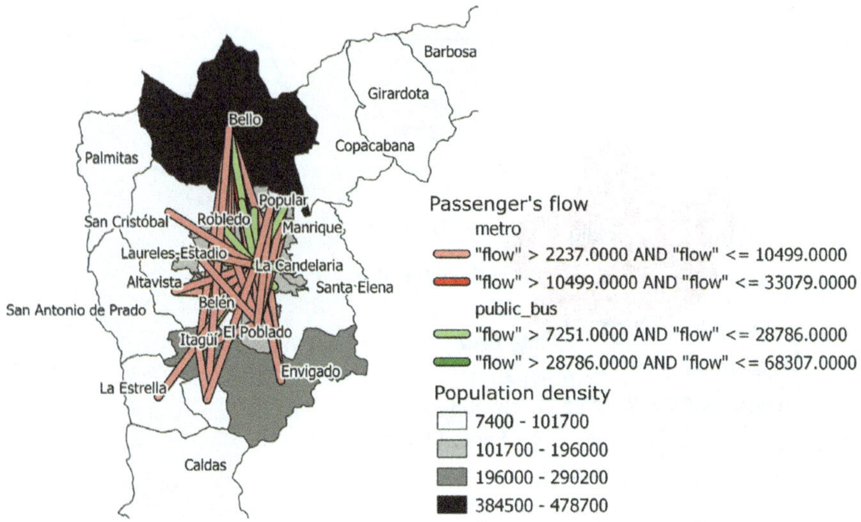

Fig. 16.2 Illustration of vehicle's flow among communes. Own elaborated from data provided by Area Metropolitana Valle de Aburrá (2019) and Alcaldía de Medellín (2019b)

16.2.1 Traffic Data Available for the Aburrá Valley Region

The data relating to the traffic demand in the Aburrá Valley can be classified into two types: open data and non-open data, as shown in Fig. 16.3.

The Open data, provided by governmental entities in the Aburrá Valley, encloses elements related to the geo-referenced road network topography, the land's use description, the public transport modes routes, the statistics about mobility (origin–destination matrix), the population density, and average traffic flow measures for the main roads.

On the other hand, the non-open data, owned by privates, contains more detailed traffic flow data than the origin–destination matrix, such as the Waze data, vehicular traffic measures from cameras, induction loops located at the traffic lights, and passenger affluence measures from cameras in the public transport vehicles.

In Fig. 16.3, we have highlighted the data sources we used to estimate the passenger's demand in the public transport system of the Aburrá Valley which are: the origin–destination matrix, the population density, and features of the public transport systems in the area (the geospatial topography, the capacity of the system, and the vehicles scheduling). These data inputs are deeply described in Sect. 16.3.1. Furthermore, as can be observed in Fig. 16.3, the developed algorithm does not include the non-open data because of the following reasons:

The acquisition of non-open data depends on the willingness of private entities to provide it. This data focuses on the traffic flow measures from private modes, i.e., own cars, own motorcycles, taxis, and freight transport; in the case of measurements generated by induction loops, cameras on traffic lights, and Waze events. Hence, it cannot be used to estimate the public transport passenger's demand.

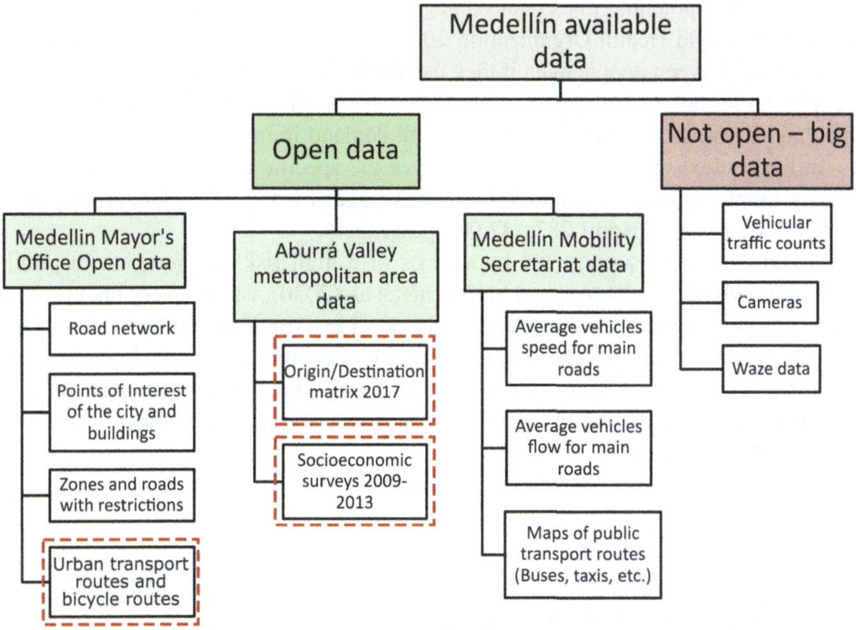

Fig. 16.3 Summary of traffic data available for Medellín city. Own elaborated from data provided by Area Metropolitana Valle de Aburrá (2019)

And finally, this data requires more complicated and computationally expensive procedures to be processed and analyzed, as those presented in reference (Acosta et al. 2021). For example, micro-services need to be developed to download the measured data for a certain period.

Therefore, to create an easily implementable tool, we used only open data sources. However, we have not limited the scope of the algorithm only to operate with them, as the trip chains for public transport systems can be obtained from another source, different than the origin–destination matrix. In any case, as we explain in Sect. 16.3.1 a file with the origin zone, destination zone, travel hour, transport mode, and travelers' amount, must be provided as input, to estimate the person density per square meter in each public transport system.

16.2.2 The Relationship Between Contagion and Public Transport Passenger's Density

Public transport systems in most of the world's cities are used massively, that is, they operate with an occupation of more than 100% of their nominal capacity. These spaces become potential places for contagion (Shen et al. 2020), since according to the WHO (World Health Organization 2021) the minimum distance that should be maintained between people even if they use bibs is 1.82 m. Therefore, the number of people should not exceed 1 person per meter square. However, this condition is not met, under normal conditions of passengers' demand in most modes of transport.

In the context of this chapter, we analyze the specific conditions of the Aburrá Valley's metro system, which mobilizes an average of 1,536,377 passengers per day, and in 1 year can transport 287,447,000 passengers.

Additionally, we support the hypothesis of a high risk of contagion in public transport with the results reported in Hörcher et al. (2020), where a very high level of spread of viral substances was found in closed spaces. In this context, a relationship usually occurs as a product of contact through proximity or physical contact with infected people, where there is direct exposure to respiratory residues generated by coughing or sneezing and even through indirect contact. For example, there is a touch with contaminated surfaces, where in both cases such confinement is the key factor that increases the probability of infection.

16.3 Software Tool Description

16.3.1 Description of the Algorithm Input Data

16.3.1.1 Origin–Destination Matrix

The Origin–Destination (OD) matrix (Area Metropolitana Valle de Aburrá 2017a) describes the travels performed for a certain population group among geospatial zones. It gives information about origin zones, destination zones, travel schedules, transport mode, traveling persons per household, household socioeconomic level, and inhabitant's economic activities.

These geospatial zones are called SIT zones in the Aburrá Valley and are illustrated in Fig. 16.4, where the population size surveyed to construct the OD matrix is also depicted. The 536 SIT zones are inside bigger areas which we denoted as communes for simplicity. The OD survey contains 87,614 samples from a total population of 3,909,676 inhabitants, which represents 2.24% of the total population. This fact implies that the OD matrix is not scaled for the total population, and only contains the travel behavior of a percentage of the inhabitants. Hence, a previous procedure of data filtering and scaling is required to obtain the travel patterns for the complete population in the zones. This pre-processing step also includes the population distribution in the same spatial zones as the OD matrix data, as we will explain in the following section.

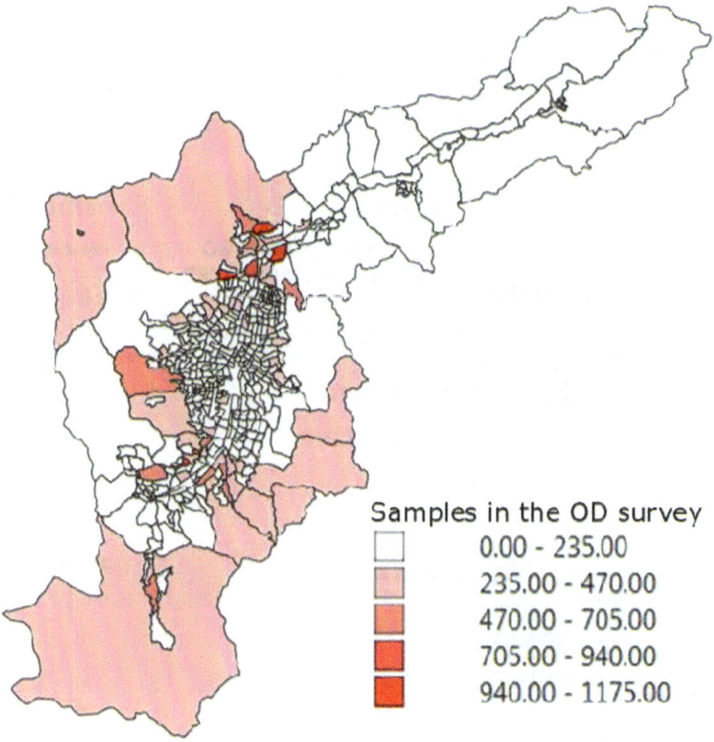

Fig. 16.4 Illustration of sample size of the OD survey. Own elaborated from data provided by Area Metropolitana Valle de Aburrá (2017a) and Alcaldía de Medellín (2019b)

16.3.1.2 Population Data

The population data of the Aburrá Valley is available in Alcaldía de Medellín (2019a) and Gobernación de Antioquia (2019) for the spatial resolution illustrated in Fig. 16.1, i.e., it is available for communes and municipalities. Note that the communes and municipalities are not the same spatial areas than the SIT zones of the OD data, as can be observed in Fig. 16.4. So, in a pre-processing step, the communes population data needs to be divided into SIT zones. In this case, we have performed this task with the land usage data, which is also open as can be observed from Fig. 16.3. Therefore, we have split the total population in the buildings tagged as residential according to its area.

16.3.1.3 Public Transport Systems Features

Massive public transport systems are a fundamental part of the cities' mobility around the world, however, the new reality derived by the COVID-19 pandemic, brings new challenges since the usual (pre-pandemic) occupancy of the transport

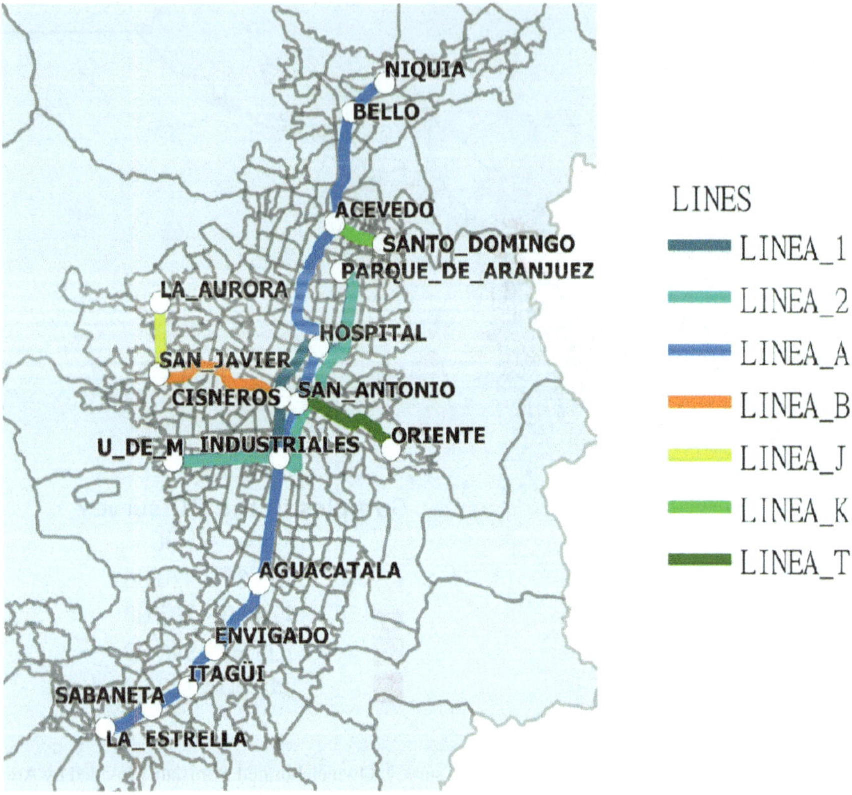

Fig. 16.5 Illustration of the Aburrá Valley METRO system

systems could lead to a considerable increase in the spread of the virus. Particularly, in the Metropolitan Area of the Aburrá Valley 2,780,000 trips, are made, daily in the public transport systems (Area Metropolitana Valle de Aburrá 2017b). Moreover, the Metro, which is the main public transport system of the region and the system focus of this chapter, transports an average of 1,536,377 passengers per day, at working days, reaching an annual value of 318,447,000 passengers (Empresa de Transporte Masivo del Valle de Aburrá Ltda 2019).

Motivated by this high passenger affluence in the Metro system, in this chapter, we analyze the passenger density in the 4 modes of the Aburrá Valley Metro mass transport, illustrated in Fig. 16.5. These modes are the railway line, the tram, the aerial cable system (Metrocable), and the integrated bus system (Metroplus).

Next, we give a detailed description of them, considering the existing infrastructure for 2017 (the year in which the origin–destination survey was performed):

- *Railway Line:* The railway line of the Medellín Metro has an approximate length of 31.3 km, a total of 28 stations, 5 of these serve as transfer nodes to other modes of the mass transport system. The main rail fleet is made up of a total of 80 trains, each train has 3 cars, which have the capacity to transport approximately 1800 passengers through two main lines interconnected at the San Antonio station:

- Line A: This line starts at the Niquia station, located in the municipality of Bello, and ends at the La Estrella station, located in the municipality of La Estrella (25.8 km).
- Line B: This line starts at the San Antonio station and ends at the San Javier station; both stations are located in Medellín city (5.5 km).

These two lines, which conform to the backbone of the Metro mass transport system, use wagons with 3.80 m of height, 3.20 meters of width, and 22.86 m of length. They have 148 seats and 4 double electronic doors on each side. And the recommended passenger density (without considering pandemic limitations) is 6 passengers per square meter, approximately.

- *Tram:* The Tram is composed of line T, which is connected to the main lines of the Metro system (A and B) at the San Antonio station, that is, its starting point, and ends at the Oriente station, having a length of 4.3 km.

  The tram system has 9 stations and 12 vehicles, each one with a transport capacity of 300 passengers. Furthermore, the vehicle's dimensions are 2.89 m in height, 2.22 meters in width, and 39 m in length. Each vehicle has 5 double electronic doors on each side, and the recommended passenger density (without considering pandemic limitations) is 6 passengers per square meter, approximately.
- *The aerial cable system (Metrocable):* This line is based on an air suspension system of cabins attached to a cable driven by a pulley. The length of the cable is 14.62 km, and the dynamics of the system is based on the movement of the cable through the stop stations. In 2017, this air transport system had 6 stations distributed in 2 lines:

  - Line K: Starting at Acevedo station and ending at Santo Domingo station.
  - Line J: Starting at San Javier station and ending at Aurora station.

The airline has a total of 212 cable cabins and each one of these cabins has dimensions 1.80 m, 2.20 m, and 1.80 m in width, interior or height, and length, respectively. The system cabins have the capacity to transport 8 persons seated and 2 standing.

- *Integrated Bus System (Metroplus):* This integrated system was developed with the intention of strengthening the Metro transport system and allowing the transportation of the population located away from the main lines.

  For the year 2017 this system had two lines:

  - Bus line 1: Starting at the Industriales station (21 stops).
  - Bus line 2: Starting at the Industriales station (23 stops).

The integrated bus lines have approximately 77 articulated gas buses and 1 articulated electric bus with a capacity of 154 passengers.

16.3.2 *Software Architecture*

We have built, in the pre-processing step, a database with the travel information of the OD survey. This database contains the origins, destination, main transport mode, travel time, and stage of the trip (departure or return). This file is exported in a csv format and taken as input by the software tool "*demand_generation_software.*" Whose operation is described below and illustrated in Fig. 16.6.

The "*demand_generation_software*" tool, programmed in Python, is used to calculate the occupation of the metro system or a public transport system considering the population traveling in a typical day. The input files required to run the tool are:

- Correspondence file between SIT zones, metro stations, buses stop. The minimum distance to the zones' centroids.
- Data from the OD matrix.
- File with coordinates of the metro stations or stops of the public transport system.
- File with frequencies of trains or vehicles of the public transport system, where the passenger capacity of them is also included.

The software is composed of the following classes: "*metro*", "*buses*", "*metro_routes*", "*buses_routes*", "*concurrences*", and "*density*".

Class "*metro*"

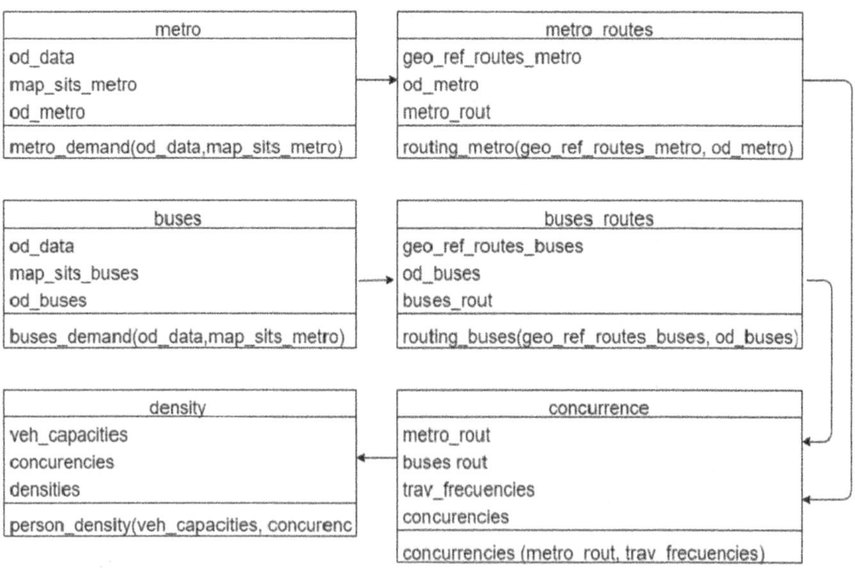

Fig. 16.6 Architecture of the proposed software tool

The class "*metro*" generates the data of the OD matrix of the metro system. This data is generated every hour.

The class attribute "*map_sits_metro*" converts trips between SIT zones to trips between metro stations.

Class "*buses*"

The "*buses*" class generates the hourly demand for buses (integrated into the Metro system) based on the OD data.

Using the classes described above, a database is generated with the number of users who travel by metro and buses with their respective SIT zones of origin and destination.

Class "*metro_routes*"

Subsequently, the class "*metro_routes*" is used to calculate the minimum distance route using the integrated systems of the metro to travel between a SIT area of origin and a SIT area of destination. And in a similar way, the class "*buses_routes*" is used to calculate the minimum distance route using public transport buses to travel between a SIT area of origin and a SIT area of destination. With this information is obtained a database that contains the number of passengers in the public transport system for a typical day.

Class "*concurrence*"

The class "*concurrence*" provides a matrix of total trips on the public transport routes.

Class "*density*"

Subsequently, with this matrix, and using the class "*density*" the occupation of the vehicles on each route is calculated, as well as the density of people on the route per m^2.

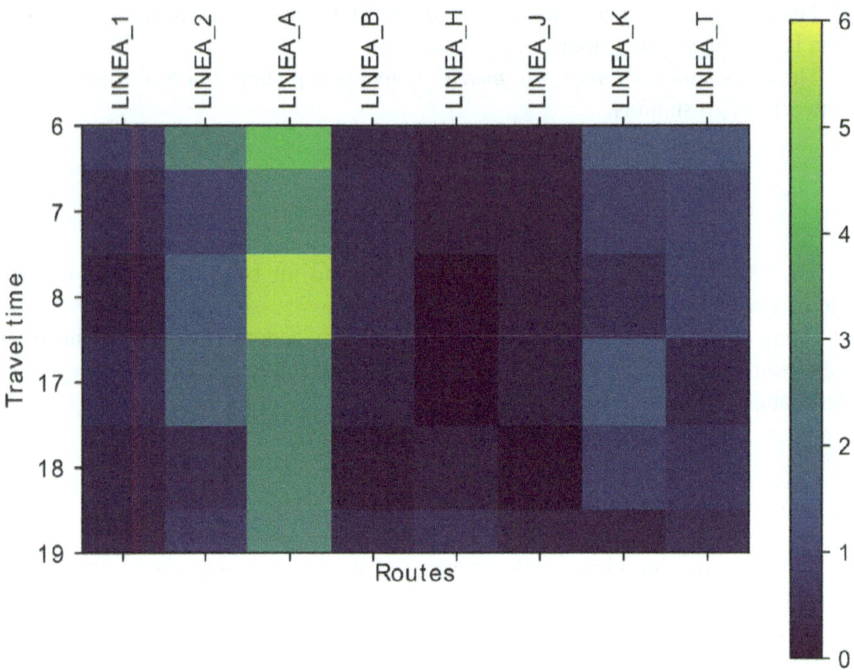

Fig. 16.7 Passenger density in the Aburrá Valley public transport system

16.4 Main Results

Figure 16.7 depicts the passenger density that the main routes of the Metro public transport system will have if the passengers' demand behaves as the origin–destination matrix indicates. Furthermore, we have included the travel time, but we have selected only the morning and afternoon peak hours.

From Fig. 16.7, it can be observed that the route called "*LINEA_A*" is the most congested in the whole day, and the hour with the highest passenger affluence is 8 A.M. At this hour, the person densities reach values of 6 persons per square meter. This implies that if any mobility restriction is imposed, the social distancing requirements will not be accomplished. Additionally, the social distancing will not be fulfilled in the routes "*LINEA_2*," "*LINEA_K*," and "*LINEA_T*." However, we highlight these results depending on the traffic behavior extracted from the origin–destination survey, which is applied only to one population sample, and which was performed in the year 2017. Hence, the estimated densities are subject to the OD sample errors, the impressions and their validity regarding the present time. As we previously mentioned, the data about mobility in the geospatial zones of the Aburrá Valley can come from other data sources different from the OD matrix, and our algorithm can be similarly applied. But the processing of this data, apart from the fact that it is not allowed for public use, because the data is non-open, will require the implementation of micro-services and other software processes more computationally complex than the processing of the OD data survey.

16.5 General Conclusions

In this research we have developed a software tool for the estimation of the passenger's densities in the public transport systems of a city; specifically, we have applied the tool to the Aburrá Valley region, located in Colombia, which contains the city of Medellín.

The main advantage of the software is that it allows a fast computation of the number of persons per square meter in the public transport systems, taken as input open and easy access data.

The passenger densities given by the algorithm, as output, are essential information for government agencies so that they can make decisions to mitigate the spread of COVID-19. Especially these results give the evidence that in order to maintain the spatial distancing between in the metro system recommended by the WHO during all the time periods, strategic logistic measures should be implemented by (local) ministries of transport.

For instance, some of the policies that could be suggested to achieve the established recommendations are:

First, the implementation of staggered schedules for the beginning of the workday discriminated with different between economic sectors, prioritizing the movement of people linked to essential activities.

Second, the constrain of movement for specific zones of the city with differentiated time schedules and with a strong emphasis on some primary places of more probable conglomeration focus.

And third, the development of campaigns motivating some of the citizens to travel among the city by bicycle or any similar medium, to reduce the number of people using the metro system.

Acknowledgments Danny A. J. Gómez-Ramírez wishes to thank Fabian Suarez, Juan Carlos Gómez, Stephanie Cuellar, and Liliana Lopera for all their friendship and support.

References

Acosta AF, Pérez E, Espinosa J, Portilla CR, Ruiz S, Noreña JP, Villa JP, Sánchez DA (2021) Plataforma de apoyo a toma de decisiones y recolección de información en tiempo real para el análisis del impacto epidemiológico de las medidas de interes público en la contingencia del COVID-19. Medellín, Colombia

Area Metropolitana Valle de Aburrá (2017a) Encuesta de Origen - Destino - Análisis de Viajes. In: Obs Metropolitan Information. https://www.metropol.gov.co/observatorio/Paginas/encuestaorigendestino.aspx

Area Metropolitana Valle de Aburrá (2017b) Encuesta Origen Destino 2017 - Datos por Hogares. In: Datos Abiertos del Área Metropolitana del Valle Aburrá. https://datosabiertos.metropol.gov.co/dataset/encuesta-origen-destino-2017-datos-por-hogares

Area Metropolitana Valle de Aburrá (2019) Área Metropolitana del Valle de Aburrá. https://datosabiertos.metropol.gov.co/. Accessed 12 Jan 2020

Brooks JHM, Tingay R, Varney J (2021) Social distancing and COVID-19: an unprecedented active transport public health opportunity. Br J Sports Med 55:411–412. https://doi.org/10.1136/bjsports-2020-102856

de Antioquia G (2019) Población. In: Gob Antioquia. https://www.dssa.gov.co/index.php/estadisticas/poblacion. Accessed 9 Feb 2019

de Medellín A (2019a) Proyecciones de Población 2016 a 2020 de Medellín. In: Alcaldía de Medellín. https://www.medellin.gov.co/irj/portal/medellin?NavigationTarget=navurl://06bdb4d911e35cb9d3de717115deedc3. Accessed 9 Feb 2019

de Medellín A (2019b) OpenData Alcaldía de Medellín. https://geomedellin-m-medellin.opendata.arcgis.com/. Accessed 26 Dec 2019

Dzisi EKJ, Dei OA (2020) Adherence to social distancing and wearing of masks within public transportation during the COVID 19 pandemic. Transp Res Interdiscip Perspect 7:100191. https://doi.org/10.1016/j.trip.2020.100191

Empresa de Transporte Masivo del Valle de Aburrá Ltda (2019) Memoria de sostenibilidad. Medellín

Gkiotsalitis K, Cats O (2020) Public transport planning adaption under the COVID-19 pandemic crisis: literature review of research needs and directions. Transp Rev 0:1–19. https://doi.org/10.1080/01441647.2020.1857886

Hörcher D, Singh R, Graham DJ (2020) Social distancing in public transport: mobilising new technologies for demand management under the COVID-19 crisis. SSRN Electron J. https://doi.org/10.2139/ssrn.3713518

Hsiao SH, Chen TC, Chien HC, Yang CJ, Chen YH (2020) Measurement of body temperature to prevent pandemic COVID-19 in hospitals in Taiwan: repeated measurement is necessary. J Hosp Infect 105:360–361

Hu M, Lin H, Wang J, Xu C, Tatem AJ, Meng B, Zhang X, Liu Y, Wang P, Wu G, Xie H, Lai S (2021) Risk of coronavirus disease 2019 transmission in train passengers: an epidemiological and modeling study. Clin Infect Dis 72. https://doi.org/10.1093/cid/ciaa1057

Shen J, Duan H, Zhang B, Wang J, Ji JS, Wang J, Pan L, Wang X, Zhao K, Ying B, Tang S, Zhang J, Liang C, Sun H, Lv Y, Li Y, Li T, Li L, Liu H, Zhang L, Wang L, Shi X (2020) Prevention and control of COVID-19 in public transportation: experience from China. Environ Pollut 266. https://doi.org/10.1016/j.envpol.2020.115291

World Health Organization (2020) Supporting healthy urban transport and mobility in the context of COVID-19. In: Social Determinants of Health. https://www.who.int/publications/i/item/9789240012554. Accessed 30 Apr 2021

World Health Organization (2021) COVID-19 weekly epidemiological update 22. World Health Org:1–3

Zhao S, Zhuang Z, Ran J, Lin J, Yang G, Yang L, He D (2020) The association between domestic train transportation and novel coronavirus (2019-nCoV) outbreak in China from 2019 to 2020: a data-driven correlational report. Travel Med Infect Dis 33:2019–2021. https://doi.org/10.1016/j.tmaid.2020.101568

Part IV
Optimization Techniques

Chapter 17
A Generalized Model for Scheduling Multi-Objective Multiple Shuttle Ambulance Vehicles to Evacuate COVID-19 Quarantine Cases

Said Ali Hassan and Ali Wagdy Mohamed

Abstract This chapter is devoted to present a generalized model for scheduling multiple ambulance vehicles from multiple ambulance centers assigned to evacuate COVID-19 patients. The proposed formulation is a multi-objective multiple 0–1 mathematical model as a new application of the multi-objective multiple 0/1 knapsack problem.

The scheduling aims at achieving the best utilization of the time shift as a planning time window. The best utilization of time is evaluated by a compromise between maximizing the number of evacuated people who might be infected with the virus to the isolation hospitals and maximizing the evacuated patients having higher relative priorities measured according to their health status. The complete mathematical model for the problem is formulated including the representation of binary decision variables, the problem constraints, and the multi-objective functions.

The proposed multi-objective multiple ambulances model is applied to an illustrated case study in Great Cairo, Egypt, the case study aims at improving the scheduling of ambulance vehicles in the back-and-forth shuttle movements between patient locations and the available multiple isolation hospitals with multiple ambulance vehicles. The solution procedure is illustrated while two efficient solutions for the case study with different numbers of evacuated patients are obtained. The proposed mathematical model is so general that it can be applied to cases covering the whole Governorates and even the whole country all over the world.

S. A. Hassan
Department of Operations Research and Decision Support, Faculty of Computers and Artificial Intelligence, Cairo University, Giza, Egypt

A. W. Mohamed (✉)
Operations Research Department, Faculty of Graduate Studies for Statistical Research, Cairo University, Giza, Egypt

Department of Mathematics and Actuarial Science, School of Sciences and Engineering, The American University in Cairo, New Cairo, Egypt

S. A. Hassan et al. (eds.), *Decision Sciences for COVID-19*, International Series in Operations Research & Management Science 320,
https://doi.org/10.1007/978-3-030-87019-5_17

Keywords Scheduling shuttle ambulance · COVID-19 · Quarantine cases · Binary programming · Multi-objective multiple knapsack problem

17.1 Introduction

Currently, the entire world is suffering from a global epidemic of COVID-19 that has infected thousands of people in almost all countries, Cleemput et al. (2020). In December last year, Wuhan in China was the origin of pneumonia of unknown cause. By Jan 2020, assured cases were detected outside Wuhan, The Lancet website (2020).

Nowadays, the new Coronavirus (COVID-19) puts humans in all countries in front of huge danger. The Center for Disease Control and Prevention (CDC) declares the major signs of COVID-19 so that any individual can discover whether or not he has such symptoms, World Health Organization (2020). The updated worldwide total cases and deaths of coronavirus are recorded daily, Worldometer website (2021).

The problem of the large number of people who feel symptoms of coronavirus infection has worsened in the recent period, which necessitates appropriate planning for the speedy transfer of these cases to quarantine hospitals. This is to be done with the best possible efficiency in the utilization of time while considering also the health and morbidity factor of those people. Patient Transport Service (PTS) have a key role in ensuring patients who are frail or vulnerable or have no other means of transport get to their health appointments safely and on time. Patient transport is a service that transfers patients to and from medical facilities, and ambulances are used to respond to emergency medical services. They can rapidly transport paramedics and other first responders to the scene, carry equipment for administering emergency care and transport patients to a hospital or other definitive care, Black (2017).

The optimal planning of the ambulance facility and the scheduling of its movements back and forth to transport suspects with the new coronavirus are of great importance. This should be considered at the level of specialists and researchers for the optimal utilization of this vital facility with the ultimate goal of reducing patient mortality and morbidity, Carlos (2017).

Scheduling of shuttle ambulance vehicles for COVID-19 quarantine cases from a single ambulance center is formulated and solved for an illustrated case study is presented, Hassan et al. (2020a, b). This research presents a generalized model for scheduling multi-objective multiple shuttle ambulance vehicles to Evacuate COVID-19 Quarantine Cases to multiple quarantine centers. The objectives to be achieved are to maximize the number of suspects evacuated patients while maximizing the relative priorities given to prospected patients according to their health conditions so that more weight is given to those who are in more dangerous status than others.

All the prospected patients are transported to the isolation hospitals to conduct the necessary analyses and treatment. The optimal scheduling of shuttle trips to transport

COVID-19 patients to isolation hospitals when receiving calls from people at risk of infection with the virus, especially those in contact with patients whose positive tests have been confirmed can save lives. It can prevent contacting with those who might be carrying the virus with their families and relatives where the infection can transfer to more people.

The structure of the chapter is as follows: The second section explains the problem of scheduling shuttle ambulance vehicles for COVID-19 patients. It Highlights the efforts done by the ambulance specialists as battlefield partners who are indifferent to the seriousness and obstacles they face. It reveals also the current plan of the ambulance facility to perform their job starting from receiving people calls, preparing and prioritizing patient lists, schedule ambulance vehicles, and then transport patients to the isolation hospitals. Finally, this section concludes the necessity of using scientific methods to ensure the best scheduling of the ambulances vehicle to do the work in the best manner, and that the statement of this problem resembles the well-known Multi-objective Multiple 0–1 Knapsack Problem.

Sect. 17.3 presents a concise review of the Knapsack Problem (KP). Important related versions of this famous problem are also mentioned like the 0–1 KP, Multi-objective Knapsack Problem (MOKP), and the Multi-objective Multiple Knapsack Problem (MOMKP).

The mathematical representation for scheduling the shuttle ambulance vehicles is designed in sect. 17.4 including all needed formulations. The proposed model is a Binary Constrained Multi-Objective Multiple Knapsack Problem, the steps of the solution procedure are also explained.

An illustrative case study in Cairo, Egypt with all relative data is explained in sect. 17.5. Its complete mathematical model is formulated and solved to obtain efficient solutions. Section 17.6 gives the main conclusions and the suggested points for future research.

17.2 Scheduling Shuttle Ambulance Problem

Suddenly, as a part of this world, we found ourselves face to face in front of a battle with COVID-19, the virus that necessitates confrontation, honing all national efforts, and raising the level of preparedness in all health sectors. Doctors and nursing make extraordinary efforts, but one must not forget the link among them and patients, they are the ambulance specialists who take nonstop daily trips to and from homes to isolation hospitals. They are battlefield partners who provide round-the-clock assistance in the event of any medical emergency and are indifferent to the seriousness and obstacles they face, Global Medical System (GMS) website (2021).

The workers in the ambulance system make great sacrifices, they work in very difficult conditions, and play an important pivotal role in facing the emerging crisis of Coronavirus to prevent the spread of this epidemic. All of them are heroes at the forefront of the first line in the face of this dreaded global epidemic that threatens all countries. The ambulance facility is the unknown soldier facing Coronavirus, where the personnel make a great effort and play a heroic role, and they deserve our support and honor, all people acknowledge the risks that they are taking and offer their gratitude, Sanders (2021).

The telephone communication rates at the ambulance facility have increased significantly during this period, as the number of phone calls that the ambulance agency receives is about 2500 calls per day from the governorates of Greater Cairo. The Egyptian Ambulance Organization (EAO) has transported between 150 and 170 people who are either infected or suspected to be infected with COVID-19 to hospitals places of isolation and medical examination on a daily basis, Egypt Independent website (2021).

The current plan of the ambulance facility is to maintain enough ambulances equipped with their crew on standby. They are ready to deal with any Corona situation that appears in the province, whether positive or suspected. The authority has also an alternative scenario to prepare other cars for each governorate in case the conditions change for the worst. There are enough drivers and paramedics, all of whom receive continuous training. It is not possible to use a paramedic for real-time interaction before he goes through entire practical training through simulated experiments, Al-Ahram Gate website (2021).

The ambulance operations room receives the reports through a contracting company, which sends the communication in writing to appear on the screen with details of the case, name, address, and all relevant data. The operations room employee immediately begins dealing with it and calls the ambulance to the location of the caller. He continues to follow the case to make sure it arrives at the hospital. Data is recorded on-screen, to calculate the report receipt time, response time, and status arrival, Boone et al. (2015).

Ambulance services are classified into these categories: pre-hospitalization transfer, inter-facility transfer, post-treatment transfer, and large-scale gatherings. During the isolation trip, they are busy with the patient following for oxygen, especially if he suffers from shortness of breath, as well as to the heartbeat through screens and devices. They are also in constant contact with the ambulance facility and recipients for any change in instructions or trip details, Little (2019).

The geographical scope of ambulance work is divided into sectors, where a hospital is designated for medical isolation of the suspected cases and several ambulances are allocated for corona only. The operating room receives the reports and a schedule is prepared according to the geographical location and the condition of the patient, Gov. UK website (2021). Ambulances from multiple centers drive in shuttle movements back and forth from the isolation hospital to the places of suspected COVID-19 infection as seen in Fig. 17.1.

Here, scientific methods must be used to ensure the best scheduling of ambulances. The ultimate objective is to perform the evacuation duties while making the best utilization of the available time. Two competing objectives arise: transporting the largest number of suspects patients, while also considering the priority of cases according to their health status. This statement of this problem resembles the well-known Multi-Objective Multiple 0–1 Knapsack Problem (MOM0/1KP).

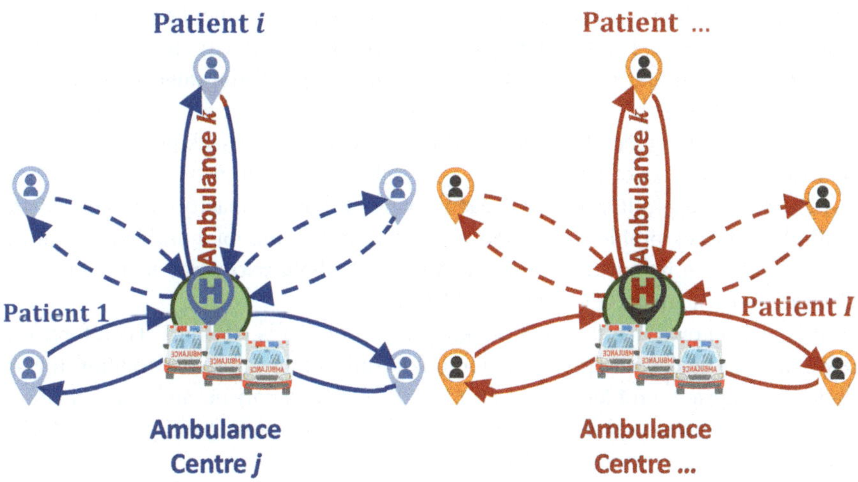

Fig. 17.1 Scheme of shuttle ambulance vehicles from multiple centers

17.3 Multi-Objective Multiple Knapsack Problem: A Literature Review

The knapsack problem (KP) is a fundamental problem in combinatorial optimization. The KP involves the maximization or minimization of a value, such as profits or costs. A knapsack can only hold a certain weight or volume that can accommodate different types of items but with limitations in total volume, weights, or both. The question is to decide the number of each item to pick up so that to minimize the total cost or to maximize the total profit (Render et al. 2020). In the Integer type KP, items can have arbitrary integer values, whereas the 0–1 KP limits the number of each type to 0 or 1. The 0–1 KP does not allow the user to put multiple copies of the same items in their knapsack. The 0–1 knapsack is a special case in which each input can be loaded or not into the knapsack.

There are other variants of KP like the Multi-dimensional 0–1 KP, Alaya et al. (2004), Fidanova (2004, 2005, 2007), or the Multiple 0–1 Knapsack Problem, Ke et al. (2010), Shahrear et al. (2010), Ji et al. (2007). The integer KP and the 0–1 KP can be solved using some dynamic programming algorithms by deriving a recurrence equation expressing a solution to an instance in terms of solutions to its small-scale instances, Vickram et al. (2016). Dynamic Programming follows the principle of optimality to reach the final solution, Anany (2003). The 0–1 KP and its versions are famous NP-hard problems, dynamic programming techniques can solve such problems in pseudo-polynomial time, Kellerer et al. (2004).

In combinatorial optimization problems like that of KP, exact methods are impractical in finding an optimal solution because the run time is exponentially increased as the problem size increases, Basheer and Algamal (2019). Therefore, interest in the application of the meta-heuristic algorithms has become necessary to solve these problems and obtain the results in a reasonable time, El-Ghazali (2009), Yang (2014, 2015). The 0–1 KP can be solved also by greedy genetic algorithm, genetic algorithm and rough set theory and ant weight-lifting algorithm, Raj and Vitthalpura (2017).

One interesting generalization of KP is the multiple knapsack problem, where a set of items and a set of knapsacks are given, Kuchta and Rynca (2019). Each item has a size and a profit value, and each knapsack has a capacity. The goal of the multiple knapsack problem is to find a subset of maximum total profit such that the picked items can be packed into the knapsacks without exceeding their capacities, Chekuri and Khanna (2006), Jansen (2009, 2012).

It is well-known that the multiple knapsack problem is NP-hard. Therefore, it is reasonable to design an approximation algorithm to find a feasible solution for the multiple knapsack problem, Li et al. (2015). Various heuristic and exact methods have been devised to solve it. Genetic Algorithm (GA) shows good performance on solving static optimization problems, Ünal (2013).

In Multi-Objective Knapsack Problem (MOKP), a major challenge is to generate efficient solutions, that have the property that no improvement on any objective is possible without sacrificing on at least another objective. A survey and an annotated bibliography about multi-objective combinatorial optimization can be found in Ehrgott and Gandibleux (2000). Many real-world applications are reported in the multi-objective case dealing with capital budgeting, selection of transportation investment alternatives, relocation issues arising in conservation biology, and planning remediation of contaminated light station sites, Chabane et al. (2005). Several exact approaches have been proposed in the literature to find the efficient set specially designed for the bi-objective case Klamroth and Wiecek (2000), Captivo et al. (2003) and Da Silva et al. (2005). Besides these exact methods, approximation algorithms Erlebach et al. (2002) and metaheuristics, Gandibleux and Freville (2000), Da Silva et al. (2006) Da Silva et al. (2007) have been proposed. Many evolutionary algorithms for solving multi-objective 0/1 knapsack problems are also proposed, Groşan et al. (2003a). In the case of multi-objective evolutionary algorithms, the outcome is usually an approximation of the Pareto-optimal set, which is denoted as an approximation set Groşan et al. (2003b) and Zitzler et al. (2003).

The Multi-Objective Multiple Knapsack Problem (MOMKP) combines the two cases of multi-objective knapsack problem (MOKP) and the multiple knapsack problem (MNP). Many approximate solution methods have been developed to solve MOMKP, Ma et al. (2018). Evolutionary algorithms (EAs) have been successfully applied to multi-objective combinatorial optimization (MOCO) problems in the last decades. EAs maintain a population of solutions and thus they can obtain multiple efficient solutions in a single run, Soylu and Köksalan (2009).

Many applications are formulated and solved as (KP) like: energy management, radio networks, resource management, mining application, wireless communication, power allocation management, production planning, mobile computing, production planning, radar systems, architectures for localization, demand response, tour conducting, waste management, workflow mapping, and network selection for mobile nodes, Biglar (2018a, b).

17.4 Mathematical Model for the Problem

The Shuttle Ambulance Scheduling (SAS) is defined as follows:

- Each patient place is approached only once by only one ambulance vehicle from one ambulance center, or he is not reached at the considered time shift.
- The shuttle ambulance performs several shuttle movements for transferring patients from their places to the isolation hospital(s), the process is performed back and forth repeatedly bounded by the time of the scheduling time shift.
- The overall goal of the problem to be achieved is the best utilization of the available time expressed by 2 competing objectives: maximizing the number of evacuated patients and maximizing efficiency through prioritizing dangerous cases.

The mathematical model is formulated as a multi-objective multiple 0–1 knapsack problem as follows:

Decision Variables:

Let:

$$x_i^{j,k}\begin{cases}1,\text{if patientiis approached from the ambulance centre } j \text{ by ambulance vehicle numberk}, i=1,2,\ldots,I; j=1,2,\ldots,J \text{ and } k=1,2,\ldots,K_j\\ 0,\text{otherwise.}\end{cases}$$

Constraints:

(1) Patients Constraints:

Each patient i is either approached by only one ambulance vehicle or not picked up in the considered time slot:

$$\sum_{j=1}^{J}\sum_{k=1}^{K_j} x_i^{j,k} \leq 1, i = 1, 2, \ldots, I. \tag{17.1}$$

(2) Time Shift Constraints:

Time shift T is the time shift for employees for planning, scheduling, and executing movements of the ambulance vehicles according to received call records. The total time spent by an ambulance vehicle in one transportation segment consists of travelling from the hospital to the place of patient, riding patient and handing him over. The total activity times for any ambulance vehicle should be within the available time shift T.

$$2\sum_{i=1}^{I} t_{ji}x_i^{j,k} + \sum_{i=1}^{I} t_i^r x_i^{j,k} + \sum_{i=1}^{I} t_i^h x_i^{j,k} \leq T, j = 1, 2, \ldots, J \text{ and } k = 1, 2, \ldots, Kj \tag{17.2}$$

where:

t_{ji} = Transportation time between the isolation hospital (center) j and patient place i, $j = 1, 2, \ldots, J$ and $i = 1, 2, \ldots, I$.

t_i^r= Riding time of patient i to an ambulance vehicle, $i = 1, 2, \ldots, I$,
t_i^h= Handing over time of patient i to a hospital, $i = 1, 2, \ldots, I$,
T = Shift time duration.

The first part is for travelling from the isolation hospital j to all the picked-up patient' places forth and back, the second part is the sum of riding time of patients to the ambulance, and the third part is the sum of handing over times of patient at the hospital.

(3) Binary Constraints:

All the decision variables are 0–1.

$$x_i^{j,k} = 0 \text{ or } 1, i = 1, 2, \ldots, I; j = 1, 2, \ldots, J \text{ and } k = 1, 2, \ldots, Kj \tag{17.3}$$

(4) Avoid the Trivial Solution

To avoid the trivial solution that all patients are assigned to one ambulance vehicle from any ambulance center, the following condition should hold:

The sum of travelling from an isolation hospital to all the picked-up patient' places forth and back, the sum of riding time of patients to the ambulance, and the sum of handing over times of patients at the hospital should be greater than the considered scheduled time shift. This condition should be checked before executing the mathematical model:

For each ambulance center:

$$\begin{aligned} &\text{The total transportation time} \\ &\quad + \text{Total riding time of patients to the ambulance} \\ &\quad + \text{Total handing over to the hospital} \\ &\quad > \text{Time slot } T \end{aligned} \tag{17.4}$$

In case of violation, then all patients can be transported in one ambulance vehicle in the corresponding ambulance center and no need to perform the scheduling process. In such a case, arranging priority should be given to cases with severe health conditions.

(5) The Objective Functions

The COVID-19 crisis operation room decided on two main competing objectives, the first is to maximize the total number of evacuated patients, and the second is to give priorities depending on the patient's conditions in terms of his age, health, and number of contacts. The problem is then a multi-objective one with two objectives. The motivation in Multi-Objective Optimization (MOO) is that it allows for a compromise (trade-off) on some contradictory issues.

The Weighted Sum or scalarization method is one of the classic (MOO) methods, it puts a set of objectives into one by adding each objective pre-multiplied by a user-supplied weight. The weight of an objective is chosen in proportion to the relative importance of the objective. The weighted sum method is simple, but it is difficult to set the weight vectors to obtain a Pareto-optimal solution in the desired region in the objective space and it cannot find certain solutions in case of a nonconvex objective space, Marler and Arora (2004). The answer is a set of solutions that define the best trade-off between competing objectives, Gunantara (2018). The weighted sum approach treats the multi-objective optimization as composite objective function, Hemamalini and Simon (2010). The composite objective function is expressed as follows:

$$Z = \text{Maximize} \sum_{i=1}^{q} w_i.f_i\ (x).$$

where w_i is the positive weight values, $f_i\ (x)$ is one of the objective functions, and q is the number of objective functions. Since the objective of this research is to provide a compromise between maximizing of Total Relative Importance (TRI) and maximizing of Total Number of Patients (TNP), the following objective functions are considered:

$$\text{Maximize } f_1(x) = (\text{TRI}) = \sum_{i=1}^{I} p_i \sum_{j=1}^{J} \sum_{k}^{K_j} x_i^{j,k}, \text{ and (a)}$$

$$\text{Maximize } f_2(x) = (\text{TNP}) = \sum_{i=1}^{I} \sum_{j=1}^{J} \sum_{k}^{K_j} x_i^{j,k} \text{ (b)}$$

where: p_i is the priority of patient No i based on his medical status.

The composite objective function will be:

$$\text{Maximize } Z = w_1 \left(\sum_{i=1}^{I} p_i \sum_{j=1}^{J} \sum_{k}^{K_j} p^k x_i^{j,k} \right) + w_2 \left(\sum_{i=1}^{I} \sum_{j=1}^{J} \sum_{k}^{K_j} x_i^{j,k} \right) \tag{17.5}$$

where: p^k = Priority assigned to ambulance vehicle number k, more priority is given to more equipped vehicles.

To assign the proper weightage values in Eq. (17.5), an investigation is carried out by varying the value w_1 from 1 to 0 with a step size of 0.1. This is done to observe the significance of each weightage set toward the objective function, Naidu et al. (2014).

Finally, we have a suggested model that contains (I. $\sum_{j=1}^{J} K_j$) binary variables and ($I + \sum_{j=1}^{J} K_j$) constraints.

Any final chosen efficient solution will produce two distinct situations:

1. If $\sum_{j=1}^{n} \sum_{k=1}^{K_j} x_i^{j,k} = 1\ x_i^{j,k}$, $\forall\ i$, then all patients are transported to the isolation hospitals during the planned time shift, the problem is completed, and the planning team will go to the next time shift.

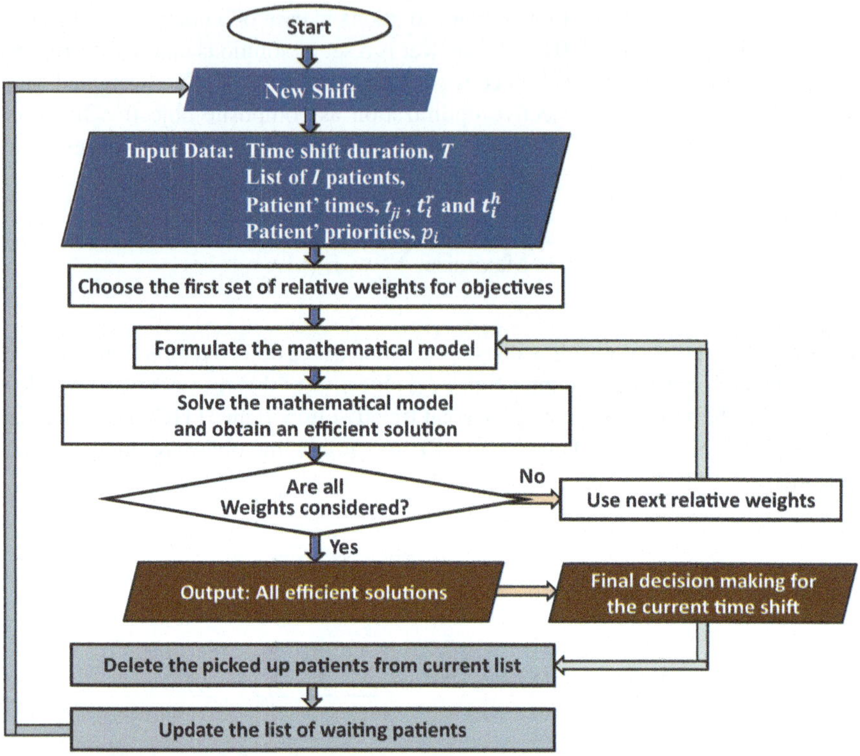

Fig. 17.2 Steps of the solution procedure

2. If $\sum_{j=1}^{n} \sum_{k=1}^{K_j} x_i^{j,k} = 0$ for any i, then the corresponding patient i is not transported to any hospital. In this case, add this patient to the updated list and repeat the procedure for another time slot. The solution procedure is presented in Fig. 17.2.

17.5 An Illustrated Case Study

The illustrated case study will take place in Cairo, Egypt. During a given time shift of 120 minutes, the ambulance operation room received 10 calls from suspicious patients to be transported to two isolation hospitals each has two ambulance vehicles for transportation of COVID-19 cases. Table 17.1 and Fig. 17.3 show names and positions of the locations of different patients. Time for riding a patient to the ambulance vehicle, t_i^r and time of handing over patient to the isolation hospital t_i^h are estimated to be equal for all patients to be as follows: $t_i^r = 8$ minutes and $t_i^h = 2$ minutes.

Table 17.1 Ambulance centers and patient locations in Cairo

	#	Location
Ambulance Center	1	Salah Salem street
	2	Shobra
Patient	1	Merry land park
	2	El-Galaa bridge
	3	Al Hayy Ath Thamin, Nasr City
	4	Port Saeed & Sekat Al Waili Streets Intersection
	5	Salah El din Al Ayouby citadel
	6	Shobra
	7	Al Sabtiah
	8	Cairo opera house
	9	Emtidad Ahmed Helmy
	10	Al Teraa Al Bolakia

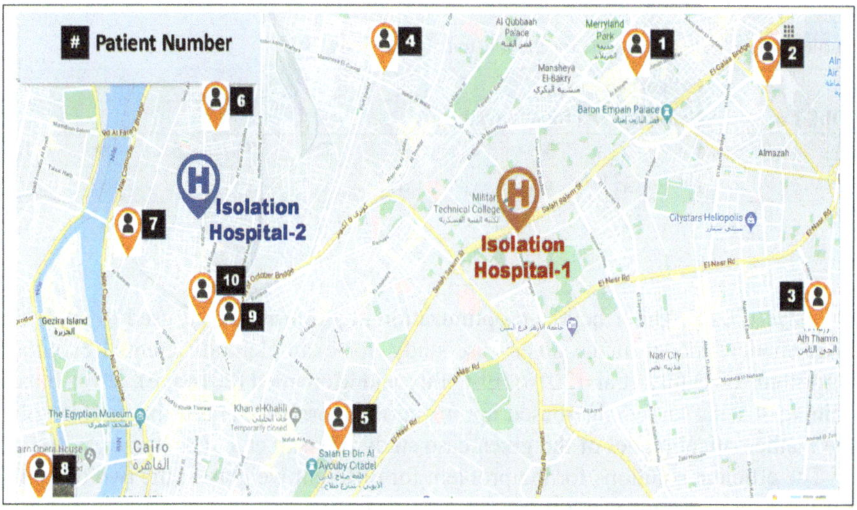

Fig. 17.3 Locations of the isolation hospital and patients in a part of Cairo map

After considering the opinion of a group of specialists practicing this work in these sites for many years, Table 17.2 shows the time required to travel from one site to another and the extension time required for each local unit.

The mathematical formulation for the given case is worked out by substituting in the previously described model for the given problem. The Excel solver is used to solve this case study, solver is a Microsoft Excel add-in program used to find an optimal value for a formula in one objective cell subject to constraints, or limits, on the values of other formula cells on a worksheet. Solver works with a group of decision variables and adjusts the values in the decision variable cells to satisfy the limits on constraint cells and produce the result for the objective cell, Microsoft

Table 17.2 Transportation time and relative priorities

	First Ambulance Centre									
Patients #	**1**	**2**	**3**	**4**	**5**	**6**	**7**	**8**	**9**	**10**
Relative priority of patient	1	1	1	1	3	3	1.5	1.5	4	4
Time t_{1i} (min.)	10	15	20	35	40	50	65	70	50	50
Total time/cycle, first vehicle	20	25	30	45	50	60	75	80	60	60
Total time/cycle, second vehicle	25	30	35	50	55	65	80	85	65	65
Relative priorities, p^1	**1.1**									
Relative priorities, p^2	1.0									
	Second ambulance Centre									
Time t_{2i} (min.)	40	55	70	30	35	25	20	35	20	20
Total time/cycle, first vehicle	50	65	80	40	45	35	30	45	30	30
Total time/cycle, second vehicle	55	70	85	45	50	40	35	50	35	35
Relative priorities, p^1	1.5									
Relative priorities, p^2	1.0									

Table 17.3 Efficient solutions for the illustrated case study

Obj. Fn.		**Patient No.**									
		Non-Zero Decision Variables									
f_1	f_2	**1**	**2**	**3**	**4**	**5**	**6**	**7**	**8**	**9**	**10**
49	9	$x_1^{1,1}$	$x_2^{2,1}$	$x_3^{1,1}$	$x_4^{1,1}$	$x_5^{1,2}$	$x_6^{1,2}$	$x_7^{2,2}$		$x_9^{2,1}$	$x_{10}^{2,2}$
52	7		$x_2^{2,1}$			$x_5^{1,1}$	$x_6^{1,1}$	$x_7^{2,2}$	$x_8^{1,2}$	$x_9^{2,1}$	$x_{10}^{2,2}$

webpage (2021). Other heuristic optimization algorithms can be used to solve the mathematical model of the given case study, for example, a differential evolution algorithm, El-Qulity et al. (2016), El-Qulity and Mohamed (2016a, b), also a binary gaining-sharing knowledge-based optimization algorithm can also be used to solve the mathematical model of the given case study, Hassan et al. (2020a, b).

The efficient solutions for the problem for different weights of the two objective functions are presented in Table 17.3. The first efficient solution has 9 transported patients in the considered time shift with total relative priorities of 49, while the second efficient solution effective solution has only 7 transported patients with a slightly higher total relative priority of 52.

The shuttle routes for all the ambulance vehicles can be seen in Fig. 17.4. In the first efficient solution: the first ambulance vehicle from the first ambulance center will transport patients 1, 3, and 4, the second ambulance vehicle will transport patients 5 and 6. The first ambulance vehicle from the second ambulance center will transport patients 2 and 9, the second ambulance vehicle will transport patients 7 and 10.

In the second efficient solution: the first ambulance vehicle from the first ambulance center will transport patients' number 3 and 6, the second ambulance vehicle will transport patient number 8. The first ambulance vehicle from the second ambulance center will transport patients' number 5 and 9, the second ambulance

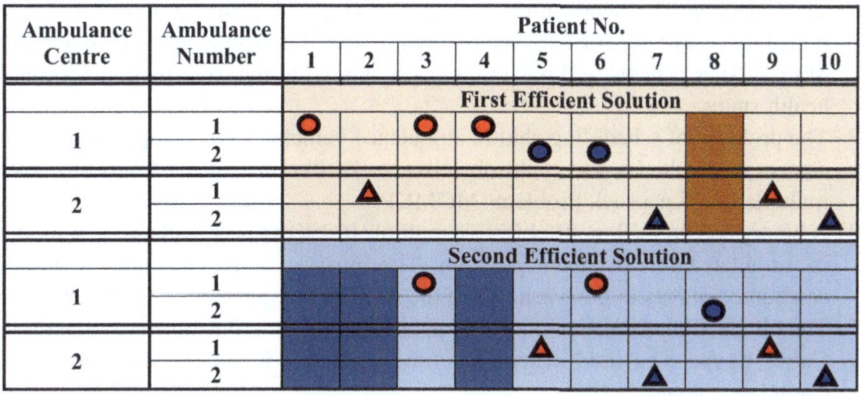

Fig. 17.4 Shuttle route for the ambulance vehicles

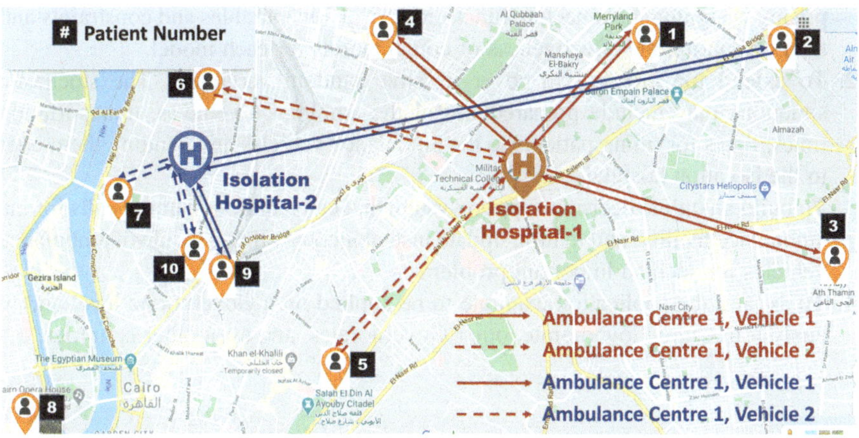

Fig. 17.5 Routes for the ambulance vehicles in the first efficient solution

vehicle will transport patients' number 7 and 10. The shuttle routes for the ambulance vehicles are shown in Fig. 17.5.

17.6 Conclusions and Points for Future Researches

The main conclusions for this chapter can be summarized as follows:

1. A generalized scheduling problem of shuttle ambulance vehicles for COVID-19 quarantine cases is presented. The problem formulation is generalized to handle multiple ambulance centers with multiple ambulance vehicles.

2. The objective is to achieve the best utilization of the available time shift determined by compromising between two objectives of maximizing the total number of evacuated patients and maximizing patients with high priorities due to their health status.
3. The problem of scheduling shuttle ambulance vehicles with multiple ambulance centers and vehicles and multi-objective looks like that of the Multi-objective multiple 0–1 Knapsack Problem (MOMKP).
4. A binary constrained multi-objective multiple 0–1 Knapsack mathematical model is formulated for the given problem. The binary decision variables represent candidate patients to be evacuated to the isolation hospitals.
5. The obtained mathematical model is applied to solve an illustrated case study in Cairo, Egypt. The used scheduling time shift is 2 hours and the number of candidate patients is 10.

The points for future research can be stated in the following points:

1. To propose other mathematical models' formulation for the same problem comprising designing the objective function(s), decision variables and constraints and then comparing the effectiveness of computations for each model.
2. To extend the problem to cover possible random variations. The stochastic situations may include preparedness of the ambulance vehicles, transportation times, times for riding patients to the ambulance vehicles and handing them over to the isolation hospital.
3. To build an online decision support system that can handle streaming calls for the ambulance facility, and in turn, update instantaneously the schedule of ambulance vehicles for such an important problem.
4. To expand the application example to be applied on a Governorate level and to apply it to Cairo Governorate, other Governorates, and even other countries.

References

Al-Ahram Gate website (2021) "The link between the patient and the doctor, rescue brigades are heroes inside ambulances". Retrieved on 9 March 2021 at: https://gate.ahram.org.eg/News/2391717.aspx

Alaya I, Solnon C, Gheira K (2004) Ant algorithm for the multi-dimensional knapsack problem, International Conference on Bioinspired Optimization Methods and their Applications, (BIOMA 2004), 2004, 63–72

Anany L (2003) The design and analysis of algorithms. Pearson Education Inc., New Jersey, p 2003

Basheer GT, Algamal ZY (2019) Nature-inspired optimization algorithms in knapsack problem: a review. Iraqi Journal of Statistical Science 30:1–18

Biglar A (2018a) Applications and solutions of knapsack problem: a literature review. https://doi.org/10.13140/RG.2.2.10921.08807. Retrieved at: file:///C:/Users/Dr%20Said/Downloads/knapsack.pdf

Biglar A (2018b) Some applications of knapsack problem preprint. https://doi.org/10.13140/RG.2.2.15115.39209

Black J (2017) Ambulance services working collaboratively with community partners, Association of Ambulance Chief Executives, 2017. Retrieved on May 15, 2020 at: https://www.kingsfund.org.uk/sites/default/files/2017-11/John%20Black%2031.101.17%20pdf_1.pdf

Boone CM Avery LW, Malone TB (2015) "A research study of ambulance operations and best practice considerations for emergency medical services personnel", Department of Homeland Security Science & Technology Directorate First Responders Group, March 2015

Captivo ME, Climaco JCN, Figueira JR, Martins EQV, Santos JL (2003) Solving bicriteria 0-1 knapsack problems using a labeling algorithm. Comput Oper Res 30(12):1865–1886

Carlos AB (2017) Emergency medical service ambulance system planning: history and models. M. Sc. Thesis, University of California, Santa Barbara, December 2017

Chabane B, Basseur M, Hao J (2005) A practical case of the multi-objective knapsack problem: design, modelling, tests and analysis. Springer International Publishing, Switzerland, pp 1–7. https://doi.org/10.1007/978-3-319-19084-623

Chekuri C, Khanna S (2006) A polynomial time approximation scheme for the multiple knapsack problem. SIAM Journal on Computing, 2006 35(3):713–728

Cleemput S, Dumon W, Fonseca V, Abdool Karim W, Giovanetti M, Alcantara LC, Deforche K, Oliveira T (2020) Genome Detective Coronavirus Typing Tool for rapid identification and characterization of novel coronavirus genomes, Bioinformatics, btaa145. https://doi.org/10.1093/bioinformatics/btaa145

Da Silva CG, Climaco JCN, Figueira JR (2005) Core problems in the bi-criteria {0,1} knapsack: new developments. Research Report 12, INESC-Coimbra

Da Silva CG, Climaco JCN, Figueira JR (2006) A scatter search method for bi-criteria {0-1}-knapsack problems. Eur J Oper Res 169(2):373–391

Da Silva CG, Climaco JCN, Figueira JR (2007) Integrating partial optimization with scatter search for solving bi-criteria {0-1}-knapsack problems. Eur J Oper Res 177(3):1656–1677

Egypt Independent website (2021) Egyptian ambulance organization transports 150–170 people per day due to coronavirus. Retrieved on March 9, 2021 at: https://egyptindependent.com/egyptian-ambulance-organization-transfers-between-150-170-persons-infected-or-suspected-with-coronavirus-president/

Ehrgott M, Gandibleux X (2000) A survey and annoted bibliography of multiobjective combinatorial optimization. OR-Spektrum 22(4):425–460

El-Ghazali T (2009) Metaheuristics from design to implementation. Wiley

Erlebach T, Kellerer H, Pferschy U (2002) Approximating multiobjective knapsack problems. Manag Sci 48(12):1603–1612

El-Qulity SA, Mohamed AW, Bafail AO, Abdelaal RMS (2016) A multistage procedure for optimal distribution of preparatory-year students to faculties and departments: a mixed integer nonlinear goal programming model with enhanced differential evolution algorithm. J Comput Theor Nanosci 13(11):7847–7863

El-Qulity SAH, Mohamed AW (2016a) A generalized National Planning Approach for admission capacity in higher education: a nonlinear integer goal programming model with a novel differential evolution algorithm. Comput Intell Neurosci 2016:5207362. https://doi.org/10.1155/2016/5207362

El-Qulity SAH, Mohamed AW (2016b) A large-scale nonlinear mixed-binary goal programming model to assess candidate locations for solar energy stations: an improved real-binary differential evolution algorithm with a case study. J Comput Theor Nanosci 13(11):7909–7921. https://doi.org/10.1166/JCTN.2016.5791

Fidanova S (2004) Ant colony optimization for multiple knapsack problem and heuristic model. Kluwer Academic Publishers

Fidanova S (2005) Ant colony optimization for multiple knapsack problem and model bias. In: Margenov S, Vulkov LG, Wasniewski J (eds) Numerical analysis and its applications, LNCS, vol 3401. Springer, Berlin Heidelberg, pp 280–287

Fidanova S (2007) Probabilistic model of ant colony optimization for multiple knapsack problem. In: Lirkov I, Margenov S, Wasniewski J (eds) LSSC 2007, LNCS 4818, Berlin, pp 545–552

Gandibleux X, Freville A (2000) Tabu search based procedure for solving the 0−1 multiobjective knapsack problem: the two objectives case. J Heuristics 6(3):361–383

Global Medical System (GMS) (2021) website, Ambulance Services. Retrieved on March 8, 2021 at: https://www.gmshospital.com/services/ambulance/

Gov. UK website (2021) Public Health England COVID-19: guidance for ambulance services, Updated 29 January 2021. Retrieved on March 9, 2021 at: https://www.gov.uk/government/publications/covid-19-guidance-for-ambulance-trusts/covid-19-guidance-for-ambulance-trusts

Groşan C Oltean M, Dumitrescu D (2003a) A new evolutionary algorithm for the multiobjective 0/1 knapsack problem, Proceedings of the International Conference on Theory and Applications of Mathematics and Informatics–ICTAMI 2003, Alba Iulia

Groşan C, Oltean M, Dumitrescu D (2003b) Performance Metrics for Multiobjective Optimization Evolutionary Algorithms, In Proceedings of Conference on Applied and Industrial Mathematics (CAIM), Oradea, 2003

Gunantara N (2018) A review of multi-objective optimization: Methods and its applications, Cogent Engineering (2018), 5: 1502242, Electrical & Electronic Engineering, Review article, https://doi.org/10.1080/23311916.2018.1502242

Hemamalini S, Simon SP (2010) Economic/emission load dispatch using artificial bee colony algorithm. In: ACEEE international journal on electrical and, power engineering, vol 1

Hassan SA, Agrawal P, Ganesh T, Mohamed AW (2020a) "Scheduling Shuttle Ambulance Vehicles for COVID-19 Quarantine Cases. A Multi-objective Multiple 0–1 Knapsack Model with A Novel Discrete Binary Gaining-Sharing knowledge-based Optimization Algorithm", in Data Science for COVID-19, Computational Perspectives, 1st. Edition, Paper No. 37, Elsevier, Academic Press, April 15, 2021

Hassan SA, Ayman YM, Alnowibet K, Agrawal P, Mohamed AW (2020b) Stochastic travelling advisor problem simulation with a case study: a novel binary gaining-sharing knowledge-based optimization algorithm. Hindawi, Complexity 2020:6692978. https://doi.org/10.1155/2020/6692978

Jansen K (2009) Parameterized approximation scheme for the multiple knapsack problem. SIAM Journal on Computing, 2009 39(4):1392–1412

Jansen K (2012) A fast approximation scheme for the multiple knapsack problem. Lect Notes Comput Sci 2012(7147):313–324

Ji J, Huang Z, Liu C, Liu X, Zhong N (2007). An ant colony optimization algorithm for solving the multidimensional knapsack problems, [in:] Proceedings of the 2007 IEEE/WIC/ACM International Conference on Intelligent Agent Technology, IEEE Computer Society, Los Alamitos, 2007, 10–16

Ke L, Feng Z, Ren Z, Wei X (2010) An ant colony optimization approach for the multi-dimensional knapsack problem. J Heuristics 16(1):65–83

Kellerer H, Pferschy U, Pisinger D (2004) Knapsack Problems. Springer-Verlag, Berlin

Klamroth K, Wiecek M (2000) Dynamic programming approaches to the multiple criteria knapsack problem. Nav Res Logist 47(1):57–76

Kuchta D, Rynca R (2019) The use of the multiple knapsack problem in strategic management of a private polish university case study. Int J Educ Manag 33(2):335–358

Li J, Li W, Wang H (2015) The multiple knapsack problem with compatible bipartite graphs, The 12th International Symposium on Operations Research and its Applications in Engineering, Technology and Management (ISORA 2015), Luoyang, China, August 21–24, 2015

Little AD (2019) Ambulance services–optimizing operations, rethinking emergency services for efficiency

Ma X, Yan Y, Liu Q (2018) A multi-objective particle swarm optimization for multiple knapsack problem with strong constraints, 13th IEEE Conference on Industrial Electronics and Applications (ICIEA), 2018. https://doi.org/10.1109/ICIEA.2018.8397892

Marler RT, Arora JS (2004) Survey of multi-objective optimization methods for engineering, structural and multidisciplinary optimization 26 (6), April 2004, pp. 369–395. https://doi.org/10.1007/s00158-003-0368-6

Microsoft webpage (2021) Define and solve a problem by using Solver. Retrieved on March 10, 2021 at: https://support.microsoft.com/en-us/office/define-and-solve-a-problem-by-using-solver-5d1a388f-079d-43ac-a7eb-f63e45925040

Naidu K, Mokhlis H, Bakar AHA (2014) Multiobjective optimization using weighted sum artificial bee Colony algorithm for load frequency control. Electr Power Energy Syst 55:657–667

Raj N, Vitthalpura J (2017) Literature review on implementing binary knapsack problem. IJARIIE 3(5):2395–4396

Render B, Stair RM, Hanna ME (2020) Quantitative analysis for management, 12th. Edition, BEARSON, 2020

Sanders D (2021) The heroism of health workers in the coronavirus crisis, The New York Times website, March 26, 2020. Retrieved on March 8, 2021 at: https://www.nytimes.com/2020/03/26/opinion/letters/coronavirus-health-care.html

Shahrear I, Faizul B, Sohel R (2010) Solving the multidimensional multi-choice knapsack problem with the help of ants. In: Dorigo M et al (eds) ANTS 2010. LNCS 6234, Berlin, pp 312–323

Soylu B, Köksalan M (2009) An evolutionary algorithm for the multi-objective multiple knapsack problem, International Conference on Multiple Criteria Decision Making, MCDM 2009: Cutting-Edge Research Topics on Multiple Criteria Decision Making, pp 1–8

The Lancet website (2020) A novel coronavirus outbreak of global health concern. Retrieved on March 09, 2021 at: https://www.thelancet.com/journals/lancet/article/PIIS0140-6736(20)30185-9/fulltext

Ünal AN (2013) A genetic algorithm for the multiple knapsack problem in dynamic environment, Proceedings of the World Congress on Engineering and Computer Science, 2013, Vol II, WCECS 2013, 23–25 San Francisco, USA

Vickram P, Krishna AS, Srinivas VS (2016) A survey on design paradigms to solve 0/1 knapsack problem. International Journal of Scientific & Engineering Research 7(11):2229–5518

World Health Organization website (2020) Guidance for health workers. Retrieved on March 1, 2021 at: https://www.who.int/emergencies/diseases/novel-coronavirus-2019/technical-guidance/health-workers

Worldometer website (2021) COVID-19 Coronavirus pandemic. Retrieved on April 15, 2020 at: https://www.worldometers.info/coronavirus/#ref-13

Zitzler E, Thiele L, Laumanns M, Fonseca CM, da Fonseca VG (2003) Performance assessment of multiobjective optimizers: an analysis and review. IEEE Trans Evol Comput 7:2

Yang X-S (2014) Cuckoo search and firefly algorithm-theory and applications. Springer International Publishing, Switzerland

Yang X-S (2015) Recent advances in swarm intelligence and evolutionary computation. Springer International Publishing, Switzerland

Chapter 18
Hyperparameters Optimization of Deep Convolutional Neural Network for Detecting COVID-19 Using Differential Evolution

Abdelrahman Ezzeldin Nagib, Mostafa Mohamed Saeed, Shereen Fathy El-Feky, and Ali Khater Mohamed

Abstract COVID-19 is one of the most dangerous diseases that appeared during the past 100 years, that caused millions of deaths worldwide. It caused hundreds of billions of losses worldwide as a result of complete business paralysis. This reason has attracted many researchers to attempt to find a suitable treatment for this dreaded virus.

The search for a cure is still ongoing, but many researchers around the world have begun to search for the safest ways to detect if a person carries the virus or not. Many researchers resorted to artificial intelligence and machine learning techniques in order to detect whether a person is carrying the virus or not.

However, many problems are arising when using these techniques, the most important problem is the optimal selection of the parameter values for these methods, as the choice of these values greatly affects the expected results.

In this chapter, Differential Evolution algorithm is used to determine the optimal values for the hyperparameters of Convolutional Neural Networks, as Differential Evolution is one of the most efficient optimization algorithms in the last two decades. The results obtained showed that the use of Differential Evolution in optimizing the hyperparameters of the Convolutional Neural Network was very efficient.

Keywords Convolutional neural network · COVID-19 · Differential Evolution · Hyperparameters Optimization

---

A. E. Nagib · M. M. Saeed · S. F. El-Feky · A. K. Mohamed (✉)
Faculty of Computer Science, October University for Modern Sciences and Arts (MSA), Giza, Egypt
e-mail: Abdelrahman.ezzeldin1@msa.edu.eg; Mostafa.mohamed52@msa.edu.eg; selfeky@msa.edu.eg; akhater@msa.edu.eg

S. A. Hassan et al. (eds.), *Decision Sciences for COVID-19*, International Series in Operations Research & Management Science 320,
https://doi.org/10.1007/978-3-030-87019-5_18

18.1 Introduction

Nowadays, COVID-19 became this ages' lesion after the new Chinese epidemic emergency which unfortunately considered from the most harmful diseases that sometimes cause death, COVID-19 is resulting from Coronaviruses that cause illnesses which are similar to flu with more severe clinical outcomes (da Costa et al. 2020). The workers of the healthcare sector have a very important role by providing healthcare for the COVID-19 patients at the front lines and make sure of prevention of the infection additional to the implementation of control (IPC) measures.

There are many clinical features of COVID-19 that are varying from case to case beginning from asymptomatic state to acute respiratory distress syndrome and multi Organ dysfunction, but there are many common clinical features as cough, sore throat, headache, fever (not in all cases), myalgia and breathlessness, in some cases, the infection can progress by the end of the first week to pneumonia then respiratory failure and death, that progress caused by the extreme rise in inflammatory cytokines (Singhal 2020).

Unfortunately, it is very hard to differentiate the COVID-19 Infection from all types of respiratory viral infections such as (Influenza, Parainfluenza, Respiratory syncytial virus, Non-COVID-19 Coronavirus, etc.) (Singhal 2020). The effective detection of COVID-19 patients is the most critical and important step towards the confrontation of the COVID-19 pandemic. Chest radiography is one of the most important detection approaches cause the radiography of the patients shows abnormalities in images which are used as characteristics of the infection (Wang et al. 2020).

The ground-glass opacity in the Chest X-rays has been observed once the COVID-19 has reached the lungs to differentiate between the COVID-19 patient and non-infected persons. Sample of CXR images for normal person, COVID-19 person, and pneumonia patient is presented in Fig. 18.1.

Although the radiological features from images are closely similar and overlapping those which have associated with SARS and MERS. The involvement of the lung bilateral is on initial imaging is more likely to be discovered as

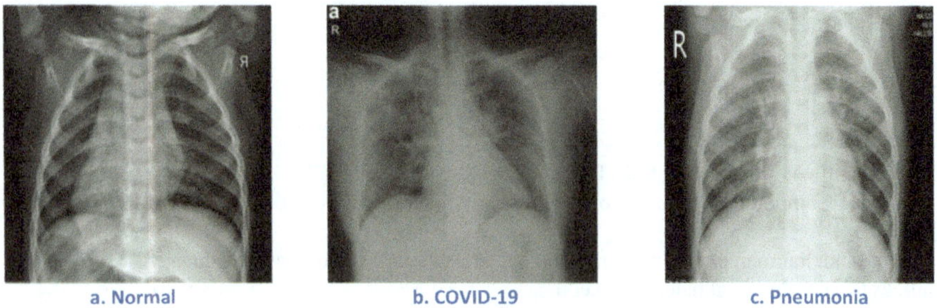

Fig. 18.1 Sample of CXR images

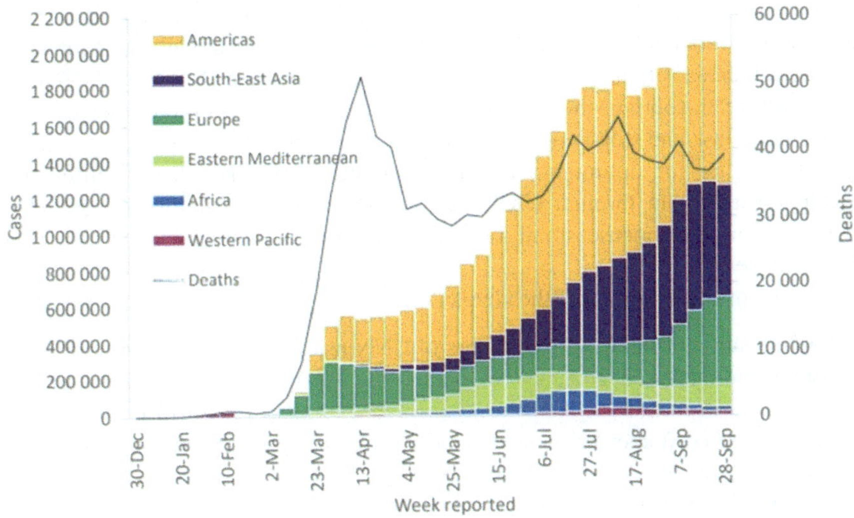

Fig. 18.2 Number of COVID-19 cases

COVID-19, as those associating SARS and MERS are more predominantly unilateral, that leads to the urgency need to expertise but need to fast diagnostic techniques that aligned with the pneumonic disease (Tahamtan and Ardebili 2020).

Over 37 million people have been infected by COVID-19 and one million deaths have been in the period by WHO Region from December 30, 2019 till October 11, 2020 as depicted in Fig. 18.2.

COVID-19 has a severe effect on all life fields as global aviation that has lost around 100 billion US dollars because of the travel restrictions which directly affect the world's economy accordingly and affect the individual's social life which causes economic crisis in addition to the health crisis, and that was the motivation for thousands of researches in all the fields after corona pandemic.

With the constant increase of infection rate, the call of the fast detection became an urgent need that led to the manifestation of the COVID-19 Auto-detection techniques which help with a rapid and automatic diagnosis through medical images processing using graphical processing units (GPUs).

Recently, the Convolutional Neural Network (CNN) presented optimistic results when used to classify radiological images. CNNs were used in many real-life applications including image classification. Therefore, this advantage was the main motive to present a CNN algorithm for diagnosing COVID-19 in the presented work.

Although CNNs networks have proposed a lot of breakthroughs but it seems like a black-box predictor as the hyperparameters of CNNs have a significant impact on the performance of the network and have a direct control on the training process.

The appropriate selection for the hyperparameters is very critical and directly affects the training process of the CNN network.

For example, the learning rate plays a vital role in network training, if it is too low, the network could lose the diversity. If it is too high, the model could converge so fast which leads to a premature convergence or stagnation (Nagib et al. 2020).

Therefore, the optimization of the hyperparameters of CNNs is considered a very challenging task and still an open area for research. A novel framework is presented in this chapter for optimizing the hyperparameters of CNN using Differential Evolution algorithm through CXR images. The proposed framework resulted in better performance metrices and avoid overfitting in the model training using the input images.

The main contributions of the chapter are:

1. Optimization of the hyperparameters of CNN using Differential Evolution algorithm for increased accuracy in diagnosing COVID-19 patients using CXR images.
2. The proposed framework is compared with other optimization algorithms used for tuning the hyperparameters of CNNs.
3. Ninety-nine percent accuracy is obtained for diagnosing COVID-19 patients using the presented framework.

The rest of the chapter is organized as follows: Section 18.2 presents related work. Differential Evolution and CNN are presented in Sect. 18.3. The material and methodology are presented in Sect. 18.4. The experiments are given in Sect. 18.5. Finally, the conclusion is presented in Sect. 18.6.

18.2 Related Work

Currently, there are many attempts that are being made to find a fast and effective way to detect the **COVID-19** virus with high accuracy using artificial intelligence techniques to end this global crisis. This section will review some of these attempts, which depend on the diagnosis of the patient through the CXR images.

Narin et al. (2020) proposed a detection method to detect COVID-19 infected patients using five pre-trained CNN-based models depends on Chest X-rays radiographs, binary classification with different three types have been implemented with an average of 99% accuracy between three datasets. El-Din Hemdan et al. (2020) have demonstrated the application of deep learning models for COVID-19 Classification based on X-ray images by using a new framework for deep learning called COVIDX-net, validation has been done on 50 X-Ray images with 25 confirmed COVID-19 positive cases, 80–20% evaluation has been done and tested on the seven different architecture of the COVIDX-net and shows that visual geometry group network and DenseNet showed similar or convergent performance with 0.89 *f*-score value for the automated COVID-19 and the normal COVID-19 is with 0.91 *f*-score value.

Khan et al. (2020) detected COVID-19 from Chest X-Ray images by using deep convolution and CoroNet Models and the trained CoroNet shows overall accuracy of

89.6%. Li et al. (2020a, b) developed deep learning model for COVID-19 autodetection, they developed convolutional neural network to extract biomarkers from chest scanning images, diagnostic performance was evaluated by the measurement of the area under the receiver operating characteristic curve and the results show the AUC (area under the receiver operating characteristic) with 0.96 value. Maghdid et al. (2020) proposed an artificial intelligence tool for COVID-19 auto-detection by implementing dataset containing the scanning results of X-rays and chest scanning and using a convolutional neural network and the mechanism of deep learning for auto-detection Process, the results shows that the accuracy up to 98% via pre-trained network and 94.1% accuracy using the modified convolutional neural network.

Mahdy et al. (2020) produced the usage of the auto classifiers for COVID-19 detection that depends on X-ray images using deep features, they used the system of multilevel thresholding and support vector machine which presented a high classification of the COVID-19 patient by using the same size of all the images in JPEG format, and the results show that the average of the accuracy of used auto-classification model is 97.48%. Rehman et al. (2020) proposed the COVID-19 auto-diagnoses deep learning method by differentiating it into three categories (viral pneumonia, bacterial pneumonia, and normal cases), the results showed that the K-fold −10 produced 98.75% accuracy by using X-Rays and chest scanning. Abbas et al. (2020) made an adaptation for the chest X-ray images irregularities by using decompose, transfer, and compose (DETRAC) for the COVID-19 classification, the results that have been introduced shows that DETRAC has a great capability in COVID-19 detection using image dataset with an accuracy of 95.12%.

Afshar et al. (2020) provided the usage of handling small datasets by using alternative modeling techniques based on capsule network, it is used in COVID-19 detection, the used technique provides an advantage over the convolutional neural network-based models because it achieved an accuracy of 95.7%. Apostolopoulos and Mpesiana (2020a) provided evaluation for the performance of the classic CNN that has been used in medical detection sector by using the transfer learning which helps in many abnormalities detection process, they build their experiment using two datasets which collected from public medical centers that include X-ray images, this experiments showed that deep learning introduces very helpful biomarks which obtain 96.78% accuracy. Apostolopoulos et al. (2020) have been investigated extracted features importance by using Convolutional Neural Network Model called Mobile Net in diseases autodetection, the results shows that Convolutional Neural Network Training from scratch produce very helpful biomarkers for diseases do not end to COVID-19. That leads to reaching 87.66 accuracy of COVID-19 detection.

Xu et al. (2020) proposed a model to differentiate between COVID-19, influenza, and normal cases by using deep learning model which has been implemented on the chest scanning images data set, the results showed an overall accuracy of 86.7% from all CT datasets. Wang et al. (n.d.) used Artificial intelligence's deep learning techniques that could extract the Biomarkers of COVID-19 disease that will save the time of clinical diagnosis and for disease control; they made internal and external validation the results showed that internal accuracy is 89.5% and external accuracy is

79.3%. Kumar Sethy et al. (2020) proposed COVID-19 auto-detection by using deep learning-based mechanism depends on X-rays images; they used RESNET50 cooperated with Support vector machine classifiers and its results showed that the detection is with an accuracy of 95.52% from the repository of GitHub, Kaggle, and Open-I.

Li et al. (2020a, b) developed a deep learning neural network for COVID-19 auto-detection by visual features extraction from chest CTs by made binary classification and made validation by using independency, specify and independency tests that ends with the fact that deep learning methods detect COVID-19 accurately. Gozes et al. (2020) developed an artificial deep learning technique to detect, quantify, and track COVID-19 disease, by using 2D and 3D deep learning models; they proposed retrospective experiments for analyzing the attitude of the system in COVID-19 detection, the result of classification is 0.996 AUC. Shan et al. (2020) implemented deep learning for auto segmentation and quantification of infection locations beside the lung from chest screening images, the convolutional neural network used to detect infection regions, they used 249 COVID-19 cases, and validated using 300 COVID-19 cases.

18.3 Theory and Methods

18.3.1 Differential Evolution Algorithm

Differential Evolution (DE) presented by Storn and Price (Storn and Price 1995, 1997) is a stochastic population-based search method. DE proved an excellent ability to solve a wide range of optimization problems with different features from many fields and many real-world applications (Kenneth et al. 2005). The evolution process of DE uses mutations, crossover, and selection operators at each generation that must be used in order to obtain the global optima. DE is recognized as one of the most efficient evolutionary algorithms (EAs) currently in use. DE has many advantages including that it is very simple to implement, reliable, speed of convergence, and it is a robust algorithm. Therefore, it is applied in a wide range of applications to solve numerous numbers of real-world applications.

A brief summary of the basic Differential Evolution (DE) algorithm is presented. In simple DE, generally known as *DE/ rand /1/bin* (Storn and Price 1997), an initial population, denoted by *P*, is randomly initialized and consists of *NP* individual. The vector $x_{i=(x_{1,i},x_{2,i},\ldots,x_{D,i})}$ is used to represent each individual, where *D* is the number of dimensions in solution space. Since the population will be varied with the running of evolutionary process, the numbers of generation in DE are expressed by $G = 0, 1\ldots,$ G_{max}, where G_{max} is the maximum number of generations. The ith individual of the population at generation number *G*, is denoted by $x_i^G = \left(x_{1,i}^G, x_{2,i}^G \ldots, x_{D,i}^G\right)$. The lower bound and upper bound in each dimension is recorded by $X_L = (x_{1,\ L}, x_{2,\ L}\cdots, x_{D,\ L})$ and $X_U = (x_{1,\ U}, x_{2,\ U}\cdots, x_{D,\ U})$. The uniform random initial population P_0 is generated within the lower and upper boundaries (X_L, X_U). After the initialization process, those individuals evolve by using DE operators (mutation and crossover) to obtain a trial vector. The parent is compared to its child in order to select the fittest vector that will survive to the next generation. Detailed steps for DE are discussed as follows.

18.3.1.1 Initialization

DE start the optimization process by generating an initial random population P_0. This process is called initialization. Typically, the value of the jth dimension ($j = 1, 2..., D$) of the ith individuals ($i = 1, 2..., NP$) in the initial population P_0 could be obtained as follows:

$$x^0_{j,i} = x_{j,L} + rand(0,1).(x_{j,U} - x_{j,L}) \tag{18.1}$$

where *rand* (0, 1) return a random number that follows uniform distribution between [0, 1].

18.3.1.2 Mutation

At generation G, a mutant vector v_i^G is generated for each target vector x_i^G using the following formula:

$$v_i^G = x_{r1}^G + F.(x_{r2}^G - x_{r3}^G), r_1 \neq r_2 \neq r_3 \neq i \tag{18.2}$$

where r_1, r_2, $r_3 \in \{1, 2, ..., NP\}$ represents three randomly chosen indices. F represents the mutation factor that is a real number that controls the amplification of the difference vector $(x_{r2}^G - x_{r3}^G)$. Storn and Price (1997) suggested that F in the range [0, 2] is a very good choice. In this work, if the value of any component of the mutant vector violates search space boundaries, then a new value is generated for this component by (1).

18.3.1.3 Crossover

There are two main crossover types, binomial, and exponential. We here elaborate on the binomial crossover. In the binomial crossover, the target vector is mixed with the mutated vector, using the following scheme, to yield the trial vectorU_i^G.

$$u_{j,i}^{G} = \begin{Bmatrix} v_{j,i}^{G}, \text{if } \left(\text{rand}_{j,i} \le \text{CR or } j = j_{\text{rand}}\right) \\ x_{j,i}^{G}, \text{otherwise} \end{Bmatrix} \tag{18.3}$$

where $rand_{j,\ i}$, $(i \in [1, NP] and\ j \in [1, D])$ is a uniformly distributed random number in [0,1], $CR \in [0, 1]$ called the crossover rate that controls how many components are inherited from the mutant vector, j_{rand} is a uniformly distributed random integer in [1, D] that makes sure at least one component of trial vector is inherited from the mutant vector.

18.3.1.4 Selection

DE adapts a greedy selection strategy. If and only if the trial vector U_i^G yields as good as or a better fitness function value than X_i^G, then U_i^G is set to X_i^{G+1}. Otherwise, the old vector X_i^G is retained. The selection scheme is as follows (for a minimization problem):

$$x_i^{G+1} = \begin{Bmatrix} u_i^G, f\left(u_i^G\right) \le f\left(x_i^G\right) \\ x_i^G, otherwise \end{Bmatrix} \tag{18.4}$$

A detailed description of standard DE algorithm is given in Fig. 18.3.

Parameters of differential evolution play a vital role in its performance and affect the convergence of the algorithm. There are three main control parameters in differential evolution:

1. Mutation factor (F).
2. Cross over rate(CR).
3. Population size (NP).

The control parameters are tuned manually, adaptively, or self-adaptively. (Mohamed and Mohamed 2019a, b) presented a novel scheme for changing the values of CR adaptively based on the probability of success, and a novel scheme for population size reduction is presented in Mohamed and Mohamed (2019a, b), Mohamed et al. (2018), Mohamed, Hadi and Mohamed et al. (2020) and it is used to solve constrained problems (Mohamed and Mohamed 2019a, b) and mixed-integer problems (Mohamed et al. 2019a, b).

From the literature, it has been shown that the mutation scheme in differential evolution has a great impact on the convergence of the algorithm. A complete study for the effect of mutation strategies on the convergence of the algorithm is presented in Mohamed, Hadi and Mohamed et al. (2021).

```
Generate an initial population
Evaluate the fitness of each individual
while the termination criteria are not satisfied do
      For i = 1 to NP do
        Select uniform randomly r_1 ≠ r_2 ≠ r_3 = i
                   j_rand = rndint(1, D)
        For j = 1 to D do
           if rndreal_j(0,1) < CR or j = j_rand then
             u_{i,j} = x_{r1,j} + F.(x_{r2,j} − x_{r3,j})
           else
             u_{i,j} = x_{i,j}
           End if
        End for
      End for
      For i = 1 to NP do
        Evaluate the offspring u_i
        If f(u_i) is better than or equal to f(x_i) then
           Replace x_i with u_i
        End if
      End for
End while
```

Fig. 18.3 Description of standard DE algorithm

18.3.2 Convolutional Neural Network (CNN)

Since the start of deep learning and the outstanding achievements achieved by the dense neural networks (ANN) that was first invented in 1958, since then dense neural networks have proved amazing results but in the 1980s the convolutional neural networks also known as CNNs or CovNet, which is a class of neural networks with a main specialty in processing the gird like topology data such as images. A digital image is a binary representation of visual data in which the images have a series of pixels that are arranged in a grid-like form that contains the pixel value to denote and show what color and how bright is that color in each pixel would be, since then the convolutional neural networks (CNN) has proved an outstanding performance in the field of deep learning in image classification. The CNN can successfully capture the temporal and spatial dependencies in an image through the application of relevant filters and the architecture performs better fitting to the dataset of images due to the reduction of parameters in the CNN; in other words, the CNN can be trained to understand the sophistication of the images really well.

The Convolutional Neural Network consists of Convolutional layers "as the name applies" in which the Convolutional layers are the layers in which filters are applied to the original image, or maybe to other feature maps in a deep convolutional neural network. This is where most of the user-specified parameters are in the network.

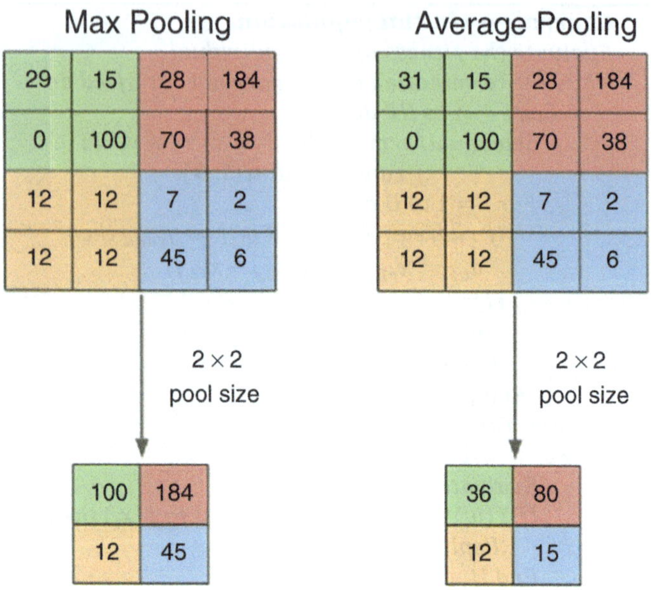

Fig. 18.4 CNN Pooling

Where the most important parameters are Features of a pooling layer and the number of kernels and the size of the kernels.

The CNN structure starts with its first secret ingredient that has made CNNs very successful; that is pooling. Pooling is a scalar transformation vector that operates on each and every local region of an image, but what makes it different from convolutions is that they do not have any filters and do not compute the dot products with the local region, they compute the average of the pixels in the region (Average Pooling) or simply select the highest intensity pixel and discard the rest of the pixels in the region (Max Pooling) (Fig. 18.4).

Another main ingredient In CNN is the kernel, the kernel is nothing but only a filter that is used to extract features from our images, we can say it is a matrix, which is sliding across the image and being multiplied with the input, in such a way that the output is improved in a such a noticeable desirable way (Fig. 18.5).

In summary, the CNN consists of 2 bases, the convolution base, and the classifier base.

The convolutional base has three main layers which are: the convolutional layers, the activation layers, and the pooling layers. These layers are used to discover the features of the input images, which is called a feature map. A feature map is constructed by performing convolution processes to the input image or prior features using a linear filter and merging a term called the bias. Then forwarding this feature map through a non-linear activation function such as Sigmoid or Rectified Linear Unit (RELU). In contrast, the classifier base has dense layers that are combined with

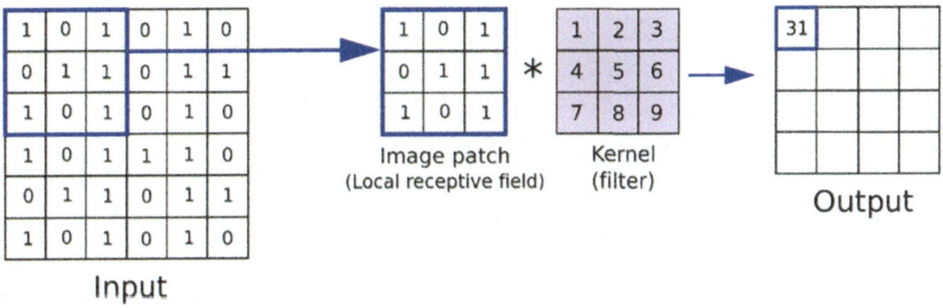

Fig. 18.5 Kernel in CNN

the activation layers to turn the feature maps into one-dimensional vectors to speed up and accelerate the task of classification using neurons.

18.4 The Proposed Framework

In image classification, it is usually better to convert images from RGB (3 color channels) to gray scale (1 color channel) in order to reduce complexity and the size of input for the artificial neural networks, so the presented framework starts by making the images of gray scale.

Another important feature in image classification is the width and height of the image, which is usually resized to smaller value to reduce the input shape for the artificial neural network and yet maintain high results because in the case of a 1080×1080 image of gray scale we get an input shape around 1,160,000. So, in the case of RGB not being gray scale then this number will be multiplied by 3.

Usually image resizing and the width and height of the image is decided randomly while creating the neural network; the presented framework in Fig. 18.6 works on using the differential evolution algorithm to find the best width and height for the images to be resized to according to given bounds that the width and height cannot be less than the minimum width nor more than the maximum height; yet in order to reduce the number of parameters that need to be optimized we decided to always make the width and height equal to each other so we do not have to consider the width as special parameter and the height as another special parameter because this will create an unimaginable size of search space given the other parameters that the Differential Evolution tries to optimize; it will be a very large combination; so for the sake of taking a step forward without including huge search space we treat the width and height as one single parameter in order to find the best number that fits them both at the same time. Then in the framework, we scale the input (the images) value to between 0 and 1.

All of these processes are done according to whether the developer wants only training or training, and testing, or training, testing and, validation.

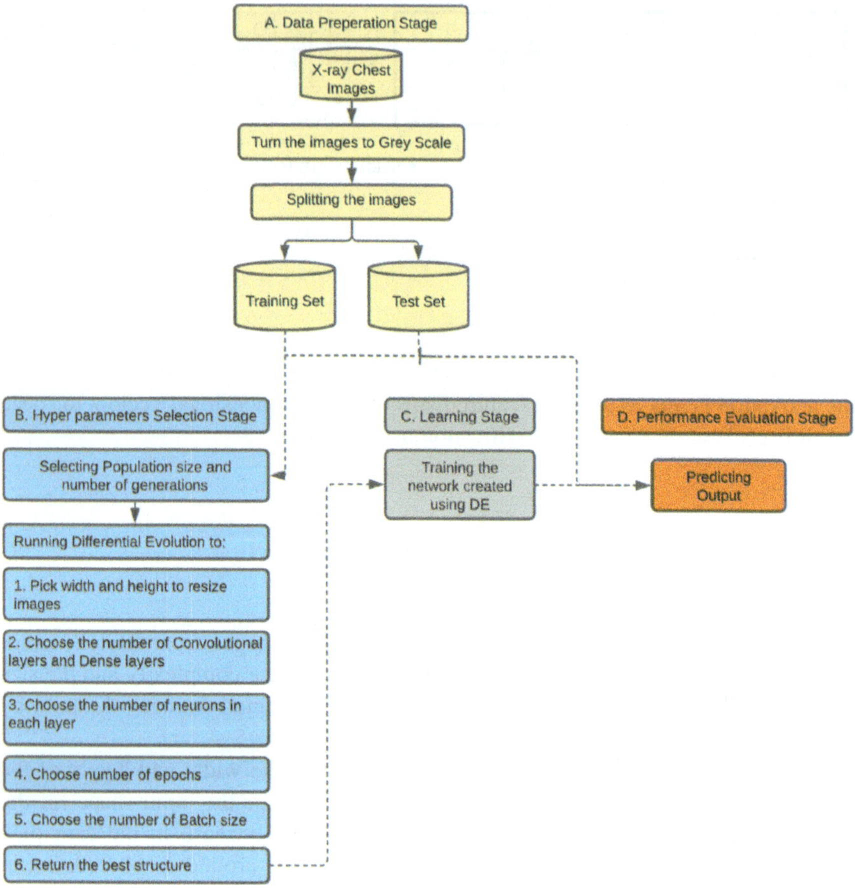

Fig. 18.6 The proposed framework

The previous process will be applied on the entire data and be split in the end according to a given parameter in the start that allows the developer to choose how he wants to split the data and to what ratios.

The differential evolution tries to optimize the number of layers given and the number of neurons in each layer as well. Since our Artificial Neural Network is a Convolutional Neural Network so we have three types of layers:

1. The Convolutional Layers.
2. The max pooling layers.
3. The dense layers.

we choose to consider the convolutional layers and the max pooling layers as one type of layers by which we create a max pooling layer of 2 * 2 after each convolutional layer (with a constant scanner of 3*3 for all the conv.) if it is possible due to the over pooling of the image might at some point cause some errors. So in order to prevent that problem from occurring we decided to handle it that once max pooling is not possible, we will stop creating convolutional layers and get on with creating the dense layers; yet this is not the way of how the number of convolutional layers and max pooling layers as picked; as the differential evolution algorithm will select a number of the convolutional layers as mentioned above and will then create one max pooling layer after each convolution layer and then create the dense layers.

In fact, the way the presented framework selects the best number of convolutional neural network for the mix of convolution layers (and max pooling) and the dense layers is by setting the number of minimum layers for the convolution layer to 1 and the minimum number of dense layers to 1 and a number is set as the maximum bounds for the total number of layers of both the convolution and dense layers combined to be set as not to exceed it.

For example, if the bounds of the number of layers for the model is 1 and 20; the layers can be 3 convolutional (with 1 max pooling after each) and 4 dense layers or 1 convolutional and 19 dense or 19 convolutional and 1 dense; so, the differential evolutional finds the best combination of the number of layers of the convolutional layers and the dense layers.

Actually, while creating the layers of each the convolutional and the dense layers, the differential evolution decides which is the best number of neurons in each layer of the convolutional and the dense layers according to some bounds that should not be less than the minimum bound and never exceed the maximum bound given in the framework.

Before creating the convolutional neural network model, the presented framework makes a few decisions according to the given data; the framework figures out the input shape for the input layer according to the images dimensions that is given, and thus the input layer is always dynamic and will work with any input shape as long as it is in gray scale, the model creates the layers and gives them an activation function of "ReLU" as the default for all the hidden layers, in the output layer we decide the number of outputs by examining the target values, if they are 2 classes then the output layer activation function will be set to 'sigmoid' and the loss function of the model will be "binary_crossentropy" but if they are more than 2 classes the output layer activation function will be set to "SoftMax" and the loss function will be set to "categorical_crossentropy" and the target data will be handled to be in the form of vectors (in case of more than 2 classes in the target data) if they are not in the form of vectors.

The presented framework handles the target data given that if its values are strings or characters, they are handled and turned into integer values for the sake of improvements and speed.

In order to measure the success and improvement of the different neural network structures and resized images created, the accuracy is considered as the default improvement measure but the improvement can be measured on any of the following as chosen while using the framework; the improvement can be done on: the training data alone, or the test alone, or the validation, or the training and test, or the training, test and validation or the training validation or the test validation in order to allow the user of the framework to focus on improving the model on all the kinds of data because one approach is to improve the test accuracy and the training accuracy will usually improve by default.

RMSE (root mean square error) is another measure that could be used instead of the accuracy. The presented framework gives the user the ability to use the accuracy and the RMSE combined as a measure of improvement.

In the end, the presented framework manages to find the best width and height that the images need to be resized to. The framework manages to find the best combination number of layers of convolutional layers to be used alongside with the dense layer as well as finding the best number of neurons for all the convolutional layers and the dense layers as well to create the convolutional neural network. Adam optimizer is used as the default optimizer in the framework.

18.5 Experimentation

This section represents the experimentation results for the presented framework presented in Sect. 18.4 and Fig. 18.6. Besides, experimentation results without using DE for hyperparameters selection is presented, in order to investigate the efficiency of using DE as a metaheuristic for hyperparameters selection.

18.5.1 CNN Using DE

The presented framework used a differential evolution algorithm as an optimization technique on Convolutional Neural Network on the COVID-19 detection using X-RAY images to optimize the hyperparameters of the network.

Data is composed of 2905 chest X-ray images, 1345 images are classified as normal cases, 1345 images are classified as Viral Pneumonia cases, 219 images are classified as COVID-19 cases, of size 1080 $\times$ 1080. Data was imported and the images were scaled to values between 0 and 1. Then, data were split to 80% for training and 20% for testing.

Differential Evolution is applied with a maximum number of iterations is set to 10 and population size =10 with the following constraints:

- Cost function is measuring the error in accuracy, so we try to reach the highest accuracy.
- Number of layers between 1 and 7.
- Number of neurons in each layer between 1 and 512.
- Batch size between 4 and 32.
- Number of epochs between 25 and 150.
- The width and height between 15 and 250.

Differential evolution algorithm returns the structure of the CNN neural network as follows:

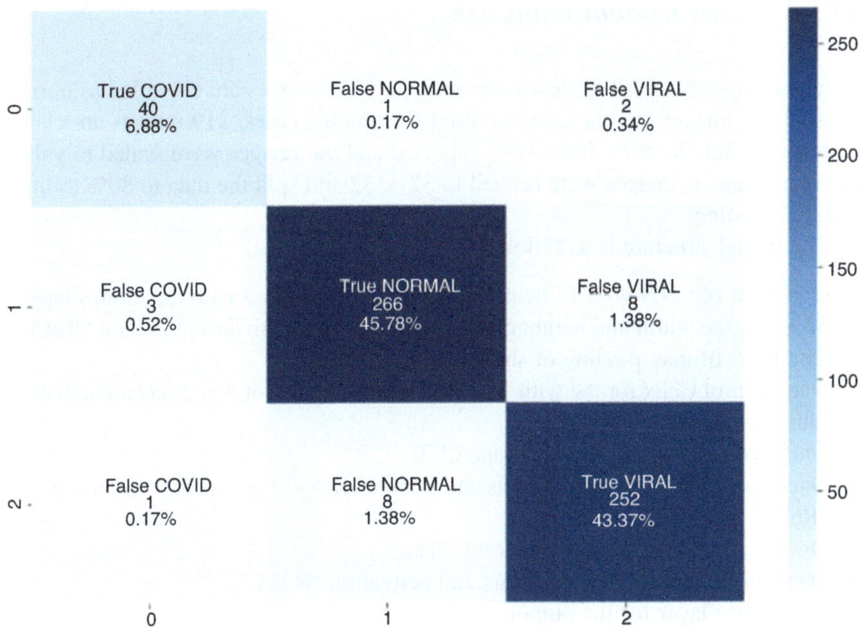

Fig. 18.7 Confusion matrix for CNN using DE

- One Conv2D layer for the input layer with shape **(width = 33 and height = 33)** with 64 Neurons with a 3 × 3 Scanner.
- One Max Pooling layer 2 × 2.
- One Conv2D Layer 134 neurons, 3x3 Scanner and activation function "ReLU".
- One Max Pooling layer 2 × 2.
- One flattens layer.
- One Dense Layer with 117 Neurons and activation function "ReLU".
- One output layer with 3 neurons and activation function SoftMax.

The total number of parameters of the current structure is 642,837 parameters, and the loss function of the model is categorical "crossentropy" and optimizer is Adam.

The Differential Evolution algorithm also returned a batch size of 72 and number of epochs equal to 125.

The resulting model was trained with the returned parameters and scored **100%** accuracy in the training and **97.25%** in the testing which proved that our differential evolution approach in the Convolutional Neural Network proved worthy by outperforming the compared various structures accuracy in the COVID-19 research area as well as it also proved to be better in finding a simpler structure (Fig. 18.7).

18.5.2 CNN without Using DE

Data is composed of 2905 chest x-ray images, 1345 images are classified as normal cases, 1345 images are classified as Viral Pneumonia cases, 219 images are classified as COVID-19 cases. Data were imported and the images were scaled to values between 0 and 1. Images were resized to 32 × 32 and split the data to 80% training and 20% testing.

The model structure is as follows:

- One layer of CNN with 15 neurons and scanner of 3 × 3 with the input shape of 32 × 32 (the width and height of the images) and an activation function "ReLU".
- One layer of max pooling of shape (2,2).
- One layer of CNN for the with 10 neurons and scanner of 3 × 3 and an activation function "ReLU".
- One layer of max pooling of shape (2,2).
- One layer of CNN with 8 neurons and scanner of 3 × 3 and an activation function "ReLU".
- One layer for flattening the neurons shape.
- One Dense layer with 16 neurons and activation "ReLU".
- One Dense layer for the output with 3 neurons.

The full structure is composed of 4353 parameters.

Class weights assigned of 1.3 weight adjustment for the COVID cases, 1 for the Normal, and 1 for the Viral Pneumonia.

"Adam" optimizer is used with learning rate equal to 0.0008, and the loss function is Sparse Categorical Crossentropy, and the model is trained for 150 epochs.

The model ends the final epoch with training accuracy of 99.14% and testing accuracy of 96.04%, where True COVID is 48, False COVID is 4, True NORMAL is 264, False NORMAL is 10, True Viral is 246, False Viral is 9, True COVID (92.30%), True Normal (96.35%), and True Viral (96.47%) (Fig. 18.8).

18.5.3 Comparison and Analysis

In this study, two models are presented to determine if the patient is affected by COVID-19 or not, one model using DE and the other without using DE. The performance of each model is evaluated and a comparative analysis with the most recent models from the literature (Khan and Aslam 2020) is presented in Table 18.1.

From Table 18.1, it could be easily observed that the proposed model outperformed other models in the literature except COVIDiagnosis-Net. Most of the previous research studies suffered from having a very limited number of COVID-19 images and imbalanced data.

The main contributions of the current research are:

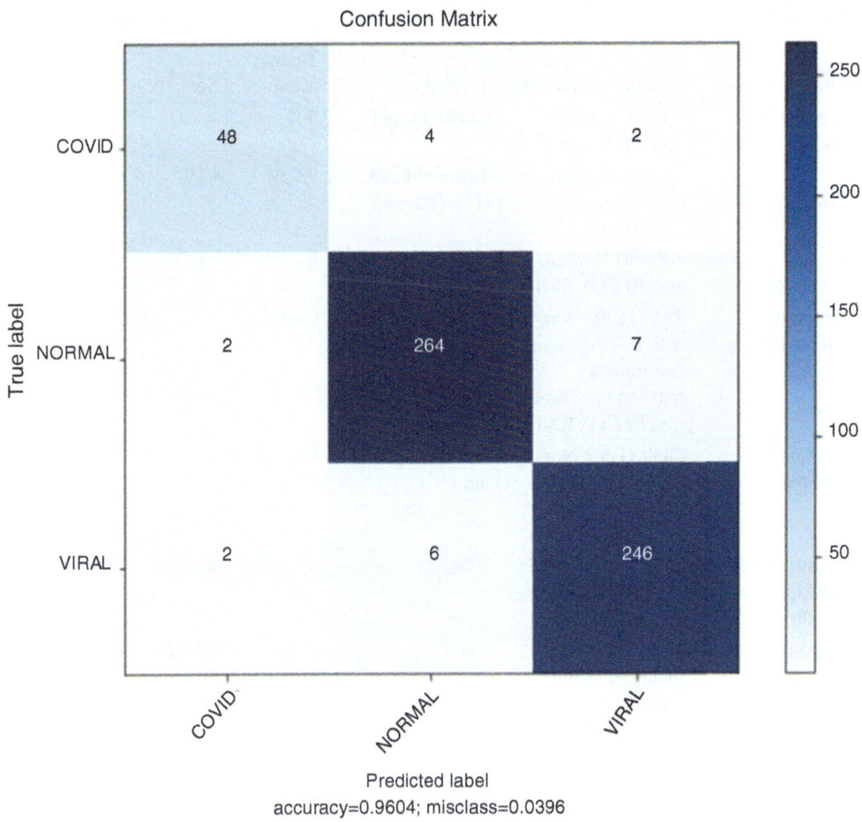

Fig. 18.8 Confusion matrix for CNN without using DE

1. The model does not suffer from imbalanced data.
2. The proposed model is a fully automated diagnosis method and does not require any separate feature extraction or annotation prior to the diagnosis.
3. Optimizing the hyperparameters of the model using Differential Evolution algorithm.
4. Less network structure as the number of parameters in the current suggested model is very simple, have very low number of parameters (296,000) compared to other complex models.
5. The model outperforms the previous models in the literature.

Table 18.1 Performance of the proposed models compared to models in the literature

Study	Number of samples	Technique	Testing accuracy	Sensitivity	Specificity
Hemdan et al. (2020)	50 (25 healthy, 25 COVID-19)	COVIDX-net	90%	–	–
Wang et al. (2020)	13,975 (normal, pneumonia, and COVID-19	Tailored CNN (COVID-net)	93.3%	91%	–
Narin et al. (2020)	100 (50 Normal, and 50 COVID-19)	ResNet50	96.1%	91.8%	96.6%
Farooq and Hafeez (2020)	2813 (1203 Normal, 931 bacterial pneumonia, 660 viral pneumonia, 19 COVID-19)	COVID-ResNet	96.23%	–	–
Ucar and Korkmaz (2020)	2839 (1203 Normal, 1591 pneumonia, and 45 COVID-19	COVIDiagnosis-net	98.30%	–	–
Apostolopoulos and Mpesiana (2020b)	1427 (224 COVID-19, 700 pneumonia, and 504 normal)	VGG19	93.48%	92.85%	98.75%
Proposed	2905 (1345 normal, 1345 viral pneumonia, and 219 COVID-19)	CNN	96.04%	99.24%	92.3%
		CNN-DE	97.25%	98.5%	93%

18.6 Conclusion

This paper presented a new approach called DE-CNN-COVID-19 that could be used to detect COVID-19 patients using chest images. The presented model starts with data preparation, followed by hyperparameters optimization using Differential Evolution algorithm, and ended with the learning phase and evaluation of performance.

Firstly, imbalanced data is handled, and the data set is divided into training and test sets. Secondly, Differential Evolution algorithm is used to optimize the hyperparameters of CNN and presenting the optimal structure of CNN. At last, the model is trained using the optimal structure obtained from Differential Evolution. The results obtained from the presented model were very promising and the model could be attaining 100% accuracy for training and 97.25% for testing.

The presented model was then compared with other models in the literature, and it proved the outperforming performance and the superiority over other models in detecting COVID-19.

References

Abbas A, Abdelsamea MM, Gaber MM (2020) 'Classification of COVID-19 in chest X-ray images using DeTraC deep convolutional neural network', medRxiv. medRxiv. Available at: http://arxiv.org/abs/2003.13815. Accessed: 16 January 2021

Afshar P et al. (2020) 'COVID-CAPS: a capsule network-based framework for identification of COVID-19 cases from X-ray Images', Pattern Recognition Letters. Elsevier B.V., 138, pp. 638–643. Available at: http://arxiv.org/abs/2004.02696. Accessed: 16 January 2021

Apostolopoulos ID, Aznaouridis S, Tzani M (2020) Extracting possibly representative COVID-19 biomarkers from X-ray images with deep learning approach and image data related to pulmonary diseases. Journal of Medical and Biological Engineering 40(3):462–469. https://doi.org/10.1007/s40846-020-00529-4

Apostolopoulos ID, Mpesiana TA (2020a) Covid-19: automatic detection from X-ray images utilizing transfer learning with convolutional neural networks. Physical and Engineering Sciences in Medicine 43(2):635–640. https://doi.org/10.1007/s13246-020-00865-4

Apostolopoulos ID, Mpesiana TA (2020b) Covid-19: automatic detection from X-ray images utilizing transfer learning with convolutional neural networks. Phys Eng Sci Med 43 (2):635–640. https://doi.org/10.1007/s13246-020-00865-4

da Costa VG, Moreli ML, Saivish MV (2020) The emergence of SARS, MERS and novel SARS-2 coronaviruses in the 21st century. Arch Virol:1517–1526. https://doi.org/10.1007/s00705-020-04628-0

El-Din Hemdan, E., Shouman, M. A. and Karar, M. E. (2020) COVIDX-Net: a framework of deep learning classifiers to diagnose COVID-19 in X-Ray images

Farooq M, Hafeez A (2020) 'COVID-ResNet: a deep learning framework for screening of COVID19 from radiographs', arXiv. arXiv. Available at: http://arxiv.org/abs/2003.14395. Accessed: 5 April 2021

Gozes O et al. (2020) 'Rapid AI development cycle for the coronavirus (COVID-19) pandemic: initial results for Automated Detection & Patient Monitoring using deep learning CT image analysis', arXiv. arXiv. Available at: http://arxiv.org/abs/2003.05037. Accessed: 16 January 2021

Hemdan EE-D, Shouman MA, Karar ME (2020) 'COVIDX-net: a framework of deep learning classifiers to diagnose COVID-19 in X-ray images', *arXiv*. arXiv. Available at: http://arxiv.org/abs/2003.11055. Accessed: 5 April 2021

Kenneth VP, Rainer MS, Jouni AL (2005) Differential evolution–a practical approach to global optimization. Springer-Verlag, Berlin, Germany

Khan IU, Aslam N (2020) A deep-learning-based framework for automated diagnosis of COVID-19 using X-ray images. Information. Multidisciplinary Digital Publishing Institute 11(9):419. https://doi.org/10.3390/info11090419

Khan AI, Shah JL, Bhat MM (2020) CoroNet: a deep neural network for detection and diagnosis of COVID-19 from chest x-ray images. Comput Methods Prog Biomed 196:105581. https://doi.org/10.1016/j.cmpb.2020.105581

Kumar Sethy P, Kumar Sethy P, Kumari Behera S (2020) Detection of coronavirus disease (COVID-19) based on deep features. https://doi.org/10.20944/preprints202003.0300.v1

Li L, Qin L, Xu Z, Yin Y, Wang X, Kong B (2020a) Artificial intelligence distinguishes covid-19 from community acquired pneumonia on chest CT. Radiology:200905

Li L et al (2020b) Using artificial intelligence to detect COVID-19 and community-acquired pneumonia based on pulmonary CT: evaluation of the diagnostic accuracy. Radiology 296(2): E65–E71. https://doi.org/10.1148/radiol.2020200905

Maghdid HS et al. (2020) 'Diagnosing COVID-19 pneumonia from X-ray and CT images using deep learning and transfer learning algorithms', arXiv. arXiv. Available at: http://arxiv.org/abs/2004.00038. Accessed: 16 January 2021

Mahdy LN et al. (2020) 'Automatic X-ray COVID-19 lung image classification system based on multi-level thresholding and support vector machine', medRxiv. medRxiv, p. 2020.03.30.20047787. https://doi.org/10.1101/2020.03.30.20047787

Mohamed AK, Hadi AA, Mohamed AW (2020) Generalized adaptive differential evolution algorithm for solving CEC 2020 benchmark problems. In: 2nd novel intelligent and leading emerging sciences conference, NILES 2020. Institute of Electrical and Electronics Engineers Inc, pp 391–396. https://doi.org/10.1109/NILES50944.2020.9257924

Mohamed AW, Hadi AA, Mohamed AK (2021) Differential evolution mutations: taxonomy, comparison and convergence analysis. IEEE Access. Institute of Electrical and Electronics Engineers Inc. 9:68629–68662. https://doi.org/10.1109/ACCESS.2021.3077242

Mohamed AK, Mohamed AW (2019a) Real-parameter unconstrained optimization based on enhanced AGDE algorithm. Studies in Computational Intelligence:431–450. https://doi.org/10.1007/978-3-030-02357-7_21

Mohamed AW, Mohamed AK (2019b) Adaptive guided differential evolution algorithm with novel mutation for numerical optimization. Int J Mach Learn Cybern 10(2):253–277. https://doi.org/10.1007/s13042-017-0711-7

Mohamed AW, Mohamed AK, Elfeky EZ, Saleh M (2019a) Enhanced directed differential evolution algorithm for solving constrained engineering optimization problems. Int J Appl Metaheuristic Comp (IJAMC) 10(1):1–28. https://doi.org/10.4018/IJAMC.2019010101

Mohamed AK et al (2018) Enhancing AGDE algorithm using population size reduction for global numerical optimization. Advances in Intelligent Systems and Computing:62–72. https://doi.org/10.1007/978-3-319-74690-6_7

Mohamed AK et al (2019b) Solving constrained non-linear integer and mixed-integer global optimization problems using enhanced directed differential evolution algorithm. In: Studies in computational intelligence. Springer Verlag, pp 327–349. https://doi.org/10.1007/978-3-030-02357-7_16

Nagib,AE et al. (2020) 'Neural Network with Adaptive Learning Rate', in 2nd Novel Intelligent and Leading Emerging Sciences Conference, NILES 2020. Institute of Electrical and Electronics Engineers Inc., pp. 544–548. https://doi.org/10.1109/NILES50944.2020.9257880

Narin A, Kaya C, Pamuk Z (2020) 'Automatic detection of coronavirus disease (COVID-19) using X-ray images and deep convolutional neural networks', arXiv. arXiv. Available at: http://arxiv.org/abs/2003.10849. Accessed: 16 January 2021

Rehman A et al. (2020) 'Improving coronavirus (COVID-19) diagnosis using deep transfer learning', medRxiv medRxiv, p. 2020.04.11.20054643. https://doi.org/10.1101/2020.04.11.20054643

Shan F et al (2020) Lung infection quantification of COVID-19 in CT images with deep learning. Med Phys. https://doi.org/10.1002/mp.14609

Singhal T (2020) A review of coronavirus Disease-2019 (COVID-19). Indian J Pediatr:281–286. https://doi.org/10.1007/s12098-020-03263-6

Storn R, Price K (1995) Differential Evolution-A simple and efficient adaptive scheme for global optimization over continuous spaces

Storn R, Price K (1997) Differential evolution - a simple and efficient heuristic for global optimization over continuous spaces. J Glob Optim 11(4):341–359. https://doi.org/10.1023/A:1008202821328

Tahamtan A, Ardebili A (2020) 'Expert Review of Molecular Diagnostics Real-time RT-PCR in COVID-19 detection: issues affecting the results'. https://doi.org/10.1080/14737159.2020.1757437

Ucar F, Korkmaz D (2020) COVIDiagnosis-net: deep Bayes-SqueezeNet based diagnosis of the coronavirus disease 2019 (COVID-19) from X-ray images. Med Hypotheses. Churchill Livingstone 140:109761. https://doi.org/10.1016/j.mehy.2020.109761

Wang L, Lin ZQ, Wong A (2020) COVID-net: a tailored deep convolutional neural network design for detection of COVID-19 cases from chest X-ray images. Sci Rep. Nature Publishing Group UK 10(1):1–12. https://doi.org/10.1038/s41598-020-76550-z

Wang S et al (n.d.) A deep learning algorithm using CT images to screen for Corona virus disease (COVID-19). https://doi.org/10.1101/2020.02.14.20023028

Xu X et al (2020) A deep learning system to screen novel coronavirus disease 2019 pneumonia. Engineering 6(10):1122–1129. https://doi.org/10.1016/j.eng.2020.04.010

Part V
Data Science

Chapter 19
Quality Design for the COVID-19 Pandemic: Use of a Web Scraping Technique on Text Comments and Quality Ratings from Multiple Online Sources

Praowpan Tansitpong

Abstract This study explores the main determinants of airline satisfaction by integrating data from two online survey sources collected via the use of a web scraping technique on text comments and quality ratings to determine service recovery procedures for the aviation industry during the COVID-19 pandemic. The text analysis technique provides information on how passengers rate service attributes (high or low) by generating clusters of the most frequent comments (WordCloud). The results suggest that satisfied passengers highlight empathy and responsive service, while negative reviews suggest frequent instances of poor operational performance, such as refund processes, rescheduling, and system breakdowns.

Keywords Web scraping · Text mining · Airline recommendation · Online review · WordCloud · COVID-19

19.1 Introduction

Due to the COVID-19 (or coronavirus) pandemic, airlines have suffered a significant decline in profit due to restrictions on global flight operations. Because of the global recession in airline operations, airlines had to deal with industry reform, including situations of flight cancelations and rescheduling. To help understand the recovery of service in this situation, direct comments or complaints from passengers are crucial. To learn quickly, airlines need to receive instant information from online websites rather than wait for complaints to be filed. In general, to survive a severe service

P. Tansitpong (✉)
NIDA Business School, National Institute of Development Administration (NIDA), Bangkok, THAILAND
e-mail: praowpan.tan@nida.ac.th

S. A. Hassan et al. (eds.), *Decision Sciences for COVID-19*, International Series in Operations Research & Management Science 320,
https://doi.org/10.1007/978-3-030-87019-5_19

situation, a service company requires an innovative approach to respond to abrupt change. New technology, such as mobile applications used to connect with passengers, may take time to develop and be hard to adopt. This study aims to solve this challenging issue for the service industry. It is important to quickly understand desired service parameters to help airlines restore satisfaction among passengers.

This study utilizes airline data from January to May 2020. Reviews from websites provide both a tangible score (e.g., seat comfort, food, magazines) and an intangible score (e.g., flight attendant performance, response to inquiries, recruitment). Because the industry is highly competitive and passengers have a wide choice of airline options, this study focuses on airlines that have still been operating during the COVID-19 pandemic, while other airlines might have ceased service due to the situation. Online review websites are one of the outlets that present airlines with service outcomes from the perspective of passengers. There are multiple websites that provide detailed ratings on various aspects of service through surveys. Each year, more than 100 detailed service items are issued to each airline on the websites. These online reviews provide instant feedback to airlines so they can identify the link between the current service situation and the misalignment of customer satisfaction and expectations. Insight from passengers' reviews is fruitful because it provides global feedback on the current service. This insight could help airlines to change and survive in the critical conditions of COVID-19; hence, this paper aims to determine the major aspects of airline services that are significant for passengers due to COVID-19.

19.2 Literature Review

Service quality design is a challenge for the airline industry due to ambiguity in passenger experience, and it is difficult for the industry to fully understand the determining factors of service satisfaction. These quality measures include a variety of tangible and intangible aspects. In the past, common measures focused on performance data, such as on-time arrivals, denied boarding, mishandled baggage, and passenger complaints (Suraratdecha et al. 2005; Pamucar et al. 2021; Gunarathne et al. 2018); however, these measures had an operational focus and did not focus on qualitative or true aspects of airline service. Previous researchers were seeking the best determinants to predict the performance of airlines (Lapré and Scudder 2004); however, the feedback was not real-time, and most of the datasets were outdated. Recently, online reviews have offered real-time feedback to airline businesses so they can identify linkages between dimensions of the services that might foster better service and what has been lacking. To improve service quality, multiple studies have determined the relationship between airline service quality, such as tangibles, schedules, services provided by ground staff, seat comfort, and passenger satisfaction (Tsikriktsis 2007; Brady et al. 2006; Chen and Hu 2013; Gentili et al. 2002), which can be derived instantly from online sources. The service recovery literature suggests that passenger attitudes toward service recovery differ according to their personal experience and that the recovery situation depends on overall satisfaction, which could be derived from both tangible (seat comfort, food, magazines, etc.) and intangible (flight attendant efficiency, response, attitude, etc.) dimensions of airline service. Service ratings are useful information to help airlines understand how they can adjust and influence others with the same service and experience (Chang and Chang 2010; Williams and Soutar 2009; Cronin and Taylor 1992). This study investigates the impact of quality of service and customer satisfaction ratings and their links to passenger comments. Moreover, the marketing and operations literature has disregarded the importance of text analysis in quantitative studies in the past, and useful information can be obtained from the qualitative approach to distinguish between reviews and comments as a predictor of satisfaction using analytical tools.

Text reviews of services can help airlines learn and better satisfy customer needs by improving two-way communication and learning from online sources. Techniques and innovations in grouping text into clusters provide direct feedback on specific topics and help determine the customer experience prior to, during, and after flight experience (Boulding et al. 1993; Mulajati and Hakim 2017; Sezgen et al. 2019; Wong and Qi 2017). Airline services are traditionally measured by individual customer survey questions. Personal satisfaction identified based on the gap between perceptions and expectations has not been analyzed in past literature. Individual customer comments do not indicate the actual performance of the airline, as each person can use different comparison criteria from different backgrounds and expect different results. Therefore, it is important to identify key aspects of service delivery during this unique crisis period and seek to find improvements in airline services based on positive and negative online reviews.

As text comments are used to examine reflections from customers' mindsets (Lucini et al. 2020; Gallagher et al. 2019; Holliman and Rowley 2014), the use of a web crawling or automated data collection technique on online websites is an effective method for researchers to reduce the time to collect large amounts of data at no cost and to create "just-in-time" knowledge. Previous studies have been conducted to determine the relationship between airline service quality, such as assets, flight schedules, ground crew services, seating facilities, and passenger satisfaction (Ghosh 2016; Park et al. 2004; Chen et al. 1994). There has been a limited focus on integrating knowledge gained from quantitative and qualitative online reviews. Hence, the goal of this study is to answer the following question: "What are attributes of airline services that could differentiate positive satisfaction from negative satisfaction during a crisis?"

19.3 Methods

19.3.1 Data Collection

The first set of airline data was obtained from Seatguru (http://www.seatguru.com) for information on aircraft type, seat configurations, and quality evaluation from passengers. The second dataset is collected from Skytrax (http://www.airlinequality.com). The database includes reviews from US-based airlines (American Airlines, Delta Airlines, JetBlue Airways, Southwest Airlines, Spirit Airlines, US Airways and United Airlines), ratings for overall satisfaction, and detailed comments written in English. The reviews are provided in response to specific questions related to seating comfort, cabin crew service, in-flight entertainment, ground service, transfer, arrival, quality of the food, cleanliness of the cabin, enthusiasm, and attitude of the staff (friendliness and hospitality), and value for money, which are ranked on a 5-point scale; in addition, an overall airline rating is provided on a 10-point scale, and a recommendation is provided as a binary value. The overall ratings were separated into "high" and "low" rating groups using a cut-off of 8 points. A rating above 8 points was categorized as the "high" rating group, and other ratings were categorized as the "low" rating group. Scores equivalent to 8 or higher have been used in multiple studies according to the net promoter score (NPS) (Suzuki et al. 2001), which confirmed a strong relationship between satisfaction and recommendation. In the airline review section, the website displays individual passenger reviews and ratings for top comments from the airline, detailed comments, flight type, passenger type, cabin flight, route, date range, seating, cabin crew service, and food and a checkmark for drinks, in-flight entertainment, ground service, value for money and recommendation to others. The detailed passenger comments also include flight experiences with short phrases such as "flight delay, missed connection" and detailed comments. The lengths of the top comment phrases were approximately 100–500 words. Seating configuration details are displayed in Fig. 19.1.

To automate the collection of data, a data crawling method was conducted by using sources written in Python to extract web information in Excel format. The crawling technique automates the collection of text and numeric data from websites. The diagram in Fig. 19.2 shows how the web scraping code detects information on the web and processes the information to the server. The web scraping process was conducted in May–June 2020, several months after airlines experienced difficulties offering services during COVID-19. Steps in web scraping include checking for the website connection, fetching or contacting the website to check for accessibility, and then retrieving and parsing the data to forward to the collection server. This web scraping process was performed separately on two websites (http://www.seatguru.com and http://www.airlinequality.com). Comment reviews were collected from the reviews passengers posted on these websites. The method of collecting this dataset was validated using Urlib 3 functional libraries for implementing HTL and TCP/IP, the host BeatifulSoup 4 (BS4) and Python's CSV library for managing connections with HTML and XML with document standards and interfaces. Numerical and text

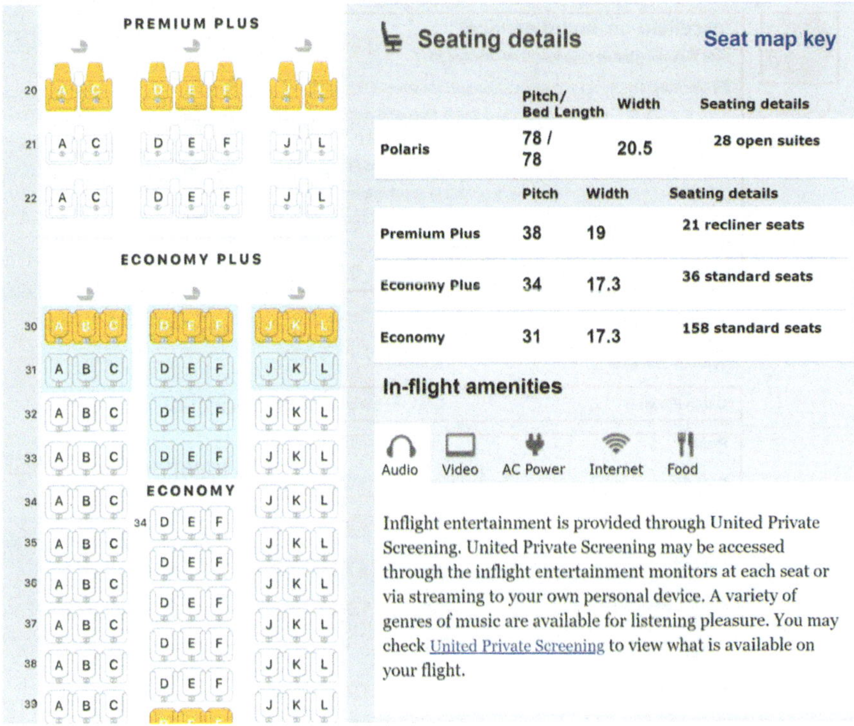

Fig. 19.1 Example of seat size data from Seatguru.com

data were merged and sent for analysis by selecting comments related to the terms "COVID-19," "Coronavirus," and "Virus."

For an analysis of text reviews, WordCloud was used to understand the frequency of words that appeared in the comment sections. Comments including terms related to "COVID-19," "Coronavirus," or "Virus" were processed into rating separation. Using text analysis, overall satisfaction scores were divided into two groups: high (more than 8) and low (less than 8 out of 10) scores. Using the nltk library provided in the text separation, the two WordCloud visualizations of the most persuasive words in the comments were divided into groups based on high and low ratings. WordCloud generation was conducted using Python packages. The diagram displays the methods for processing data collection and the analysis of text comments (Fig. 19.3).

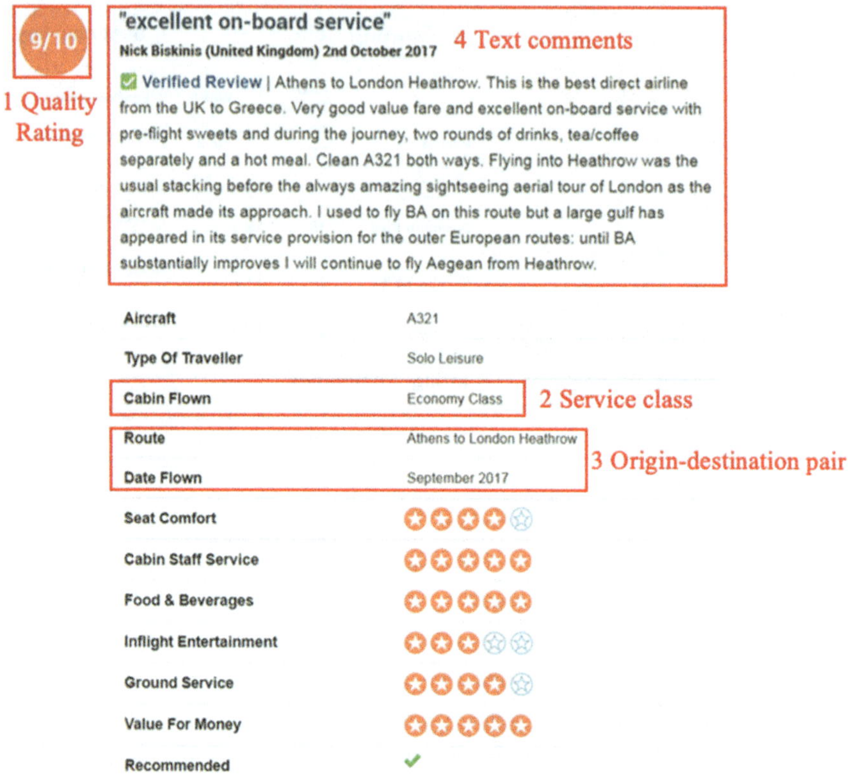

Fig. 19.2 Example of text and ratings data scraped from airlinequality.com before COVID-19 pandemic

19.4 Results

19.4.1 Passenger Reviews

Detailed service quality data include information on airline operations, on-board products and flight attendant services. Each category had its own list of services and ratings. A descriptive summary of the service parameter results and rating groups is presented in Table 19.1, which displays a scaled comparison of the seven parameters of service ratings. The top three highest ratings were found for cabin staff service, ground service, and value for money; on the other hand, the scores were low for the aspects of food and beverages and in-flight entertainment (Fig. 19.4).

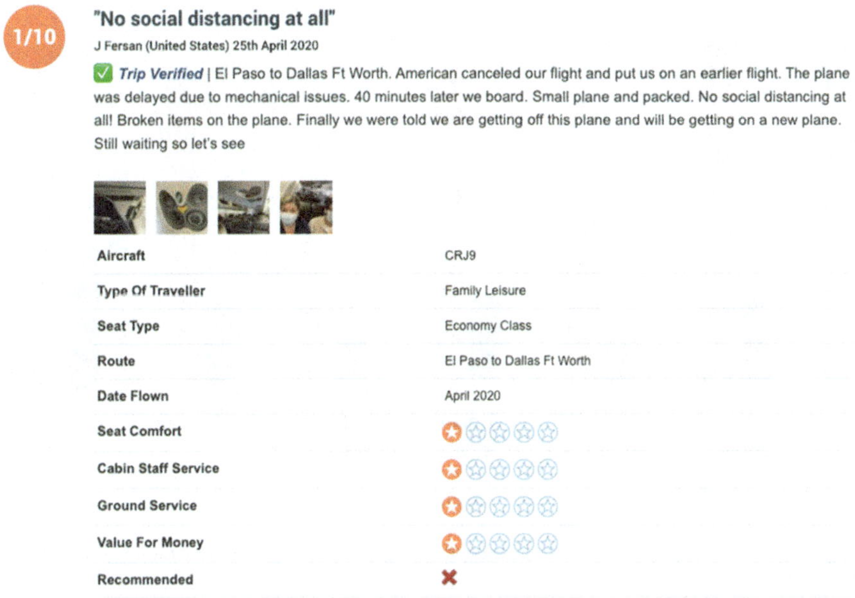

Fig. 19.3 Example of negative comments scraped from airlinequality.com during COVID-19 situation

Table 19.1 Linkage between logistic regression results and Word Cloud

Low ratings	High ratings
Refunded	Experience
Disappointed	Service
Rescheduled/Cancelled	Pleasant
Scam/robbed/rip-off	Amazing
Heartless	Excellent
Denied	Good/best/better
Keeping/waiting	Helped/helpful
Penalty/Glitch	Friendly/kind

19.4.2 Text Analysis

Frequently used word lists were collected from WordCloud to extract the most commonly used functions in the comments. Customers often used the words "Nice," "Great," and "Excellent" to describe the service they experienced and mentioned "Service" and "Experience" when commenting on flight attendants. In low-rated reviews, the words "Refund" and "Canceled/Postponed" were most often used to explain services. Other words that were associated with the poor performance of service operations included "disappointed," "refund," "cancel," and "system failure" as part of descriptions of experiences with the airport, staff courtesy, cleanliness of the airport and a comfortable environment, and the close proximity of

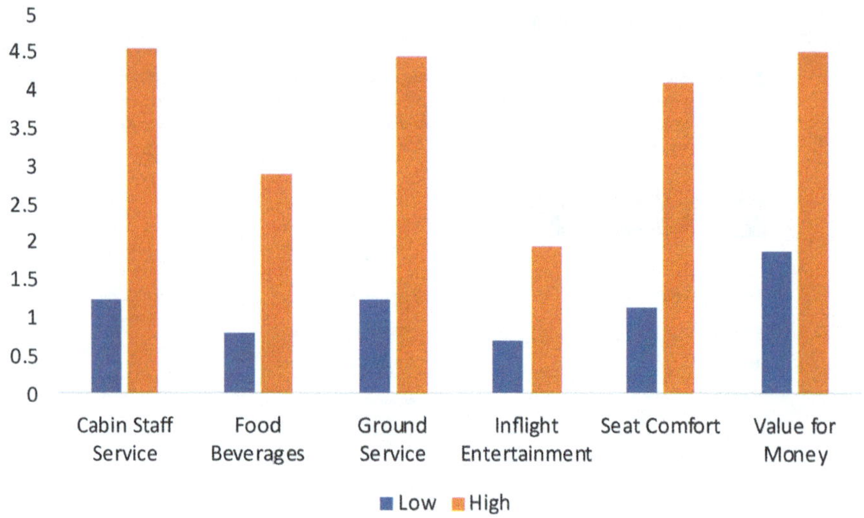

Fig. 19.4 A comparison of service dimension scores between high and low rating groups during COVID-19

passengers. The results showed that helpful airport personnel is important for helping passengers cope with difficult situations such as delayed or canceled flights. Using web scraping techniques, airlines can use online reviews to improve their current services. As the results confirm, airlines should focus on flight attendants, ground staff service and in-flight entertainment. The frequent word lists of the two highest- and lowest-rating groups are shown in Figs. 19.5 and 19.6. The difference between the two sets of word lists shows that passengers submitting high ratings were satisfied with the service related to the flight attendant service, while passengers submitting low ratings gave comments related to operational management and empathy of service.

19.4.3 Text Analysis for Quality Differentiation: Post-Pandemic Results

By using a rating score separator, frequent word lists were gathered from WordCloud to extract features that were the most commented on during and after the spread of the COVID-19 situation. Frequently used word lists were collected from the word cloud to extract the most commonly used functions in comments. Customers often used the words "Nice," "Great," and "Excellent" to describe the service they experienced and mentioned "Service" and "Experience" when commenting on flight attendant service. In low-rated reviews, the words "Refund" and "Canceled/Postponed" were most often used to explain services. Other words

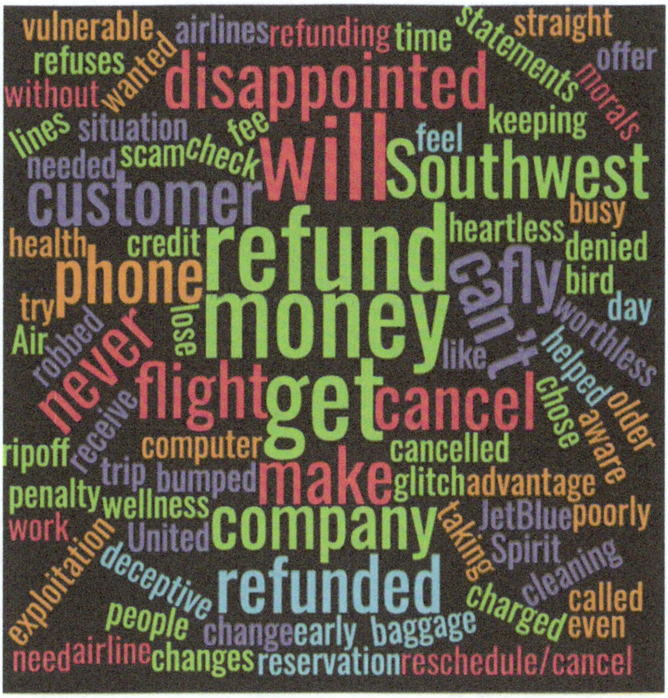

Fig. 19.5 WordCloud results from low rating scores

that were associated with the poor performance of service operations included "disappointed," "refund," "cancel," and "system failure" as part of descriptions of experiences with the airport, staff courtesy, cleanliness of the airport and a comfortable environment, and the close proximity of passengers. The results showed that helpful airport personnel is important for helping passengers cope with difficult situations such as delayed or canceled flights. Using text mining techniques, airlines can use online reviews to improve their current services in several aspects, such as flight attendant service, food and beverages, ground handling, in-flight entertainment, seating comfort and value for money. As the results confirm, airlines should focus on flight attendants, ground handling and in-flight entertainment. The frequent word lists of the two highest- and lowest-rating groups are shown in Figs. 19.5 and 19.6.

WordCloud results from low- and high-rating reviews visually characterize an intuitive representation derived from passengers' comments. The importance attached to desired in-flight and ground services can help airlines determine factors that are meaningful for passengers. Many frequent keywords were related to the operational aspect of airline services, such as ground staff, seat comfort, and value for money (Siering et al. 2018). However, some evaluations of traditional operations, such as in-flight entertainment and food/beverages, did not show significant ties to the text comment reviews. This confirms that by using information from text

Fig. 19.6 WordCloud results from high-rating scores

comments, airlines can derive multiple dimensions of desired qualities from the point of view of passengers. The missing operational aspects of money transaction and processing, such as refunds for tickets, extra charges, credit card and penalty charges, are important for customers in regard to critical situations during COVID-19 when flights were normally canceled or postponed due to international and domestic travel restrictions. This dimension cannot be captured using traditional surveys. On the other hand, the positive reviews showed that sentiment feedback was related to empathy (Pakdil and Aydın 2007), related by comments with words such as excellent, great, amazing, thankful, helpful, friendly, kind, and satisfied passengers, which were previously difficult to detect in traditional rating surveys. These positive comments confirm that passengers have received services beyond their expectations. This could help airlines train flight attendants to improve their willingness to help passengers experience difficult situations and could help change passengers' evaluations of the overall trip experience.

19.5 Conclusion

This study empirically demonstrates that service quality design is an essential component of product line strategy. The difference in quality can be determined in two ways. Through the use of customer ratings and text comments, the airline industry could use this study to examine whether a service strategy is suitable to counteract the effect of severe situations and whether a range of seats and other services that affect the customer experience would be offered differently than during a normal situation. This study confirms that airlines operating under pandemic conditions due to limited capabilities and resources can experience service failures. Hence, the airline should adjust operations and policies to improve their services given this situation. This paper has examined the relationship between service practice and practical results based on quality ratings and text comments. Based on previous literature indicating that changing service quality should have some impact on performance, the results of this study confirm that airlines gain better ratings by emphasizing a range of operational dimensions, such as on-time service and reliable computer systems.

A summary of text analysis indicates another aspect of service design, where positive reviewers are satisfied with human dimensions such as personality and friendly service, while negative comments often indicate aspects of poor communication, such as personal communication, timeliness, and seating. This study shows how website crawling for data can be useful for measuring the performance of services. A comparison between pre-COVID-19 and post-COVID-19 showed differences in expectations and evaluations of air services. According to a previous study on the determinants of recommendations in the airline industry prior to COVID-19 (Aydin and Yildirim 2012), comments frequently focused on the professionalism and personality of flight attendants, such as surprised, helpful, impressed, efficient, polite, and courteous, while low-rating reviews suggested operational difficulties, such as system breakdowns, downgrading, and limited access to alcohol beverages. This comparison between pre- and post-COVID confirms that the point of view regarding desirable service has shifted from the operational dimension to the personal dimension, as passengers frequently express emotional and touching terms such as "pleasant" and "experience" for positive comments and "heartless" and "keeping/waiting" after experiencing the COVID-19 situation.

The COVID-19 pandemic has created a difficult service environment for global airlines. This study explores the determinants of the airline rating system from online reviews during this critical period. Using information from online web scraping, this study distinguishes between positive and negative comments from online reviews to understand what significant service factors need improvement to meet passenger expectations in this critical situation. The text analysis technique provides information to characterize how passengers decide to recommend other people and the attributes of the service that differentiate the positive and negative feedback of the generated Word Cloud. This study demonstrates how web crawling could be useful in measuring service industry performance and predicting recommendation patterns, which is a rare service context. While all dimensions of service are considered important determinants of air services based on common perceptions and evidence provided by websites, key determinants of airline recommendations can narrow the scope of service in regard to choosing priorities. The results suggest that positive recommenders are satisfied with human dimensions such as personality and courteous treatment, while negative comments suggest frequent complaints about poor dimensions of operations such as punctuality and seating comfort.

The results from this study confirm the linkage between rating scores and text comments obtained from two online sources. Passengers usually compare the cost of the flight with the service they receive. It is important to improve specific airline products (such as in-flight entertainment and seats) and services (such as cabin staff service and ground service). Given a short window to improve air service, there is an urgent need to identify service parameters in times of crisis to maintain customer satisfaction and to survive the critical conditions of COVID-19.

References

Suraratdecha C, Saithanu S, Tangcharoensathien V (2005) Is universal coverage a solution for disparities in health care?: findings from three low-income provinces of Thailand. Health Policy 73(3):272–284

Pamucar D, Yazdani M, Montero-Simo MJ, Araque-Padilla RA, Mohammed A (2021) Multi-criteria decision analysis towards robust service quality measurement. Expert Syst Appl 170:114508

Gunarathne P, Rui H, Seidmann A (2018) When social media delivers customer service: differential customer treatment in the airline industry. MIS Q 42:489–520

Lapré MA, Scudder GD (2004) Performance improvement paths in the US airline industry: linking trade-offs to asset frontiers. Prod Oper Manag 13:123–134

Tsikriktsis N (2007) The effect of operational performance and focus on profitability: a longitudinal study of the US airline industry. Manuf Serv Oper Manag 9:506–517

Brady MK, Voorhees CM, Cronin JJ, Bourdeau BL (2006) The good guys don't always win: the effect of valence on service perceptions and consequences. J Serv Mark 20:83–91

Chen PT, Hu HHS (2013) The mediating role of relational benefit between service quality and customer loyalty in airline industry. Total Qual Manag Bus Excell 24:1084–1095

Gentili GB, Tesi V, Linari M, Marsili M (2002) A versatile microwave plethysmograph for the monitoring of physiological parameters. IEEE Trans Biomed Eng 49:1204–1210

Chang YW, Chang YH (2010) Does service recovery affect satisfaction and customer loyalty? An empirical study of airline services. J Air Transp Manag 16:340–342

Williams P, Soutar GN (2009) Value, satisfaction and behavioral intentions in an adventure tourism context. Ann Tour Res 36:413–438

Cronin JJ, Taylor SA (1992) Measuring service quality: a reexamination and extension. J Mark 56:55–68

Boulding W, Kalra A, Staelin R, Zeithaml VA (1993) A dynamic process model of service quality: from expectations to behavioral intentions. J Mark Res 30:7–27

Mulajati M, Hakim RF (2017) Sentiment analysis on online reviews using Naïve Bayes classifier method and text association (case study: Garuda Indonesia airlines passengers reviews on tripadvisor site). Indian J Sci Res:274–281

Sezgen E, Mason KJ, Mayer R (2019) Voice of airline passenger: a text mining approach to understand customer satisfaction. J Air Transp Manag 77:65–74

Wong CUI, Qi S (2017) Tracking the evolution of a destination's image by text-mining online reviews-the case of Macau. Tour Manag Perspect 23:19–29

Lucini FR, Tonetto LM, Fogliatto FS, Anzanello MJ (2020) Text mining approach to explore dimensions of airline customer satisfaction using online customer reviews. J Air Transp Manag 83:101760

Gallagher C, Furey E, Curran K (2019) The application of sentiment analysis and text analytics to customer experience reviews to understand what customers are really saying. Int J Data Warehous Min (IJDWM) 15:21–47

Holliman G, Rowley J (2014) Business to business digital content marketing: marketers' perceptions of best practice. J Res Interact Mark 8:269–293

Ghosh M (2016) Case study: text-mining customers view point and perceived value about brand. Int J Bus Anal Intell 4:1

Park JW, Robertson R, Wu CL (2004) The effect of airline service quality on passengers' behavioural intentions: a Korean case study. J Air Transp Manag 10:435–439

Chen IJ, Gupta A, Rom W (1994) A study of price and quality in service operations. Int J Serv Ind Manag 5:23–33

Suzuki Y, Tyworth JE, Novack RA (2001) Airline market share and customer service quality: a reference-dependent model. Transp Res A Policy Pract 35:773–788

Siering M, Deokar AV, Janze C (2018) Disentangling consumer recommendations: explaining and predicting airline recommendations based on online reviews. Decis Support Syst 107:52–63

Pakdil F, Aydın Ö (2007) Expectations and perceptions in airline services: an analysis using weighted SERVQUAL scores. J Air Transp Manag 13:229–237

Aydin K, Yildirim S (2012) The measurement of service quality with servqual for different domestic airline firms in Turkey. Serb J Manag 7:219–230

Chapter 20
Data Science Models for Short-Term Forecast of COVID-19 Spread in Nigeria

Ijegwa David Acheme, Olufunke Rebecca Vincent, and Olaniyi Mathew Olayiwola

Abstract The study presents data science models for a real-time forecast of COVID-19 size and spread in Nigeria. Firstly, an exploratory and comparative study of the disease spread in Nigeria and some other African nations are carried out. Then variants of support vector machine (SVM) using the Gaussian kernel and regression machine learning models suitable for modeling count data variables are built to estimate a 15-day prediction of infection cases. The data science models built in this research give a short-term forecast of the disease's spread which is useful in better understanding the spread patterns of the disease as well as enabling future preparedness and better management of the disease by the government and relevant authorities. The research outcome can therefore serve as an effective decision support system. This work can also serve as an alternative to the mathematical-based epidemiological models for the forecast of COVID-19 spread because of their inherent advantages of learning from historical datasets and generalizing with new sets of data which promises better results.

Keywords Coronavirus spread · COVID-19 spread · Machine learning · Poisson regression · Support vector

---

I. D. Acheme (✉)
Department of Computer Science, Edo State University Uzairue, Edo State, Nigeria
e-mail: acheme.david@edouniversity.edu.ng

O. R. Vincent
Department of Computer Science, Federal University of Agriculture, Abeokuta, Nigeria
e-mail: vincentor@funaab.edu.ng

O. M. Olayiwola
Department of Statistics, Federal University of Agriculture, Abeokuta, Nigeria

S. A. Hassan et al. (eds.), *Decision Sciences for COVID-19*, International Series in Operations Research & Management Science 320,
https://doi.org/10.1007/978-3-030-87019-5_20

20.1 Introduction

The coronavirus was first discovered in Africa on February 14, 2020, in Egypt. The COVID-19 disease was first reported in Nigeria on February 28th, 2020. It was an infected Italian who had travelled from the city of Milan in Italy to Nigeria. Since then, the World Health Organization (WHO) has expressed concerns over the growing number of cases in Africa owing to the continent's poor households, weak health systems, and living conditions that do not allow for the strict social distancing required to control the disease spread (Nkengasong and Mankoula 2020). The United Nations has warned that unless urgent steps are taken, the disease will kill over three hundred thousand Africans and push over 30 million to extreme poverty conditions. (Sumner et al. 2020).

The enormity of the economic and health challenges posed by the disease outbreak is inexplicable. This disease outbreak has led to unprecedented disruptions of markets and economies globally. Nigeria particularly faces a greater challenge than most other countries from the COVID-19 pandemic as a result of being the largest economy in Africa, with an economy that solely relies on crude oil exports which have been severely affected by a huge drop in prices. The collapse in oil prices by over 60% since the start of the year to below $30 per barrel currently has affected government revenues, which could fall by as much as 45%. Other key economic sectors that have been badly hit by the pandemic globally include transportation, tourism, education, agriculture, production, and global supply chains.

Several research efforts, therefore, have been reported in efforts to curtail it. Efforts toward sequencing the generic configuration for the provision of reliable vaccines are in top gear and reported (Sun et al. 2020). Other research efforts have focused on studying the economic effects of the disease outbreak (Guerrieri et al. 2020), while several others have focused on the disease's spread patterns and building mathematical-based epidemiological models aimed at understanding the disease spread and hence propose better strategies for curtailing it (Fanelli and Piazza 2020; Peng et al. 2020; Roosa et al. 2020; Atangana 2020).

The statistical and mathematical models for the COVID-19 disease reported in recent studies (Tuiteand and Fisman 2020; Russell et al. 2020; Zhao et al. 2020; Li et al. 2020; Wu et al. 2020b, b) were largely based on the Survival Infected and Recovery (SIR) models and their extended versions. Though, these models offer great insights into the estimation and transmission dynamics of the disease so as to propose interventions strategies to the authorities, the models are limited by the need to estimate initial parameters and rely on several assumptions (Hu et al. 2020). In this work machine learning models are deployed as alternatives to the SIR-based models since machine learning models may yield better results because of their inherent capability to learn from the training datasets and generalize with new datasets.

Hence, this study is motivated by the need to deploy machine learning algorithms for the prediction of the disease's spread as this could yield better results owing to their inherent abilities. The study utilizes the convergence of variants of two well-known regression and support vector machine algorithms. Considering the nature of the training dataset as containing count data, the poison regression model with log-loss and the multinomial dirichlet model were used as variants of the regression and support vector machine algorithms, respectively. The prediction outputs are used for real-time forecast of COVID-19 spread in Nigeria to predict the disease trend. The study examines and analyses the infection rate in Nigeria and compares the spread to some other African countries.

The rest of the chapter is organized as follows: Section 20.2 presents related works. In Sect. 20.3, the procedure for data collection, methodology, and comparative analysis is presented. Section 20.4 presents the results and discussion, while sect. 20.5, presents the concluding remarks are presented.

20.2 Related Work

Several research efforts have reported the use of epidemiological models for the prediction of COVID-19 disease. Fanelli and Piazza (2020) presented an analysis and forecast of the disease outbreak in China, Italy, and France with the time frame of January 22, 2020, to March 15, 2020. The study analyzed a simple day-lag map that revealed the universal nature of the disease spread, indicating that simple mean-fields models can help predict the epidemic peak and the number of infections at the peak. A susceptible-infected-recovered-deaths (SIRD) model was used to forecast the peak periods in the countries of interest as well as predicting the likely numbers of infected, recovered, and fatalities. However, Roberts et al. (2015) outline the limitations of the SIR epidemiological models to include its endemic equilibrium, instability, and the inability to handle time-varying infections in cases where super-infections are possible.

Villela (2020) presented a discrete-time phenomenological model for epidemic forecasting. The model utilized disease infection data that are reported in discrete-time batches of days, weeks, and months. The infection data of the severe acute respiratory syndrome (SARS) was utilized and compared with simulated data. The research showed promising results with the forecast following the time dynamics and resilience to the noise in the dataset. However, a delay effect was also shown as a result of the segmented units of time. Peng et al. (2020) presented dynamical modeling of COVID-19 spread in China using the generalized susceptible-exposed-infected-recovered (SEIR) model with infection data of January 20th, 2020, to February ninth, 2020, obtained from China's National Health Commission. Using estimated parameters, the model forecasted the epidemic's peak as well as the ending date for the Beijing and Shanghai regions. Details of several cities with peak infection rates were also forecast by the model. However, Yang et al. (2012) state that the underlying assumptions of a homogeneous population and disease environment in SIER models treat communicable diseases spread as a linear function, hence the need to utilize models based on machine learning with the ability for more effective results in nonlinear disease patterns as well as heterogeneous environments.

Roosa et al. (2020) presented a model for the real-time forecast of the COVID-19 spread in China from fifth February to 24th February, 2020. The research utilized a phenomenological model built with the cumulative daily number of confirmed cases from the Hubei region from where the virus originated in order to determine the spread trend across mainland China. The research presented a 5, 10, and 15-day consecutive forecast with a result showing the mean estimates and uncertainty bounds being relatively stable within the period of the forecast. Furthermore, the model showed that the disease spread had reached its peak within the period of the study indicative of the fact that the Government's containment efforts had been successful in curtailing the transmission rate. Hirose & Wang (2012) proposed a technique of predicting the future spread pattern of influenza disease using a dataset obtained from Twitter which was used to build a social network system (SNS). Using a method of multiple linear regression utilizing messages on Twitter and the Center for Disease Control's data of influenza-like illness, a regression model was built to predict the future spread patterns of the disease.

While the majority of the previous works have demonstrated the predictability of COVID-19 spread in selected countries particularly China and parts of Europe, which were the epicenters of this disease, little work has been done to study and understand the disease spread patterns in Africa where cases are beginning to rise, hence the motivation for this study. A huge threat is posed in the continent to the essential workers, especially the health workers with Africa's peculiar challenges of lack of resources for effective tackling of highly infectious diseases especially within the rural areas, acute shortage of health facilities, and settlement patterns that are impossible to impose social distancing (Farmer 2001). The World Health Organisation (WHO) has already warned that Africa could become the new epicenter of the disease and predicted that several thousands of lives would be lost if urgent steps are not taken (Kapata et al. 2020). This study, therefore, aims at providing an analysis of the spread patterns of COVID-19 in selected African countries in order to understand better and forecast the expected spread enabling better planning and management. Though, previous research efforts have focused on mathematical and statistical-based models which heavily rely on assumptions of key parameters for predictions, the machine learning models presented in this study provides a comparative alternative as they are self-learning and generalization models (Hu et al. 2020).

20.3 Methodology

This study sets out to better understand the spread patterns of COVID-19 in Africa with the aim of providing short-term prediction of infection cases so as to provide a knowledge-based decision support system for government and health authorities. This will improve the preparedness by government and relevant authorities as they will be better able to anticipate the expected number of hospital spaces and medical supplies for the coming days ahead, furthermore, the social demographic factors

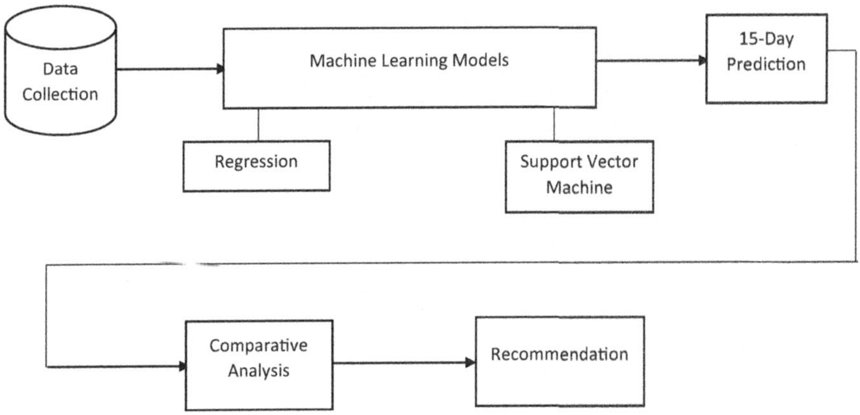

Fig. 20.1 The proposed system

affecting the disease's spread pattern will be better understood, also the containment measures put in place by the government can be evaluated.

The research utilized the dataset of COVID-19 cases, which is daily tallied by the World Health Organization (WHO) of reported cases of COVID-19 for the selected countries of interest. The dataset contained 21,900 multivariate instances with attributes related to date, country/region, last update, number of confirmed cases, number of deaths, and number of recoveries. The study follows the well-known data science research methodology, as proposed by Wickham and Grolemund (2017). Figure 20.1 shows the workflow of the proposed system. It comprises the following phases: Data collection, exploratory analysis, building machine learning predictive models, and comparative analysis of the models.

20.3.1 Data Collection

The dataset utilized for this study is the daily recorded incidence of COVID-19 cases across the world as captured by the World Health Organization (WHO), which contained the following features: country, date, number of confirmed cases, number of recoveries, and number of fatalities. These features are the predictor variables used in this study while the target variable is the number of confirmed cases. The machine learning models will be trained with this dataset so as to forecast the possible number of confirmed cases for a 15-day period. The dataset was found at https://www.who.int/emergencies/diseases/novel-coronavirus-2019/situation-reports

Table 20.1 Correlation matrix of COVID-19 confirmed cases

	World	Egypt	Ethiopia	Kenya	Nigeria	Senegal	South Africa	Component1
World	1	−0.119	−0.129	−0.124	−0.109	−0.100	−0.121	−0.142
Egypt	−0.119	1	0.970	0.999	0.993	0.974	0.994	0.999
Ethiopia	−0.129	0.970	1	0.979	0.940	0.892	0.989	0.973
Kenya	−0.124	0.999	0.979	1	0.987	0.962	0.997	0.998
Nigeria	−0.109	0.993	0.940	0.987	1	0.992	0.979	0.993
Senegal	−0.100	0.974	0.892	0.962	0.992	1	0.947	0.972
South Africa	−0.121	0.994	0.989	0.997	0.979	0.947	1	0.995

20.3.2 Explorative Analysis of the Dataset (Table 20.1)

The correlation matrix among confirmed COVID-19 cases in Africa countries and the world show that the number of COVID-19 cases in the world is negatively correlated with the number of cases in the African countries. There is a clear increase in these African Countries cases and a decrease in the number of the world cases (Table 20.2).

It is noticeable that the sharp drop is after the first component. Therefore, we can based on selecting only one component, as it explains 84.03% of the variation in the number of confirmed COVID-19 cases (Table 20.1 and Screen plot (Fig. 20.2)). The first component explains 84.03% of the amount of variation that exists in the cases of COVID-19 in Africa countries and the world, which is why the number of confirmed cases of COVID-19 in African countries clustered around the first principal component.

First Principal Component Analysis—PCA1.

The first principal component is strongly correlated with the African countries. The first principal component increases with COVID-19 confirm cases in Egypt, Ethiopia, Kenya, Nigeria, Senegal, and South Africa. This suggests the cases in these countries vary together. If the cases in one of the countries increase, then the remaining ones tend to increase as well. In fact, we could state that based on the correlation of 0.999 that this principal component is primarily a measure case in Egypt

$$Y_1 = -0.142\ \text{World} + 0.999\ \text{Eqypt} + 0.973\ \text{Ethiopia} + 0.993\ \text{Nigeria} \\ + 0.972\ \text{Senegal} + 0.995\ \text{SA}$$

The magnitudes of the coefficients give the contributions of each variable to that component.

Table 20.2 Total variance from principal component analysis

	Initial eigenvalues			Extraction sums of squared loadings		
Component	Total	% of Variance	Cumulative %	Total	% of Variance	Cumulative %
1	5.882	84.029	84.029	5.882	84.029	84.029
2	0.984	14.054	98.083			
3	0.130	1.854	99.937			
4	0.003	0.044	99.981			
5	0.001	0.010	99.991			
6	0.000	0.006	99.997			
7	0.000	0.003	100.000			

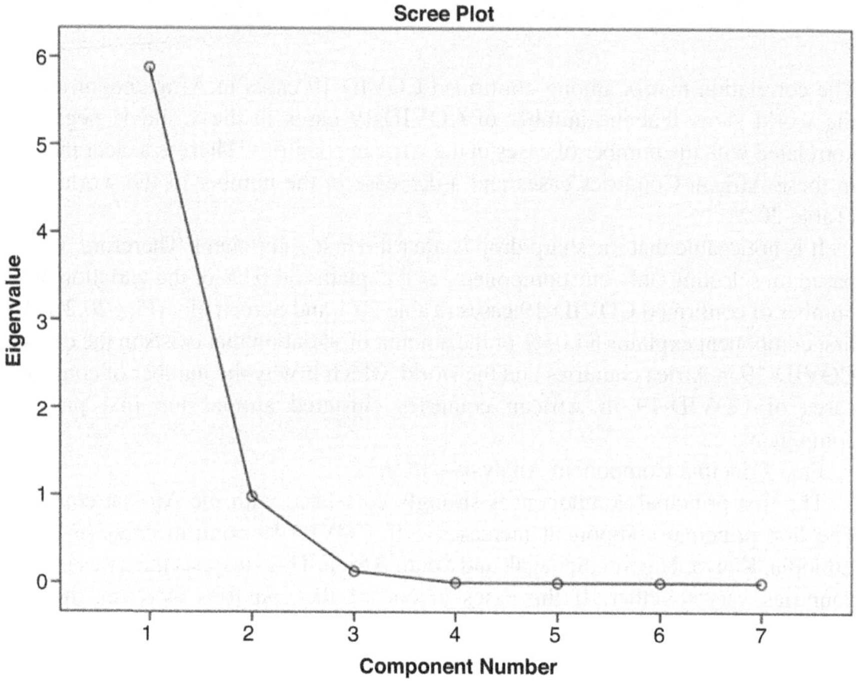

Fig. 20.2 Principal component plot

20.3.3 Machine Learning Models

In machine learning (ML), Artificial Intelligence is applied through the use of statistical, probabilistic, and tools for optimization that learn from patterns in training datasets to classify new data presented after training (Alpaydin 2020; Mitchell 1999). While this technique has been applied to statistical problems for analysis and interpretation of data, ML extends statistical methods by the usage of programming constructs such as Boolean logic, conditional statements if...else, and conditional probabilities for optimization, classification, and clustering problems. Therefore, despite the foundation of Machine learning being firmly rooted in statistics and probability, it offers more robust results as it allows inferences and decisions to be drawn from models that may not be possible with conventional techniques (Ijegwa et al. 2018; Alpaydin 2020). Statistical methods used in multivariate regression and correlation analysis, for example, assumes variable independence hence, a strict statistical model will only build linear combinations of such variables, which is not realizable in most real-world systems, hence, statistical models are limited by non-linear, inter-dependent, and conditional variables characteristic of most real-life systems, in this kinds of situation, ML models offer better results (Michie, et al. 1994). The success of a good machine learning model depends on the understanding of the problem and the data used, understanding the assumptions and limitations of the chosen algorithms as the best models are dependent on the quality of the training dataset (Kotsiantis et al. 2007). Other problems are classified under the dimensionality of variables, overtraining, and overfitting of models (Ayodele 2010).

20.3.3.1 Regression Model

Regression analysis is a machine learning technique that is used for prediction. It models the relationship between a **dependent** (target) and **an experimental variable (s)** (predictor). This modeling technique is used for forecasting and time series analysis so as to find the causal effect relationship between variables. It has the general form, $(x_1, y_1), (x_2, y_2), \ldots, (x_n, y_n)$. The regression line of best fit is given as

$$y = a_0 + a_1x + \ldots + a_mx^m. \tag{20.1}$$

In Eq. (20.1), y is the dependent (target) variable, in this case, the number of infections, a_0, a_1, and a_m are the parameters representing the cost, which is minimized The accuracy of the polynomial regression model is evaluated with the Root Mean Squared Error (RMSE). The RMSE is a quadratic value that reveals the size of the predicted value from the actual value of infections. This is given as

$$RMSE = \sqrt{\sum_{t=1}^{n} \frac{(Predicted\ (t) - actual\ (t))^2}{n}}. \tag{20.2}$$

However, a model as Eq. 20.1 will not work well for Poisson data (Agresti 2002). It is certain to yield negative values for certain xi, but y_i can only take on values from 0 to infinity. Also, the assumption of equal variance in linear regression inference is violated because as the mean rate for a Poisson variable increase the variance also increases. Hence, this work considers the poison regression model described in Eq. 20.2 which models $\log(y_i)$ rather than y_i as a function of the covariates. The $\log(y_i)$ takes on values from $-\infty\ to \infty$.

$$log\ (y_i) = a_0 + a_1x \tag{20.3}$$

where the observed values $y_i \sim$ Poisson with $y = y_i$ for a given x_i. Hence, for every state, i can potentially have a different y which is dependent on its value of x_i. Poisson regression models, however, contain no separate error term like classical regression models.

The summary of the parameters and hyperparameters utilized in building the binomial regression model are: nine features, twenty-three thousand training examples, 80:20 training/test, hyperparameters refer to parameters whose values are set before the learning process begins. These parameters are tunable during the training and learning process and can directly affect how well a model train. While the following assumptions of the Poisson regression models—Poisson response, independence, mean/variance, and linearity were verified to assure the validity of the forecasts.

20.3.3.2 Support Vector Machine

The Support Vector Machine (SVM) belongs to the family of supervised machine learning algorithms used for regression and classification problems. The SVM algorithm plots each data item as a point in an n-dimensional space with the value of each feature represented by the value of a coordinate and n refers to the number of features in the dataset. Thus, classification is performed by finding the hyper-plane that most distinguishes the two classes. The study utilized the COVID-19 dataset in the form $\{(x_k, y_k)\}_{k=1}^{n}$. The x are the features, and y is a $1 \times n$ vector. Since there is n training dataset. Given that training variable x and output variable y are defined by

$$x = \{x_1, x_2, x_3 \ldots\} \in R^N \tag{20.4}$$

and

$$y = \{y_1, y_2, y_3 \ldots\} \in R. \tag{20.5}$$

Eq. (20.3) represents the training variable while Eq. (20.4) is the output variable. The SVM algorithm is the function $y(x)$ that minimizes the error for all the learning features x_i, hence the objective is to; $minimize : \frac{1}{2}||w||^2 + C\sum_{i=1}^{l}\left(\xi_i + \widehat{\xi}_i\right)$,

$$subject\ to : w.x_i + b - y \leq \epsilon + \xi_i,$$

$$y_i - w.x_i - b < \varepsilon + \widehat{\xi}_i.$$

where, w is an n-dimensional vector representing the weight, $i = 1, \ldots, l$, $C > 0$ determines the trade-off between the differences in the decision function and $\xi_i, \widehat{\xi}_i \geq 0$.

The classical support vector machine algorithm as described in Eqns. 20.4 and 20.5 have been noted for yielding poor results with count data variables. Hence a variant given by Bouguila and Ziou (2007) has been utilized for the forecast in this work. Eqn 20.5 describes the multinomial dirichlet model.

$$p\left(\vec{X}|\xi_j\right) = \frac{\Gamma\left(\sum_{v=1}^{V} X_v + 1\right)\Gamma\left(\sum_{v=1}^{V} \alpha_{jv}\right)}{\Gamma\left(\sum_{v=1}^{V} \alpha_{jv} + \sum_{v=1}^{V} X_v\right)} \times \Pi_{v=1}^{V} \frac{\Gamma(X_v + \alpha_{jv})}{\Gamma(\alpha_{jv})\Gamma(X_v + 1)} \tag{20.6}$$

where $\xi_j = (\{\alpha_{jv}\})$ are the parameters associated with the component density. $\alpha_{jv} > 0$ and $\Gamma(.)$ is the Gamma function satisfying $\Gamma(X + 1) = X\Gamma(X)$ and $\Gamma(1) = 1$, ($C = 1$, degree = 6, kernel = "guass").

20.4 Implementation and Results

The exploratory data analysis is presented in order to gain insights and understand the trend of the disease spread, comparative analysis of the cases incidences in Nigeria, and some selected African countries are also presented. The retrieved dataset required very little cleaning and pre-processing as it contained no missing values, however, careful features' selections and transformation were carried out to obtain the relevant features for the work which included date, country_region, last_update, number_of_confirmed_cases, number_of_ deaths, and number_of _recoveries. The target variable is number_of_confirmed_cases and the goal is to train the models with these features so as to predict this target variable.

20.4.1 Analysis COVID-19 in Nigeria

The initial spread of the disease in the first month was very slow; however, starting in the month of April, 2020 there has been a spike in the number of confirmed cases and deaths. Figures 20.3 and 20.4 show the disease spread in Nigeria from February 28th, 2020.

Figures 20.3 and 20.4 indicate the growth of confirmed, active, and closed cases at almost exponential rates from April 2020, indicating that infection rates were beginning to rise. As can be seen in Figs. 20.3 and 20.4 the first reported cases was on February 28th with very slow infection rate in the month of March and April. From February to the end of April, a period of over 60 days the total infection rates stood at less than 2000, however, beginning in the month of May to June another period of about 60 days, the number of infections has risen to over thirteen thousand.

20.4.2 Comparative Analysis COVID-19 Nigeria and Selected African Countries

The comparative analysis of Nigeria with six other African countries revealed that, among the six countries under consideration, as shown in Table 20.3, Ethiopia has the lowest cases (2506.0), while South Africa has reported the highest number of cases followed closely by Egypt. Senegal has the lowest mortality rate of 1.12%,

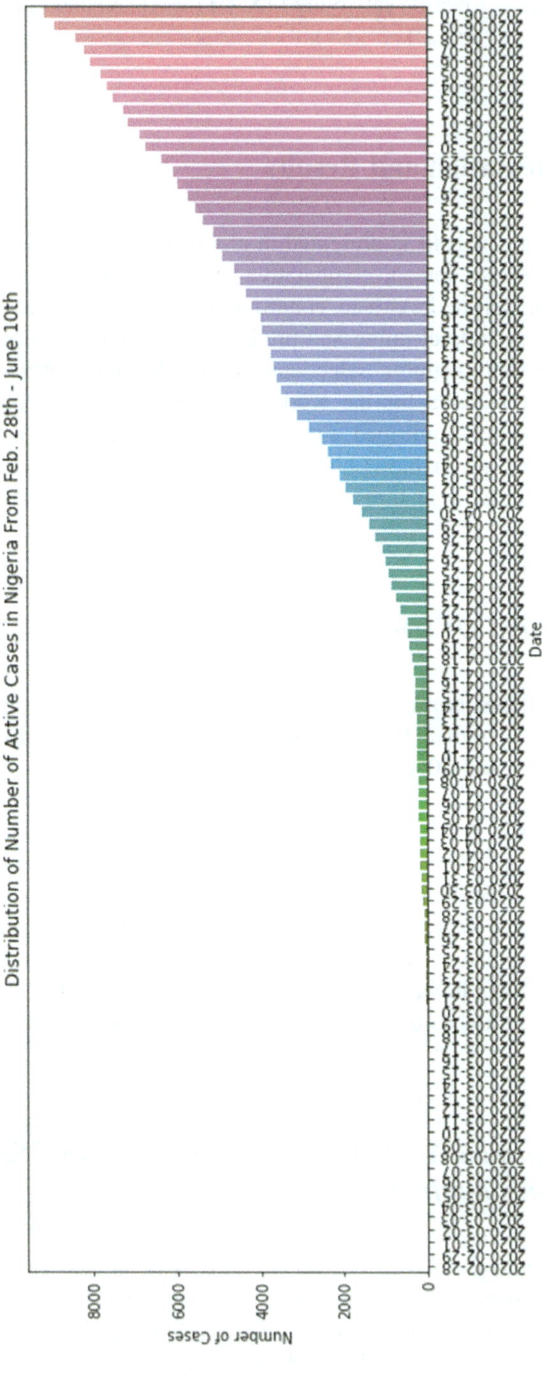

Fig. 20.3 Distribution of active cases in Nigeria

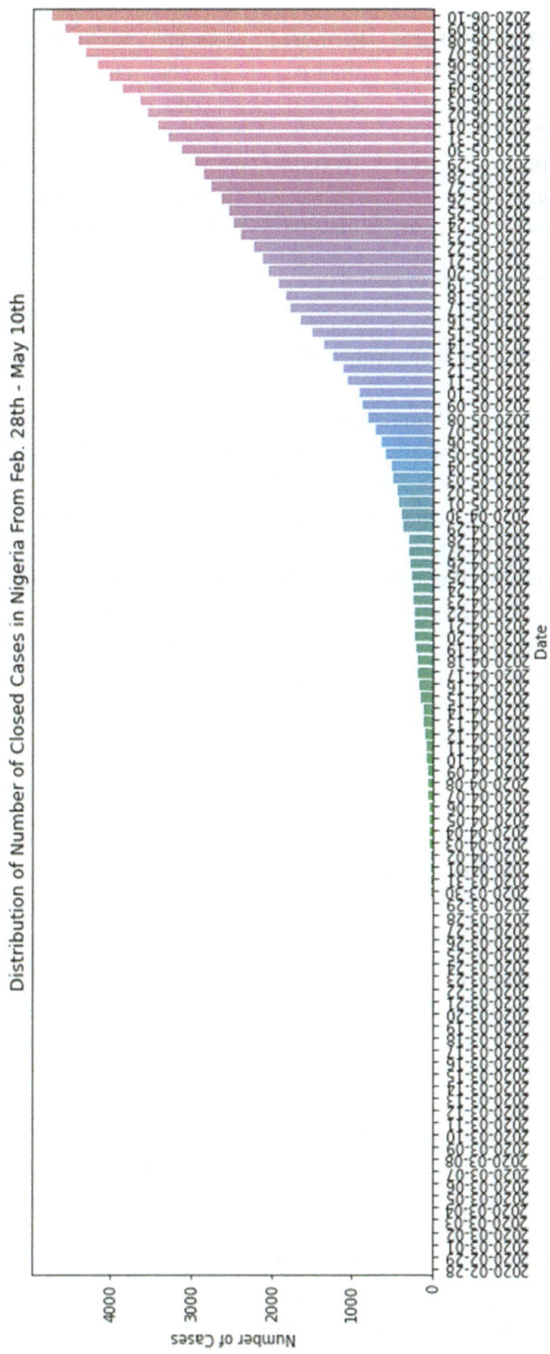

Fig. 20.4 Distribution of closed cases in Nigeria (Deaths and Recoveries)

Table 20.3 Comparison of cases of Nigeria and other African nations

Country	Confirmed	Deaths	Recovered	Mortality	Recovery	Survival probability
South Africa	55421.0	1210.0	31505.0	2.183288	56.846683	97.816712
Egypt	38284.0	1342.0	10289.0	3.505381	26.875457	96.494619
Nigeria	13873.0	382.0	4351.0	2.753550	31.363079	97.246450
Ghana	10201.0	48.0	3755.0	0.470542	36.810117	99.529458
Senegal	4640.0	52.0	2885.0	1.120690	62.176724	98.879310
Kenya	3094.0	89.0	1048.0	2.876535	33.872010	97.123465
Ethiopia	2506.0	35.0	401.0	1.396648	16.001596	98.603352

while Egypt reports the highest number of fatalities (3.5%). In Nigeria, the death rate is 2.7%.

Figure 20.5 shows that South Africa is the worst affected country among the selected African nations followed closely by Egypt, while Ethiopia is the least affected. In Fig. 20.6, the low mortality and recovery rates are shown the cases in Nigeria appears to be in a manageable range. However, exponential increase can be observed for all the countries under study beginning in the month of April 2020.

Figure 20.6 presents the mean mortality and recovery rates for the selected African countries, from the figure it is evident that Senegal has performed best in keeping the disease under control, with the lowest mortality rate and highest recoveries, on the other hand, Egypt and Kenya are the worst hit in terms of mortality and recoveries, while the mortality and recovery rates in Nigeria has been within limits considered to be acceptable.

20.4.3 Results of the Models

The data exploration and machine learning models were implemented using python programming language in jupyter notebook. The standard scaler is used for transforming the time series data to the form suitable for machine learning modeling. This subtracts the mean and divides the result by the standard deviation of the data sample, which transforms the data to have a mean of zero, or centered, with a standard deviation of 1. This resulting distribution is called a standard Gaussian distribution, or a standard normal, hence the name of the transform.

Using the parameters described in Sect. 20.3.1 simulations are carried out and the results of predictions are presented in Figs. 20.7 and 20.8 and Table 20.4.

Figure 20.7 presents the results of the support vector machine for the 15-day prediction of COVID-19 spread in Nigeria. The Poisson regression model forecasts a cumulative count case of 19,814 (Table 20.4) infected persons on the fifteenth day (June 26th, 2020). The support vector machine predicts cumulative count cases of 31,844 infected persons on the same day.

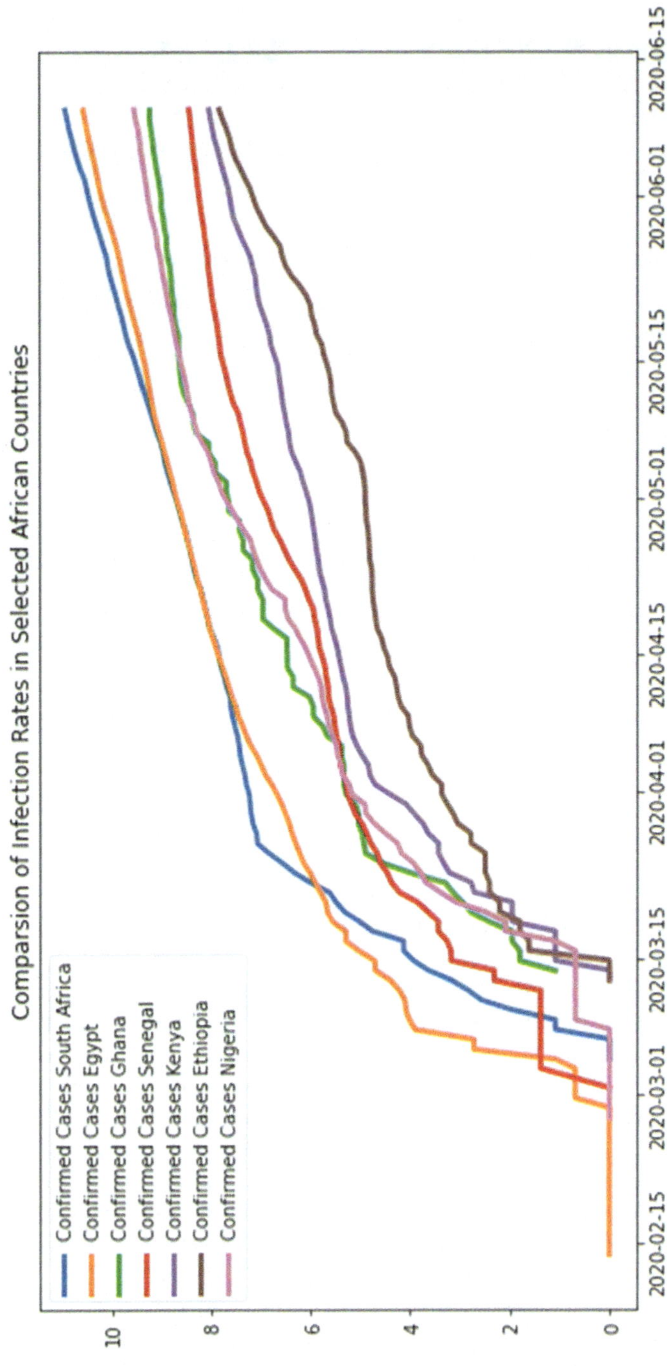

Fig. 20.5 Comparison of confirmed cases (Nigeria, South Africa, Senegal, Ethiopia, Kenya, Ghana, South Africa, and Egypt)

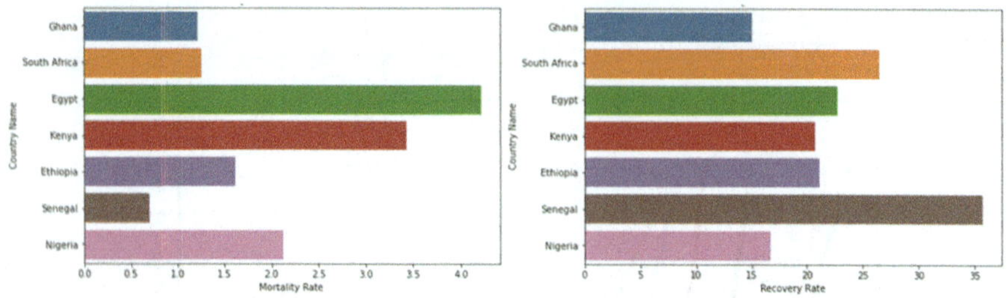

Fig. 20.6 Mean mortality and recovery rates in some African countries

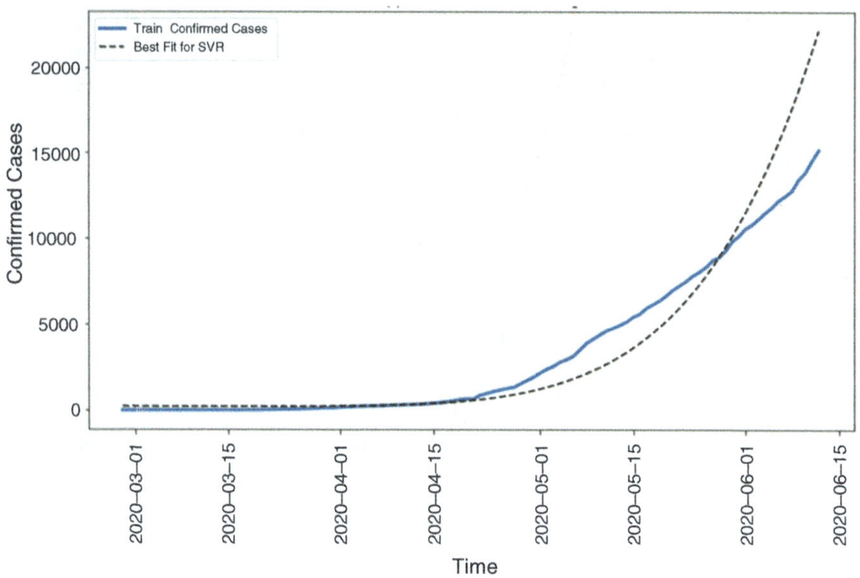

Fig. 20.7 Result of support vector machine model

Table 20.4 shows the prediction results for the Poisson regression and a variant of the support vector machine as well as the average prediction of the two models which is adopted as the system's output. Order to establish the effectiveness of the approach used in this work we compared the model's predicted values against the actual confirmed cases on the particular dates as shown in Fig. 20.8 and Table 20.5.

Table 20.5 and Fig. 20.8 is a day-by-day comparison of predicted values against the actual values reported by the Nigerian Center for Disease Control (NCDC). This comparison helps to reveal the efficiency of the approach employed in this study as the error margin (Table 20.5) is within acceptable ranges. Hence, it is shown that the result of the model proposed in this study is quite efficient and reliable.

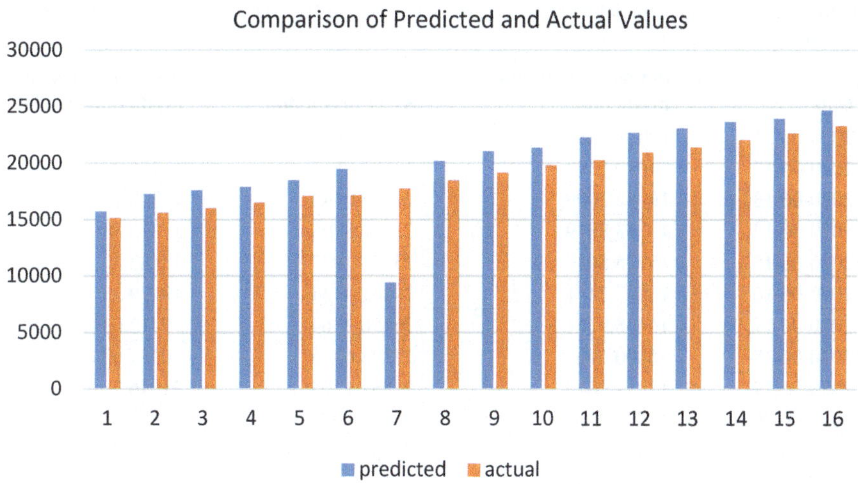

Fig. 20.8 Predicted versus confirmed cases in Nigeria

20.4.4 Discussion

In the experimentation carried out in this study, the models built from the reported number of cases of the COVID-19 dataset for Nigeria gave a short-term prediction as shown in Table 20.4. The average cumulative values of all the models are plotted against the actual observed cases, as confirmed by the Nigerian Center for Disease Control (NCDC), in order to validate the accuracy of the model. Table 20.5 presents these values, while Fig. 20.8 is the plotted values (Predicted Vs. Actual Values). Furthermore, individual comparisons of the two chosen models revealed that the Poisson regression model yielded better results in terms of prediction accuracy than the support vector.

These values, Table 20.4 and Fig. 20.8 reveal a high prediction accuracy of over 90% in the period under study, this is easily seen in the small error values, which could be attributed to the containment measures put in place by the authorities. The second reason that could be attributed to the observed error values could be mis/disinformation report of infection cases, as observed by Ngengasong & Mankoula (2020).

Recent papers have also reported the forecast of COVID-19 spread in different countries. A short-term forecast of COVID-19 cases in France, Canada, India, South Korea, and the UK is reported in Chakraborty & Ghosh (2020) with their prediction yielding very high values of root mean square error (RMSE) values of 631.91 and 149 for France and Canada, respectively, indicative of poor prediction accuracy. Other forecast models are presented in Ribeiro et al. (2020) for the short-term forecast of COVID-19 infections in Brazil with error rates of 6.90% on the sixth day, also Chimmula & Zhang (2020) show a time series forecast of COVID-19 spread in Canada with an RMSE value of 34.83. While Ribeiro et al. (2020) and

Table 20.4 Average cumulative prediction of COVID-19 in Nigeria (Polynomial Regression and Support Vector Machine)

Date	Poisson regression prediction	SVM prediction	A convergence of predictions models
2020-06-11	14038.8387	17454.5685	15746.7
2020-06-12	14610.5198	19902.0614	17256.29
2020-06-13	14982.2009	21214.0142	17598.11
2020-06-14	15353.8819	22692.8705	17873.38
2020-06-15	15725.563	23241.1427	18483.35
2020-06-16	16097.2441	24861.4134	19479.33
2020-06-17	16468.9251	26556.3364	9412.631
2020-06-18	16840.6062	27328.6385	20184.62
2020-06-19	17212.2873	27881.1203	21046.7
2020-06-20	17583.9683	28116.6577	21350.31
2020-06-21	17955.6494	28638.2035	22296.93
2020-06-22	18327.3305	30248.7884	22688.06
2020-06-23	18699.0115	30451.5226	23075.27
2020-06-24	19070.6926	30949.5969	23660.14
2020-06-25	19442.3737	31446.2844	23944.33
2020-06-26	19814.0547	31844.9417	24679.5

Table 20.5 Predicted values, actual confirmed values, and difference

Date	Predicted	Actual	Difference (Error)
6/11/2020	15746.7	15,166	581
6/12/2020	17256.29	15,609	1647
6/13/2020	17598.11	16,011	1587
6/14/2020	17873.38	16,500	1373
6/15/2020	18483.35	17,101	1382
6/16/2020	19479.33	17,148	2331
6/17/2020	9412.631	17,735	−8322
6/18/2020	20184.62	18,480	1705
6/19/2020	21046.7	19,147	1900
6/20/2020	21350.31	19,808	1542
6/21/2020	22296.93	20,244	2053
6/22/2020	22688.06	20,919	1769
6/23/2020	23075.27	21,371	1704
6/24/2020	23660.14	22,020	1640
6/25/2020	23944.33	22,614	1330
6/26/2020	24679.5	23,298	1382

Chimmula & Zhang (2020) reports error ranges and RMSE values similar to this work, however, a strict comparison may not be reasonable, as noted by Fanelli and Piazza (2020), a significant factor responsible for influencing forecast models of this current disease outbreak using the dataset of different countries is the containment measures put in place by individual countries in terms of strictness which affects the rate of spread in those countries; hence comparisons will be more relevant when different methods and models are applied to the same dataset within same time frames.

20.5 Conclusion

This study has presented short-term forecasts models of COVID-19 infections in Nigeria and a comparative study of cases in some other African countries using two machine learning prediction models, namely, Poisson Regression and Support Vector Machine with the chosen models trained and evaluated using the datasets of confirmed cases as reported by the World Health Organization.

The objective of the study was to better understand the spread patterns of the disease as well as to build models that can serve as decision support systems for government and health authorities for better management of the pandemic. In order to achieve this, the relevant datasets were retrieved from the WHO, detailed data preprocessing for the selection of the relevant features as well as the target variable was carried out, with the resulting shaped dataset used to train the chosen machine learning algorithms from which a 15-day prediction of infection cases was obtained.

As this disease continues to spread, this work can serve as a handy tool to tell the epidemic's direction so as to guide decision-making by the relevant authorities. The short-term forecasts based on machine learning models presented in this study can be useful for preparedness by government and relevant authorities in Nigeria, such as anticipating the expected number of hospital spaces and medical supplies for the coming days ahead thereby generally enhancing better management of the pandemic.

The COVID-19 outbreak in Nigeria presents an emerging challenge for researchers, because of limited datasets on the initial growth trajectory of the disease as well as other epidemiological characteristics of the disease that are yet to be well understood. Also, delays associated with identification, isolation, and reporting of cases in Nigeria are challenges that could have influenced the models' predictions. As a future work, ensemble techniques and deep neural network models could be employed for the prediction of the cases so that comparative analysis can be carried out.

References

Alpaydin E (2020) Introduction to machine learning. MIT press

Atangana A (2020) Modeling and forecasting of epidemic spreading: The case of Covid-19 and beyond. Chaos Solitons and Fractals. https://doi.org/10.1016/j.chaos.2020.109794

Ayodele TO (2010) Types of machine learning algorithms. New advances in machine learning:19–48

Bouguila N, Ziou D (2007) Unsupervised Learning of a Finite Discrete Mixture: Applications to Texture Modeling and Image Databases Summarization. J Vis Commun Image Represent 18 (4):295–309

Chakraborty T, Ghosh I (2020) Real-time forecasts and risk assessment of novel coronavirus (COVID-19) cases: A data-driven analysis. Chaos, Solitons Fractals 109850

Chimmula VKR, Zhang L (2020) Time Series Forecasting of COVID-19 transmission in Canada Using LSTM Networks. Chaos, Solitons Fractals 109864

Fanelli D, Piazza F (2020) Analysis and forecast of COVID-19 spreading in China, Italy and France *Chaos Solitons and Fractals* Chaos. Solitons and Fractals 134(2020):109761. https://doi.org/10.1016/j.chaos.2020.109761

Farmer P (2001) Infections and inequalities: The modern plagues. Univ of California Press

Guerrieri V, Lorenzoni G, Straub L, Werning I (2020) Macroeconomic Implications of Covid-19: Can Negative Supply Shocks Cause Demand Shortages. The National Bureau of Economic Research Paper No 26918 April 2020

Hirose H, Wang L (2012) Prediction of Infectious Disease Spread Using Twitter: A Case of Influenza, 2012 Fifth International Symposium on Parallel Architectures. Algorithms and Programming, Taipei 2012:100–105. https://doi.org/10.1109/PAAP.2012.23

Hu Z, Ge Q, Li S, Jin L, Xiong M, (2020) Artificial Intelligence Forecasting of Covid-19 in China Department of Biostatistics and Data Science, School of Public Health, The University of Texas Health Science Center at Houston, Houston, TX 77030, USA

Ijegwa AD, Olufunke VR, Folorunso O, Richard JB (2018) A Bayesian-based system for evaluating customer satisfaction in an online store. In: Proceedings of SAI Intelligent Systems Conference. Springer, Cham, pp 1047–1061

Kapata N, Ihekweazu C, Ntoumi F, Raji T, Chanda-Kapata P, Mwaba P et al (2020) Is Africa prepared for tackling the COVID-19 (SARS-CoV-2) epidemic. Lessons from past outbreaks, ongoing pan-African public health efforts, and implications for the future. Int J Infect Dis 93:233–236

Kotsiantis SB, Zaharakis I, Pintelas P (2007) Supervised machine learning: A review of classification techniques. Emerging artificial intelligence applications in computer engineering 160:3–24

Kucharski A, Russell T, Diamond C, Liu Y, (2020) CMMID nCoV working group, Edmunds J, Funk S, Eggo R. Analysis and projections of transmission dynamics of nCoV in Wuhan. 2020. https://cmmid.github.io/ncov/wuhan_early_dynamics/index.html

Li Q, Guan X, Wu P (2020) Early transmission dynamics in Wuhan, China, of novel coronavirus–infected pneumonia. N Engl J Med. https://doi.org/10.1056/NEJMoa2001316

Michie D, Spiegelhalter DJ, Taylor CC (1994) Machine learning. Neural and Statistical Classification 13(1994):1–298

Mitchell TM (1999) Machine learning and data mining. Commun ACM 42(11):30–36

Nkengasong JN, Mankoula W (2020) Looming threat of COVID-19 infection in Africa: act collectively, and fast. Lancet 395(10227):841–842

Peng L, Yang W, Zhang D, Zhuge C, Hong L (2020) Epidemic analysis of COVID-19 in China by dynamical modeling. Beijing Institute for Scientific and Engineering computing, College of Applied Sciences, Beijing University of Technology, Beijing. arXiv:2002.06563v1 [q-bio.PE] 16 Feb 2020

Ribeiro MHDM, da Silva RG, Mariani VC, dos Santos Coelho L (2020) Short-term forecasting COVID-19 cumulative confirmed cases: Perspectives for Brazil. Chaos, Solitons Fractals 109853

Roberts M, Andreasen V, Lloyd A, Pellis L (2015) Nine challenges for deterministic epidemic models. Epidemics 10:49–53

Roosa K, Lee Y, Lou R, Kirpich A, Rothenberg R, Hyman JM, Yan P, Chawell G (2020) Real-time forecasts of the COVID-19 epidemic in China from February 5th to February 24th, 2020. Infectious Disease Modelling 5(2020):256e263. https://doi.org/10.1016/j.idm.2020.02.002

Sumner A, Hoy C, Ortiz-Juarez E (2020) Estimates of the Impact of COVID-19 on Global Poverty. *UNU-WIDER, April*, 800–809

Sun J, He WT, Wang L, Lai A, Ji X, Zhai X, Li G, Suchard MA, Tian J, Zhou J, Viet M, Su S (2020) COVID-19: Epidemiology, Evolution, and Cross-Disciplinary Perspectives. Trends Mol Med 26:5. https://doi.org/10.1016/j.molmed.2020.02.008

Tuite AR, Fisman DN (2020) Reporting, epidemic growth, and reproduction numbers for the 2019 novel coronavirus (2019-nCoV) epidemic. Ann Intern Med (2020 Feb 5). https://doi.org/10.7326/M20-0358

Villela DAM (2020) Discrete-Time Forecasting of Epidemics. Infectious Disease Modelling 5 (2020):189–196. https://doi.org/10.1016/j.idm.2020.01.002

Wickham H, Grolemund G (2017) R for Data Science: *Journal of Statistical Software* ISBN 978–1–4919-1039-9. 522 pp. http://r4ds.had.co.nz/

Wu YC, Chen CS, Chan YJ (2020a) The outbreak of COVID-19: An overview. J Chin Med Assoc 83(3):217

Wu JT, Leung K, Leung GM. (2020b) Now casting and forecasting the potential domestic and international spread of the 2019-nCoV outbreak originating in Wuhan, China: a modelling study. Lancet. 2020. pii: S01406736(20)30260-9. doi: https://doi.org/10.1016/S0140-6736(20)30260-9

Yang Q, Jiang D, Shi N, Ji C (2012) The ergodicity and extinction of stochastically perturbed SIR and SEIR epidemic models with saturated incidence. J Math Anal Appl 388(1):248–271

Zhao ML (2020) Estimating the unreported number of novel coronavirus (2019-nCoV) cases in China in the first half of January 2020: A Data-driven modelling analysis of the early outbreak. J Clin Med 9(2):388

Part VI
COVID-19 Detection

Chapter 21
Attention-Based Residual Learning Network for COVID-19 Detection Using Chest CT Images

R. Karthik, R. Menaka, Sundar Anand, Annie Johnson, and Kanagala Srilakshmi

Abstract COVID-19 is a deadly viral infection that is highly contagious and has brought a great loss of human lives and economic resources. Hence, it is very critical to detect the virus at an early stage with high accuracy. Deep learning algorithms are very effective in learning the discriminative features of medical images. It can facilitate the rapid diagnosis of the disease. In this research, different deep learning architectures have been evaluated on a balanced CT image dataset for COVID-19. From the analysis performed, this research proposes a deep learning architecture that performs substantially better than the reviewed models. This research aims to analyze and identify an effective baseline deep architecture on which a model can be built for advanced COVID-19 detection. The architectures analyzed in this research include, AlexNet, DenseNet, GoogLeNet, InceptionV4, ResNet, ShuffleNet, SqueezeNet, and Visual Geometric Group (VGG16). Experimental results indicated that GoogLeNet based network was able to detect COVID-19 with an accuracy of 83.27%. This backbone model was further modified to design a better performing network exclusively for COVID-19. This work proposes an attention-based residual learning block that is integrated with the GoogLeNet backbone. The proposed model performed obtained an accuracy of 91.26%.

Keywords COVID-19 · Deep Learning · CNN · Detection · CT

R. Karthik (✉) · R. Menaka
Centre for Cyber Physical Systems, Vellore Institute of Technology, Chennai, India
e-mail: r.karthik@vit.ac.in; menaka.r@vit.ac.in

S. Anand · A. Johnson · K. Srilakshmi
Present Address: School of Electronics Engineering, Vellore Institute of Technology, Chennai, India
e-mail: a.sundar2016@vitstudent.ac.in; annie.johnson2016@vitstudent.ac.in; kanagala.srilakshmi2016@vitstudent.ac.in

S. A. Hassan et al. (eds.), *Decision Sciences for COVID-19*, International Series in Operations Research & Management Science 320,
https://doi.org/10.1007/978-3-030-87019-5_21

21.1 Introduction

The first case of COVID-19 was reported in December 2019, in the capital of China's Hubei province, Wuhan (Ye et al. 2020). Wuhan was the epicenter of the global pandemic disease. At the end of 2019, several patients were admitted to Wuhan on account of pneumonia.

The World Health Organization (WHO) declared the outbreak of a global pandemic on 30th January 2020 (Wang et al. 2020b). As of February 25, 2020, out of 77,779 confirmed cases, 2666 were reported dead due to the COVID-19 infection (Wang et al. 2020b). The number of active cases had increased to 168,826 cases, and 6503 deaths were reported on 16th March 2020 (Zhai et al. 2020).

The virus had spread across many parts of the world, including Thailand, Korea, and the United States of America (Zhao et al. 2020a). Thus giving rise to concern on a global scale.

The common symptoms of COVID-19 are fever, cough, shortness of breath, headache, and sore throat, loss of appetite, loss of smell, and taste (Udugama et al. 2020; Liang et al. 2020). COVID-19 has similar symptoms to that of Influenza-A, Influenza-B, common flu, and other types of common cold (Ai et al. 2020). This high similarity ratio resulted in the spreading of the virus rapidly. The genome sequences of the respiratory samples of affected patients were analyzed. It suggested a new type of beta coronavirus that was similar to Severe Acute Respiratory Syndrome Coronavirus (SARS-CoV) (Udugama et al. 2020; Zhou et al. 2020). Thus, careful and thorough identification processes were required, making the entire process time demanding.

Early detection of COVID-19 is very crucial, as it can help to cure the disease at an initial stage and reduce the chances of spreading. The availability of detection kits is minimal, and a lot of time is consumed for producing the results. The detection of COVID-19 with Computed Tomography (CT) helps to get accurate results, overcome the disadvantage of fewer detection kits, and improves the clinical diagnosis of the disease. Hence, it has become a priority to find efficient COVID-19 detection methods from CT images.

The review is organized as follows. In Sect. 21.2, the details of the existing research work in COVID-19 detection from CT scans are presented. The research gaps, motivation, and the contribution of this paper are examined in this section. The section also highlights the importance of using deep learning methods to obtain better results. Section 21.3 discusses the dataset and the deep learning models employed in the experiments performed in this study. The architecture of each of these popular deep learning models has been explained in detail. Section 21.3 also details the architecture of the proposed system. The results of the experiments conducted with the aforementioned dataset and deep learning models have been explored in Sect. 21.4. It presents a consolidated review of the setup used in comparing the various deep learning architectures. The performance of each network is evaluated based on different performance metrics such as accuracy, precision, recall, F1-Score and specificity. Finally, Section 21.5 provides a conclusion to contemporary research in COVID-19 detection from CT scans and includes the future scope in this domain.

21.2 Related Works

The research in COVID-19 detection from CT scans has emerged only within the past few months. Most of the work proposed in COVID-19 detection involves the use of X-radiation (X-ray) images rather than CT scans. However, X-ray image-based datasets may pose a few drawbacks. The accuracy is limited, and false-positive might arise (Apostolopoulos and Mpesiana 2020). Therefore, using a dataset of CT images would result in better performance. As the research in this area is only budding, most of the works deal with manual or statistical analysis performed on these images.

Fang et al. (Fang et al. 2020) employed the McNemar Chi-squared test on chest CT scans of 51 patients. A significance at the $p < 0.05$ level was used in the statistical analysis. The study recorded a sensitivity of 98% for COVID-19 detection with CT. The sensitivity measure determines how well true positives can be recognized by a test. The Chi-square test and Fisher's exact test were utilized by Wang et al. (Wang et al. 2020b) in their statistical analysis. Liang et al. (Liang et al. 2020) applied a one-way analysis of variance (ANOVA) and the Chi-square test in their work.

Many studies used software packages to perform statistical analysis. Shi et al. (Shi et al. 2020) performed a statistical analysis based on the Statistical Analysis System (SAS) software. They concluded that combining analysis of imaging features with laboratory findings could expedite the process of early diagnosis of COVID-19 pneumonia.

However, a majority of the work published in manual COVID-19 detection has made use of the Statistical Package for the Social Sciences (SPSS) software (Zhao et al. 2020a; Ai et al. 2020; Zhou et al. 2020; Huang et al. 2020; Liu et al. 2020; He et al. 2020; Ding et al. 2020; Zhao et al. 2020b; Pan et al. 2020b; Dai et al. 2020; Guan et al. 2020; Wang et al. 2020a; Long et al. 2020; Luo et al. 2020; Pan et al. 2020a; Li et al. 2020a). Various versions of the software were used. Also, several tests, including the ANOVA and the Pearson chi-square test for normality, were employed on the categorical CT data. Li et al. (Li et al. 2020a) achieved a sensitivity of 82.6% on the COVID-19 CT scans. The Chi-square test and the Fisher exact test were employed in this work to compare the number of lobes involved in several categories of CT images. Yet, the study faced challenges due to the imbalance in the distribution of CT images obtained that were of a particular clinical type. Ai et al. (Ai et al. 2020) were able to produce a sensitivity of 97%. This analysis made use of SPSS 21.0 and compared the performance of CT images with Reverse transcription-polymerase chain reaction (RT-PCR). It concluded that CT scans performed better in the detection of COVID-19. Amongst all these works, Long et al. (Long et al. 2020) made use of the SPSS 17.0 software to accomplish statistical analysis and achieved the highest sensitivity of 97.2% for COVID-19 detection from chest CT scans.

Manual methods have several limitations, including slower results and less efficient performance. Most of the datasets used in the works have few images to train on and are unbalanced. This limited dataset tends to affect the results and give biased interpretations (Liang et al. 2020; Li et al. 2020a). To overcome these issues, automation in image analysis and classification is essential. Hence deep learning algorithms were employed to overcome these downsides.

Butt et al. (Butt et al. 2020) compared the performance of two CNN-based three-dimensional architectures. A traditional Residual Network (ResNet) model was evaluated on CT scan images with no infection, Influenza viral pneumonia, or COVID-19. The ResNet18 model utilized in this work was used for extracting image features. To further improve the performance of the baseline architecture, a location-attention mechanism was added in the output fully connected layers. This improvisation reduces the overfitting of the model and increases the overall accuracy attained. However, other deep learning approaches reported better results than this study. Li et al. (Li et al. 2020b) proposed a deep learning model called the COVID-19 detection neural Network (COVNet). This model is similar to that of Butt et al. (Butt et al. 2020) as the baseline model on which improvements are made the ResNet model. But the ResNet model used in this work has more layers and hence performs better than that of Butt et al. (Butt et al. 2020). Initially, the images are segmented using a U-net. This step is done to set the Region of Interest (ROI) as the lung region in the CT scan. This pre-processed image is fed into the COVNet. COVNet architecture is based on the ResNet50 model. The features generated from the images using ResNet50 are fed into a fully connected layer and a softmax activation function to produce a probability score for each category. In this way, the network classifies the dataset into COVID-19, Community-Acquired Pneumonia (CAP), and non-pneumonia. The study also performed statistical analysis to compare the efficiency of manual and deep learning-based methods. Li et al. (Li et al. 2020b) concluded that compared to statistical methods, deep learning models could accurately and quickly detect COVID-19 and distinguish it from other lung infections. Yet, a lighter deep learning architecture would result in even faster results and efficient performance. In the research presented by Ucar et al. (Ucar and Korkmaz 2020), a Deep Bayes-SqueezeNet based model is used in the detection of COVID-19. This proposed system consists of three main stages: offline augmentation of the dataset, training the network, and classification with the testing phase. This work performed the best, yielding an accuracy of 98.26%. This model overcomes the challenge of working on an imbalanced dataset. It proposes a lightweight and easily executable network that can be applied in mobile remote systems. This network could facilitate a faster and timely diagnosis of the COVID-19 infection in patients, thus aiding in the control of the current epidemic. The metrics achieved by this model depict the superiority of deep learning architectures over other machine learning and traditional approaches in COVID-19 detection.

21.2.1 Research Gaps and Motivation

The research of COVID-19 detection from radiological images has only come into being very recently. Hence many drawbacks need to be addressed in the upcoming works. Automating the entire process will facilitate better results in this research.

1. The majority of research in COVID-19 detection utilizes datasets that have fewer COVID-19 images than healthy images. This imbalance in the dataset causes the classifier to have a bias towards the class, having more training samples. Hence, the performance of the model cannot be accurately evaluated on such datasets (Zhao et al. 2020a).
2. Most of the works published so far involve the use of X-ray images. Though X-rays are used as the first-line imaging modality for COVID-19 diagnosis, there is a possibility that false positives may arise (Ren et al. 2019; Li et al. 2019). On the other hand, CT scans produce high-quality, detailed annotation of the anatomical structures. This modality makes it easier for algorithms to differentiate between symptoms. In the early stage of infection, CT shows scattered opacities or a cluster of opacities. This observation is seen predominantly in the lower and middle regions of the lung and can be detected in CT scans (Long et al. 2020).
3. Though CT-based research was reported, it primarily involved the use of manual analysis to differentiate between target and non-target images. Tests including the Chi-square test, the Fisher's exact test, and ANOVA were carried out to estimate parameters like Sensitivity and Specificity of the classification. Although most works use software tools such as SAS, R, and SPSS, the overall efficiency of manual methods is much lower than that of machine learning and deep learning systems. Traditional methods provide less accurate results and are slower than automated systems (Liang et al. 2020; Li et al. 2020a).
4. Very few researches were reported using deep learning. But it is in the beginning stages of development (Butt et al. 2020; Li et al. 2020b; Ucar and Korkmaz 2020). Extensive research on the available deep architectures needs to be done to improve the diagnostic accuracy of the COVID-19 detection.

Taking the above requirements into consideration, the proposed study intends to address the following research gaps.

1. An exhaustive experimental study involving the use of popular deep learning architectures to extract key features from CT images is performed. Subsequently, these features are applied for classification.
2. A balanced dataset containing an equal distribution of target and non-target images must be used to overcome the class imbalance issue and to provide reliable results with the model. Moreover, class weight assignments can be performed to address the imbalance.
3. Deep learning architectures are better equipped at classifying image data accurately and in lesser duration of time. Several architectures have been compared and contrasted against each other in this research. This process has to be

performed to determine which model performs best for interpreting COVID-19 samples.

21.2.2 Research Contributions

The key contributions of this research are reported as follows:

1. A comprehensive analysis involving the state-of-the-art deep architectures for medical image classification was performed in the proposed research. Models including DenseNet, VGGNet, AlexNet, SqueezeNet, ShuffleNet, GoogLeNet, InceptionV4 Net, and ResNet variants were evaluated based on various performance metrics.
2. From the thorough analysis of these models, it was concluded that GoogLeNet yielded optimal results on k-fold cross-validation. To enhance the classification accuracy further, Attention-based residual learning mechanism was integrated with the GoogLeNet backbone. An Attention-based Residual Block that uses the skip connection concept is used, skips connection between layers to make it less deep and thus making it computationally less expensive. This Residual Block consists of convolutional layers which do convolution operation over the input to extract crucial features by utilizing different sizes of kernel filters.

21.3 Materials and Methods

Deep learning models are robust architectures that are capable of learning discriminative features for classification. It also contributes to the better categorization of data and yields very high-performance metrics. In this research, a publically open, labeled CT scan-based dataset is used in COVID-19 detection. The data consists of two types of CT scans; COVID-19 infected and non-COVID-19 samples. These images are then augmented to provide a larger dataset to the network for training. The training is performed with models like AlexNet, DenseNet, ResNet50, ResNet101, ResNet152, ShuffleNet, SqueezeNet, GoogLeNet, and the InceptionV4 network. Then the trained model is subjected to the testing process for evaluation. The evaluation metrics of each model are recorded and tabulated for discussion and comparison. The overall workflow of the proposed experimentation is depicted in Fig. 21.1.

21.3.1 Data Acquisition

CT scans provide a fast, accurate, and affordable method for testing COVID-19 among subjects. The publically accessible COVID-CT-Dataset was collected by

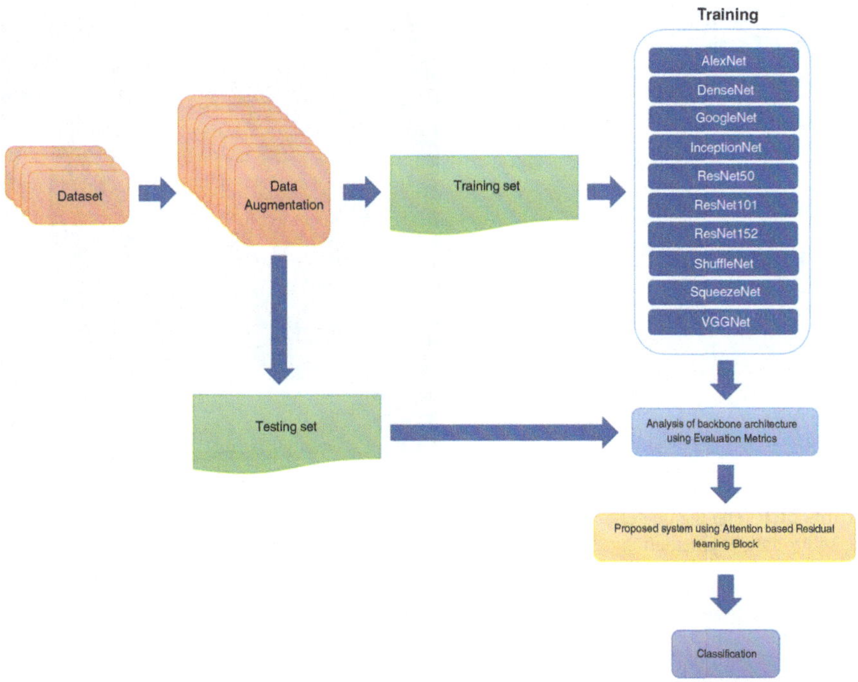

Fig. 21.1 The block diagram of the proposed system

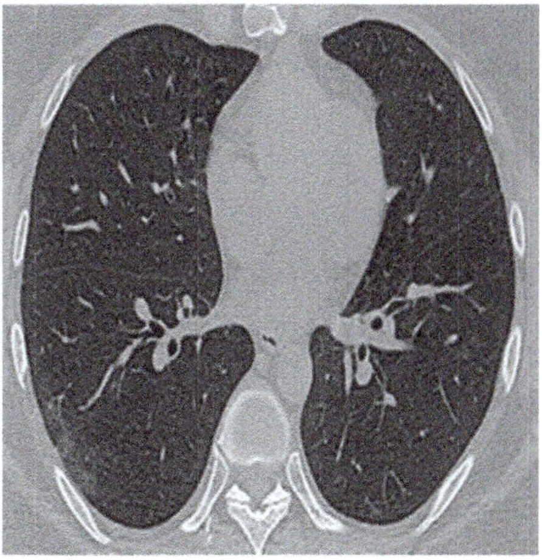

Fig. 21.2 COVID-19 positive CT scan

Zhao et al. (Zhao et al. 2020), and it is used in this research. This dataset has 349 chest CT images collected from 216 patients diagnosed with COVID-19. These images were collected by crawling the images from 760 preprints related to COVID-19. To obtain a balanced dataset, 398 non-COVID CT scans of healthy patients were included. A total of 747 images were present in this dataset. Figure 21.2

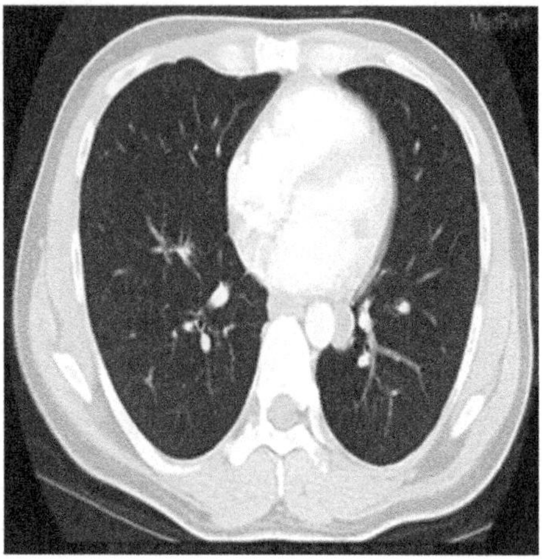

Fig. 21.3 COVID-19 negative CT scan

Table 21.1 The parameters used for data augmentation

S.No	Transformation	Range
1	Shear	± 20%
2	Room	± 20%
3	Shearing	± 5%
4	Flip	Horizontal
5	Brightness	50% to 150%

indicates a COVID-19 positive CT scan, and Fig. 21.3 shows a COVID-19 negative CT scan.

21.3.2 Data Augmentation

Discriminate patterns can be generated by CNNs irrespective of the geometric representation of the object image. Geometric modifications such as translation, rotation, vertical and horizontal flip, shearing, zooming, etc., change the perspective of the image, image pixels are moved to new locations altogether within the image while some pixels may be preserved in the same location, thus creating a new image. These geometric modifications are handled efficiently by CNNs. Due to this advantage, even though a considerable amount of CT images were available, data augmentation can be done to enhance the dataset size. Thus the performance of the trained network can also be improved. The parameters that were modified to perform geometric transformation and, in turn, augment the data are tabulated in Table 21.1.

The augmented data samples are presented in Fig. 21.4. A total of 747 images were initially provided as a part of the dataset. Data augmentation was done on those

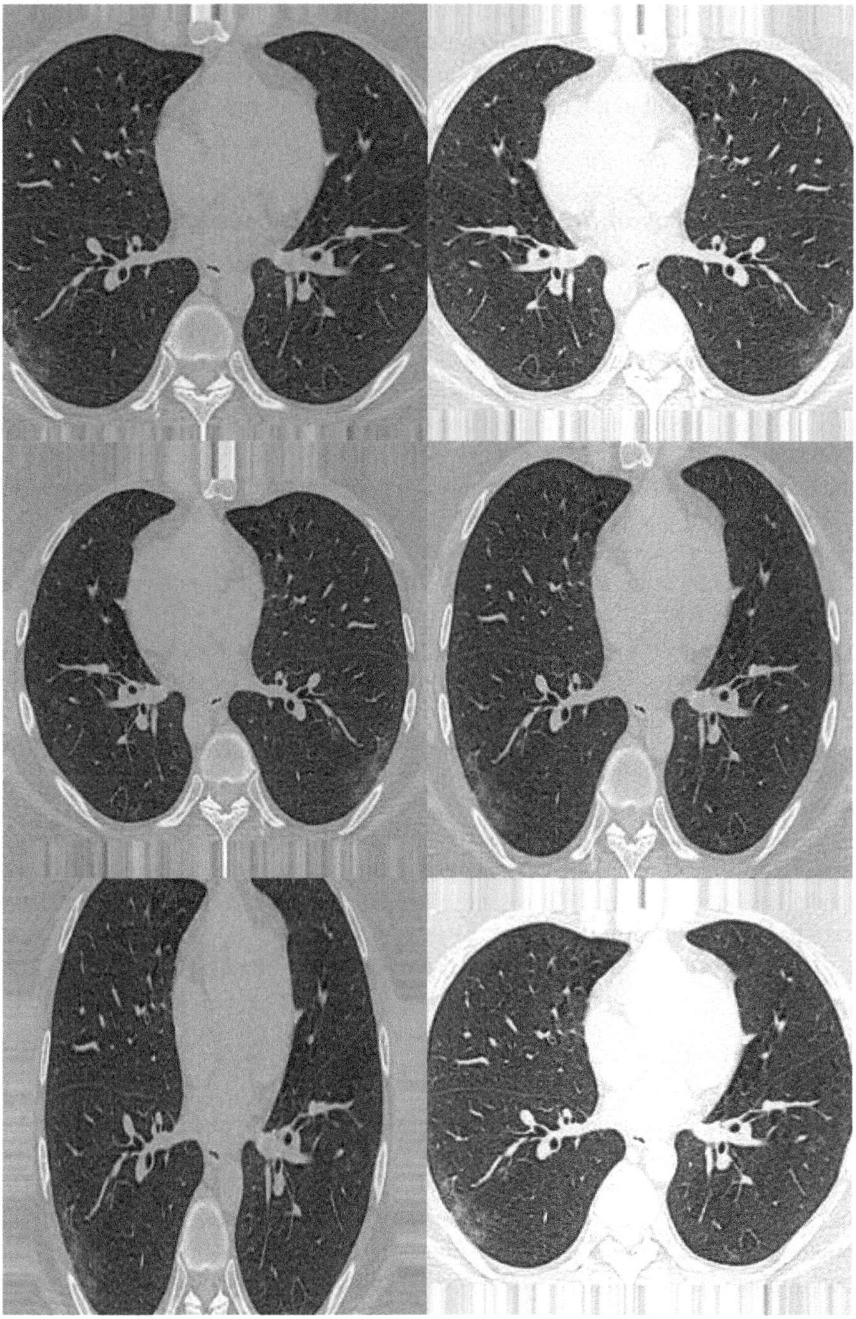

Fig. 21.4 Data augmentation sample images

images to increase the count by a factor of "6." The total number of images obtained for training was "4474." Class weight assignment was also performed to make sure that one class does not dominate another. In this way, the dataset size was increased, and an improvement in the model performance was analyzed.

21.3.3 Analysis of the Architecture of Deep Learning Models

The COVID-CT-Dataset images are trained on ten different models with a fivefold cross-validation process. Each fold was iterated for 50 epochs, and the metrics such as Accuracy, Precision, Recall, Specificity, and F1 score were recorded.

Accuracy is a measure that gives insight into how well the model learned and is producing a liable result. It is the fraction of predictions that were provided correctly by the model. The accuracy of a model is the ratio of correctly predicted samples to the number of input samples. The number of correctly predicted samples is the sum of the number of true positives and false negatives.

Precision is a measure of how precise or accurate the model is in terms of positive classifications. In other words, it measures the number of true positives out of all the predicted positives.

Recall, or sensitivity is a measure of how many positives the model managed to capture out of all the correctly predicted positive values by labeling the samples as positive.

Specificity is a measure of how many negatives the trained model managed to capture out of the entire set of correctly predicted negative values by labeling the samples as negative.

F1-score is a measure of the balance between the precision and recall of a model. It is used to perform a statistical analysis of the test accuracy. The F1-score of a model lies between 0 and 1. It is said to be very good if its value lies near one and very bad if it is near zero.

The architecture of these models is discussed in this section.

21.3.3.1 AlexNet

AlexNet is a deep learning architecture consisting of 5 convolutional layers, which are followed by three fully connected layers. In total it has 61 million parameters. The first layer in the network is used to filter the 224×224 input image with 11×11 filters and a stride factor of 4. A total of 96 such filters are used in this layer. The number of channels of the image is estimated from the depth of the filters used. The output obtained is passed on to the second layer that performs convolution using 256 filters, each having a dimension of $5 \times 5 \times 96$. This operation is done after carrying out normalization and max-pooling. The third layer has performed convolution with 384 filters, and the fourth layer has 384 filters for convolution. The final convolution layer operates on the input with 256 filters. This output is then sent as input to 2 fully connected layers having 4096 neurons each. The final output layer, which is also a dense layer, is used to categorize the data and perform classification.

The deep network uses a Rectified Linear Unit (ReLU) activation function. Unlike the sigmoid and tanh activations, the ReLU activation is less computationally expensive. The sigmoid function saturates for large values, but the ReLU function implements non-saturating nonlinearity. This activation function, in turn, speeds up the training phase of the network. It can differentiate between inputs better, and classification is improved significantly. The usage of a nonlinear activation function also helps to rectify the vanishing gradient problem (Krizhevsky et al. 2017). Another specification of the AlexNet model is the dropout regularization that is used in the three fully connected layers to overcome the problem of overfitting (Sharma et al. 2018).

21.3.3.2 DenseNet

The DenseNet architecture is comprised of dense blocks and transition layers. The dense blocks have layers that are densely connected to each other. In total it has 0.8 million parameters. Each layer performs batch normalization and a 3×3 convolution with ReLU activation. The transition layer is present between two dense blocks to ensure down-sampling. This layer consists of three operations: batch normalization followed by a 1×1 convolution and an average pooling layer. Each layer of each dense block is connected to its subsequent layers, thus making the architecture heavy. Hence it is computationally more expensive than most other architectures and may also tend to overfit for few problems (Huang et al. 2017).

This model encourages feature reuse and reduces the number of parameters involved in training. It utilizes the depth of the network to provide the listed advantages. In this work, the input dataset is analyzed to see if deeper networks provide better results on COVID-19 detection.

21.3.3.3 GoogLeNet

Over the past years, various versions and upgrades on the Inception network have been made (Szegedy et al. 2016a; Szegedy et al. 2016b). In total it has five million parameters. Inception networks are built up with dimension-reduced inception modules. This inclusion was done to tackle the issue of varying sizes of the significant objects in an image. The inception module performs convolution on the input image with 1×1, 3×3, and 5×5 filters, which are followed by a max-pooling layer. Once these outputs are concatenated, they are sent as input to the next inception module. To reduce the number of input channels, a 1×1 filter convolution is performed before performing the 3×3 and 5×5 filter convolutions. This decreases the computational expense of the model. The Inception v1 network, more commonly known as the GoogLeNet, has nine such inception modules which are stacked linearly. It consists of 22 layers, and at the end of the last module, it uses the global average pooling (Szegedy et al. 2015).

Compared to heavier models like DenseNet, GoogleNet trains much faster. This model is built not only depthwise but also width wise. A single layer performs several types of feature extraction. Thereby allowing the network to make better classification choices.

21.3.3.4 InceptionV4

Another salient inception network is the Inception v4 architecture, which was introduced to make the modules more uniform. In total it has 23 million parameters. This was achieved by modifying the operations performed initially on the input before sending it to the inception blocks. The Inception V4 architecture mainly consists of 3 blocks of inception modules. These blocks are composed of an average pooling layer and multiple convolution operations with filters of different dimensions.

Specialized blocks called reduction blocks are incorporated to change the width and height parameters of the network grid. These reduction blocks eliminated all the unnecessary complications in the modules. The reduction blocks consisted of a max-pooling layer and multiple convolutional layers that use filters of different sizes. Although a similar functionality was implemented in earlier Inception networks, explicit blocks were not added to the previous networks (Szegedy et al. 2016b).

21.3.3.5 ResNet

The ResNet was developed as an improvement made on the GoogLeNet and the Inception networks. In total, it has 25 million parameters. It is composed of layers that learn residual functions for the inputs given. In turn, the network is a very deep one that contains many layers stacked on top of another. Usually, deeper networks are harder to train due to the vanishing gradient problem, but the ResNet introduces a new concept to deal with this issue. The fundamental concept involved in the construction of the ResNet is the identity shortcut connection. These connections allow the network to skip one or more layers. This stabilizes the network and prevents the network from getting saturated as the model gets deeper. It also prevents rapid degradation during training as the number of layers increases in the network. ResNet well balances the computational cost and representation capability trade-off.

Three of the most common ResNet architectures used in Deep Learning research are ResNet50, ResNet101, and ResNet152. The architecture of the ResNet50 can be divided into four stages. The model uses 7×7 and 3×3 filters to perform the initial convolution and max-pooling operations on the input. The output of these layers is sent to stage 1 of the network. Stage 1 of the architecture contains three-layered residual blocks that perform convolution with kernels of size 64, 64, and

128, respectively. The residual block operations have a stride of 2, and hence the output will be half the size of the input. But the channel width will be doubled. This happens as the model advances from one stage to another. Deeper networks like the ResNet101 and the ResNet152 use a bottleneck design to increase the performance of the network. 1×1, 3×3, 1×1 convolutions are stacked one over another in every residual function F to achieve this bottleneck design. Finally, an average pooling layer and a fully connected layer having 1000 neurons are used as the output layer (He et al. 2016). The performance of ResNet50, ResNet101, and ResNet152 were evaluated in this research.

21.3.3.6 ShuffleNet

The ShuffleNet is a small network and contains many stacks of ShuffleNet units grouped into three stages. A ShuffleNet unit is made up of group convolutions and channel shuffles (Zhang et al. 2018). Initially, group convolutions were introduced in AlexNet. Group convolutions aimed at distributing the model over Graphical Processing Units (GPU) to increase the overall efficiency of the model. A group convolution consists of multiple convolutions, each covering a part of the input channels. Group convolutions help achieve a balanced trade-off between the representation capability and computational cost of the model. However, on stacking multiple group convolutions together, the outputs from an individual channel are only considered from all the available input channels. This leads to the blockage of information between group convolutions. To overcome this consequence of using multiple group convolution units, the channel shuffle operation was utilized. The channel shuffle process performs a random channel shuffle followed by a group convolution. Initially, the design principle of a bottleneck unit is used in the design of the shuffle unit. A 3×3 depthwise convolution is applied on the bottleneck feature map. The first 1×1 layer is then replaced with a pointwise group convolution, after which a channel shuffle operation is performed. These operations form a ShuffleNet unit.

21.3.3.7 SqueezeNet

The SqueezeNet is composed of a convolution layer, followed by eight fire modules and a final convolutional layer. A fire module is made of a squeeze convolution layer comprised of only 1×1 filters followed by an expanding layer consisting of 1×1 and 3×3 convolution filters (Iandola et al. 2016). The total number of parameters in the network is decreased by reducing the number of filters used in the "squeeze" layer. It substantially minimizes the number of connections entering the 3x3 filters present in the expanded layer. The fire module serves as the basic building block of the entire architecture. With the limited number of parameters available to the SqueezeNet for training, the model takes a very different approach to achieve desirable results. The stride parameter is reduced in the last few convolutional layers of the fire module, which creates a larger feature map in the network. Thus, the classification accuracy of the network increases tremendously. The fire modules learn the features of the data. As compared to AlexNet, the SqueezeNet achieves a 50X reduction in network size while performing better than AlexNet. By applying the deep compression mechanism on the SqueezeNet, a smaller model than AlexNet was constructed.

21.3.3.8 VGGNet

VGG was developed based on CNN to enable better extraction of images. A filter size of 3×3 is used in VGG (Sharma et al. 2018). In total it has 138 million parameters. This small filter size helps to reduce the number of parameters and improve accuracy (Simonyan and Zisserman 2014). There are two types of VGG in terms of architecture, VGG19 (19 weight layers) and VGG16 (16 weight layers). The VGG16 architecture consists of 13 convolutional layers, five max-pooling layers, and three dense layers. Thus adding up to a total of 21 layers, but only 16 are weight layers. The pooling layers following the convolution layers are responsible for the narrowing of the layers. This architecture performs well on benchmarking tasks, but there are still many drawbacks to this model. However, the network takes a very long time to train, as there are many weight parameters involved in the training. This may lead to running out of memory before completing the training process. VGG19 is a deeper network and has more layers than VGG16, thus increasing the computational expense and time consumed in training (Sharma et al. 2018). This work made use of the VGG16 model to analyze the performance of VGG for the other standard networks discussed here.

21.3.4 Proposed System

Figure 21.5 presents the architecture of the proposed system. The GoogLeNet architecture is the main backbone of the proposed system in this work. To this backbone architecture, Attention-based residual blocks have been added to obtain a better performing network for COVID-19. The architecture and significance of the various components of the proposed system have been explained in this section.

The GoogLeNet model is entirely built up of two types of blocks, the inception blocks with dimensionality reduction and the auxiliary network blocks. The nine inception blocks and the two auxiliary network blocks used in the architecture have been depicted in Fig. 21.5. In this work, the attention-based residual learning blocks (indicated on the left side of the architecture) in Fig. 21.5, were proposed as additions to the GoogLeNet architecture to improve the performance of the model.

The GoogLeNet architecture consists of nine inception modules stacked up linearly. Several layers and techniques have been used in the proposed architecture, and the operations performed by each of these layers and techniques have been

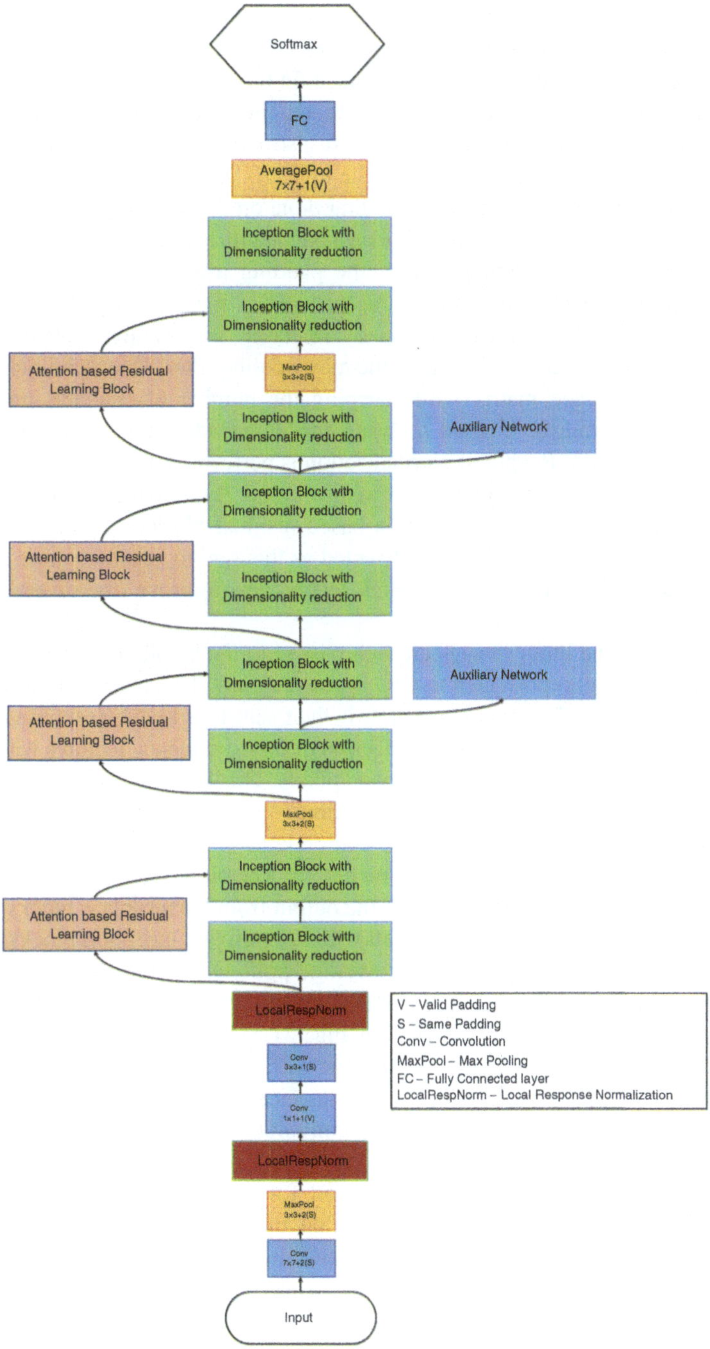

Fig. 21.5 Architecture of the proposed system

discussed here. The two types of padding used in the experiments conducted in this paper are the same zero padding and valid zero padding. Padding is a technique used to preserve the dimensions of the input image. In this technique, layers of zeroes are added around the entire image to prevent the input data from shrinking when performing convolution operations on it. It also conserves the information in the corners and borders of the image while the image is processed. When no zeroes are added around an image, there is no padding done on the image. This is called valid padding. After performing an operation on a padded image, if the input image has the same dimension as the output image, the padding used is called the same zero padding. Another technique used in this network is max-pooling. Max-pooling is used to downsample an image matrix by selecting the maximum element in the region of the image matrix covered by the max-pooling filter. This technique reduces the size of the image matrix while keeping the number of channels intact. The normalization technique is used to bring the data distribution of the input to a common scale. Local Response Normalization is performed in a local neighborhood by square-normalizing the pixel values in this region. By utilizing this technique, the ReLU activations used in the attention-based residual blocks are normalized. The activations having larger values are detected as this region becomes more sensitive than its neighbors. These layers and techniques improve the overall model performance.

In Fig. 21.5, the input images are first passed to a convolution layer having the same zero padding, represented as "S" in the diagram. This layer is followed by a max-pooling layer. The resulting output is then sent to a Local Response Normalization layer. This layer enables the detection of high-frequency features with a large response by normalizing the neighborhood pixels. This layer is then followed by two convolution layers having valid zero padding, "V" and same zero padding "S," respectively. A Local Response Normalization layer is used on top of the preceding layer. It is then followed by the inception blocks. The output from the four attention-based residual blocks is combined with the output from the inception blocks. Two of the residual block outputs are combined with not only the inception blocks but also the auxiliary network output. Finally, average pooling is performed on the output obtained from the ninth inception block, and this response is passed to a fully connected layer represented as "FC" in Fig. 21.5. The final classification of the image as COVID positive or healthy is done with a softmax classifier.

Figure 21.6 depicts the architecture of an inception block with dimensionality reduction. An inception block performs a convolution operation on the input image with a 1×1, 3×3 and a 5×5 filter, which is then followed by a max-pooling layer. The 1×1 convolutions were mainly used for dimensionality reduction in the network. These convolutions make sure that the depth of the input and output matched. As the dimensions of the input and the output do not differ drastically, there is almost no loss in information and hence better performance of the neural network.

The auxiliary block in Fig. 21.7 consists of an average pooling operation being performed on the output from the inception module with a valid zero padding. This

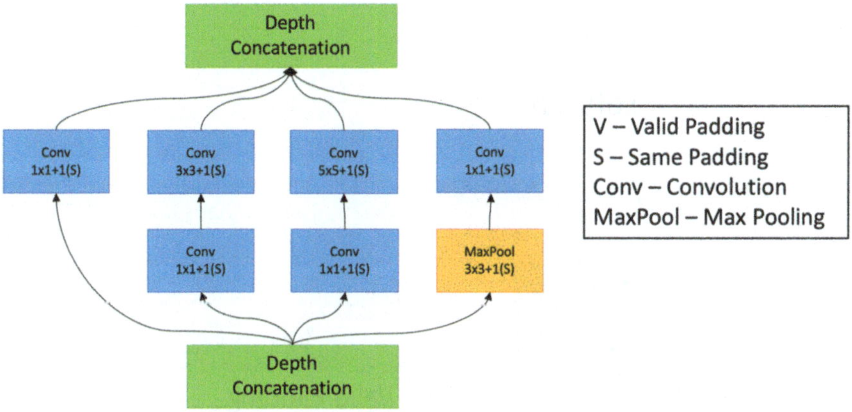

Fig. 21.6 Inception block with dimensionality reduction

layer is followed by a convolution layer with the same padding, followed by two fully connected layers and a softmax classifier.

In total, the GoogLeNet model consists of 22 layers, and at the end of the last module, a global average pooling operation is performed (Szegedy et al. 2015). To further improve the results and performance of this 22 layered architecture, attention mechanisms and residual blocks are added to the network, as shown in Fig. 21.5. A separate block called Attention-based residual block is incorporated into the architecture to enhance the network's performance.

Attention-based residual blocks play a significant role in the network employed in this work. Four residual blocks are used in the proposed model. Each block contains three convolution layers with the same zero padding and a max-pooling layer. The convolution layer convolves distinct kernels with the zero-padded input feature map. This layer is followed by an activation function, the Rectified Linear Unit (ReLU). ReLU is used to replace all the negative values from the output of the convolution layer with zeros, thus adding stability to the learning.

The output from the previous inception block represented as "X" in Fig. 21.8 is passed on to two functional branches. The first branch is an inception block, and the second branch is a residual block followed by the attention block, which learns and adds weights to the significant features of the image data. This attention embedded output is then passed on to the next function in the network or the output layer of the model, as depicted in Fig. 21.8 (Ramamurthy et al. 2020).

Attention mechanisms enable the network to focus on only the most important features of the image data, thus allowing for researchers to compromise on complicated feature engineering. It provides a balance between feature learning and region localization. By prioritizing features, marginal differences in visual representation can be detected. The use of these localized features in the model leads to expert image-level recognition (Fu et al. 2017). An attention block learns attention weights and four such attention blocks are used in between the inception modules of GoogLeNet.

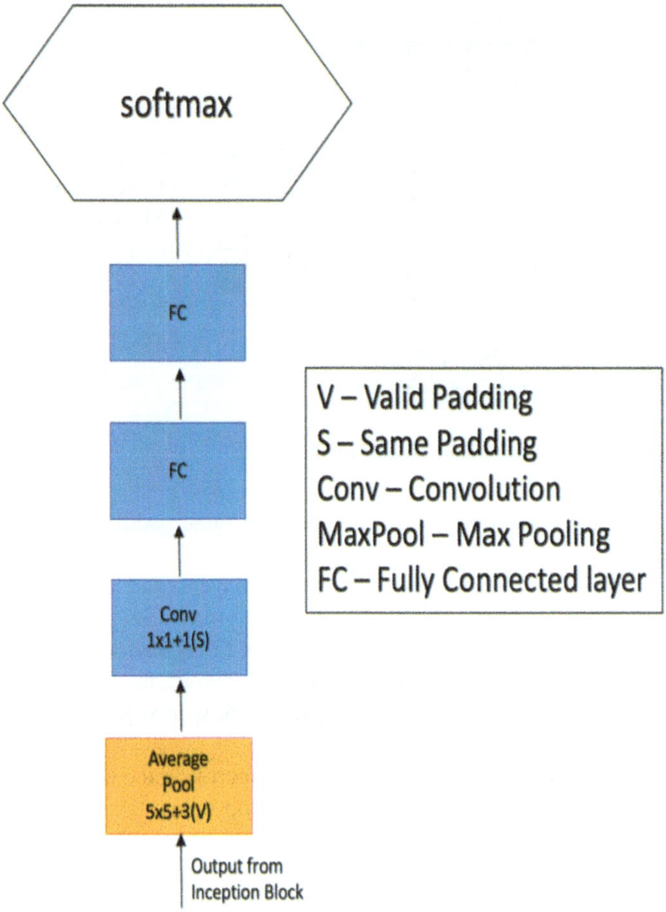

Fig. 21.7 Auxiliary network

Another contributing factor to the proposed model is the use of a residual network in the model. The model was implemented with the use of nonlinear functions such as ReLU. The network also skipped two or three layers in the network in order to facilitate faster and better performance. Skipping simplifies the network by using very few layers in the initial stage. This in turn, helps to increase the learning rate of the model as there are comparatively fewer layers to propagate through in the network. It also reduced the vanishing gradient problem of the network. The skipped layers were restored with the help of feature space. Batch normalization was used in between the layers. Residual neural networks are easy to optimize. The use of a residual network increased the learning rate of the model. Instead of learning unreferenced functions that were not beneficial to the network, residual learning in the layers enabled the learning of residual functions with reference to the layer inputs. This improved the overall performance of the network.

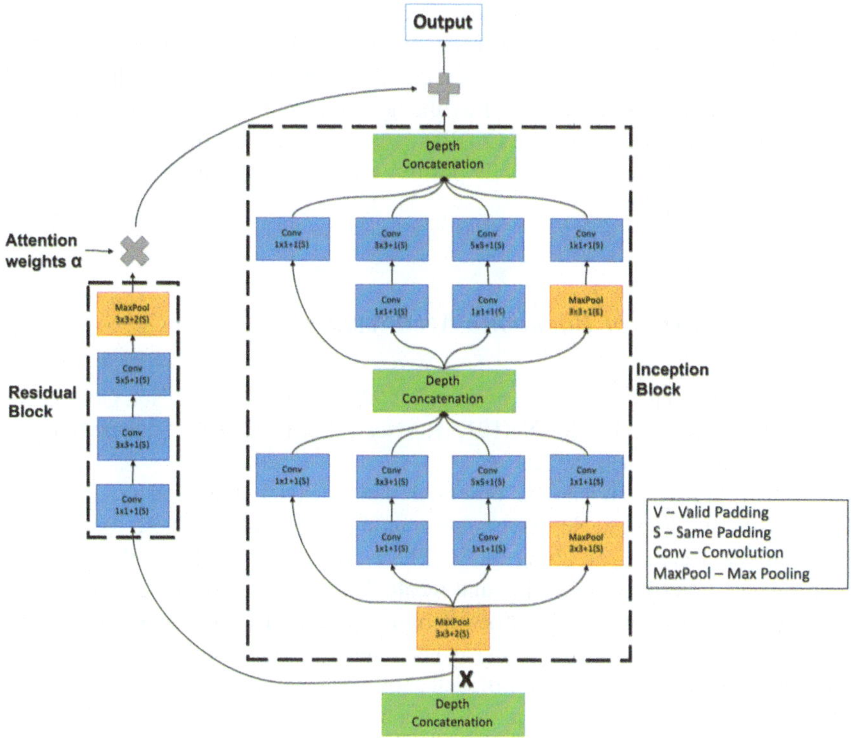

Fig. 21.8 Interaction between Attention-based residual learning block and Inception block (Ramamurthy et al. 2020)

21.4 Results and Discussion

A comprehensive experimental analysis of multiple models has been performed in this research. The observations and performance metrics of these experiments have been utilized to propose a deep learning architecture with improved performance on the COVID-CT dataset. The results and analysis of all the networks trained in this work have been explained in detail in this section.

21.4.1 Experimental Setup

The deep learning architectures were trained in the Tensorflow Deep Learning programming framework with the Tesla K80 GPU. A RAM of 25GB was provided for the execution. The source codes for training these models were written using the

Keras library. The deep networks were trained on a total of 4474 augmented CT scans obtained from the 747 scans present in the COVID-CT-Dataset. The objective of each model was to perform binary classification. To carry out this task, the features of the images were learned by the networks and were used to differentiate between the COVID-positive and COVID-negative scans. The numeric value 1 indicated that the image was of a COVID-positive type, and the number 0 represented all healthy chest scans.

21.4.2 Analysis of Backbone Architecture

The architectures analysed in this research include, AlexNet, DenseNet, GoogLeNet, InceptionV4, ResNet50, ResNet101, ResNet152, ShuffleNet, SqueezeNet, and Visual Geometric Group (VGG16).

From Table 21.2, it can be observed that AlexNet, GoogLeNet, ResNet50, ResNet101, and SqueezeNet have good training performance. These models have obtained an accuracy of above 80%. These models have also exhibited good performance in terms of precision and recall. They have the right balance between precision and recall, which could be inferred from their F1 score values since all of them are close to 1. In specific, AlexNet and GoogleNet have performed consistently with a very high average accuracy of well above 80% and have precision above 0.80. They have good F1 scores; hence these models are balanced.

From the analysis that was conducted in the series of experiments detailed above, it was inferred that the GoogLeNet model yielded good performance. The overall

Table 21.2 Analysis of performance measures for different deep learning models

Network	Training Accuracy (in %)	Testing Accuracy (in %)	Precision	Recall (Sensitivity)	F1-Score	Specificity
AlexNet	87.039	80.685	0.816	0.846	0.823	0.762
DenseNet	78.547	66.464	0.776	0.717	0.638	0.606
GoogLeNet	84.787	83.278	0.877	0.801	0.836	0.868
InceptionV4	54.972	50.77	0.384	0.436	0.336	0.593
ResNet50	80.67	78.628	0.827	0.763	0.79	0.814
ResNet101	81.991	73.508	0.742	0.787	0.755	0.676
ResNet152	75.083	70.646	0.784	0.631	0.69	0.794
ShuffleNet	77.852	73.42	0.761	0.71	0.714	0.761
SqueezeNet	80.67	75.095	0.808	0.715	0.748	0.789
VGG16Net	65.698	59.021	0.568	0.985	0.72	0.741

accuracy of 83.27% was achieved with GoogLeNet on the COVID-19 dataset used. Hence, GoogLeNet architecture was used as the base model to improve the performance for COVID-19 detection.

21.4.3 Cross-Validation Performance for Proposed Network

The proposed deep network with the GoogLeNet backbone was trained on the 4474 augmented CT scans obtained from the COVID-CT-Dataset. The scans were labeled with the value 1 and 0, indicating that the image was of a COVID-positive type and a healthy chest scan, respectively. 60% of the dataset was used in training, 20% for validation and 20% for testing the model. Fivefold cross-validation was performed on the data where each fold ran for 50 epochs. The fivefold cross-validation metrics have been shown in Table 21.3.

The performance metrics of the proposed architecture have been compared with the performance of the existing deep learning research works in this section. Figure 21.9 and Fig. 21.10 presents the observations of the trained work based on GoogleNet backbone with and without attention-based residual learning blocks.

The original GoogLeNet model performed with an accuracy of 83.27%. By adding Attention mechanism and residual network blocks into the architecture, the performance of the network was increased by 9.598% in terms of classification accuracy. This improvement in the performance reveals the efficiency of the residual block and attention mechanisms towards precise classification. The attention mechanism improved the model by enabling it to focus on the significant features of the image. All the performance metrics observed in the training of the proposed system have been tabulated in Table 21.4.

Table 21.3 Results of fivefold cross-validation for the GoogLeNet model and the proposed system

	Accuracy in Percentage					
Model	**Fold-1**	**Fold-2**	**Fold-3**	**Fold-4**	**Fold-5**	**Average**
Traditional GoogLeNet	83.3464	83.9106	83.5642	83.7875	83.7808	83.2779
Proposed system with GoogleNet backbone and attention-based residual learning blocks	91.1620	91.0503	91.0280	91.5973	91.4720	91.2619

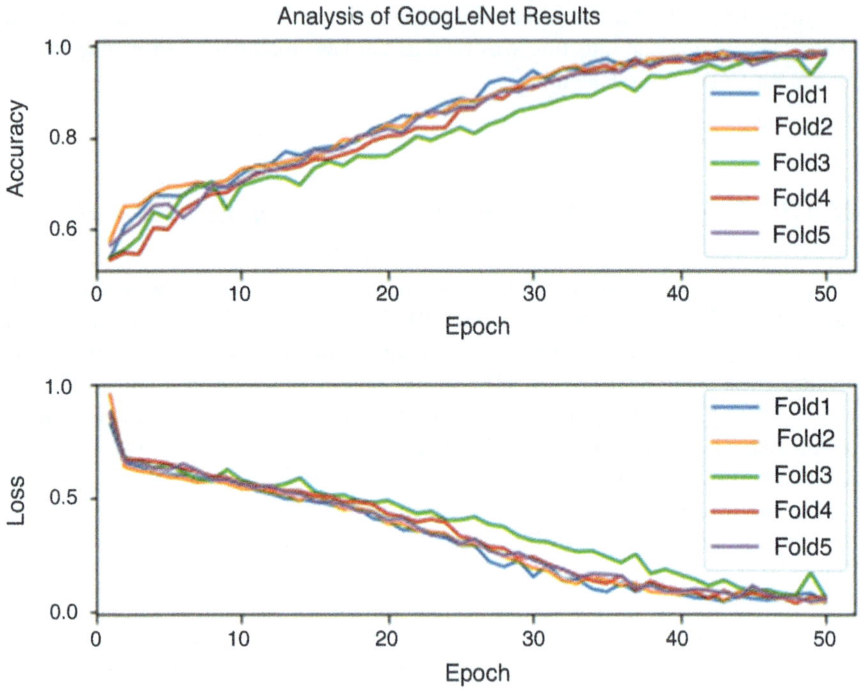

Fig. 21.9 Observations for traditional GoogLeNet backbone architecture

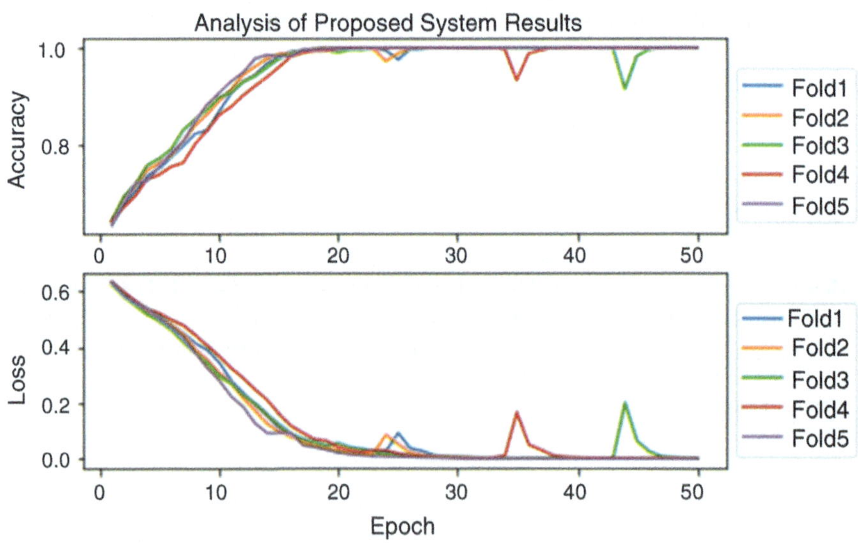

Fig. 21.10 Observations of the proposed system

Table 21.4 Analysis of performance measures for the proposed system

Network	Training Accuracy (in %)	Testing Accuracy (in %)	Precision	Recall (Sensitivity)	F1-Score	Specificity
Proposed system	91.322	91.262	0.869	0.914	0.891	0.842

21.5 Conclusion

A performance evaluation of various deep learning models was carried out for the detection of COVID-19 using CT scan images. AlexNet, DenseNet, GoogLeNet, InceptionV4Net, ResNet, ShuffleNet, SqueezeNet, and VGG16 were the state-of-the-art CNN architectures that are analyzed in this study. The results obtained using the various evaluation metrics indicated that GoogLeNet is an effective learning model that gave the highest accuracy of 83.27%. Based on the performance of these networks, a new architecture based on GoogLeNet backbone was proposed in this work. The proposed work introduced Attention-based residual learning blocks that were integrated with GoogLeNet backbone to improve the accuracy of the model. The proposed model obtained an accuracy of 91.26%. Future researchers can incorporate various additions to these standard architectures, which will substantially improve the performance of the model. COVID-19 datasets obtained from several sources can be utilized to train and test a superior network that would greatly influence the spread of the disease and help mitigate the impact of the virus.

References

Ai T, Yang Z, Hou H et al (2020) Correlation of chest CT and RT-PCR testing for coronavirus disease 2019 (COVID-19) in China: a report of 1014 cases. Radiology. https://doi.org/10.1148/radiol.2020200642

Apostolopoulos ID, Mpesiana TA (2020) Covid-19: automatic detection from X-ray images utilizing transfer learning with convolutional neural networks. Phys Eng Sci Med 43:635–640. https://doi.org/10.1007/s13246-020-00865-4

Butt C, Gill J, Chun D, Babu BA (2020) Deep learning system to screen coronavirus disease 2019 pneumonia. Appl Intell. https://doi.org/10.1007/s10489-020-01714-3

Dai H, Zhang X, Xia J et al (2020) High-resolution chest CT features and clinical characteristics of patients infected with COVID-19 in Jiangsu, China. Int J Infect Dis 95:106–112. https://doi.org/10.1016/j.ijid.2020.04.003

Ding X, Xu J, Zhou J, Long Q (2020) Chest CT findings of COVID-19 pneumonia by duration of symptoms. Eur J Radiol 127:109009. https://doi.org/10.1016/j.ejrad.2020.109009

Fang Y, Zhang H, Xie J et al (2020) Sensitivity of chest CT for COVID-19: comparison to RT-PCR. Radiology. https://doi.org/10.1148/radiol.2020200432

Fu J, Zheng H, Mei T (2017) Look closer to see better: recurrent attention convolutional neural network for Fine-Grained Image Recognition. 2017 IEEE Conference on Computer Vision and Pattern Recognition (CVPR). https://doi.org/10.1109/cvpr.2017.476

Guan CS, Lv ZB, Yan S et al (2020) Imaging features of coronavirus disease 2019 (COVID-19): evaluation on thin-section CT. Acad Radiol 27:609–613. https://doi.org/10.1016/j.acra.2020.03.002

He J-L, Luo L, Luo Z-D et al (2020) Diagnostic performance between CT and initial real-time RT-PCR for clinically suspected 2019 coronavirus disease (COVID-19) patients outside Wuhan, China. Respir Med 168:105980. https://doi.org/10.1016/j.rmed.2020.105980

He K, Zhang X, Ren S, Sun J (2016) Deep residual learning for image recognition. 2016 IEEE Conference on Computer Vision and Pattern Recognition (CVPR). https://doi.org/10.1109/cvpr.2016.90

Huang L, Han R, Ai T et al (2020) Serial quantitative chest CT assessment of COVID-19: deep-learning approach. Radiology: Cardiothoracic Imaging. https://doi.org/10.1148/ryct.2020200075

Huang G, Liu Z, Maaten LVD, Weinberger KQ (2017) Densely connected convolutional networks. 2017 IEEE Conference on Computer Vision and Pattern Recognition (CVPR). doi: https://doi.org/10.1109/cvpr.2017.243

Iandola FN, Moskewicz MW, Ashraf K, Han S, Dally WJ, Keutzer K (2016) SqueezeNet: alexnet-level accuracy with 50x fewer parameters and <1MB model size. ArXiv, abs/1602.07360

Krizhevsky A, Sutskever I, Hinton GE (2017) ImageNet classification with deep convolutional neural networks. Commun ACM 60:84–90. https://doi.org/10.1145/3065386

Li K, Fang Y, Li W et al (2020a) CT image visual quantitative evaluation and clinical classification of coronavirus disease (COVID-19). Eur Radiol 30:4407–4416. https://doi.org/10.1007/s00330-020-06817-6

Li B, Kang G, Cheng K, Zhang N (2019) Attention-guided convolutional neural network for detecting pneumonia on chest X-rays. In: 2019 41st annual international conference of the IEEE engineering in medicine and biology society (EMBC). https://doi.org/10.1109/embc.2019.8857277

Li L, Qin L, Xu Z et al (2020b) Using artificial intelligence to detect COVID-19 and community-acquired pneumonia based on pulmonary CT: evaluation of the diagnostic accuracy. Radiology. https://doi.org/10.1148/radiol.2020200905

Liang T, Liu Z, Wu CC et al (2020) Evolution of CT findings in patients with mild COVID-19 pneumonia. Eur Radiol. https://doi.org/10.1007/s00330-020-06823-8

Liu H, Liu F, Li J et al (2020) Clinical and CT imaging features of the COVID-19 pneumonia: focus on pregnant women and children. J Infect. https://doi.org/10.1016/j.jinf.2020.03.007

Long C, Xu H, Shen Q et al (2020) Diagnosis of the coronavirus disease (COVID-19): rRT-PCR or CT? Eur J Radiol 126:108961. https://doi.org/10.1016/j.ejrad.2020.108961

Luo Z, Wang N, Liu P et al (2020) Association between chest CT features and clinical course of coronavirus disease 2019. Respir Med 168:105989. https://doi.org/10.1016/j.rmed.2020.105989

Pan Y, Guan H, Zhou S et al (2020a) Initial CT findings and temporal changes in patients with the novel coronavirus pneumonia (2019-nCoV): a study of 63 patients in Wuhan, China. Eur Radiol 30:3306–3309. https://doi.org/10.1007/s00330-020-06731-x

Pan F, Ye T, Sun P et al (2020b) Time course of lung changes at chest CT during recovery from coronavirus disease 2019 (COVID-19). Radiology 295:715–721. https://doi.org/10.1148/radiol.2020200370

Ramamurthy K, Hariharan M, Anand S et al (2020) Attention embedded residual CNN for disease detection in tomato leaves. Appl Soft Comput 86:105933. https://doi.org/10.1016/j.asoc.2019.105933

Ren J, Ren R, Green M, Huang X (2019) Defect detection from X-ray images using a three-stage deep learning algorithm. In: 2019 IEEE Canadian conference of electrical and computer engineering (CCECE). https://doi.org/10.1109/ccece.2019.8861944

Sharma N, Jain V, Mishra A (2018) An analysis of convolutional neural networks for image classification. Procedia Computer Science 132:377–384. https://doi.org/10.1016/j.procs.2018.05.198

Shi H, Han X, Jiang N et al (2020) Radiological findings from 81 patients with COVID-19 pneumonia in Wuhan, China: a descriptive study. Lancet Infect Dis 20:425–434. https://doi.org/10.1016/s1473-3099(20)30086-4

Simonyan K, Zisserman A (2014) Very deep convolutional networks for large-scale image recognition. CoRR, abs/1409.1556

Szegedy C, Ioffe S, Vanhoucke V (2016b) Inception-v4, inception-resnet and the impact of residual connections on learning. arXiv:1602.07261

Szegedy C, Liu W, Jia Y, et al (2015) Going deeper with convolutions. 2015 IEEE Conference on Computer Vision and Pattern Recognition (CVPR). https://doi.org/10.1109/cvpr.2015.7298594

Szegedy C, Vanhoucke V, Ioffe S, et al (2016a) Rethinking the Inception Architecture for Computer Vision. 2016 IEEE Conference on Computer Vision and Pattern Recognition (CVPR). https://doi.org/10.1109/cvpr.2016.308

Ucar F, Korkmaz D (2020) COVIDiagnosis-net: deep Bayes-SqueezeNet based diagnosis of the coronavirus disease 2019 (COVID-19) from X-ray images. Med Hypotheses 140:109761. https://doi.org/10.1016/j.mehy.2020.109761

Udugama B, Kadhiresan P, Kozlowski HN et al (2020) Diagnosing COVID-19: the disease and tools for detection. ACS Nano 14:3822–3835. https://doi.org/10.1021/acsnano.0c02624

Wang K, Kang S, Tian R et al (2020a) Imaging manifestations and diagnostic value of chest CT of coronavirus disease 2019 (COVID-19) in the Xiaogan area. Clin Radiol 75:341–347. https://doi.org/10.1016/j.crad.2020.03.004

Wang H, Wei R, Rao G et al (2020b) Characteristic CT findings distinguishing 2019 novel coronavirus disease (COVID-19) from influenza pneumonia. Eur Radiol. https://doi.org/10.1007/s00330-020-06880-z

Ye Z, Zhang Y, Wang Y et al (2020) Chest CT manifestations of new coronavirus disease 2019 (COVID-19): a pictorial review. Eur Radiol 30:4381–4389. https://doi.org/10.1007/s00330-020-06801-0

Zhai P, Ding Y, Wu X et al (2020) The epidemiology, diagnosis and treatment of COVID-19. Int J Antimicrob Agents 55:105955. https://doi.org/10.1016/j.ijantimicag.2020.105955

Zhang X, Zhou X, Lin M, Sun J (2018) ShuffleNet: an extremely efficient convolutional neural network for mobile devices. 2018 IEEE/CVF Conference on Computer Vision and Pattern Recognition. https://doi.org/10.1109/cvpr.2018.00716

Zhao J, Zhang Y, He X, Xie P (2020) COVID-CT-dataset: a CT scan dataset about COVID-19. arXiv preprint arXiv:2003.13865

Zhao W, Zhong Z, Xie X et al (2020a) Relation between chest CT findings and clinical conditions of coronavirus disease (COVID-19) pneumonia: a multicenter study. Am J Roentgenol 214:1072–1077. https://doi.org/10.2214/ajr.20.22976

Zhao W, Zhong Z, Xie X et al (2020b) CT scans of patients with 2019 novel coronavirus (COVID-19) pneumonia. Theranostics 10:4606–4613. https://doi.org/10.7150/thno.45016

Zhou Z, Guo D, Li C et al (2020) Coronavirus disease 2019: initial chest CT findings. Eur Radiol 30:4398–4406. https://doi.org/10.1007/s00330-020-06816-7

Chapter 22
COVID-19 Face Mask Detection Using CNN and Transfer Learning

Cecilia Ajowho Adenusi, Olufunke Rebecca Vincent, Jesufunbi Abodunrin, and Bukola Taibat Adebiyi

Abstract The trauma produced by the COVID-19 sickness, which was proclaimed by the World Health Organization (WHO) in 2020, has impacted the entire world. WHO has recommended several recommendations and precautions to effectively prevent the spread of the deadly disease, including social distance, hand sanitizer, and the use of a face mask or face shield. Most particularly in crowded settings, which is what inspired this investigation into one of the WHO recommended preventive measures, namely the use of a face mask. This research used a Convolutional Neural Network and a Transfer Learning Model to determine whether or not a citizen wears a mask. This suggested model is trained and tested on the Face Masked Dataset, then image augmentation on limited available data for improved training and testing, with a 98 percent accuracy rate during training and testing.

Keywords COVID-19 · Face mask · Transfer learning · Dataset · Pandemic · Image augmentation · CNN · WHO

22.1 Introduction

The COVID-19 pandemic has had a long-term impact on numerous places throughout the world since December 2019. Wuhan, China, was the origin of it all. On March 11, 2020, the World Health Organization (WHO) declared it a dangerous disease after it swept across the globe and severely impacted 114 countries. Medical professionals, healthcare organizations, medical practitioners, and researchers are all working on efficient vaccinations and treatments to tackle this terrible disease, but there has been no breakthrough to date (Megahed and Ghoneim 2020). The virus

C. A. Adenusi (✉)
Linux Professional Institute, Nigeria Master Affiliate, UI, Abeokuta, Nigeria

O. R. Vincent · J. Abodunrin · B. T. Adebiyi
Department of Computer Science, Federal University of Agriculture, Abeokuta, Nigeria

S. A. Hassan et al. (eds.), *Decision Sciences for COVID-19*, International Series in Operations Research & Management Science 320,
https://doi.org/10.1007/978-3-030-87019-5_22

spreads through the air channel when an infected person sneezes or speaks to another person, as water droplets from their nose or mouth disperse through the air channel.

Due to the COVID-19 pandemic, which requires people to wear face masks, maintain social distance, and wash their hands with hand sanitizers, face mask detection is now a useful feature. While other social distancing and sanitization issues have been addressed in the past, face mask detection yet to be properly addressed. Wearing a mask during this pandemic is a key preventive measure (Rahmani and Mirmahaleh 2020) and is an especially important step in times when maintaining social distance is difficult. A mask is required for anyone who is at risk of serious illness from COVID-19 diseases, especially those who are at higher risk. COVID-19 is found to spread primarily among individuals who are close to one another (nearly 6 feet), but it can also be spread by people who do not have symptoms and are unaware that they are infected (Ge et al. 2020).

The Centers for Disease Control and Prevention (CDC) advises all people aged 2 years and above to wear a mask in public places, particularly when other social distancing measures are difficult to maintain. As a result, lowering the risk of transmission of this lethal virus from an infected person to a healthy person will greatly limit the spread of the virus and the severity of the disease. In the field of image processing and computer vision, face mask detection has proven to be a difficult problem. Face detection has a variety of applications, ranging from face recognition to capturing facial motions, the latter of which necessitates revealing the face with extreme accuracy.

As a result, many nations have adopted the rule of "no mask, no entry." Face mask detection is a critical problem in COVID-19 prevention and security. In the medical field, a mask eliminates the risk of infection for an infected person, whether or not they have any symptoms. Face mask detection is used in a variety of settings, including airports, hospitals, offices, and educational institutions. The virus spreads through close contact and densely populated areas. The coronavirus outbreak has resulted in unparalleled levels of international scientific collaboration.

Multitasking is a learning model. MTL has now been referred to by a variety of names, including joint learning, learning to learn, and learning with auxiliary activities, to name a few. Multitask learning has been successfully used across all applications of machine learning, from natural language processing (Collobert and Weston 2008) and speech recognition (Deng et al. 2013) to computer vision (Girshick 2015) and drug discovery (Girshick 2015) (Ramsundar et al. 2015). MTL has been referred to by a variety of names, including joint learning, learning to learn, and learning with auxiliary activities, to name a few.

When you find yourself optimizing more than one loss function, multitask learning is a success. MTL (Multi-Task Learning) is a methodology for learning multiple tasks at once to improve the performance of the main task or all activities. This study intends to design for face mask detection using a convolutional neural network and transfer learning, after taking into account models that have previously been used by other researchers.

22.2 Literature Review

Many datasets for face detection have been created in the past to form an impression of face mask detection models. Wider Face (Yang et al. 2016), IJB-A (Klare et al. 2015), MALF (Tian et al. 2015), and CelebA (Tian et al. 2015) are examples of recent datasets that were created by taking online pictures (Wen et al. 2016). Annotations are given for current faces rather than previous ones in all of these datasets. Large datasets are required for better training and testing data, as well as for executing real-world applications more simply. This needs several deep learning algorithms capable of reading faces and masks directly from the user's data.

In Deformable Part Model-based categorization, the structure and orientations of numerous distinct faces are modeled using DPM. In 2006, Ramanan proposed a random forest tree model for face mask identification that accurately predicts facial shapes and poses. One of the well-known specialists (Zhang and Zhang 2014) developed a DPM-based face mask detector that used approximately 30, 000 faces divided into masks and non-mask categories. His work was extremely precise, with a precision of 97.14 percent. Sun created several face mask detection models. Face mask detection models based on DPM will often achieve fantastic precision, but they may not be tolerant of the extremely high computational cost due to the usage of DPM.

Face detector models learn directly from the user's data and then apply some deep learning algorithms to it in Convolutional Neural Network-based classification (Ren et al. 2015). Cascade CNN was created in the year 2007 (Li et al. 2015). The concept of feature aggregation of faces in the face detection model was proposed by (Tian et al. 2015) in 2015. (Ojala et al. 2002) Improved the Alex Net architecture for fine-tuning the image dataset in subsequent studies. (Howard et al. 2017) proposed a Contextual Multi-Scale Region-based Convolutional Neural Network (CMS-RCNN) for uninhibited situations, which had a substantial effect on face detection models. A grid loss layer was created by (Opitz et al. 2016) to reduce the errors in the replacement layers of the CNN layers and to deal with the biased obstructions produced in the mask detection models. As technology progressed, more CNN-based 3D models emerged, one of which was suggested by (Li et al. 2015). It was the learning structure of a face mask detection model. Several other projects were completed in the areas of pose recognition, gender estimation, and landmark localization.

Models for detecting face masks are available in a variety of shapes and sizes. These can be divided into a variety of categories. The Viola-Jones face detector (Viola and Jones 2001) was used to boost cascades in boosting-based classification using simple features. The Multiview face mask detector was then inspired by the Viola-Jones detector model. A face mask detector model was also created using decision tree methods. Face mask detectors in this category were excellent at detecting masks.

The most extensively utilized technology in the realm of security is closed-circuit television (CCTV), sometimes known as video surveillance. CCTV cameras can now be found in a wide range of places, from public spaces to private homes. The quantity of space necessary for footage storage is one of the most significant and hardest components of deploying CCTV cameras on a wide scale. The majority of the footage is stored on secondary storage methods such as hard disk drives. To reduce the amount of storage space needed, compression techniques are applied. A storage optimization method is being developed, especially for CCTV cameras (Andrew et al. 2013).

Complex algorithms such as feature-based algorithms and learning-based algorithms are dealt with by general mask detection or scarf detection algorithms (Wright et al. 2008). Face-based methods use information from facial characteristics such as the mouth (Wen and Ding 2005) or skin color (Lin and Liu 2006) to determine whether or not there is occlusion on the face. (Jia and Martinez 2009) Propose occluded facial recognition using support vector machines (Min et al. 2011). Gabor wavelets, principal components analysis, and support vector machines are used to find occluded faces. These methods, too, are computationally intensive.

22.3 Dataset Used

This study makes use of secondary data called Face Mask Detection (FMD) by OMKAR GURAV (2020) (https://www.kaggle.com/omkargurav/face-mask-dataset), which consists of 7553 images that include both masked and unmasked facial images, with images of face masks accounting for 3732 and those without accounting for 3828. 6083 Photos from both groups were used for training, while the remaining 1510 photos were utilized to test the model. Figure 22.1 shows some examples of sample photographs.

22.4 Proposed Methodology

The dataset description entails limited samples that were allowed due to government concerns about individual privacy and security. Whereas the ResNet50 CNN model manages to learn due to the limited samples available. There are two types of proposed methods. One is to deal with the oversample using image augmentation and the second is to deal with the face mask detection using CNN and transfer learning.

22.4.1 Augmentation of Images

This technique is used to artificially modify images in the dataset to increase the training dataset size via augmentation. The training images are being augmented

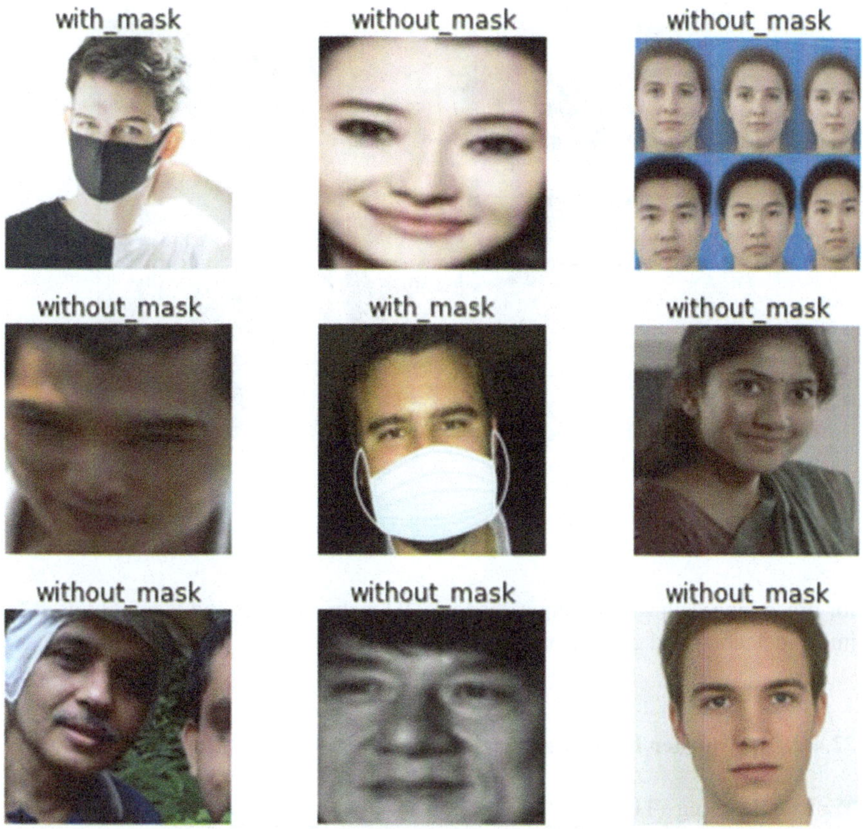

Fig. 22.1 Sample image from FMD dataset

using eight (8) operations which are rotating, blurring, zooming, contrasting, flipping horizontally, and the whole dataset will then be rescaled to $224 \times 224,224 \times 224$ and also converted to a single grayscale representation, as shown in Fig. 22.2.

22.4.2 Convolutional Neural Network

Convolutional neural networks (CNNs) are a class of deep artificial neural networks. CNN is used to identify images, cluster them by resemblance and execute object recognition within scenes.

CNN can be used to identify faces, tumors, street signs, platypuses, and other aspects of visual data. The CNN's core building block is the convolutional layer. The layer's parameters consist of a set of learnable filters that have a small receptive field but extend to the full depth of the input volume. This is a specialized neural network

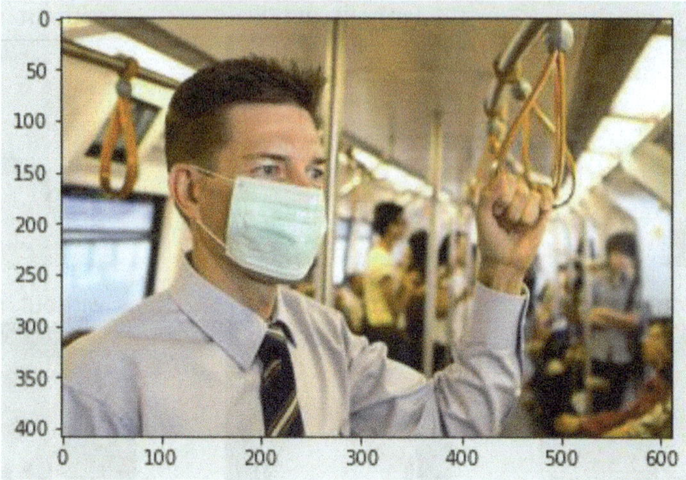

Fig. 22.2 Augmentation of training images

for processing data with an input shape of 2D matrix images and is mainly used for image detection and classification.

22.4.2.1 Transfer Learning

Transfer learning (TL Model) is the advancement of learning new tasks via the transfer of knowledge from a related work that has already been learned. Transfer learning is in the context of inductive learning and includes expanding known inference and classification algorithms such as Bayesian networks, neural networks, and Markov Logic Networks. Another major area is in the context of reinforcement learning and involves algorithms like policy search and Q-learning.

A transfer can help students learn in three ways. The first way is the amount of time it takes to fully learn the target work given the transferred knowledge compared to the number of times you learn it from the start. The second is the incipient performance achieved in the target work using only the transferred knowledge before further learning is completed, compared to the initial performance of an ignorant agent. The final performance level is achievable in the target work compared to the final level with no transfer.

This utilizes the ResNet50 CNN pre-trained model for classifying and detecting people without a face mask. The architecture of the proposed model is shown in Fig. 22.3.

Fig. 22.3 explains the whole process of work done in the research, commencing from how the dataset is obtained to the training and testing aspect, down to the visualization and interpretation of results, to the methods used for classification and detection of people wearing face masks.

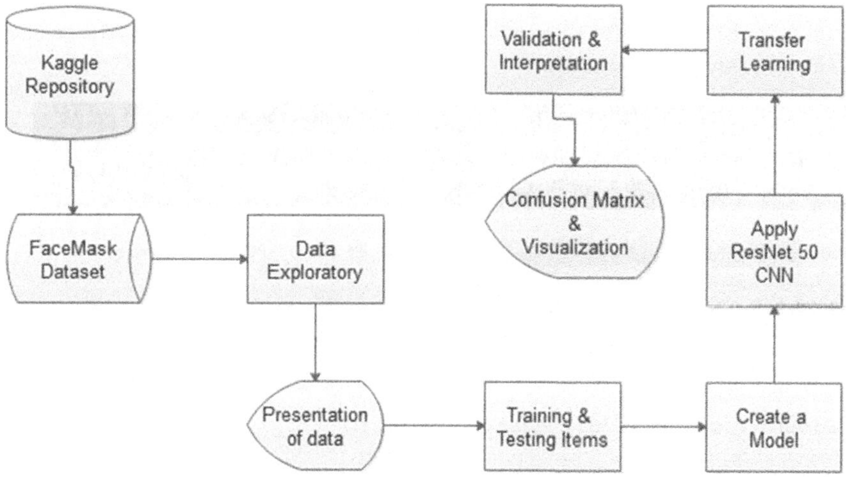

Fig. 22.3 Proposed face mask detection model

The dataset used was gotten from the Kaggle repository named FaceMask dataset (OMKAR GURAV, 2020) and it was downloaded to a local computer for analysis and detection. The data exploratory, augmentation, and presentation were carried out, followed by the training and testing of the model used via ResNet50 CNN and the loading the CNN used and used the transfer learning model for a better detection system via the learning rate function applied and the transfer learning model was also trained and validation and visualization results were carried out with the confusion matrix, which is shown in Fig. 22.3.

22.5 Results

To perform the face masked detection, this component of the work was carried out utilizing Kaggle Notebook and a face masked dataset obtained from its repository. Because transfer learning (TL) starts with an already-trained neural network for image identification, we used a pre-trained ResNet50 CNN and TL to learn weights of only the last network layer. To achieve reasonable neural network training, around 300,000 picture samples must be supplied, and millions of photos will be required for good performance. This was used to develop the CNN function, created by CNN, which allowed a pre-trained ResNet50 network to be loaded, which had been trained on almost a million photos from the FaceMask Detection database.

Fig. 22.4 shows how the method used to load the pre-training model needed for the path works on the Kaggle notebook. After the data exploration and argumentation, the Resnet50 module was downloaded to the Jupyter notebook and used on the training data.

```
learn = cnn_learner(data, models.resnet50, metrics=[accuracy], model_dir =
Path('../kaggle/working'),path = Path("."))
```

```
Downloading: "https://download.pytorch.org/models/resnet50-19c8e357.pt
h" to /root/.cache/torch/checkpoints/resnet50-19c8e357.pth
```

```
100% 97.8M/97.8M [01:18<00:00, 1.31MB/s]
```

Fig. 22.4 ResNet50 network loaded

```
learn.lr_find()
learn.recorder.plot(suggestions=True)
```

```
0.00% [0/2 00:00<00:00]

epoch  train_loss  valid_loss  accuracy  time

91.49% [86/94 01:03<00:05 1.6725]
```

Fig. 22.5 Finding the learning rate model

22.6 Using Learning Rate

The quantity of weight loaded during training is referred to as the step size or learning rate. The slower learning rate necessitates more training epochs, resulting in smaller increases in the weight for each update, whereas the faster learning rate necessitates fewer training epochs.

Fig. 22.5 depicts the weights loaded during training, which are 91.49% accurate. The TL module was also loaded and applied to the CNN model results, resulting in the above accuracy.

22.7 Output

The goal of this research is to create an automated system to assist organizations, individuals, and government officials in monitoring people who wear or do not wear masks in public places.

Figure 22.6 is the outcome of TL on the CNN model used earlier for training of dataset used.

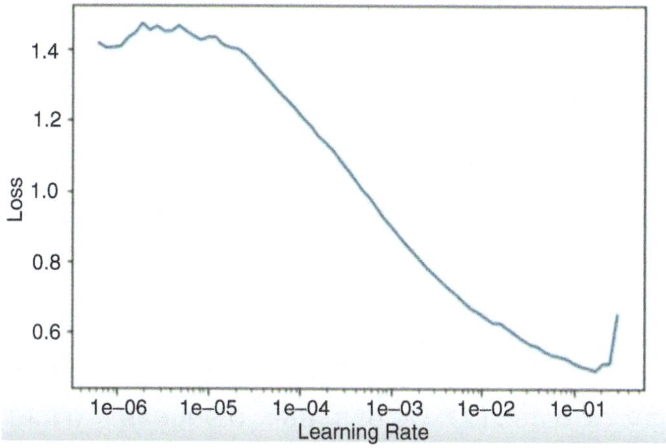

Fig. 22.6 Learning rate model output

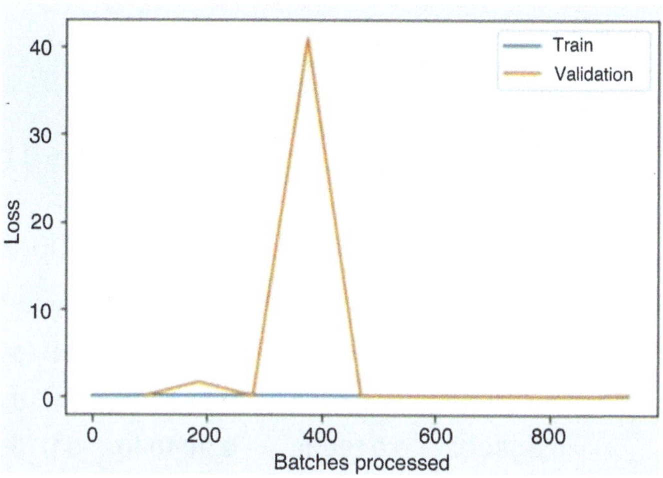

Fig. 22.7 Batches processed visualization

The graph in Fig. 22.7 shows the training and testing processes in batches and the loss encountered during the batch processes of the dataset used.

Fig. 22.8 shows that the learning rate cycle is four, denoting (epoch 0–3) with an accuracy of each of the epochs increasing as the iteration increases.

In Fig. 22.9, iterations were done in 10 places, with the accuracy also increasing as the epoch value increases, while other variables involved are not fixed either but keep changing.

Fig. 22.10 shows how face masks are detected by considering four variables, which are prediction, actual, loss, and probability of occurrence are put in place.

```
lr1 = 1e-3
lr2 = 1e-1
learn.fit_one_cycle(4,slice(lr1,lr2))
```

epoch	train_loss	valid_loss	accuracy	time
0	0.317080	2.066353	0.914570	01:18
1	0.463127	0.058465	0.996689	01:15
2	0.155862	0.004193	0.999338	01:16
3	0.048622	0.008202	0.998676	01:18

Fig. 22.8 Training model result

```
learn.unfreeze()
learn.fit_one_cycle(10,slice(1e-4,1e-3))
```

epoch	train_loss	valid_loss	accuracy	time
0	0.057610	0.029165	0.995364	01:19
1	0.074849	1.584362	0.991391	01:18
2	0.068002	0.019006	0.994040	01:19
3	0.046452	41.017384	0.969536	01:18
4	0.032881	0.031075	0.988742	01:20

Fig. 22.9 Hyper parameter tuning result

22.8 Result Interpretation

The model's performance can be validated by a variety of methods, one of which is by the use of a confusion matrix, which employs diagonal values to signify correct predictions for each class and inaccurate predictions for the other cell values.

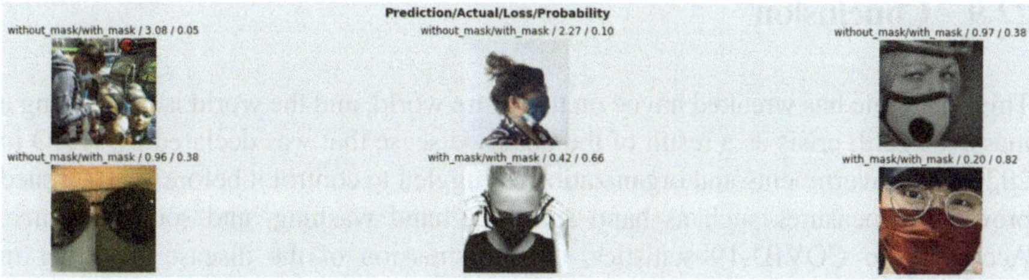

Fig. 22.10 Face mask detection and prediction

Fig. 22.11 Confusion matrix

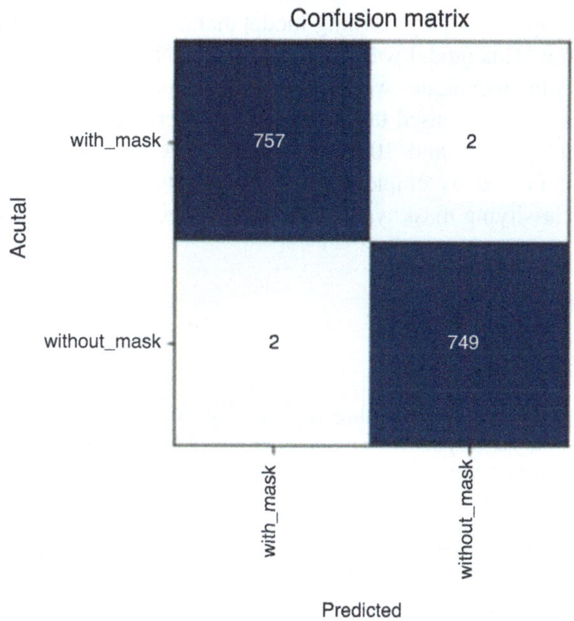

Fig. 22.11 justifies the accuracy of the models used for the face mask detection system, showing the actual value and predicted value. After the model has been trained and the required results have been obtained, the model can be deployed into production, and the entire model structure must be saved with the trained parameters.

22.9 Conclusion

This pandemic has wreaked havoc on the entire world, and the world is now facing a massive health crisis as a result of the deadly disease that was declared by WHO in 2020, and governments and organizations struggled to control it before WHO issued preventive measures such as hand sanitizer, hand washing, and social distance. According to COVID-19 statistics, the transmission of the disease is higher in crowded areas, and many researchers have determined that wearing a mask in public spaces will immediately minimize the disease's spread. As a result, governments in various countries have mandated the wearing of masks in public and crowded areas, making it extremely difficult to monitor people's compliance with government rules, particularly in crowded areas. This is what prompted this research work, and we proposed a deep learning model that detects whether a person is wearing a mask or not. This model was created using CNN and transfer learning, and image augmentation techniques were employed to improve the model's performance in this study, as they increased the difference between training and test data with an accuracy of 98 percent and 100 percent during testing on the FMD. This study could be improved by employing larger datasets, and it could subsequently be applied to classifying mask types using a facial recognition system.

References

Andrew G, Arora R, Bilmes J, Livescu K (2013) Deep canonical correlation analysis. In: International conference on machine learning. PMLR, pp 1247–1255

Collobert R, Weston J (2008) A unified architecture for natural language processing: deep neural networks with multitask learning. In: Proceedings of the 25thinternational conference on machine learning, pp 160–167

Deng L, Hinton G, Kingsbury B (2013) New types of deep neural network learning for speech recognition and related applications: an overview. In: 2013 IEEE international conference on acoustics, speech and signal processing. IEEE, pp 8599–8603

Ge XY, Pu Y, Liao CH, Huang WF, Zeng Q, Zhou H, Huang X (2020) Evaluation of the exposure risk of SARS-CoV-2 in different hospital environment. Sustain Cities Soc 61:102413

Girshick R (2015) Fast r-CNN. In: Proceedings of the IEEE international conference on computer vision, pp 1440–1448

Howard AG, Zhu M, Chen B, Kalenichenko D, Wang W, Weyand T, Adam H (2017) Mobilenets: efficient convolutional neural networks for mobile vision applications. arXiv preprint arXiv:1704.04861

Jia H, Martinez AM (2009) Support vector machines in face recognition with occlusions. In: 2009 IEEE conference on computer vision and pattern recognition. IEEE, pp 136–141

Klare BF, Klein B, Taborsky E, Blanton A, Cheney J, Allen K, Jain AK (2015) Pushing the frontiers of unconstrained face detection and recognition: Iarpa janus benchmark a. In: Proceedings of the IEEE conference on computer vision and pattern recognition, pp 1931–1939

Li H, Lin Z, Shen X, Brandt J, Hua G (2015) A convolutional neural network cascade for face detection. In: Proceedings of the IEEE conference on computer vision and pattern recognition, pp 5325–5334

Lin DT, Liu MJ (2006) Face occlusion detection for automated teller machine surveillance. In: Pacific-rim symposium on image and video technology. Springer, Berlin, Heidelberg, pp 641–651

Megahed NA, Ghoneim EM (2020) Antivirus-built environment: lessons learned from Covid-19 pandemic. Sustain Cities Soc 61:102350

Min R, Hadid A, Dugelay JL (2011) Improving the recognition of faces occluded by facial accessories. In: Face and gesture 2011. IEEE, pp 442–447

Ojala T, Pietikainen M, Maenpaa T (2002) Multiresolution gray-scale and rotation invariant texture classification with local binary patterns. IEEE Trans Pattern Anal Mach Intell 24(7):971–987

Opitz M, Waltner G, Poier G, Possegger H, Bischof H (2016) Grid loss: detecting occluded faces. In: European conference on computer vision. Springer, Cham, pp 386–402

Rahmani AM, Mirmahaleh SYH (2020) Coronavirus disease (COVID-19) prevention and treatment methods and effective parameters: a systematic literature review. Sustain Cities Soc 102568

Ramsundar B, Kearnes S, Riley P, Webster D, Konerding D, Pande V (2015) Massively multitask networks for drug discovery. arXiv preprint arXiv:1502.02072

Ren S, He K, Girshick R, Sun J (2015) Faster r-CNN: towards real-time object detection with region proposal networks. arXiv preprint arXiv:1506.01497

Tian Y, Luo P, Wang X, Tang X (2015) Deep learning strong parts for pedestrian detection. In: Proceedings of the IEEE international conference on computer vision, pp 1904–1912

Viola P, Jones M (2001) Rapid object detection using a boosted cascade of simple features. In: Proceedings of the 2001 IEEE computer society conference on computer vision and pattern recognition. CVPR 2001, vol 1. IEEE, p I-I

Wen D, Ding Y (2005) Formulation of nanofluids for natural convective heat transfer applications. Int J Heat Fluid Flow 26(6):855–864

Wen Y, Zhang K, Li Z, Qiao Y (2016) A discriminative feature learning approach for deep face recognition. In: European conference on computer vision. Springer, Cham, pp 499–515

Wright J, Yang AY, Ganesh A, Sastry SS, Ma Y (2008) Robust face recognition via sparse representation. IEEE Trans Pattern Anal Mach Intell 31(2):210–227

Yang S, Luo P, Loy CC, Tang X (2016) Wider face: a face detection benchmark. In: Proceedings of the IEEE conference on computer vision and pattern recognition, pp 5525–5533

Zhang C, Zhang Z (2014) Improving multiview face detection with multi-task deep convolutional neural networks. In: IEEE winter conference on applications of computer vision. IEEE, pp 1036–1041

Chapter 23
LASSO-DT Based Classification Technique for Discovery of COVID-19 Disease Using Chest X-Ray Images

Roseline Oluwaseun Ogundokun, Joseph Bamidele Awotunde, Paul Onawola, and Taye Oladele Aro

Abstract Early recognition of COVID-19 can aid in the development of a treatment plan and disease containment decisions. This research aims to use the LASSO+DT technique to automatically diagnose COVID19 lung-related individuals employing automated chest X-ray imageries while optimizing detection accuracy. The study employed machine learning (ML) based techniques, specifically feature selection (FS) and classification models were used to classify COVID-19 and normal chest X-ray imageries. For the FS, Least Absolute Shrinkage and Selection Operator (LASSO) technique was employed and for the classification, a Decision Tree (DT) classifier was used. A dataset containing 372 instances of chest X-rays was employed in this research for the investigation. The classification accuracy, detection rate (DR), and false-positive rate (FPR) were employed for the performance evaluation of the investigation. The study implemented DT alone as well as LASSO+DT and the results of both classifiers were compared. The result of the study recognized that the proposed LASSO+DT outperformed that of the DT alone. The investigation demonstrates that the proposed LASSO+DT is efficient and effective for the identification of COVID-19 ailment and further lung diseases because it has higher accuracy, DR, and lower FAR when compared with DT technique and few state of the art.

Keywords COVID-19 · LASSO · Decision tree · Feature selection · Chest X-ray

---

R. O. Ogundokun · P. Onawola
Department of Computer Science, Landmark University, Omu-Aran, Nigeria
e-mail: Ogundokun.roseline@lmu.edu.ng; onawola.paul@lmu.edu.ng

J. B. Awotunde (✉)
Department of Computer Science, University of Ilorin, Ilorin, Nigeria
e-mail: awotunde.jb@unilorin.edu.ng

T. O. Aro
Department of Computer Science, Koladaisi University, Onidudun, Nigeria
e-mail: taye.aro@koladaisiuniversity.edu.ng

S. A. Hassan et al. (eds.), *Decision Sciences for COVID-19*, International Series in Operations Research & Management Science 320,
https://doi.org/10.1007/978-3-030-87019-5_23

Abbreviations

ML	Machine learning
DL	Deep learning
DCNN	Deep convolutional neural network
FS	Feature selection
LASSO	Least Absolute Shrinkage and Selection Operator
DT	Decision tree
SVM	Support Vector Machine
TLT	Transfer learning technique
CNN	Convolutional Neural Networks
FE	Feature Extraction

23.1 Introduction

COVID-19 has been conveyed from Wuhan to the entire region in China and several other countries since December 2019 (Lukman et al. 2020). Due to its rapid dissemination from one individual to another in population, the modern coronavirus (similarly recognized as COVID-19) has instigated a worldwide infection problem (Zaim et al. 2020; Ogundoku et al. 2020). COVID-19's terrifying spread is the largest threat humankind has faced since WWII. In March 2020, the World Health Organization (WHO) declared COVID-19 a global pandemic (Rubin et al. 2020; Daniyal et al. 2020). COVID-19 is characterized by a dry cough, sore throat, and fever (Wang et al. 2020a). Septic shock, pulmonary edema, acute respirational distress disorder, and multi-organ dysfunction are all possible complications of a serious case of pneumonia (Zaim et al. 2020), unfortunately, clinical characteristics alone do not allow for the analysis of COVID-19, particularly in an affected individual who are experiencing symptoms for the first time. The Reverse Transcription Polymerase Chain Reaction (RT-PCR) examination was employed as the "gold standard" for establishing COVID-19 positively infected individuals among nucleic acid-based studies (Li et al. 2020). Nevertheless, RT-PCR can fail to identify many infected individuals, resulting in those individuals not receiving timely care (Wang et al. 2020a). Patients that are unaware of the dangers are especially hazardous and they are a direct source of contamination due to the virus's high contagiousness. The precision of the RT-PCR test is high, but the sensitivity is poor. As a response, a negative RT-PCR outcome does not ignore the risk of contamination with COVID-19 (Wang et al. 2020a). Medical care professionals have recently opted to utilize imagery scans to detect COVID-19, which is often performed with a CT scan or chest X-ray. The latest findings, nevertheless, warn against using an imagery examination to identify COVID-19 since it has a high rate of false positives (FP) and negatives (Li et al. 2020b). Such undiagnosed cases may have devastating consequences due to COVID-19's exponential distribution (Bertone et al. 2020). For immediate identification of COVID-19 persons, new diagnostic techniques must be created (Awotunde et al. 2021a).

COVID-19 identification that is fast and effective is becoming increasingly important for preventing infection and assisting patients in preventing ailment advancement. Soft Computing (SC) approaches like fuzzy logic, neural networks, and genetics be effective in ailment identification (Aydin and Yurdakul 2020; Gayathri and Satapathy 2020). They will help with decision-making by allowing for instantaneous seclusion and care of patients (Pegoraro et al. 2020). Numerous methods for detecting COVID-19 infections have been suggested, but the desired detection sensitivity is yet to be achieved. Several methods for diagnosing COVID-19 have recently been proposed, but none of them take into account the effect of feature weight on the classifier judgment (El Asnaoui and Chawki 2020; Barstugan et al. 2020a; Brunese et al. 2020; Islam et al. 2020; Salehi et al. 2020; Apostolopoulos and Mpesiana 2020a). Since both methods handle all functions similarly, consistency is degraded. The epidemic will propagate pandemic if COVID-19 cases are incorrectly diagnosed (Guo et al. 2020). Allocating a weight or a rank to every element, on the other hand, would aid the classifier in making correct judgments, thus improving diagnostic accuracy.

Several groups have recently identified COVID 19 pneumonia detection techniques based on deep learning (Wang et al. 2020b, 2021). Wang et al. (2021) employed DL approaches on CT imageries to scan COVID 19 persons and had an 89.5%, 88%, 87% for accuracy, precision, and sensitivity correspondingly. Wang et al. (2020b) introduced a deep convolutional neural network (DCNN) called COVID Net that had an accuracy of 83.5% in identifying COVID 19 persons from chest X-ray photographs. Ayrton (Ghoshal & Tucker, 2020) employed a limited dataset of 339 imageries for training and testing. They registered a validation accuracy of 96.2% using a ResNet50-built deep transfer learning approach. Peng (Salman et al. 2020) employed an ML detection approach to categorize chest X-ray images using DCNN. Since accuracy is regarded as the most important factor in their study, the DCNN accuracy was increased by taking a large number of imageries for training the system and increasing the number of iterations. A significant number of confirmed cases must be tested for proper isolation and care to curb the dissemination of COVID-19. Pathogenic research facility monitoring is the highest performing standard, but it is time overwhelming and has a high rate of FN results. To combat the disease, fast and precise analytic techniques are urgently needed (Awotunde et al. 2021b).

Given COVID-19 radiographic improvements in X-ray images, the study intends to use an enhanced ML technique for extracting COVID-19's imagery attributes to deliver a medical study before the pathogenic examination, thereby redeeming a serious period for ailment curbing (Folorunso et al. 2022). Generally, in this study, we attempted to identify the best appropriative algorithm for early COVID-19 discovery built on chest X-ray datasets. We analyzed clinical data of 372 patients who were confirmed as COVID-19 or other respiratory infectious diseases from the Kaggle database repository. The study used a classification technique which is

DT. The study implemented DT alone as well as LASSO+DT and the results of both classifiers were compared.

The rest of the paper is structured as thus:

23.2 Literature Review

ML and DL methods have recently been employed to examine chest X-rays in a limited amount of time. This has favored the chest CT scan because of the small ionizing radiation and transferability of X-rays (Singh et al. 2020). Wang and Wong (2020) established a DCNN for the recognition of COVID-19 cases from chest X-rays. An overall of 13,975 chest X-ray imageries was employed to execute their model. The model achieved a classification accuracy of 98.9%. The COVIDX-Net technique was created by Hemdan et al. (2020). This was employed for automated recognition of coronavirus diseased people employing chest X-ray photographs. The COVIDX-Net technique was implemented on a dataset comprising 50 normal and 25 confirmed COVID-19 cases. For COVID events, COVIDX-Net had a classification accuracy of 91%. Narin et al. (2020) proposed 3 related CNN methods: ResNet-50, Inception-ResNetV2, and Incep-tionV3 for the characterization of COVID-19 from chest X-ray images. In comparison to the other ones, ResNet50 had a classification accuracy of 98%. Sethy and Behera (2020) employed a pre-trained transfer approach known as ResNet-50 for removing imageries features from contaminated patients. These characteristics were used to aid classification using Support Vector Machine (SVM). The built model produced a classification accuracy of 95%. Farooq and Hafeez (2020) proposed a multiple-phase fine-tuning structure for ResNet-50 architectures that have been pre-trained. COVIDRes-Net is the name of the newly created model. COVIDRes-Net had a 96.23% accuracy rating. Asnaoui et al. (2020a) conducted a comparative analysis of 8 transfer learning approaches for the identification of COVID-19 pneumonia. The method was trained on 5856 chest X-rays. The classification accuracy of MobileNet-V2 and Inception-V3 was 96%. Abbas et al. (2020) proposed a DCNN technique called Decompose, Transfer, and Compose (DeTraC) for separating COVID-19 indications using chest X-rays. Using the decomposition technique, they looked into the anomalies in class limits. The DeTraC method achieved 95.12% accuracy and 97.91% sensitivity. For the identification of COVID-19 disease using chest X-ray photographs, Chowdhury et al. (2020) suggested an image line of reasoning strategy combined with a transition procedure. AlexNet, ResNet-18, DenseNet-201, and SqueezeNet are four well-known pre-trained strategies used for classification. SqueezeNet's classification accuracy was 98.3%, with 99% precision and 96.7% sensitivity.

Alqudah et al. (2020a) employed ML techniques such as SVM and RF for immediate recognition of COVID-19 signs in people. For feature extraction (FE), they used the CNN model. The derived features are then employed to distinguish COVDI-19 and non-COVID-19 instances using ML techniques. The SVM and RF accuracy rate was 90.5% and 81% correspondingly. By using a chest X-ray, Ghoshal and Tucker (2020) used Bayesian CNN (BCNN) to diagnose the COVID-19. They looked at the importance of BCNN weight loss. It was looked into the relationship between doubt and prognosis accuracy. BCNN had a classification accuracy of 90%. For identifying COVID-19 ailment on chest X-ray photographs, Salman et al. (2020) used a professional CNN. The model's sensitivity and precision were also 100% accurate. Using chest X-ray files, Li and Zhu (2020) created a DCNN for the extraction of imagery features. COVID-Xpert is the name of the built model. To distinguish between COVID-19 and virus-related pneumonia cases, a DenseNet-based transfer learning technique (TLT) was used. Karim et al. (2020) created a DeepCOVIDExplainer to diagnose COVID-19 signs from an affected person's chest X-ray automatically. They employed image recognition and TLT in an ensemble methodology. For COVID-19 cases, the projected technique attained a classification accuracy of 96.12%. For the identification of coronavirus signatures from an affected individual chest X-ray, Apostolopoulos and Mpesiana (2020b) employed a TLT. They used 224 COVID-19 confirmed photographs, 714 virus-related pneumonia images, and 504 natural imageries. The binary class accuracy attained from this method was 98.75%. Ozturk et al. (2020) created a DarkNet Model that uses chest X-ray images to automatically identify an infected human. The COVID-19 binary and multiple class glitches were classified using the DarkNet model. For binary and multiple class complications, the classification accuracies were 98.08% and 87.02%, respectively. Deep CNN was used by Jamil et al. (2020) to identify and diagnose COVID-19 disease by utilizing chest X-rays. They put their approach to the test using 150 verified cases from the Kaggle dataset. Their model yielded a 93% accuracy rate. Using chest X-ray scans, Asif and Wenhui (2020) suggested an automated COVID-19 recognition method. For noticing contagion in an infected individual chest, they employed Inception V3 with TLT. A total of 1341 natural, 1345 virus-related pneumonia, and 864 COVID-19 photos were used to test their approach. This approach had a classification accuracy of 96%. Loey et al. (2020) employed a GAN-built DL approach for COVID-19 recognition in chest X-rays. They looked at three different models: AlexNet, GoogleNet, and ResNet-18. For four and two class instances, the GoogleNet model achieved 80.6% and 99.9% accuracy, respectively. A comparative analysis of 7 distinct DL structures for noticing COVID-19 warning signs in chest X-ray imageries was proposed by Elasnaoui and Chawki (2020). 6087 photographs were used to train these models. The classification accuracy of Inception-ResNetV2 was 92.18%. COVID-19 has over 24,000 academic papers from recognized publications including bioRxiv, arXiv, and medRxiv. About 1500 of these articles have undergone peer review (Loey et al. 2020). Two recent research articles (Farooq and Hafeez 2020; Jamil et al. 2020) discuss the use of AI approaches in COVID-19 identification. We concentrate on COVID-19 detection research that employs deep learning. Wang et al. (2018) used InceptionNet to construct a classifier from 1065 chest CT scan photographs of COVID-19 affected persons. They claim 89.5% accuracy, 88% specificity, and 87% sensitivity. Xu et al. (2020) recorded an accuracy of 86.7% by using a 3D CNN. Chen et al. (2020) use UNet++ to segment contaminated areas in CT scans. TLT and predefined structures have also been used to characterize COVID-19 in CT scans, for example, using DenseNet (Alqudah et al. 2020a; Salman et al. 2020), ResNet (Hemdan et al. 2020; Ozturk et al. 2020), and CNN (Asnaoui et al. 2020b; Gozes et al. 2020; Li et al. 2020c). Traditional FE techniques and traditional ML procedures for classification were similarly employed. Mucahid et al. (2020b) employed the grey-level co-occurrence matrices (GLCM), local directional pattern (LDP), grey-level run-length matrix (GLRLM), and discrete wavelet transform (DWT) FE techniques, as well as an SVM to categorize the mined features. In the best configuration conditions, they claim a 99.68% accuracy. Alqudah et al. (2019) used a variety of ML technique, including SVM and RF, achieved a 95% accuracy rate.

23.3 Material and Method

This section outlined the tools used for this study, such as a chest X-ray, and the methods used, such as the ML technique. The DT and LASSO + DT ML techniques used in this study are listed in this section.

23.3.1 Dataset

The dataset is a chest X-ray dataset downloaded from the Kaggle database repository. The dataset can be found in this link https://www.kaggle.com/bachrr/covid-chest-xray. Figure 23.1 and 23.2 shows sample datasets. Fig. 2.1 depicts a sample of

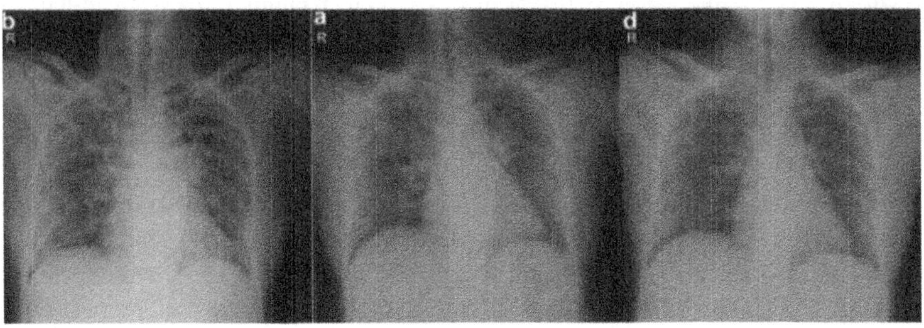

Fig. 23.1 Sample of X-ray image for an infected person

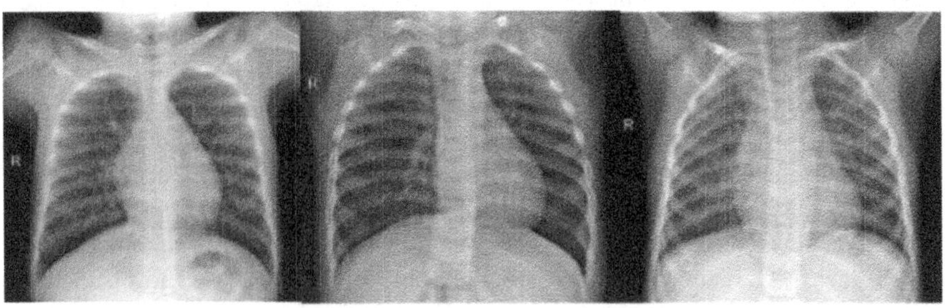

Fig. 23.2 Sample of X-ray of Normal person

Table 23.1 COVID-19 chest X-ray dataset image distribution

Disease	Training	Testing
ARDS	4	1
COVID-19	296	102
Chlamydophila	2	1
E. Coli	4	3
Legionella	2	2
Pneumocystis	15	4
SARS	16	7
Streptococcus	17	6
Klebsiella	1	0

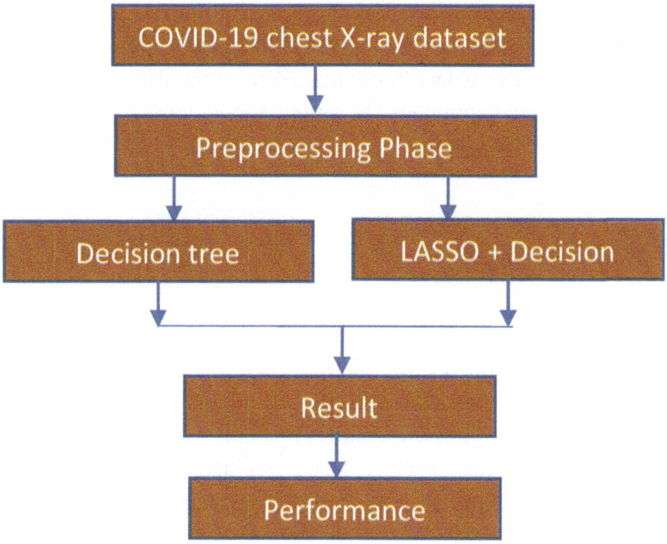

Fig. 23.3 Proposed system block flow diagram

chest X-ray images for people afflicted with COVID-19, while Fig. 2.2 depicts a sample of chest X-ray images for people who do not have COVID-19 or who test negative for the disease. The preparation phase included 357 chest X-ray datasets, while the testing phase included 126 chest X-ray datasets. The method used a 65:35 dataset split, which means that 65% of the dataset was used for preparation and 35% for research. The distribution of chest X-ray dataset images used in the proposed system's execution is seen in Table 23.1.

23.3.2 *Proposed System*

The proposed system was executed and tested using a python programming language. The system employed chest X-ray datasets for the detection of COVID-19 disease. The dataset was passed into the feature selection technique LASSO to determine relevant features from the dataset and also rank the features according to their importance. The DT technique which is the supervised ML classification technique was employed for disease detection. The system implementation is divided into two phases as shown in Fig. 2.3. The figure also shows the projected system flow diagram implemented. In the first phase, the chest X-ray dataset is loaded into the system and the supervised ML classifier DT was implemented on the dataset while in the second phase of the system execution, the chest X-ray dataset is loaded into the system after which the LASSO feature selection technique was employed on it. The relevant features extracted from the dataset are thereby passed into the DT classification technique. The system thereafter classifies the datasets and the result gathered are evaluated using the confusion matrix measures.

23.3.3 *Feature Selection*

Lasso

Feature selection (FS) is a machine learning approach for selecting a subset of specific features for use as explicit variables in models. The FS approach aims to exclude unnecessary or meaningless features that are closely associated with the data without sacrificing too many details. It is also used to make the model more understandable and to improve generalization by reducing variance (Fonti and Belitser 2017). In the feature selection procedure, LASSO is a technique for reducing dimensionality. It presents a method for reducing the number of features in a rational model. It is used to train a model that can recognize the most relevant features in a dataset (Fonti and Belitser 2017) while discarding the ones that are not.

23.3.4 *Classification Technique*

The classification phase was performed after the FS phase. A supervised ML technique which is DT was used to perform the classification.

23.3.4.1 Decision Tree

DT is one of the regularly employed ML approach in classification owing to their rapidity and proficiency. Classification with DT is performed in two phases. In the

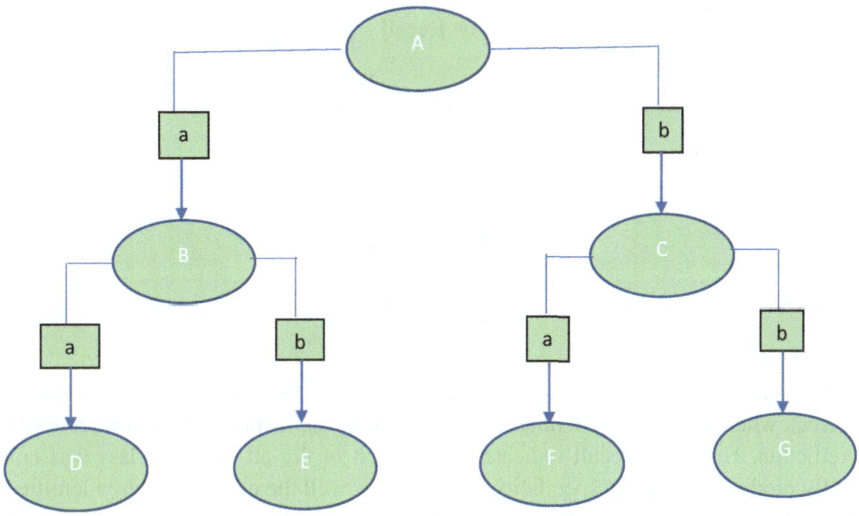

Fig. 23.4 A simple decision tree structure (Kilincer et al. 2021)

first phase, the tree is fashioned, while in the second phase, classification rules are attained from the tree fashioned in the first phase. DT mainly comprises roots, branches, and leaves. The decision-making procedure is from root to leaf and in the decision procedure, branches are followed (Xin et al. 2018; Friedl and Brodley 1997). A simple decision structure is shown in Fig. 23.4.

23.3.5 Performance Analysis

To assess the performance of a classification model in DL, the confusion matrix is most commonly used. The amount of correct and incorrect outcomes in a classification problem was summed and the results were compared with the reference results. Accuracy, Precision, Recall, Specificity, and F1-score are just a few of the most common matrices. True positive (TP), true negative (TN), false positive (FP), and false-negative (FN) statistical indices were calculated to solve the confusion matrix, as shown in Eq. (23.1–23.5).

$$\text{Accuracy} : \frac{TP + TN}{TP + FP + FN + TN} \tag{23.1}$$

$$\text{Precision} : \frac{TP}{TP + FP} \tag{23.2}$$

$$\text{Sensitivity or Recall} : \frac{TP}{TP + FN} \tag{23.3}$$

$$\text{Specificity} : \frac{TN}{TN + FP} \tag{23.4}$$

$$\text{F1} - \text{score} : \frac{2 * Precision * Recall}{Precision + Recall} \tag{23.5}$$

$$\text{TPR} = \frac{TP}{TP + FN} \tag{23.6}$$

$$\text{FPR} = \frac{FP}{FP + TN} \tag{23.7}$$

From the above Eqs. (23.1–23.5), accuracy denotes how often the classification is correct, whereas precision denotes how often the classification will be correct during prediction. However, recall indicates how much of the all-positive class was correctly predicted, whereas specificity assesses how well the negatives were identified. The F1-score is a combination of precision and recall. The number of accurate negative predictions divided by the total number of negative predictions is known as specificity. The True Positive Rate (TPR) is defined as the percentage of correctly identified records over the total number of COVID-19 records, as seen in Eq. 23.6. The TPR stands for Detection Rate. The False Alarm Rate (FAR) is calculated by dividing the number of records wrongly denied by the total number of normal records. Eq. 23.7 defines the FAR evaluation metric. As a result, in the medical industry, the impetus for disease diagnosis is to achieve a higher accuracy and detection rate (DR) with a lower false alarm rate.

23.4 Result and Discussion

A chest X-ray was used to identify COVID-19 in this proposed investigation, which included both compromised and healthy people. The tests were carried out on an Apple Macintosh laptop with a Core i5, 8GB of RAM, and the ML technique DT was implemented in the Python programming language. ML procedures such as DT were used to train and validate the dataset on chest X-ray images. Figure 23.5 shows the confusion matrix plot for DT, while Fig. 23.6 shows the confusion matrix plot for LASSO+DT. Table 23.2 shows the confusion matrix for the two classifiers employed in this study. The results of the two ML classifiers are shown in Table 23.3 and it was shown that the LASSO + DT technique outperformed that of DT alone having a higher percentage value for all the performance metrics used for evaluation in this study having 84.31% accuracy, 87.10% sensitivity, 55.56% specificity, 95.29% precision, and 91.01% f1-score over DT having 81.37%% accuracy, 86.02% sensitivity, 33.33% specificity, 93.02% precision, and 89.39% f1-score. Figure 23.7 shows the graphical representation of the proposed ML classification techniques and it was demonstrated that the proposed LASSO + DT

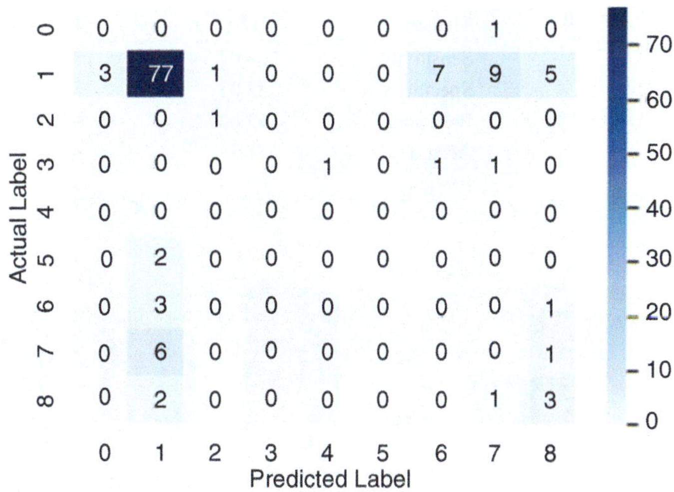

Fig. 23.5 Confusion matrix plot for DT

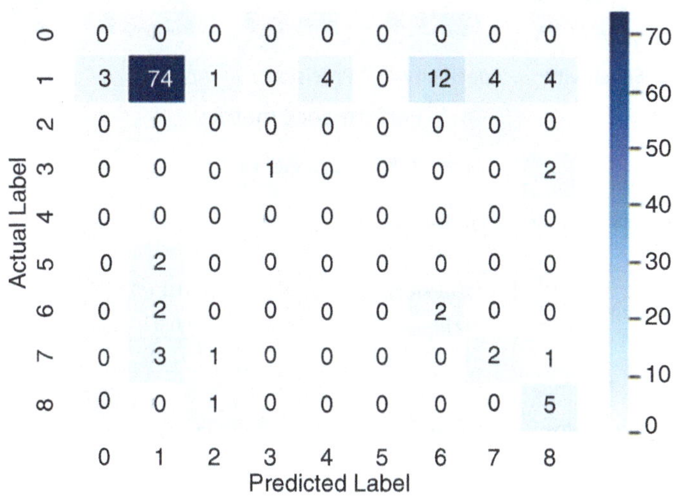

Fig. 23.6 Confusion matrix plot for LASSO+DT

Table 23.2 Confusion matrix for the classifiers

Classifier	TP	TN	FP	FN
DT	80	3	6	13
LASSO + DT	81	5	4	12

technique outperformed the DT classifier in terms of accuracy, sensitivity, specificity, precision, and f1-score.

The DR and FAR for COVID-19 detection from chest X-ray are seen in Table 23.4, and the inspiration for disease detection is to achieve a higher accuracy

Table 23.3 Comparative analysis between the classifiers

Measure	DT (%)	LASSO + DT (%)
Sensitivity	86.02	87.10
Specificity	33.33	55.56
Precision	93.02	95.29
Accuracy	81.37	84.31
F1-score	89.39	91.01

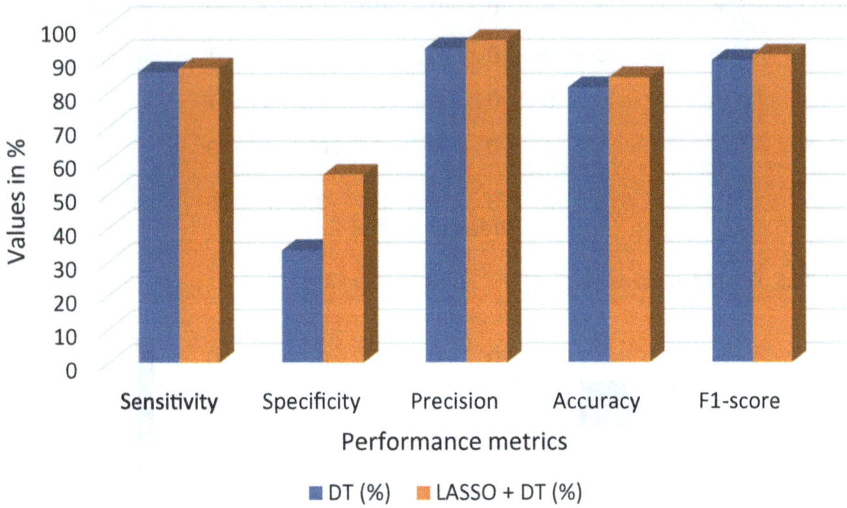

Fig. 23.7 Comparative analysis of the classifier

Table 23.4 DR and FAR of proposed techniques for COVID-19 Detection

Classifiers	DR (%)	FAR
DT	86.02	0.667
LASSO + DT	87.10	0.444

Table 23.5 Comparative analysis with state of the art

Authors	Algorithms	Accuracy (%)
Jiang et al. (2020).	DT	70%
Alqudah et al. (2020b).	RF	81%
Proposed system	**DT** **LASSO + DT**	**81.4%** **84.3%**

and detection rate with a lower FAR. When LASSO + DT classifier is used for COVID-19 disease identification, it has a lower FAR and a higher DR than when the DT classifier is used alone, as seen in the table. This means that LASSO + DT classifier outperformed the DT classifier alone with a lower FAR of 0.444 to 0.667 for DT and a higher DR of 87.10% for LASSO + DT over 86.02% for DT alone. Table 23.5 shows a comparison of the potential classifiers with similar research that has already been done. Both proposed classifiers outperformed the three state-of-the-art experiments in terms of accuracy, sensitivity, memory, and precision, according to the findings.

23.5 Conclusion

It is important to predict COVID-19 affected person early to prevent the disease from conveying to other individuals. In this research, we suggested an FS with ML technique for automated identification of COVID-19 disease by employing chest X-ray imageries collected from COVID-19 patients, normal and virus-related pneumonia. COVID-19 was detected with greater than 65% percent accuracy using the proposed LASSO+DT classification approach. It is widely assumed that the proposed LASSO+DT ML classification achieved the highest accuracy, detection rate and lowest FPR outperformed that of the DT ML implemented alone. This study would assist medical doctors in making decisions in scientific practice. This study explains how the LASSO+DT technique should be employed to detect COVID-19 at the initial phase. COVID-19 has now posed a risk to the global medical infrastructure, with thousands of patients having to die as a result. Death was caused by a lack of oxygen in the lungs, which led to the loss of other organs. Since doctors' time is scarce due to the vast number of affected persons been attended to out-of-hospital or emergency treatment, computer-assisted research will save lives by early screening and appropriate care.

It is assumed that using this computer-assisted diagnostic method would increase the rapidity and precision of identifying COVID 19. This may be very helpful in the case of a contagion, where the disease burden and the demand for prevention action do not meet the resources available.

Deep learning algorithms can be employed for the diagnosis and detection of the COVID-19 disease and other features such as body temperature and knowledge about the prevalence of chronic conditions such as diabetes and heart disease would be merged with the system, in addition, to picture data to make it more efficient and accurate to help healthcare practitioners. Such medical uses for the proposed work include breast cancer screening, tumor detection, and so on.

References

Abbas A, Abdelsamea MM, Gaber MM (2020) Classification of COVID-19 in chest X-ray images using detract deep convolutional neural network. arXiv preprint. arXiv:2003.13815

Alqudah AM, Qazan S, Alquran H et al (2019) COVID-2019 detection using X-ray images and artificial intelligence hybrid systems. Biomedical Signal and Image Analysis and Project; Biomedical Signal and Image Analysis and Machine Learning Lab: Boca Raton, FL, USA

Alqudah AM, Qazan S, Alquran H et al (2020a) COVID-2019 detection using X-ray images and artificial intelligence hybrid systems. https://doi.org/10.5455/jjee.204-1585312246

Alqudah AM, Qazan S, Alquran H et al (2020b) COVID-19 detection from x-ray images using different artificial intelligence hybrid models. Jordan J Electrical Eng 6(6):168

Apostolopoulos ID, Mpesiana TA (2020a) Covid-19: automatic detection from x-ray images utilizing transfer learning with convolutional neural networks. Phys Eng Sci Med 43 (2):635–640

Apostolopoulos ID, Mpesiana TA (2020b) COVID-19: automatic detection from X-ray images utilizing transfer learning with convolutional neural networks. Phys Eng Sci Med 1

Asif S, Wenhui Y (2020) Automatic detection of COVID-19 using X-ray images with deep convolutional neural networks and machine learning. medRxiv

Asnaoui KE, Chawki Y, Idri A (2020a) Automated methods for detection and classification pneumonia based on X-ray images using deep learning. arXiv preprint. arXiv:2003.14363

Asnaoui, KE, Chawki Y, Idri A (2020b) Automated methods for detection and classification pneumonia based on x-ray images using deep learning. arXiv preprint arXiv:2003.14363

Awotunde JB, Jimoh RG, Oladipo ID et al (2021a) Big data and data analytics for an enhanced COVID-19 epidemic management. Stud Sys Decis Control 358:11–29

Awotunde JB, Folorunso SO, Jimoh RG et al (2021b) Application of artificial intelligence for COVID-19 epidemic: an exploratory study, opportunities, challenges, and future prospects. Stud Sys Decis Control 358:47–61

Aydin N, Yurdakul G (2020) Assessing countries' performances against COVID-19 via WSIDEA and machine learning algorithms. Appl Soft Comput 97:106792

Barstugan M, Ozkaya U, Ozturk S (2020a) Coronavirus (covid-19) classification using CT images by machine learning methods. arXiv preprint arXiv:2003.09424

Barstugan M, Ozkaya U, Ozturk S (2020b). Coronavirus (covid-19) classification using CT images by machine learning methods. arXiv preprint arXiv:2003.09424

Bertone E, Luna Juncal MJ, Prado Umeno RK et al (2020) Effectiveness of the early response to COVID-19: data analysis and modeling. Systems 8(2):21

Brunese L, Mercaldo F, Reginelli A et al (2020) Explainable deep learning for pulmonary disease and coronavirus COVID-19 detection from X-rays. Comput Methods Prog Biomed 196:105608

Chen J, Wu L, Zhang J et al (2020) Deep learning-based model for detecting 2019 novel coronavirus pneumonia on high-resolution computed tomography. Sci Rep 10(1):1–11

Chowdhury ME, Rahman T, Khandakar A et al (2020) Can AI help in screening viral and COVID-19 pneumonia? arXiv preprint. arXiv:2003.13145

Daniyal M, Ogundokun RO, Abid K et al (2020) Predictive modeling of COVID-19 death cases in Pakistan. Infect Dis Modelling 5:897–904

El Asnaoui K, Chawki Y (2020) Using X-ray images and deep learning for automated detection of coronavirus disease. J Biomol Struct Dyn:1–12

Elasnaoui K, Chawki Y (2020) Using X-ray images and deep learning for automated detection of coronavirus disease. J Biomol Struct Dyn:1–22

Farooq M, Hafeez A. (2020) COVID-ResNet: a deep learning framework for screening of COVID19 from radiographs. arXiv preprint. arXiv:2003.14395

Folorunso SO, Awotunde JB, Adeboye NO, Matiluko OE (2022) Data classification model for COVID-19 pandemic. Stud Sys Decis Control 378:93–118

Fonti V, Belitser E (2017) Feature selection using a lasso. VU Amsterdam Research Paper in Business Analytics 30:1–25

Friedl MA, Brodley CE (1997) Decision tree classification of land cover from remotely sensed data. Remote Sens Environ 61(3):399–409

Gayathri GV, Satapathy SC (2020) A survey on techniques for prediction of asthma. In smart intelligent computing and applications. Springer, Singapore, pp 751–758

Ghoshal B, Tucker A (2020) Estimating uncertainty and interpretability in deep learning for coronavirus (COVID-19) detection. arXiv preprint. arXiv:2003.10769

Gozes O, Frid-Adar M, Sagie N, et al (2020) Coronavirus detection and analysis on chest ct with deep learning. arXiv preprint arXiv:2004.02640

Guo L, Ren L, Yang S et al (2020) Profiling early humoral response to diagnose novel coronavirus disease (COVID-19). Clin Infect Dis 71(15):778–785

Hemdan EE-D, Shouman MA, Karar ME (2020) COVIDX-net: a framework of deep learning classifiers to diagnose COVID-19 in X-ray images. arXiv preprint. arXiv:2003.11055

Islam MZ, Islam MM, Asraf A (2020) A combined deep CNN-LSTM network for the detection of novel coronavirus (COVID-19) using X-ray images. Informatics in Medicine Unlocked 20: 100412

Jamil M, Hussain I et al (2020) Automatic detection of COVID-19 infection from chest X-ray using deep learning. medRxiv

Jiang X, Coffee M, Bari A et al (2020) Towards an artificial intelligence framework for data-driven prediction of coronavirus clinical severity. Computers, Materials & Continua 63(1):537–551

Karim M, Dohmen T, Rebholz-Schuhmann D et al (2020) Deepcovidexplainer: explainable COVID-19 predictions based on chest X-ray images. arXiv preprint. arXiv:2004.04582

Kilincer IF, Ertam F, Sengur A (2021) Machine learning methods for cybersecurity intrusion detection: datasets and comparative study. Comput Netw 188:107840

Li L, Qin L, Xu Z et al (2020b) Using artificial intelligence to detect COVID-19 and community-acquired pneumonia based on pulmonary CT: evaluation of the diagnostic accuracy. Radiology 296(2):E65–E71

Li L, Qin L, Xu Z et al (2020c) Artificial intelligence distinguishes COVID-19 from community-acquired pneumonia on chest CT. Radiology 200905

Li Y, Yao L, Li J et al (2020) Stability issues of RT-PCR testing of SARS-CoV-2 for hospitalized patients clinically diagnosed with COVID-19. J Med Virol 92(7):903–908

Li X, Zhu D. (2020) COVID-Xpert: an AI-powered population screening of COVID-19 cases using chest radiography images. arXiv preprint. arXiv:2004.03042

Loey M, Smarandache F, M-Khalifa NE (2020) Within the lack of chest COVID-19 X-ray dataset: a novel detection model based on GAN and deep transfer learning. Symmetry 12(4):651

Lukman AF, Rauf RI, Abiodun O et al (2020) COVID-19 prevalence estimation: four most affected African countries. Infect Dis Modelling, 5:827–838

Narin A, Kaya C, Pamuk Z (2020) Automatic detection of coronavirus disease (COVID-19) using X-ray images and deep convolutional neural networks. arXiv preprint. arXiv:2003.10849

Ogundoku RO, Lukman AF, Kibria GBM et al (2020) Predictive modelling of COVID-19 confirmed cases in Nigeria. Infect Dis Modelling 5:543–548

Ozturk T, Talo M, Yildirim EA et al (2020) Automated detection of COVID-19 cases using deep neural networks with X-ray images. Comput Biol Med 103792

Pegoraro F, Santos EAP, Loures EDFR (2020) A hybrid model to support decision-making in emergency department management. Knowl-Based Syst 203:106148

Rubin GD, Ryerson CJ, Haramati LB et al (2020) The role of chest imaging in patient management during the COVID-19 pandemic: a multinational consensus statement from the Fleischner society. Chest 158(1):106–116

Salehi AW, Baglat P, Gupta G (2020) Review on a machine and deep learning models for the detection and prediction of Coronavirus. Materials Today: Proceedings 33:3896–3901

Salman FM, Abu-Naser SS, Alajrami E et al (2020) COVID-19 detection using artificial intelligence. Int J Acad Eng Res 4(3):18–25

Sethy PK, Behera SK (2020) Detection of coronavirus disease (COVID-19) based on deep features. Preprints :2020030300

Singh D, Kumar Vaishali V, Kaur M (2020) Classification of COVID-19 patients from chest CT images using multi-objective differential evolution-based convolutional neural networks. Eur J Clin Microbiol Infect Dis 39:1–11. https://doi.org/10.1007/s10096-020-03901-z

Wang Y, Kang H, Liu X, Tong Z (2020a) A combination of RT-qPCR testing and clinical features for diagnosis of COVID-19 facilitates management of SARS-CoV-2 outbreak. J Med Virol 92 (6):538–539

Wang S, Kang B, Ma J et al (2021) A deep learning algorithm using CT images to screen for Corona virus disease (COVID-19). Eur Radiol:1–9

Wang L, Lin ZQ, Wong A (2020b) Covid-net: a tailored deep convolutional neural network design for detection of covid-19 cases from chest x-ray images. Sci Rep 10(1):1–12

Wang X., Lu Y., Wang Y et al (2018). Diabetic retinopathy stage classification using convolutional neural networks. In 2018 IEEE International Conference on Information Reuse and Integration (IRI):pp. 465–471. IEEE

Wang L, Wong A (2020) COVID-Net: a tailored deep convolutional neural network design for detection of COVID-19 cases from chest X-ray images. arXiv preprint. arXiv:2003.09871

Xin Y, Kong L, Liu Z et al (2018) Machine learning and deep learning methods for cybersecurity. Ieee Access 6:35365–35381

Xu X, Jiang X, Ma C et al (2020) A deep learning system to screen novel coronavirus disease 2019 pneumonia. Engineering 6(10):1122–1129

Zaim S, Chong JH, Sankaranarayanan V, Harky A (2020) COVID-19 and multi-organ response. Curr Probl Cardiol 100618

Part VII
Economy

Chapter 24
Economic Policies for the COVID-19 Pandemic: Lessons from the Great Recession

Guillermo Peña

Abstract This chapter provides possible policies for overcoming the economic effects of the current COVID-19 pandemic on the light of the mistakes and achievements of the Great Recession (2007–2009). It can be useful to policy makers, providing theoretical and empirical evidence of the effects of monetary policy on lending behavior and banking crises in a context of a deep economic crisis where policy makers might avoid its expansion to a banking crunch. The work also focuses on the mechanisms and channels of monetary transmission to financial cracks. The impact of monetary policy on credit raises the likelihood of a banking crisis, both in quantities and in prices. According to our results, there is at least one channel of monetary transmission to prevent crisis, via bank capitalization. Our estimations show the monetary policy of the previous years to the Great Recession implicated an increase in the risk of a banking crunch, with interferences among financial and economic variables, in contrast to the Great Moderation, without any interference. Results show that currently, in the post-COVID era, but also in the Between-Crises period (2010–2020), there have been such interferences again, even more than previously. Finally, policy implications for the current crisis are also provided.

Keywords COVID-19 · Decision-making · Lending · Monetary Policy · Banking Crisis · Great Recession · Great Moderation

24.1 Introduction

Currently, the COVID-19 pandemic has driven the world into a sanitary, economic, and social crisis without precedents. Nowadays it is more important than ever knowing the mechanisms of lending, monetary policy, and banking crises in order to avoid the prolongation of this economic crisis into an additional banking crisis,

G. Peña (✉)
Department of Economic Analysis and IEDIS, University of Zaragoza, Zaragoza, Spain
e-mail: gpena@unizar.es

S. A. Hassan et al. (eds.), *Decision Sciences for COVID-19*, International Series in Operations Research & Management Science 320,
https://doi.org/10.1007/978-3-030-87019-5_24

which could lead to a New Great Depression. The best way, up to our knowledge, to evaluate possible measures for avoiding these fatidic mechanisms is focusing on the previous times to the recent Great Recession (GR) and to learn from that. Learning from past is the main objective of this essay.

This paper deepens on the effects of monetary policy on lending behavior and the risk of a banking crisis (Peña 2015). In Peña (2017), I show that monetary policy measured as lending interest rates impacts on the triggering of banking crises, but this paper is focused mainly in quantity amounts. In fact, last decades Central Banks have employed more quantitative policies (i.e., Quantitative Easing, QE) than interest-policies in a context of low or null interest rates. Lee and Werner (2018) analyze these phenomena and conclude that policy makers might focus on quantity variables after observing that in most cases interest rates follow economic growth.

We aim to answer the following questions. First, does the impact of monetary policy quantities, that is, money supply growth, on credit rise the likelihood of a banking crisis? Second, is there any channel of monetary transmission to prevent crisis? Third, did the monetary policy amounts of the previous years to the Great Recession implicate an increase in the risk of a banking crunch? In this paper we will see that the answer for first and third questions is yes, and for the second one is also yes, via banking capitalization.

As Keynes (1923) pointed out in *A Tract on Monetary Reform*, central banks aim to control the amount of money, especially now that the interest rate is so low. We are uncertain whether the conventionally used monetary policy instrument, the interest rate, is the best variable to explain monetary policy. Money growth could be a better instrument to analyze credit than the interest rate, as Werner (1994) highlights on the importance of quantities.

A dynamic panel data model estimated by the GMM method in two steps is handled to answer the questions. Concretely, it is used a no-balanced panel from 1961 to 2011 for 36 countries, all the EU (27) and OECD countries with the exceptions of Switzerland, Cyprus, Romania, and Malta. One of the main results is the achievement of a better explanation of lending behavior with the use of money supply growth rather than the use of the lending interest rate, the existence of a banking sector lending channel, and the effectiveness of the monetary policy on lending, with a significant impact of a positive monetary shock on bank lending in the Euro area for the previous years of the crisis. Our results show that the lending behavior previous and during the GR, provoked by a lax monetary policy, could encourage a banking crisis. The interest rates, but also the growth of money supply, were responsible of this fact, as will be seen.

In order to answer the above-mentioned questions, we follow the next section with the relevance of the study and previous literature, Section 3 includes the contribution of the theoretical expectations about the topic, in Sect. 4 methods are explained, Section 5 exposes data and the specification of the econometric models, Section 6 exposes the results, in Sect. 7 the results are discussed and possible mechanisms are developed, and Sect. 8 provides evidence and policies for the current COVID-19 crisis. Finally, in Sect. 9 concluding remarks are provided.

24.2 Relevance of the Study and Previous Literature

The relevance of this paper yields on the analysis of the banking channels of monetary policy before the Great Recession, taking into account quantity amounts as recently did other authors (Lee and Werner 2018), in order to better understand possible roots of the previous crisis, the Global Financial Crisis (GFC) of 2007–2009. The paper also takes into account other papers as Peña (2020a, b) and considers the possible interferences among financial, monetary, and economic variables in order to check the possible good economic performance during periods as the Great Moderation (GM) or others, with the difference of taking monetary quantities into account in a wide sample. Additionally, the paper performs the same last exercise but with current monthly data in order to include recent periods as the COVID times and learn lessons from the past for this new economic crisis provoked by the pandemic, with no immediate precedent in the human being's history.

As Davis and Kahn (2008) state, Great Moderation implied a significant decrease in the volatility of the main economic variables, mainly inflation, inflation volatility, output, and employment in aggregate level and across most manufacturing sectors and categories of expenditure. Many studies claim the correct monetary policy of those years was responsible of this low volatility and uncertainty (Clarida, Cali, and Gertler 2000; Taylor 2012). Other authors suggested financial innovation as one of the main causes, while others claimed the possible "good luck" due to small exogenous shocks (Stock and Watson 2003; Benati 2008), or the "ignorance," instead of the good luck (Giannone et al. 2008). Nonetheless, reality is complex and each view could have part of the truth (Bernanke 2004). Maybe performing monetary policy is easier when exogenous shocks are smoother (Davis and Kahn 2008). These authors also highlight that the lower volatility mostly took place in aggregate output rather than in other variables, and mainly in durable goods. Davis and Kahn's (2008) argue that it is not in contradiction with the hypothesis of the correctly applied monetary policy during the Great Moderation because higher volatility in durable goods could also be influenced, as Peña (2020a) suggests, by non-ruled or discretionary monetary policies as before the Great Recession.

During the Global Financial Crisis (GFC) and after that, during the GR, there was a global deteriorating economy and interest rates in most advanced countries were reduced to virtually zero. Many central banks turned to unconventional monetary policies (e.g., the QE), where central banks set strategies to buy securities and debt, expanding their balance sheet. This led to, for instance, a five-fold expansion in the balance sheet of the federal reserve (Fed) and policy statements began to use explicit references to the trend of the official interest rates, which started to be known as "forward guidance," as Kuttner (2018) shows. This author found that these policies were effective in lowering long-term interest rates and having repercussion on firms and financial institutions as intermediaries. Nonetheless, this author sustains that benefits of unconventional policy may outweigh the costs.

The Great Moderation was a period of certainty and monetary rules were followed, while the "Ad Hoc Era" is the current period of uncertainty and discretional monetary policy rules. These are the "two tales" of part of the "two eras" (Taylor 2012), but what happened in the "middle point?" What did this change in monetary policy lead to provoke such a huge crisis? The answer may be formulated by stating that the main initial impact of monetary policy was the lending behavior before the GR, at the beginning of the "Ad Hoc Era" (2003-present, after the "Ruled-Based Era," 1985–2003), and may due to incorrect monetary policies. The beginning of this "Ad Hoc Era" is known in this paper as the "laxity period" (2002–2006), which is the "middle point" between the two eras. In this period, there was a lending boom may be derived from lax monetary policy that does not follow monetary rules but communicative or discretionary ones. There is a debate about its causes, regarding financial innovation, incorrect monetary policy, etc. This section does not focus on the origins of this lending boom (there may be many possible causes apart from monetary policy[1]), but in the effects of this lending boom on the likelihood of a banking crisis.

Some authors consider the Great Moderation (GM) ended on 2007, with the GR starting as a consequence of the Economic Policy Uncertainty (EPU, Prüsser and Schlösser 2020). Therefore, these authors clearly distinguish between three main Eras: the Great Inflation (1965–82), the GM (1982–2007), and the GR (2007-present). Nonetheless, uncertainty was not the only driver of the crisis. For instance, Bean (2010) points the optimistic assessment of risk made in the GM, which jointly with low interest rates, loose monetary policy and high Asian saving rates encouraged the construction of an excessive leverage in the banking system that could lead to the GR. According with our proposal, this mispricing of risk would correspond to the lax period.

The study on the credit channel of monetary policy is not new in the literature. The importance of monetary policy on loans is related with some economy- and bank-specific issues. Among others, the degree of development of a country, with a higher effectivity on developed countries, overall after financial crises (Sanfilippo-Azofra et al. 2018), liquidity, with opposite findings: higher liquidity makes less effective monetary policy (Matousek and Sarantis 2009) or more sensitive (Sanfilippo-Azofra et al. 2018), banking competition (Fungacova et al. 2014), size of the bank (Altunbas et al. 2010) or capitalization, where higher capitalized banks are less influenced by monetary policy (Bubeck et al. 2020).

Previous literature related to this paper also includes Peña (2017), where there is an exercise of estimating the impact of monetary policy on lending and banking crises, and its transmitting mechanisms. The main difference with the models of the present paper is that the former only considers monetary prices, that is, interests, instead of quantity amounts. Additionally, this paper also takes Granger-causality into account, in contrast to the other one. Literature close to these tests is Lee and

[1]There is, at least, one causal-relationship corroborated in Peña (2020a) between the growth of loans, interest rates, and GDP growth, generating so a vicious circle.

Werner (2018) consider a full sample where they focus on the impact of the monetary amounts on national income, but the present paper divides between GM and lax periods. Peña (2020a, b) develops an indicator of good economic performance given by the economies whether the Granger-causality among financial, monetary, and economic variables is not significant, so it can be deducted that the variables can reach their natural trends, behaving efficiently, if there are no interferences among these variables. Nonetheless, the former author only considers monetary prices (interests), not amounts. This paper also takes monetary amounts for this exercise.

Additionally, the present paper applies these tests also for the COVID times for the case of Spain, establishing a division between different economic periods. So, it is checked whether the mistakes previous to the GR were repeated again for the previous years to the COVID crisis or whether these mistakes and incorrect policies are still in force. So, summing up, in addition to providing a brief theoretical framework to the empirical analysis, this paper proposes applying tests and empirical analysis from other authors to quantities environment as the current one where the following issues are studied. First, banking channels with money amounts are analyzed and compared with prices models, second, the correct application of monetary policies before the GR is checked dividing into subsamples, and finally, this analysis is extended to recent COVID times for Spain.

24.3 What Lessons Can Be Learned from the Lending Behavior before the Great Recession? A Theoretical Approach

Next, a short theoretical explanation of the unchain of the GFC (2007–2009) by the lending behavior in the previous moments to the GR is proposed, starting from the credit expansion and following on how this could impact on a banking crisis.

Lent credit at year t can be considered to depend on variables of supply and demand, as the following:

$$C_t = C(M, i, Y \ldots). \tag{24.1}$$

It depends from supply and demand variables as the monetary supply (M), the official interest rate (r), the GDP (Y). In addition, profits from banks (Peña 2020b) can be expressed as, in the short-term (st):

$$\Pi_{st} = r \cdot C_{st} - R \cdot D - II - wL, \tag{24.2}$$

where "*II*" represents the intermediary inputs, "r" the lending interest rate, "R" is the deposit interest rate, and D reflects the deposits. So, even taking into account that an increase on C would initially rise, in the short-term, the profits of the banking sector in the same quantity as the lending interest rate $\partial\Pi_{st}/\partial C_{st} = r > 0$ in the economy, nonetheless, in the mid-term (*mt*) the loan defaults would be increased due to the excess of credit that the society cannot afford. Hence, they would have to provide provisions (*P*), leading to the following profits:

$$\Pi_{mt} = r \cdot C_{mt} - R \cdot D - II - wL - P_{mt}(C_{st}). \tag{24.3}$$

These provisions are a function that positively depends on the credits of some years ago, in the short-term, because this increases the defaults, reducing the banking profits in the mid-term. According to the previous equation, this relation could be expressed as follows:

$$\frac{\partial P_{mt}}{\partial C_{st}} = P_{mt}'(C_{st}) > 0; \frac{\partial \Pi_{mt}}{\partial P_{mt}} < 0 \Rightarrow \frac{\partial \Pi_{mt}}{\partial C_{st}} < 0, \tag{24.4}$$

where credit increases provisions in the mid-term and this reduces profits in the mid-term. Taking into account, as Peña (2020b), that the probability of a bank crisis (Φ) negatively depends on, among others, the profits of the banking sector, the next result would be obtained:

$$\frac{\partial \Pi_{mt}}{\partial C_{st}} < 0; \frac{\partial \Phi}{\partial \Pi} < 0 \Rightarrow \frac{\partial \Phi_{mt}}{\partial C_{st}} > 0. \tag{24.5}$$

So, a decrease on banking profits and a raise on the probability of a banking crisis arise after an increase of credit in the mid-term. These theoretical expectations will be confirmed with data in the empirical section. This is also confirmed by Haan et al. (2020), who develop theoretical models in which while banks are in a boom period, they accelerate loan growth by shifting portfolios to high profit assets from more liquid assets with less returns in that moment (such as public bonds). So, banks increase profits because the return now is higher during the boom period. Nonetheless, losses are expectable after the crash.

There is another way to see this, by considering that there is a banking crisis in the mid-term if the "pure" interest, i.e. the interest without fees, related to the intertemporal interest rate, is not the optimal, expressed mathematically:

$$\varepsilon_0' \neq \varepsilon_0^* \Rightarrow \Phi > 0, \tag{24.6}$$

considering by optimal the specification given in López-Laborda and Peña (2018), that we will see after adapting previous equations. After considering again (24.2), but rearranged for differentiating between loan and deposit services in the mid-term, by considering no defaults will be:

$$\begin{array}{l} \Pi_{mt}^{C} = r \cdot C_{mt} - \varepsilon_0 B - II_{mt}^{C} - wL_{mt}^{C} \\ \Pi_{mt}^{D} = \varepsilon_0 B - R \cdot D - II_{mt}^{D} - wL_{mt}^{D} \end{array}, \tag{24.7}$$

where C and D indicate the value added of credit and deposits, respectively, ε_0 the pure interest rate, and B the bonds amount. The previous equation reflects the profits obtained by considering the profits, input costs, and wages of credits and deposits, taking them separately into account. According to the previous authors, (24.7) can be presented differently as the financial value added by direct (left side) and indirect way (right):

$$\begin{array}{l} VA_{mt}^{C} : \rho \cdot r \cdot C_{mt} = r \cdot C_{mt} - \varepsilon_0 B \\ VA_{mt}^{D} : \rho \cdot R \cdot D = \varepsilon_0 B - R \cdot D \end{array}, \tag{24.8}$$

according to this, the optimal values for the marginal productivity of financial services ρ and for the pure interest rate ε_0 can be obtained:

$$\varepsilon_0^* = \frac{2 \cdot r \cdot C_{mt} \cdot R \cdot D}{(r \cdot C_{mt} + R \cdot D)B}, \rho^* = \frac{r \cdot C_{mt} - R \cdot D}{r \cdot C_{mt} + R \cdot D}. \tag{24.9}$$

So, considering now a mid-term case with default from the short-term, the mid-term provisions would affect to the pure interest rate in this way, ceteris paribus:

$$\begin{array}{l} VA_{mt}^{C} : \rho \cdot r \cdot C_{mt} = r \cdot C_{mt} - \varepsilon_0' B - \gamma^{C} P_{mt}(C_{st}) \\ VA_{mt}^{D} : \rho \cdot R \cdot D = \varepsilon_0' B - R \cdot D - \gamma^{D} P_{mt}(C_{st}) \end{array} \tag{24.10}$$

where γ for C and D is the share of provisions allocated to the value added of loans and deposits, respectively. Finally, solving the pure interest would lead to, where $P = P_{mt}(C_{st})$:

$$\begin{array}{l} \varepsilon_0' = \dfrac{2 \cdot r \cdot C_{mt} \cdot R \cdot D + P(r \cdot C_{mt} - R \cdot D)}{(r \cdot C_{mt} + R \cdot D)B} \neq \varepsilon_0^* = \dfrac{2 \cdot r \cdot C_{mt} \cdot R \cdot D}{(r \cdot C_{mt} + R \cdot D)B}, \\ \Rightarrow \Phi > 0 \end{array} \tag{24.11}$$

which is higher than the optimal and would encourage banking crises according to (6) via distortions in intertemporal decisions. So, there are at least two ways, via lower profits and via intertemporal distortions via higher pure interest rate for raising loans could lead to banking crises through defaults and provisions.

According to Laubach and Williams (2003, p. 1), the distance with the natural (pure) interest rate marks whether the economy stays in an expansionary or contractionary cycle:

> Since Wicksell (1936), the natural rate of interest—the real short-term interest rate consistent with output equaling its natural rate and constant inflation—has played a central role in macroeconomic and monetary theory. The natural or "equilibrium" real interest rate provides

a benchmark for measuring the stance of monetary policy, with policy expansionary (contractionary) if the short-term real interest rate lies below (above) the natural rate.

This can be the theoretical support for considering that loan booms lead to banking crises, after the crash, effective interest rates are higher than the pure rate, as in Eq. (24.11), where the provisions clearly make the effective interest rates higher than the optimal because loan interest is usually higher than deposits. Taylor (1993) proposed a simple rule for monetary policy according to which the real interest rate has to exceed the natural one when the inflation exceeds the proposed objective and vice versa, ceteris paribus. Peña (2017) empirically obtained evidence that it was good for the economy decreasing reference interest rates, related with the pure or optimal interest rates, when there is a banking crisis, and rising them when there are expansionary periods.

24.4 Methodology

The methods of this paper consist on two main different procedures for two kinds of models, following Peña (2016, 2017) for the first methodology and Peña (2020a, b) for the second one. The first one develops a methodology that consists on estimating panel data models. First, it is evaluated the impact of the two main instruments of monetary policy on loans, which are the lending interest rates as a proxy of the reference rates (Peña 2017), and monetary amounts (M2) in order to take into account the supply quantities. The two models are estimated in OLS in order to check which one benefits from a higher R^2, and the monetary instrument corresponding to the model with the highest R^2 will be the one chosen as a target variable for the rest of models and estimates. The second step is to consider the impact on loan supply of the control variables (economy- and banking-specific terms) and the target variables (the instrument of monetary policy, and its interaction with periodical variables and banking-(sector-)specific variables, these last variables are involved in the channel of monetary policy for the banks). The final step in this first type of models is the estimation of the impact of loans and the rest of control and target variables on the likelihood of banking crises, again by only considering quantities (M2) as target variable. These final models are estimated by logit panel data models.

The methodology is the same for the two last kinds of problems: first, a complete model with all the control and target variables where no evidence of autocorrelation is estimated, being variance robust as will be explained below. Second, the econometric properties are analyzed. If they can be improved, then additional models are estimated by consecutively eliminating from the estimates the non-significant variables (with the 10% p-value as a threshold). First, the variables with the least significant coefficients are suppressed. At the end, an econometric model with good econometric properties is reached, and then the procedure stops.

The second kind of models uses the same panel data, in addition to temporal monthly data from Spain, but in a time series analysis, so an estimation is performed

Table 24.1 Simple regression explaining lending

Dependent variable: *lnloan*	*Coefficient*	*p-value*	*Coefficient*	*p-value*
M2	−0.0102***	0		
interest			−0.0078***	0
constant	25.3087***	0	25.6847***	0
R-squared	0.0328		0.0283	

per each country, concretely, a Vector Auto Regressive (VAR) estimate, in order to, with this result, apply additionally the Granger-causality test, which considers as null hypothesis the non-Granger-causality (GC) between the two variables taken into account. So, these tests are performed among the selected variables and after that it is checked whether there is GC or whether there is any interference by checking, first, if the average percentage of countries with rejections in at least one of the two directional GCs is higher or equal to 80%, and second, if the lowest p-values between the two directional GCs for each country are lower than 10% on average.

24.5 Econometric Model and Data

Data are available in World Bank data. To use a good monetary policy indicator, we compare the most explanatory variable with the analyzed indicators (lending interest rate, *interest,*[2] and money supply (M2) growth, *M2*; both variables are available in World Bank data). We chose the indicator of monetary policy with the highest R-squared obtained in a simple regression between the indicator and the lending growth rate. In Table 24.1 we can see that the chosen variable is *M2* because it is the highest explanatory variable (an R-squared of 0.0328 compared with an R-squared of 0.0283 of the variable *interest*). Therefore, we can answer the first question in the introduction: in this case, money supply growth is a better explanatory variable of lending growth.

The specification of the econometric model follows the GMM system developed by Arellano and Bover (1995) and Blundell and Bond (1998) as in Peña (2017), but also considers money supply quantities as an indicator of monetary policy in addition to prices measured by the lending interest rate as an indicator of the official interest rate:

$$\Delta \ln\left(L_{i,t}\right) = \gamma * \Delta \ln\left(L_{i,t-j}\right) + \beta_{MP} MP_t + \beta_X X_{i,t} + \beta_{MPX} MP_t X_{i,t} + \beta_Y Y_{i,t} + \beta_{MPZ} MP_t Z_{i,t} + \varepsilon_{i,t} \tag{24.12}$$

with $i = 1, \ldots,$ N the countries and $t = 1, \ldots, T$ the years, $L_{i,\ t}$ is the domestic credit, MP_t is the indicator of monetary policy, alternatively the growth rate of the monetary

[2]This is a proxy of the official interest rate as in Peña (2017).

Table 24.2 Countries and years in the sample

Years: 52	*Countries: 36*
1961–2011	Australia, Austria, Belgium, Bulgaria, Canada, Chile, Czech Republic, Denmark, Estonia, Finland, France, Germany, Greece, Hungary, Ireland, Iceland, Israel, Italy, Japan, Korea, Luxembourg, Latvia, Mexico, Lithuania, Netherlands, New Zealand, Norway, Poland, Portugal, Slovak Republic, Slovenia, Spain, Sweden, Turkey, United Kingdom, United States.

supply or the lending interest rate, $X_{i,\ t}$ denotes the bank-specific characteristics of the country, and $Y_{i,\ t}$ reflects the economy-specific characteristics of the country (GDP per capita growth, value of the stock exchange, inflation, and investment). $Z_{i,\ t}$ are the temporal or regional dummy variables.

Following Peña (2017), the presence of a bank lending channel should be seen in a significant coefficient for the interaction of bank-specific characteristics (liquidity, capitalization, or bank market power), with shifts in money supply performed by the Central Bank. The variable *liquid* measures the proportion of liquid bank reserves in total bank assets; *capital,* capitalization, is the ratio of bank capital and reserves to total bank assets; and the Lerner index, *lerner,* is a measurement of banking competition comparing output pricing and marginal costs (that is, the mark-up) in the banking market (a rise in the Lerner index shows a deterioration in the competitive conduct of financial intermediaries). These variables are lagged one year to avoid endogeneity and simultaneity problems. This allows us to include cyclical movements, being useful for controlling effects of demand and for isolating the economic policies. In addition, some dummies regarding time and region are included to handle potential impacts of the banking crisis. Some interferences are included between the indicator of monetary policy, *euroarea* (countries where the Euro is the common currency), *GM* (period of the Great Moderation: 1986–2001), *GR* (period of the Great Recession: 2007–2009), or *laxity* (years when a lax monetary policy was established: 2002–2006).

The second-type models follow Nocetti (2006). This author claims the relevance of the use of logit models to forecast crises, in fact, Beutel et al. (2019) consider these models as good predictors of banking crisis as the GFC. The variable *crisis* is our target variable to explain. This dependent variable takes the value 1 whether the country suffers a banking crisis and 0 otherwise. The model is estimated following a panel logit model with population-averaged effects, as Büyükkarabacak and Valev (2010).[3]

$$\text{logit}\,\Pr(Y_{it} = 1|X_{it}) = \beta^* X_{it}. \tag{24.13}$$

The data used are shown in Table 24.2, including 36 countries from the OECD and the EU (27) for 1961–2011. We handle non-balanced panel data.

[3]These authors refer to Zeger et al. (1988), Neuhuas et al. (1991), and Wooldridge (2002) for a detailed description of the population-averaged model.

Table 24.3 Summary statistics

Variable	Observations	Mean	Standard Deviation	Minimum	Maximum
lnloan	1547	25.0057	2.365934	18.61231	31.25849
crisis	1836	0.082789	0.275638	0	1
gdppc	1587	2.65145	3.62974	−31.17752	17.55749
lnc2	1485	24.92688	2.34787	18.61231	31.19756
surplus	575	−1.54559	4.35130	−29.42016	20.00958
infl	1518	11.16206	39.33546	−4.47994	1058.37400
cons	1486	57.85461	7.840445	31.41163	95.25039
M2	1206	20.17612	43.05358	−28.6298	829.307
stock	832	54.8372	46.55807	0.0198936	479.8116
inv	1527	23.59319	5.197639	2.388235	53.31139
liquid	425	7.83940	9.14341	0.22961	60.94282
cap	415	7.007952	2.442035	2.7	17.4
lerner	535	0.18510	0.11775	−1.60869	0.50311

The interest variables are *M*, the growth rate of the money supply (M2) lagged two years; and *interest*, the lending interest rate as a proxy of the official interest rate. Logarithm of domestic credit and the growth rate of M2 are lagged two periods to minimize simultaneity problems (Büyükkarabacak & Valev 2010). In Table 24.3 are shown the variable characteristics. Data are available in World Bank data, with the exceptions of the temporal and geographical dummies, which are by the author.

The explanatory variables used in this paper are the main determinants of banking crises that do not present multicollinearity problems, in addition to our objective variables. The most relevant determinants of banking crises are included as explanatory variables. In some models we also capture the different channels of monetary transmission to output. The main mechanisms to be tested are the interactions between the monetary policy indicator with credit, liquidity, and bank competition. In model 3 we also include the interaction of a dummy which reflects the years of the Great Recession (2007–2009) with the monetary policy of previous years. This variable is included due to monetary shocks affect output with delay (Douch 2010) and we see the monetary behavior in those years.

The following variables are included as determinants of financial crises: the growth rate of Gross Domestic Product (GDP) per capita, *gdppc*; the logarithm of domestic credit lagged two periods, *lnc2*; the public cash surplus or deficit, *surplus*; inflation, *infl*, measured as the annual growth rate of the cost of a shopping basket for a customer; household final consumption expenditure as share of GDP is measured by variable *cons*; the growth rate of the money supply (M2) is *M*; *liquid* measures the proportion of liquid bank reserves in total bank assets; the Lerner index, *lerner,* a measurement of banking competition that compares output pricing and marginal costs (that is, the mark-up) in the banking market: a raise in the Lerner index shows a deterioration in the competitive conduct of financial intermediaries. The estimation also includes some economy-specific characteristics of the country (the value of the stock exchange, *stock*; and *investment*, which is gross capital growth, obtained by the sum of all the increases in fixed assets and the net variation in stocks).

The expected sign of *cons* is negative, because a crisis reduces the aggregated demand, and therefore, the consumption. If the variable was lagged, then we would expect a positive sign as Hardy and Pazarbasioglu (1999) suggests. We also expect a negative effect of *M2*, because in most of the studied period central banks applied a rule-based monetary policy that avoided financial crises by rising the interest rate (Olson and Young 2015). In spite of that, Österholm (2005) only found compatibility between Taylor rule and US data from 1960 to 1979.

The correlation matrix among independent variables is shown in Table 24.4. All correlations are lower than 0.5, therefore we can state there are not multicollinearity problems. If it was not the case, the variables with high correlation had not been included in the models and hence, in the matrix.

24.6 Results

Table 24.5 shows the results of the paper, where they can be seen the effects of monetary policy on the risk of a banking crisis and its mechanisms.

The results highlight the effects of the variables on bank lending. Model (1b) is the definitive model; it was estimated following the GMM System method in two steps developed by Arellano and Bover (1995) and Blundell and Bond (1998). The model they suggested ensures efficiency and consistency, and it is robust to heteroscedasticity. Absence of serial correlation of order two and the validation of the used instruments are tested with the Arellano and Bond test (A-B) and the Sargan test, respectively (Peña 2015).

Model 1a is the complete dynamic model for money supply, estimated by the GMM System with the WC-robust estimators of Windmeijer (2005), which is a bias-corrected robust estimator for two-step VCEs (variance–covariance matrix estimators) from GMM estimators. After eliminating the non-significative variables, we have models 1b and 1c, considering money supply or the interest rate as monetary policy variables in the starting initial models.[4]

The three models estimate the impact on lending growth and have good econometric properties. The first model includes *M2* without its interactions with other variables, and we can see a non-significant negative effect of its coefficient on the dependent variable. The second model includes the interactions of the monetary policy indicator with the logarithm of credit lagged two years, and the liquidity, capitalization, and bank competition indicator variables. We observe that including these variables M2 turns out to have a significant effect, and the interaction of M2 with lending also has a significant impact on banking crises, being a positive sign. Finally, in the third model we capture the interaction between years 2007 and 2009

[4]The initial starting model with interest rate is not shown in Table 25.5.

Table 24.4 Correlation matrix of independent variables

	gdp	*M*	*stock*	*liquid*	*capital*	*lerner*	*interest*	*investment*
gdp	1.000							
M	0.327	1.000						
Stock	0.032	−0.016	1.000					
liquid	0.185	−0.036	−0.491	1.000				
capital	0.077	0.073	−0.354	0.232	1.000			
lerner	0.087	0.118	−0.020	−0.008	0.023	1.000		
interest	−0.059	0.075	−0.171	0.192	0.306	0.062	1.000	
investment	0.534	0.279	−0.142	0.248	0.093	0.172	0.145	1.000
euroarea	−0.139	−0.131	0.154	−0.376	−0.403	−0.197	−0.270	−0.299
GM	0.072	0.229	0.086	−0.076	−0.025	−0.034	0.167	−0.084
laxity	0.311	0.068	0.050	0.072	−0.077	−0.027	−0.134	0.062
GR	−0.371	−0.158	−0.082	−0.038	−0.033	0.010	0.094	0.131
gdppc	0.979	0.319	−0.100	0.236	0.153	0.095	−0.040	0.536
surplus	0.431	0.345	0.326	−0.187	−0.179	0.037	0.076	0.417
inflation	0.135	0.112	−0.309	0.244	0.324	0.126	0.565	0.431
consumption	0.062	0.025	−0.140	−0.008	0.497	0.135	0.183	0.072
	euroarea	*GM*	*laxity*	*GR*	*gdppc*	*surplus*	*inflation*	*consumption*
euroarea	1.000							
GM	0.170	1.000						
laxity	−0.013	−0.434	1.000					
GR	−0.031	−0.235	−0.565	1.000				
gdppc	−0.192	0.057	0.293	−0.359	1.000			
surplus	0.060	0.185	0.133	−0.092	0.377	1.000		
inflation	−0.242	0.094	−0.138	0.182	0.170	0.127	1.000	
consumption	−0.319	−0.047	−0.010	0.014	0.142	−0.216	0.167	1.000

Table 24.5 Estimated models

Dependent variable: *dlnloan*	**Initial model (1a)**		**Definitive model (1b)**		**Check model (1c)**	
	Coefficient	*p-value*	*Coefficient*	*p-value*	*Coefficient*	*p-value*
dlnloan (t-1)	0.318**	0.013	0.398***	0.000	0.140***	0.000
gdppc	0.017***	0.001	0.021***	0.000	0.020***	0.000
M (t-1)	−0.008**	0.028	−0.006**	0.001		
interest					−0.018***	0.000
stock	0.000	0.188	0.001***	0.034		
liquid (t-1)	0.003	0.108	0.001	0.698		
capital (t-1)	0.002	0.875				
lerner (t-1)	−0.037	0.659				
M (t-1) liquid (t-1)*	0.000	0.438				
M (t-1) capital (t-1)*	0.001*	0.088	0.001**	0.024		
M (t-1) lerner (t-1)*	0.010	0.238				
interest liquid (t-1)*					0.001***	0.000
investment	0.008	0.234				
*euroarea*M (t-1)*GM*	−0.005	0.235				
*M (t-1)*laxity*	0.001	0.606				
interest laxity*					0.007**	0.011
*euroarea*M (t-1)*GR*	0.003*	0.064	0.003**	0.031		
*euroarea*interest*GR*					0.020***	0.004
constant	−0.177	0.328	−0.023	0.478	0.084***	0.000
No obs	315		345		248	
No inst	37		30		39	
Sargan test (second step; p-value)	0.476		0.128		0.898	
A-B test MA (1), MA (2) (p-value)	0.000	0.357	0.000	0.468	0.007	0.913

* Significance level of 10%, ** significance level of 5%, *** significance level of 1%

and the money supply growth lagged two years. We obtain the same significant coefficients of model 1b but in addition to the coefficients of the interactions of *M2* with *capital* and *GR*. Both coefficients are with a positive sign.

The main result that derives from the previous table is the presence of a capitalization monetary transmission channel with the quantity instrument (M), meaning that higher capitalizes are less affected by monetary policy, because of the opposite sign of its coefficient respect to the one of M, which is negative. This result is consistent with Bubeck et al. (2020). The same happens with liquidity as a channel of the transmission of monetary policy, with higher liquidity the effect of monetary policy is lower, as in Matousek and Sarantis (2009) or Peña (2017) but the opposite than Sanfilippo-Azofra et al. (2018). Furthermore, there is a countercyclical effect of monetary policy in the lax period (1c) and in the Great Recession (1a-c), which confirms the existence of differences in the periods during and before the GR respect

Table 24.6 Policy measures

Dependent variable: *crisis*	**Initial model (2a)**		**Initial model (2b)**		**Check model (2c)**	
	Coefficient	*p-value*	*Coefficient*	*p-value*	*Coefficient*	*p-value*
gdppc	−0.343***	0.000	−0.236*	0.082	−0.213***	0.009
dlnloan (t-2)	0.239**	0.044	0.232	0.328		
surplus	−0.460***	0.000	−0.387*	0.079	−0.344**	0.034
inflation	0.593***	0.000	0.310**	0.037	0.319**	0.018
consumption	−0.179***	0.007	−0.170	0.155	−0.051**	0.011
M (t-2)	−1.440***	0.001				
interest			−2.224**	0.025	−2.153**	0.005
M (t-2) dlnloan (t-2)*	0.047***	0.001				
M (t-2) liquid*	−0.003	0.295				
M (t-2) capital*	0.005	0.565				
M (t-2) lerner*	0.545**	0.020				
interest lnloan (t-2)*			0.044*	0.078	0.057**	0.020
interest liquid*			0.001	0.922		
*interest*capital*			0.124***	0.001	0.064***	0.000
*interest*lerner*			−0.163	0.337		
*Euroarea*M (t-2)*GR*	0.112***	0.009				
*Euroarea*interest*GR*			0.720***	0.001	0.587***	0.000
No obs	278		193		258	
No groups	31		27		32	
Wald (p-value)	0		0		0	

* Significance level of 10%, ** significance level of 5%, *** significance level of 1%

to the GM. Finally, other control variables as *stock* or *gdppc* are also significative at least sometimes.

It is worth to avoid the typical argue about whether the monetary supply is endogenous or exogenous and focus more on data, because, ultimately, all economic variables can be considered to be influenced among themselves sometimes. Nonetheless, an endogeneity test is performed with the most problematic variable, money supply, in the logit model, because GMM system models are to avoid endogeneity. This is performed in order to ensure the results, and the test is passed.[5] In contrast, endogeneity in GMM System models is not a problem because it is removed by internal transformation of the data (Ullah et al. 2018), that makes the model more efficient but with less observations, using in this paper a two-step procedure in order to reduce this last problem. Table 24.6 shows the same models, with the same methodology, but where crisis as dependent variable and the econometric method

[5] In order to test the endogeneity of M in the logit model, a probit model with instrumental variables has been developed in order to check the exogeneity of the full sample respect to the reduced sample with the instrumental variable as null hypothesis. The chosen instrumental variable for *M* has been the variable *interest*, and the null hypothesis was accepted with a Wald test of p-value higher than 0.6.

is a populated-averaged logit model as stated in the previous section. In this case, the contribution is the model 2a while models 2b and 2c are shown from the Peña's (2017) paper in order to check the previous result.

In the short-term and according to Eq. (24.1), what mainly influences on credit, and hence in the banking profits, are monetary policy and GDP. They can positively impact by the reduction of the interest rates and of the money supply M2 and the increase of the per capita GDP. This is shown by the robustness of the significative and negative impact of the coefficients of the variables M (t-1) and interest and positive of gdppc in Table 24.5. On the other hand, in the long term the interaction of the increase of the credit with the accumulation of reductions of the money supply or the interest rates, and the separately effects leads to an increase of the probability of banking crises. Furthermore, this is shown by the positive and significative coefficients of the variables *M (t-2)* lnloan (t-2)* y *lnloan (t-2)*, in addition to the negative sign of the coefficients M(t-2) and gdppc.

The robustness of the coefficients is tested by estimating the same model but changing the variable M2 with interest and applying the same elimination process from model 2a to model 2b. As we can see in Table 24.6, with this monetary policy indicator we can see that the aggregated banking sector lending channel (the coefficient of *interest*capital* is significative) is also significant in the impact on banking crises, so monetary policy in more capitalized banks is more sensible if central bankers want to avoid banking crises by shifting interest rates. Furthermore, the money supply is better monetary policy indicator than the lending interest rate, due to the R2 with money supply as indicator is 0.9008 and with interest rate is 0.8875. Our expectations shown in Table 24.1 are confirmed: money supply explains better lending behavior.

Additional results are the confirmation of the lending channel directly to banking crises due to the positive and significant coefficient of the *dlnloan* variable in the first model, and the credit channel of monetary policy for the three models, as the positive and significant sign of the coefficient of the interaction of the *dlnloan* variable and the instrument of monetary policy shows the three times. Furthermore, in the three models the coefficients of the monetary instruments are negative and significative. So, a decrease in reference interest rates during lax periods can influence through a double effect on banking crises likelihood: directly, by rising the probability due to the negative and significant coefficient of the policy indicator in Table 24.6, and at least another indirectly, by the following steps. First, an increase on loan supply, as the negative and significant coefficient shows in Table 24.5, and second, this loan supply also positively influences on the likelihood of a banking crisis, rising its probability. Nonetheless, the transmission channels (by lending, 2a-c, competition, 2a, and capitalization, 2b-c) mitigate these perverse effects that generate such a vicious circle.

24.7 Mechanisms and Discussion

This section shows that the years of the lax policy and the Great Recession in the Euro area led to an expansion in lending. In the laxity period, the interest rate was the responsible of a raise in the loan growth, as the positive and significant coefficient of interest*laxity shows, but money supply did not impact, and the non-significant positive coefficient of *M(t-1)*laxity* points in Table 24.5. On the other hand, the monetary policy indicator, both in quantities and in prices, significantly impacted on a higher loan growth during the GR for countries in the Eurozone, as the significant and positive coefficients of *euroarea*M(t-1)*GR* and *euroarea*interest*GR* suggest in Table 24.5. Additionally, the coefficient of these variables took a positive and significant sign in Table 24.6, which means that during the GR the money supply and the interest rate led to even a higher likelihood of a banking crisis. In addition, the interacted effect of monetary policy and the growth of loans led to the same result, as the positive and significant coefficients *M(t-2)*loan(t-2)* and *interest*loan (t-2)* suggest in the same table. This implies that there was a feedback in the negative effects of monetary policy after the Great Moderation: before the GR and during it, monetary policy increased loan growth, and this growth led to a higher probability of a banking crisis, both in quantities (money supply) and in prices (interest rates). So, our results are in contrast to Lee and Werner (2018), who believe quantities were the main driver of monetary policy, but these results show that even in the GR prices (interest rates) also had an impact.

On the other hand, isolated effects of monetary policy led to a reduction on the growth of loan and of the probability of a banking crisis, which suggests that the detrimental monetary policy that permitted a vicious circle between lending and money that unchain banking crisis was after the Great Moderation, as Peña (2020a) shows with Granger-causalities and now is corroborated with correlations regarding money supply and interest rates as indicators of monetary policy. Finally, it can be said that the perverse lending behavior derived from the monetary policy in prices and quantities just before the GR could be correlated with the immediately posterior GR.

On the sight of the results, while the monetary policy of 1961–2011 was in general restrictive and ruled-based, the effect reduced the risk of banking crises. In spite of this, it is seen in model 1c of Table 24.5 that the monetary policy of the previous years to the Great Recession (discretionary instead of rule-based) was harmful to banking by rising loans. Other effect prejudicial to the economy in addition to the impact of monetary policy on lending is that the encouragement of credit increases the likelihood of a banking crisis.

In this section the three initial questions of sect. 1 are answered. First, the impact of monetary policy on credit rises the likelihood of a financial crisis, as we see in the significant and positive coefficient of the interaction of lending and M2 in model 2a of Table 24.6. Second, there is a channel of monetary transmission to prevent crisis, and would be via a higher bank capitalization, as we find in the significant and positive coefficient of the interaction of *M2* and *capital* on loan growth in models 1a

Table 24.7 Sum up of the Granger-causality tests

	H0: *M* does not Granger-cause *dlnc*		**H0: *dlnc* does not Granger-cause *M***			
	p-value	Countries with reject	*p*-value	Countries with reject	**Average of the lowest *p*-value for each country**	**Countries with interferences**
Full sample	0.381	0.389	0.387	0.361	0.242	0.583
Lax period before the GR	0.135	0.778	**0.082**	**0.821**	**0.006**	**0.964**

and 1b in Table 24.5 and the impact of *interest* and *capital* on banking crises in models 2b and 2c in Table 24.6. Third, the monetary policy of the previous years to the Great Recession implicated an increase in the risk of a banking crunch, as shows the significant and positive sign of the coefficient of the interaction of monetary policy, Eurozone and GR in the same Table 24.6. Finally, a positive impact of the interest rate and money supply on loan growth was found for the full sample, and a negative and significant coefficient in the impact on banking crises, showing the differences between the ruled-based era and the laxity period before and during the GR.

Lee and Werner (2018) showed that interest rates follow nominal GDP, while Peña (2020a, b) states that, for the interest rates used as monetary policy it was seen that in the Great Moderation there were not interferences among loans, GDP, and interest rates, but there were interferences in the period of lax monetary policy. This paper shows two additional issues. First, the previous statement is also corroborated for *dlnc* and quantities, and second, that choosing 2001–2006 rather than 2002–2007[6] the Granger-causality is even more robust and significant in the lax period. Table 24.7 presents the summing up of the results presented at the Appendix I where the average p-value and percentage of countries with reject of the null hypothesis is collected, jointly with the percentage of countries where at least one of the Granger-causalities is obtained.

Previous results show that the Peña (2020a, b) test for good economic performance is fulfilled for the full sample period (less than 80% of countries with interferences and an average lowest p-value for each interaction higher than the 10%). Nonetheless, the indicator is not passed for the lax period immediately previous to the GR, with a 96.4% of countries with interferences (the 96.4% of the countries have a rejection of the non-GC between two of the variables at the level of *p*-value lower than the 10% at least in one of the two directions). The results follow

[6]In both articles, there is not enough sample to obtain complete data for the "official" Great Moderation period (2002–2006). In this paper we propose to including 2001 in these calculus because at the end of this year there were already lax monetary policies, and because 2007 was inside the GR, but not before.

Table 24.8 Sum up of the Granger-causality tests for the different Eras (*p*-values)

	loans GC *unem*	*unem* GC *loans*	*loans* GC *infl*	*infl* GC *loans*	*infl* GC *unem*	*unem* GC *infl*	No of obs.
Full sample	**0*****	**0*****	0.707	**0*****	**0.054***	0.456	287
GM	0.4	0.427	**0.057***	0.191	**0.026****	0.724	54
LP	**0.052***	0.463	0.575	0.569	0.889	0.8	75
GR	**0*****	0.334	**0.014****	0.106	**0.032****	0.317	18
BC	**0*****	**0.004*****	0.762	0.191	0.29	0.688	129
CC	0.566	**0*****	**0.054***	**0.001*****	0.802	0.725	11

Note: *: *p*-value <=0.1,**: *p*-value <=0.05,***: *p*-value <=0.01, obs.: observations

Lee and Werner (2018) but for loans instead of output: our results show that loans follow money amounts during the lax periods, while they reject the traditional view that interest rates affect economic growth for a wider period that included GM. However, Peña (2020a, b) effectively confirmed that GDP followed reference interest rates during the lax period of 2002–2007 as an average of 36 countries of the European Union and the OECD.

All these findings also suggest some policies: first, a rule-based policy would be better from a discretionary one. Second, a monetary policy more independent from (or encouraging less) lending behavior would mitigate the probability of a crisis. Third, as monetary tightening is more effective in well-capitalized banks, policy makers should stimulate capitalization.

24.8 Results from the COVID Times and Proposed Economic Policies

This section uses updated data from the Bank of Spain for analyzing the evolution of the financial and economic effects of the COVID-19 pandemic on inflation (*infl*), loans (*loans*), and unemployment rate (*unem*) of Spain between January of 1997 and January of 2021 (maximum length of the available data for being balanced). The sample is also divided into five "Eras" in order to compare with the pandemic: Great Moderation (GM, -August 2001), Lax period (LP, September 2001–November 2007), Great Recession (GR, December 2007–March 2009), "Between-Crises period" (BC, April 2009–February 2020), and "COVID-19 Crisis" (CC, March 2020-). The data is collected in Appendix II.

For analyzing the interrelations among the variables, a Granger-causality has been applied. There is one financial variable, *loans*, and one real, *unem*, so it is expected a non-interference between them as a fact of good economic performance (Peña 2020a), as it is usual with the GM, a period of stability and without uncertainty where rules-based monetary policies were applied. Results are provided in Table 24.8.

As we can see, in the GM period, considered an example of good performance, there were no interferences of loans (in thousands of Euros) and unemployment (growth rate). Despite, loans affected inflation, showing that monetary policy channels work because a higher supply of credit increases the inflation, as we can see with the significance of p-value for *loans* Granger Causes (GC) *infl*. Furthermore, inflation (Harmonized Index of Consumer Price in growth rate) also affects unemployment, so there is no sticky inflation, allowing the economy to be flexible in order to achieve an economic equilibrium, as we can see with the significance of the p-value of *infl* GC *unem*. In spite of not directly causing unemployment, credit affects inflation and inflation affects unemployment, so inflation correctly plays the transmission role of monetary policy by linking the financial with the real economies. This is thanks to, overall, the rules-based monetary policy (Taylor 2009, 2012; Peña 2017).

Other studies find a positive relationship between inflation and unemployment only for the long-term, as it is the case of Serbia for 2009 M1 to 2019 M6 (Veselinović 2020), with a positive and significant impact of inflation on the unemployment rate in the long-run, but without relationship with the key monetary policy rate in the long-run neither Granger-causality in the short run. According to our results, almost all variables present interferences, with similarities with the previous author, because no policy rate is considered. The reason for the interactions between financial (loans) and real (unemployment) variables could be that the full sample only considers few years from the GM and the rest from the extended lax period. In contrast, for the full GM period there are no interactions between such variables as explained above.

Regarding existing theories, Marjanović (2010) points out that Keynesian theory conceived economy as unstable, applying economic policy (mainly the fiscal one) for achieving full employment. In spite the previous theory believes monetary policy may impact on labor, according to monetarism this policy only influences on inflation (Ng'andwe 2020). According to Lucas and Sargent (1979), one of the common points in Keynesian models is the procyclical trend of national income, in contrast to the inflation rate, achieving a countercyclical reaction with unemployment. Nonetheless, here was a sharp decrease of output during the 1970s, jointly with extremely high inflation rates. Thus, previous theories failed to solve the denominated stagflation. These circumstances taught that inflation was a limiting factor of economic growth (Đorđević & Lojanica 2015). Monetary policy resulted to be the most effective instrument for economic and price stabilization. This turned out the neoclassical postulates as the new consensus in macroeconomics (Mihajlović 2018 and Veselinović 2020).

For the rest of Eras there were interferences between the real and financial variables, overall in the Between-Crisis period, with a bidirectional causality. Monetary policy channels did not correctly work in the lax and Between-Crisis periods, while the non-sticky inflation only worked, apart from the GM, in the Great Recession. It is worth to remind that, apart from the GM period, the rest of Eras can be considered as an extended lax period, since the correct use of simple economic policy rules was not in force maybe since the year 2001, when policy started to be discretionary. So, the only Era with a correct monetary policy transmission was when monetary policy was rules-based, that is, the Great Moderation. These results can provide us a first economic policy rule: following simple rules-based monetary policies, and non-discretionary ones, obtains better stability and economic performance. So, for this pandemic following simple rules for monetary policy might be basic.

Other economic policies, more closed to the sanitary issues, could be to digitalize the medical services by using, for instance, tele-medicine, as Gil et al. (2021) show with the economic advantages of using telephonic calls instead of direct and face-to-face medical assistance in some cases of a primary healthcare sanitary center.

24.9 Concluding Remarks

As seen in the absence of effect on the banking crash during the years when that rule was a guideline for Central Bankers (the "Great Moderation") and the prejudicial impact (directly or indirectly) on the risk of a crisis by the years when or where Central Banks were not guided by that rule (the years of lax monetary policy, and, in the Euro area, the years of the Great Recession), a ruled-based policy could be better than a discretionary one, which can also generate uncertainty. Furthermore, a better indicator of monetary policy than lending interest was found to explain lending: money supply growth. We have seen that in current periods discretional monetary policies continue to be applied, as the results on the interference between financial and economic variables in Spain show. For the pandemic, and in order to make monetary policy effective and neutral, with working money channels, simple rules of monetary policy should be followed.

This paper studies the impact of monetary policy on crises and its mechanisms. The main conclusions of the paper are the following. First, the impact of monetary policy on credit rises the likelihood of a financial crisis. Second, there is a channel of monetary transmission to loan growth and to prevent crisis, via a bank capitalization. Third, the monetary policy of the previous years to the Great Recession implicated an increment in the risk of a banking crunch. Fourth, simple rules-based monetary policies have not employed since the end of the Great Moderation at September of 2001, leading to uncertainty, financial instability, and bad economic performance.

The main contributions are theoretical and empirical. The first one is the theoretical expectation of the results illustrated after that, where it is shown that credit booms in the short-term, may be due to lax monetary policy, raises loan default growth and provisions in the medium term and this could lead to higher likelihood of banking crises via lower banking profits or higher pure interest rates. The second is the empirical corroboration that, although monetary prices also matter, monetary models could be better explained by money supply. There is also contribution of corroborating some policies: a rule-based policy would be better from a discretionary one. A monetary policy that discourages lending behavior would reduce the probability of a crisis. Additionally, we think policy makers should stimulate bank capitalization. Finally, we think this chapter is the first in showing the perverse economic effects of discretionally monetary policy in the current COVID-19 pandemic, with the shown interaction between financial and real economic variables without being the inflation the intermediary.

Appendix I

Table 24.9 Granger-causality between dlnc and M for the full sample

Country	H0: M does not Granger-cause dlnc		H0: dlnc does not Granger-cause M		Countries with interferences
	p-value	Countries with reject	*p*-value	Countries with reject	
Australia	0.241	0	0.198	0	0
Austria	0.914	0	0.523	0	0
Belgium	0.417	0	0	1	1
Bulgaria	0.006	1	0	1	1
Canada	0.414	0	0.933	0	0
Chile	0.504	0	0.02	1	1
Czech republic	0.733	0	0.372	0	0
Denmark	0.577	0	0.009	1	1
Estonia	0.05	1	0.276	0	1
Finland	0.137	0	0.061	1	1
France	0.536	0	0.458	0	0
Germany	0.772	0	0.872	0	0
Greece	0.042	1	0.099	1	1
Hungary	0.786	0	0.804	0	0
Ireland	0.008	1	0.487	0	1
Island	0.151	0	0.079	1	1
Israel	0.453	0	0.103	0	0
Italy	0.78	0	0.835	0	0
Japan	0.045	1	0.918	0	1
Korea	0	1	0.179	0	1
Luxembourg	0.1	1	0.98	0	1
Latvia	0.009	1	0.143	0	1
Mexico	0.009	1	0.541	0	1
Lithuania	0.587	0	0.541	0	0
Netherlands	0.003	1	0.006	1	1
New Zealand	0.772	0	0.798	0	0
Norway	0.086	1	0.879	0	1
Poland	0	1	0.068	1	1
Portugal	0.31	0	0.581	0	0
Slovak republic	0.711	0	0.705	0	0
Slovenia	0.752	0	0.523	0	0
Spain	0.094	1	0.004	1	1
Sweden	0.069	1	0	1	1
Turkey	0.918	0	0.801	0	0
United Kingdom	0.943	0	0.053	1	1
United States	0.788	0	0.089	1	1
Average	**0.381**	**0.389**	**0.387**	**0.361**	**0.583**

Table 24.10 Granger-causality between dlnc and M for the lax period 2001–2006, before the GR

Country	H0: M does not Granger-cause dlnc		H0: dlnc does not Granger-cause M		Countries with interferences
	p-value	Countries with reject	*p*-value	Countries with reject	
Australia	0	1	0.526	0	1
Austria					
Belgium					
Bulgaria	0	1	0	1	1
Canada	0	1	0	1	1
Chile	0.003	1	0	1	1
Czech republic	0.001	1	0.006	1	1
Denmark	0	1	0.002	1	1
Estonia	0.009	1	0.027	1	1
Finland	0.272	0	0.103	0	0
France			0	1	1
Germany					
Greece					
Hungary	0.679	0	0	1	1
Ireland	0	1	0	1	1
Island	0	1	0.805	0	1
Israel	0.723	0	0	1	1
Italy	0	1	0.021	1	1
Japan	0.049	1	0.139	0	1
Korea	0.003	1	0	1	1
Luxembourg					
Latvia	0.813	0	0	1	1
Mexico	0.195	0	0	1	1
Lithuania	0.004	1	0	1	1
Netherlands					
New Zealand	0	1	0.076	1	1
Norway	0.867	0	0	1	1
Poland	0	1	0	1	1
Portugal					
Slovak republic	0	1	0.582	0	1
Slovenia	0	1	0	1	1
Spain					
Sweden	0	1	0	1	1
Turkey	0.001	1	0	1	1
United Kingdom	0.015	1	0	1	1
United States	0	1	0	1	1
Average	**0.134**	**0.778**	**0.082**	**0.821**	**0.964**

Appendix II

Table 24.11 Data for the analysis of the economic effects of the COVID-19 pandemic in Spain

Month	infl	loans	unem	Month	infl	loans	unem	Month	infl	loans	unem	Month	infl	loans	unem
Jan-97	2.8	7.60E+08	2.80E+06	Jan-03	3.8	1.30E+09	2.20E+06	Jan-09	0.8	3.10E+09	3.30E+06	Jan-15	−1.5	2.80E+09	4.50E+06
Feb-97	2.5	7.60E+08	2.80E+06	Feb-03	3.9	1.30E+09	2.20E+06	Feb-09	0.7	3.10E+09	3.50E+06	Feb-15	−1.2	2.80E+09	4.50E+06
Mar-97	2.2	7.60E+08	2.80E+06	Mar-03	3.7	1.30E+09	2.20E+06	Mar-09	−0.1	3.10E+09	3.60E+06	Mar-15	−0.8	2.80E+09	4.50E+06
Apr-97	1.6	7.60E+08	2.70E+06	Apr-03	3.2	1.40E+09	2.10E+06	Apr-09	−0.2	3.10E+09	3.60E+06	Apr-15	−0.7	2.70E+09	4.30E+06
May-97	1.3	7.70E+08	2.60E+06	May-03	2.7	1.30E+09	2.00E+06	May-09	−0.9	3.10E+09	3.60E+06	May-15	−0.3	2.70E+09	4.20E+06
Jun-97	1.4	7.90E+08	2.60E+06	Jun-03	2.8	1.40E+09	2.00E+06	Jun-09	−1	3.20E+09	3.60E+06	Jun-15	0	2.70E+09	4.10E+06
Jul-97	1.5	8.00E+08	2.50E+06	Jul-03	2.9	1.40E+09	2.00E+06	Jul-09	−1.3	3.10E+09	3.50E+06	Jul-15	0	2.70E+09	4.00E+06
Aug-97	1.7	7.80E+08	2.50E+06	Aug-03	3.1	1.40E+09	2.00E+06	Aug-09	−0.7	3.10E+09	3.60E+06	Aug-15	−0.5	2.70E+09	4.10E+06
Sep-97	1.9	7.90E+08	2.60E+06	Sep-03	3	1.40E+09	2.00E+06	Sep-09	−0.9	3.10E+09	3.70E+06	Sep-15	−1.1	2.70E+09	4.10E+06
Oct-97	1.8	7.90E+08	2.60E+06	Oct-03	2.7	1.40E+09	2.10E+06	Oct-09	−0.6	3.10E+09	3.80E+06	Oct-15	−0.9	2.70E+09	4.20E+06
Nov-97	1.9	8.00E+08	2.60E+06	Nov-03	2.9	1.40E+09	2.10E+06	Nov-09	0.4	3.10E+09	3.90E+06	Nov-15	−0.4	2.70E+09	4.10E+06
Dec-97	1.9	8.10E+08	2.60E+06	Dec-03	2.7	1.40E+09	2.20E+06	Dec-09	0.9	3.10E+09	3.90E+06	Dec-15	−0.1	2.60E+09	4.10E+06
Jan-98	1.9	8.30E+08	2.60E+06	Jan-04	2.3	1.50E+09	2.20E+06	Jan-10	0.7	3.10E+09	4.00E+06	Jan-16	−0.4	2.70E+09	4.20E+06

Feb-98	1.7	8.40E+08	2.60E+06	Feb-04	2.2	1.50E+09	2.20E+06	Feb – 10	0.4	3.10E+09	4.10E+06	Feb – 16	-1	2.70E+09	4.20E+06
Mar-98	1.7	8.30E+08	2.50E+06	Mar-04	2.2	1.50E+09	2.20E+06	Mar-10	2.7	3.10E+09	4.20E+06	Mar-16	-1	2.60E+09	4.10E+06
Apr-98	1.9	8.20E+08	2.40E+06	Apr-04	2.7	1.50E+09	2.20E+06	Apr-10	2.4	3.10E+09	4.10E+06	Apr-16	−1.2	2.60E+09	4.00E+06
May-98	2	8.20E+08	2.40E+06	May-04	3.4	1.50E+09	2.10E+06	May-10	2.5	3.20E+09	4.10E+06	May-16	−1.1	2.60E+09	3.90E+06
Jun-98	2	8.50E+08	2.30E+06	Jun-04	3.5	1.60E+09	2.10E+06	Jun-10	2.1	3.20E+09	4.00E+06	Jun-16	−0.9	2.70E+09	3.80E+06
Jul-98	2.3	8.60E+08	2.30E+06	Jul-04	3.3	1.60E+09	2.00E+06	Jul-10	1.8	3.10E+09	3.90E+06	Jul-16	−0.7	2.60E+09	3.70E+06
Aug-98	2.1	8.50E+08	2.20E+06	Aug-04	3.3	1.60E+09	2.00E+06	Aug-10	1.6	3.20E+09	4.00E+06	Aug-16	−0.3	2.60E+09	3.70E+06
Sep-98	1.5	8.50E+08	2.30E+06	Sep-04	3.2	1.60E+09	2.10E+06	Sep-10	2.8	3.10E+09	4.00E+06	Sep-16	0	2.60E+09	3.70E+06
Oct-98	1.5	8.60E+08	2.30E+06	Oct-04	3.6	1.60E+09	2.10E+06	Oct-10	2.5	3.10E+09	4.10E+06	Oct-16	0.5	2.60E+09	3.80E+06
Nov-98	1.4	8.70E+08	2.30E+06	Nov-04	3.5	1.60E+09	2.10E+06	Nov-10	2.3	3.10E+09	4.10E+06	Nov-16	0.5	2.60E+09	3.80E+06
Dec-98	1.3	8.70E+08	2.20E+06	Dec-04	3.3	1.70E+09	2.10E+06	Dec-10	2.9	3.10E+09	4.10E+06	Dec-16	1.4	2.50E+09	3.70E+06
Jan-99	1.4	8.80E+08	2.30E+06	Jan-05	3.1	1.70E+09	2.20E+06	Jan-11	3	3.10E+09	4.20E+06	Jan-17	2.9	2.50E+09	3.80E+06
Feb-99	1.8	8.80E+08	2.20E+06	Feb-05	3.3	1.70E+09	2.20E+06	Feb-11	3.4	3.10E+09	4.30E+06	Feb-17	3	2.50E+09	3.80E+06
Mar-99	2.1	8.90E+08	2.20E+06	Mar-05	3.4	1.70E+09	2.10E+06	Mar-11	3.3	3.10E+09	4.30E+06	Mar-17	2.1	2.50E+09	3.70E+06
Apr-99	2.3	8.70E+08	2.10E+06	Apr-05	3.5	1.80E+09	2.10E+06	Apr-11	3.5	3.10E+09	4.30E+06	Apr-17	2.6	2.50E+09	3.60E+06

(continued)

Table 24.11 (continued)

Month	infl	loans	unem	Month	infl	loans	unem	Month	infl	loans	unem	Month	infl	loans	unem
May-99	2.1	8.80E+08	2.10E+06	May-05	3.1	1.80E+09	2.00E+06	May-11	3.4	3.10E+09	4.20E+06	May-17	2	2.50E+09	3.50E+06
Jun-99	2.1	9.00E+08	2.00E+06	Jun-05	3.2	1.90E+09	2.00E+06	Jun-11	3	3.20E+09	4.10E+06	Jun-17	1.6	2.50E+09	3.40E+06
Jul-99	2.1	9.10E+08	2.00E+06	Jul-05	3.3	1.90E+09	2.00E+06	Jul-11	3	3.20E+09	4.10E+06	Jul-17	1.7	2.50E+09	3.30E+06
Aug-99	2.3	9.00E+08	2.00E+06	Aug-05	3.3	1.90E+09	2.00E+06	Aug-11	2.7	3.20E+09	4.10E+06	Aug-17	2	2.50E+09	3.40E+06
Sep-99	2.5	9.20E+08	2.00E+06	Sep-05	3.8	2.00E+09	2.00E+06	Sep-11	3	3.20E+09	4.20E+06	Sep-17	1.8	2.50E+09	3.40E+06
Oct-99	2.4	9.20E+08	2.00E+06	Oct-05	3.5	2.00E+09	2.10E+06	Oct-11	3	3.20E+09	4.40E+06	Oct-17	1.7	2.60E+09	3.50E+06
Nov-99	2.7	9.50E+08	2.10E+06	Nov-05	3.4	2.00E+09	2.10E+06	Nov-11	2.9	3.20E+09	4.40E+06	Nov-17	1.8	2.60E+09	3.50E+06
Dec-99	2.8	9.60E+08	2.00E+06	Dec-05	3.7	2.10E+09	2.10E+06	Dec-11	2.3	3.30E+09	4.40E+06	Dec-17	1.2	2.50E+09	3.40E+06
Jan-00	2.9	9.50E+08	2.10E+06	Jan-06	4.2	2.10E+09	2.20E+06	Jan-12	2	3.30E+09	4.60E+06	Jan-18	0.7	2.50E+09	3.50E+06
Feb-00	3	9.50E+08	2.10E+06	Feb-06	4.1	2.10E+09	2.20E+06	Feb-12	1.9	3.30E+09	4.70E+06	Feb-18	1.2	2.50E+09	3.50E+06
Mar-00	3	9.70E+08	2.00E+06	Mar-06	3.9	2.10E+09	2.10E+06	Mar-12	1.8	3.30E+09	4.80E+06	Mar-18	1.3	2.50E+09	3.40E+06
Apr-00	3	9.80E+08	2.00E+06	Apr-06	3.9	2.20E+09	2.10E+06	Apr-12	2	3.30E+09	4.70E+06	Apr-18	1.1	2.50E+09	3.30E+06
May-00	3.2	9.80E+08	1.90E+06	May-06	4.1	2.20E+09	2.00E+06	May-12	1.9	3.40E+09	4.70E+06	May-18	2.1	2.50E+09	3.30E+06
Jun-00	3.5	9.90E+08	1.90E+06	Jun-06	4	2.20E+09	2.00E+06	Jun-12	1.8	3.40E+09	4.60E+06	Jun-18	2.3	2.50E+09	3.20E+06

Jul-00	3.7	1.00E+09	1.90E+06	Jul-06	4	2.30E+09	2.00E+06	Jul-12	2.2	3.30E+09	4.60E+06	Jul-18	2.3	2.50E+09	3.10E+06
Aug-00	3.6	1.00E+09	1.90E+06	Aug-06	3.8	2.20E+09	2.00E+06	Aug-12	2.7	3.30E+09	4.60E+06	Aug-18	2.2	2.50E+09	3.20E+06
Sep-00	3.7	1.00E+09	1.90E+06	Sep-06	2.9	2.30E+09	2.00E+06	Sep-12	3.5	3.30E+09	4.70E+06	Sep-18	2.3	2.50E+09	3.20E+06
Oct-00	4	1.00E+09	1.90E+06	Oct-06	2.6	2.30E+09	2.00E+06	Oct-12	3.5	3.30E+09	4.80E+06	Oct-18	2.3	2.50E+09	3.30E+06
Nov-00	4.1	1.00E+09	2.00E+06	Nov-06	2.7	2.40E+09	2.00E+06	Nov-12	3	3.30E+09	4.90E+06	Nov-18	1.7	2.50E+09	3.30E+06
Dec-00	4	1.10E+09	1.90E+06	Dec-06	2.7	2.40E+09	2.00E+06	Dec-12	3	3.30E+09	4.80E+06	Dec-18	1.2	2.50E+09	3.20E+06
Jan-01	2.9	1.10E+09	2.00E+06	Jan-07	2.4	2.40E+09	2.10E+06	Jan-13	2.8	3.20E+09	5.00E+06	Jan-19	1	2.50E+09	3.30E+06
Feb-01	2.7	1.10E+09	2.00E+06	Feb-07	2.4	2.50E+09	2.10E+06	Feb-13	2.9	3.20E+09	5.00E+06	Feb-19	1.1	2.50E+09	3.30E+06
Mar-01	3	1.10E+09	2.00E+06	Mar-07	2.5	2.50E+09	2.10E+06	Mar-13	2.6	3.20E+09	5.00E+06	Mar-19	1.3	2.50E+09	3.30E+06
Apr-01	3.6	1.10E+09	1.90E+06	Apr-07	2.5	2.50E+09	2.00E+06	Apr-13	1.5	3.20E+09	5.00E+06	Apr-19	1.6	2.50E+09	3.20E+06
May-01	3.8	1.10E+09	1.90E+06	May-07	2.4	2.60E+09	2.00E+06	May-13	1.8	3.10E+09	4.90E+06	May-19	0.9	2.50E+09	3.10E+06
Jun-01	3.7	1.10E+09	1.80E+06	Jun-07	2.4	2.60E+09	2.00E+06	Jun-13	2.2	3.10E+09	4.80E+06	Jun-19	0.6	2.50E+09	3.00E+06
Jul-01	2.4	1.10E+09	1.80E+06	Jul-07	2.3	2.70E+09	2.00E+06	Jul-13	1.9	3.00E+09	4.70E+06	Jul-19	0.6	2.50E+09	3.00E+06
Aug-01	2.1	1.10E+09	1.90E+06	Aug-07	2.2	2.70E+09	2.00E+06	Aug-13	1.6	3.00E+09	4.70E+06	Aug-19	0.4	2.60E+09	3.10E+06
Sep-01	2.3	1.10E+09	1.90E+06	Sep-07	2.7	2.70E+09	2.00E+06	Sep-13	0.5	3.00E+09	4.70E+06	Sep-19	0.2	2.50E+09	3.10E+06
Oct-01	2.5	1.20E+09	1.90E+06	Oct-07	3.6	2.80E+09	2.00E+06	Oct-13	0	3.00E+09	4.80E+06	Oct-19	0.2	2.50E+09	3.20E+06

(continued)

Table 24.11 (continued)

Month	infl	loans	unem	Month	infl	loans	unem	Month	infl	loans	unem	Month	infl	loans	unem
Nov-01	2.5	1.20E+09	2.00E+06	Nov-07	4.1	2.80E+09	2.10E+06	Nov-13	0.3	2.90E+09	4.80E+06	Nov-19	0.5	2.60E+09	3.20E+06
Dec-01	2.5	1.20E+09	2.00E+06	Dec-07	4.3	2.80E+09	2.10E+06	Dec-13	0.3	2.90E+09	4.70E+06	Dec-19	0.8	2.50E+09	3.20E+06
Jan-02	3.1	1.20E+09	2.10E+06	Jan-08	4.4	2.80E+09	2.30E+06	Jan-14	0.3	2.90E+09	4.80E+06	Jan-20	1.1	2.50E+09	3.30E+06
Feb-02	3.2	1.20E+09	2.10E+06	Feb-08	4.5	2.90E+09	2.30E+06	Feb-14	0.1	2.90E+09	4.80E+06	Feb-20	0.9	2.50E+09	3.20E+06
Mar-02	3.2	1.20E+09	2.10E+06	Mar-08	4.6	2.90E+09	2.30E+06	Mar-14	−0.2	2.90E+09	4.80E+06	Mar-20	0.1	2.60E+09	3.50E+06
Apr-02	3.7	1.20E+09	2.10E+06	Apr-08	4.2	2.90E+09	2.30E+06	Apr-14	0.3	2.90E+09	4.70E+06	Apr-20	−0.7	2.70E+09	3.80E+06
May-02	3.7	1.20E+09	2.00E+06	May-08	4.7	2.90E+09	2.40E+06	May-14	0.2	2.90E+09	4.60E+06	May-20	−0.9	2.70E+09	3.90E+06
Jun-02	3.4	1.20E+09	2.00E+06	Jun-08	5.1	3.00E+09	2.40E+06	Jun-14	0	2.80E+09	4.40E+06	Jun-20	−0.3	2.80E+09	3.90E+06
Jul-02	3.5	1.20E+09	2.00E+06	Jul-08	5.3	3.00E+09	2.40E+06	Jul-14	−0.4	2.80E+09	4.40E+06	Jul-20	−0.7	2.80E+09	3.80E+06
Aug-02	3.7	1.20E+09	2.00E+06	Aug-08	4.9	3.00E+09	2.50E+06	Aug-14	−0.5	2.80E+09	4.40E+06	Aug-20	−0.6	2.70E+09	3.80E+06
Sep-02	3.5	1.30E+09	2.00E+06	Sep-08	4.6	3.00E+09	2.60E+06	Sep-14	−0.3	2.80E+09	4.40E+06	Sep-20	−0.6	2.70E+09	3.80E+06
Oct-02	4	1.30E+09	2.10E+06	Oct-08	3.6	3.10E+09	2.80E+06	Oct-14	−0.2	2.80E+09	4.50E+06	Oct-20	−0.9	2.70E+09	3.80E+06
Nov-02	3.9	1.30E+09	2.10E+06	Nov-08	2.4	3.10E+09	3.00E+06	Nov-14	−0.5	2.80E+09	4.50E+06	Nov-20	−0.8	2.70E+09	3.90E+06
Dec-02	4	1.30E+09	2.10E+06	Dec-08	1.4	3.10E+09	3.10E+06	Dec-14	−1.1	2.80E+09	4.40E+06	Dec-20	−0.6	2.70E+09	3.90E+06
												Jan-21	0.4	2.70E+09	4.00E+06

Source: Bank of Spain (2021)

References

Altunbas Y, Gambacorta L, Marques-Ibanez D (2010) Bank risk and monetary policy. J Financ Stab 6(3):121–129

Arellano M, Bover O (1995) Another look at the instrumental variable estimation of error-components models. J Econ 68:29–51

Bean C (2010) The great moderation, the great panic, and the great contraction. J Eur Econ Assoc 8 (2–3):289–325

Benati L (2008) The "great moderation" in the United Kingdom. J Money Credit Bank 40 (1):121–147

Bernanke B (2004) "The Great Moderation." Speech at the meetings of the Eastern Economic Association, Washington, D.C., February 20. http://www.federalreserve.gov/boarddocs/speeches/2004/20040220/default.htm

Beutel J, List S, von Schweinitz G (2019) Does machine learning help us predict banking crises? J Financ Stab 45:100693

Blundell R, Bond S (1998) Initial conditions and moment restrictions in dynamic panel data models. J Econ 87:115–143

Bubeck J, Maddaloni A, Peydró JL (2020) Negative monetary policy rates and systemic Banks' risk-taking: evidence from the euro area securities register. J Money Credit Bank 52 (S1):197–231

Büyükkarabacak B, Valev NT (2010) The role of household and business credit in banking crises. J Bank Financ 34(6):1247–1256

Clarida R, Gali J, Gertler M (2000) Monetary policy rules and macroeconomic stability: evidence and some theory. Q J Econ 115(1):147–180

Davis SJ, Kahn JA (2008) Interpreting the great moderation: changes in the volatility of economic activity at the macro and micro levels. J Econ Perspect 22(4):155–180

Đorđević M, Lojanica N (2015) Kontroverze o statistički značajnoj povezanosti inflacije i ekonomskog rasta: Primer Republike Srbije. In V. Lekovic (Ed.): Institucionalne promene kao determinanta privrednog razvoja Srbije (pp. 149–164) Kragujevac: Ekonomski fakultet Univerzitet u Kragujevcu. Retrieved from http://www.ekfak.kg.ac.rs/sr/nir/nir-naucni-skupovi/naucni-skup-instit-promene

Douch M (2010) The macroeconomic effects of monetary policy and financial crisis. Cankaya University Journal of Humanities and Social Sciences 7(1):1–35

Fungacova Z, Solanko L, Weill L (2014) Does competition influence the Bank Lending Channel in the euro area? J Bank Financ 49:356–366

Gil Giménez N, Peña Blasco G, Bartolomé Lalanza ML, Atance Melendo E, Buil Langarita S, Blasco Pérez-Aramendia MJ (2021) Economic evaluation of e-mail consulting and its impact on COVID-19. A five-year experience. Rev Pediatr Aten Primaria 23(89):43–51

Giannone D, Lenza M, Reichlin L (2008) Explaining the great moderation: it is not the shocks. J Eur Econ Assoc 6(2–3):621–633

de Haan J, Fang Y, Jing Z (2020) Does the risk on banks' balance sheets predict banking crises? New evidence for developing countries. International Review of Economics & Finance 68:254–268

Hardy DC, Pazarbaşioğlu C (1999) Determinants and leading indicators of banking crises: Further evidence", IMF staff papers, pp. 247–258

Keynes JM (1923) A tract on monetary reform. Macmillan and co., London

Kuttner KN (2018) Outside the box: unconventional monetary policy in the great recession and beyond. J Econ Perspect 32(4):121–146

Laubach T, Williams JC (2003) Measuring the natural rate of interest. Rev Econ Stat 85 (4):1063–1070

Lee KS, Werner RA (2018) Reconsidering monetary policy: an empirical examination of the relationship between interest rates and nominal GDP growth in the US, UK, Germany and Japan. Ecol Econ 146:26–34

López-Laborda J, Peña G (2018) A new method for applying VAT to financial services. Natl Tax J 71(1):155–182
Lucas RE, Sargent TJ (1979) After Keynesian macroeconomics. Federal Reserve Bank of Minneapolis Quarterly Review 3(2):295–319. https://www.minneapolisfed.org/research/quarterly-review/after-keynesian-macroeconomics
Marjanović G (2010) Uticaj ekonomske krize na glavni tok ekonomske misli. Ekonomski horizonti 12(2):5–20. http://horizonti.ekfak.kg.ac.rs/sites/default/files/Casopis/2010_2/1_Gordana_Marjanovic.pdf
Mihajlović V (2018) Implikacije novog konsenzusa u makroekonomskoj teoriji za ekonomsku politiku u Republici Srbiji. In: Ivanović, V., Nikolić, J., Jovković, B., Zlatanović, D., Kalinić, Z., Janković, N. (Eds.): Implikacije ekonomije znanja za razvojne procese u Republici Srbiji (pp. 263–275) Kragujevac: Ekonomski fakultet Univerzitet u Kragujevcu. http://www.ekfak.kg.ac.rs/images/Nir/ProjekatFakulteta/Zbornik2018-Implikacije_ekonomije_znanja_za_razvojne_procese_u_Republici_Srbiji.pdf
Matousek R, Sarantis N (2009) The Bank Lending Channel and monetary policy transmission in central and eastern European countries. J Comp Econ 37:321–334
Neuhuas JM, Kalbfleisch JD, Hauck WW (1991) A comparison of cluster-specific and population averaged approaches for analyzing correlated binary data. Int Stat Rev 59(1):25–35
Ng'andwe C (2020) Inadequacies of monetarism for developing countries with mass unemployment: a theoretical framework. Journal of Economics and Development Studies 8(4):50–67
Nocetti D (2006) Central bank's value at risk and financial crises: an application to the 2001 argentine crisis. J Appl Econ 9(2):381–402
Olson E, Young AT (2015) Discretionary monetary policy, quantitative easing and the decline in US labor share. Economics and Business Letters 4(2):63–78
Österholm P (2005) The Taylor rule: a spurious regression? Bull Econ Res 57(3):217–247
Peña G (2015) "Does monetary policy affect lending? Evidence from some panel data", Pre-print: http://www.revecap.com/encuentros/trabajos/p/pdf/078.pdf
Peña, G. (2016) "The determinants of banking crises: Further evidence", Working Paper No. 70093, MPRA
Peña G (2017) Money, lending and banking crises. Economic Papers: A journal of applied economics and policy 36(4):444–458
Peña G (2020a) Monetary policy after the great moderation. Journal of Central Banking Theory and Practice 9(3):5–26
Peña G (2020b) VAT experience as an influence on banking crises. National Accounting Review 2 (1):53–65
Prüser J, Schlösser A (2020) On the time-varying effects of economic policy uncertainty on the US economy. Oxf Bull Econ Stat 82(5):1217–1237
Sanfilippo-Azofra S, Torre-Olmo B, Cantero-Saiz M, López-Gutiérrez C (2018) Financial development and the bank lending channel in developing countries. J Macroecon 55:215–234
Stock JH, Watson MW (2003) Has the business cycle changed and why? In: Gertler M, Rogoff K (eds) NBER macroeconomics annual 2002. MIT Press, Cambridge, MA, pp 159–218
Taylor, J. B. (1993) Discretion versus policy rules in practice. In Carnegie-Rochester conference series on public policy (Vol. 39, December, pp. 195–214) North-Holland
Taylor JB (2012) Monetary policy rules work and discretion doesn't: a tale of two eras. J Money Credit Bank 44(6):1017–1032
Taylor JB (2009) The Financial crisis and the policy responses: an empirical analysis of what went wrong", National Bureau of Economic Research Working Paper 14631
Ullah S, Akhtar P, Zaefarian G (2018) Dealing with endogeneity bias: the generalized method of moments (GMM) for panel data. Ind Mark Manag 71:69–78
Veselinović N (2020) Monetary policy and unemployment in the republic of Serbia. Industrija 48 (2):73–88
Werner RA (1994) Japanese foreign investment and the "land bubble". Rev Int Econ 2(2):166–178

Wicksell K (1936) Interest and Prices (London: Macmillan) Translation of 1898 edition by R. F. Kahn

Windmeijer F (2005) A finite sample correction for the variance of linear efficient two-step GMM estimators. J Econ 126(1):25–51

Wooldridge J (2002) Econometric analysis of cross section and panel data. The MIT Press, Cambridge

Zeger S, Liang K, Albert P (1988) Models for longitudinal data: a generalized estimating equation approach. Biometrics 44(4):1049–1060

Chapter 25
QOL Barometer for the Well-being of Citizens: Leverages during Critical Emergencies and Pandemic Disasters

Arindam Chakrabarty, Uday Sankar Das, and Saket Kushwaha

Abstract Improving the quality of life for its citizens has been the focal point of any governmental system across the globe. Every state is committed to providing good governance to its countrymen. Society is moving through the Fourth Industrial Revolution (4IR) where the e-governance ecosystem has become the priority need of the hour. The days of mechanistic bureaucracy have become unpopular and outdated. The modern democracies desire an organic, citizen-friendly governmental system where information needs to be collected from the people at the bottom of the pyramid so that the state could ensure delivery of improvised and augmented public goods and services effectively and efficiently keeping in view its commitments for achieving all the UN-SDGs by 2030. This chapter has devised a dedicated model based on an e-governance framework. This QOL Barometer would be designed using the 4IR ecosystem. The innovative QOL Barometer or the "CARE-Protocol" may be developed and implemented for improving the quality of life of its citizens. This protocol would be conceptualized, based on inputs and insights from secondary sources. The benefits of this model can be leveraged during critical emergencies and pandemic disasters.

Keywords e-Governance · Fourth Industrial Revolution · QOL Barometer · UN-SDGs Emergencies · Pandemic

---

A. Chakrabarty (✉)
Department of Management, Rajiv Gandhi University (Central University), Doimukh, Arunachal Pradesh, India
e-mail: arindam.chakrabarty@rgu.ac.in

U. S. Das
Department of Management & Humanities, National Institute of Technology Arunachal Pradesh (INI under MHRD Government of India), Jote, Arunachal Pradesh, India

S. Kushwaha
Banaras Hindu University, India & Vice-Chancellor, Rajiv Gandhi University (Central University), Doimukh, Arunachal Pradesh, India
e-mail: saket.kushwaha@bhu.ac.in

S. A. Hassan et al. (eds.), *Decision Sciences for COVID-19*, International Series in Operations Research & Management Science 320,
https://doi.org/10.1007/978-3-030-87019-5_25

25.1 Introduction

Qualities of life for the citizens are the real measures in the efficacy of governance. In the era of automation and knowledge and economy, it is imperative to provide prompt and quality solutions to the countrymen. Society deserves a dedicated comprehensive and robust mechanism so that the state can render appropriate services to its target audience. The fourth industrial revolution has been strengthening E-Governance systems to energize for the superior momentum of public service and distribution framework. It is observed that the developed states have been prioritizing creating an improvised E-Governance ecosystem to fulfill the commitment toward good governance.

25.1.1 Idea of Quality of Life

Mankind has been on a constant quest for the betterment of his living conditions since the inception of humanity. As humans grouped themselves under the societal banners many pre-scientific philosophers, thinkers, and administrators attempted at understanding the basics of quality and how it might be applied for improving the human condition.

These frugal attempts found a base when psychiatrist Viktor Emil Frankl conceptualized "Logotherapy" inspired by the works of Sigmund Freud & Alfred Adler. The idea of "Logotherapy" was simply an inquiry into the three basic psychological philosophical pointers, i.e., (a) Freedom of Will, (b) Will to Meaning, (c) Meaning in Life (VFI/Logotherapy and Existential Analysis n.d.). These ideas were derived from the experiences of his struggle to remain hopeful for a better life while imprisoned in a German Nazi Camp; the ordeal of which later turned out to be a best seller (Frankl 1963). The works of Viktor Emil Frankl silently conceptualized the idea called "Quality of Life" and cemented it in academics, literature, polity, healthcare, and global policies from the 1970s onward often without landing any credits to his works or perhaps simply denying it as an idea pulled out of thin air.

Quality of Life became the catchphrase of decision makers in determining the qualitative aspect of human life. Several symposiums derived and concluded several measurement tools to quantify the dimensions into measurable terms (Environmental Protection Agency USA 1973). Quality of Life (QOL) measurement is sometimes looked upon as a means of needs satisfaction for the holistic distribution of goods and services to propagate social justice (McCall 1975). Different academic streams tried to utilize the idea of QOL and formulate measurement tools for various fields like economics, healthcare, marketing, and organizational psychology (Michalos et al. 2006).

QOL is often modeled around a three-dimensional aspect of an individual's being, belonging, and becoming (Quality of Life Research Unit n.d.). Over the years QOL has turned out to be a favorite tool to assesses and quantify the qualitative aspect of human life for both researchers and policymakers across the globe, pitching it up as a de-facto choice.

25.1.2 QOL as a Part of Holistic Development

The answer to the quest for refining the human condition is answered with improving the quality of life to achieve holistic development. Several indices and indexes are either formulated or incorporated in existing policy reports to measure the quality of life for a very long time.

One of the early attempts to measure the QOL in this direction was possibly the Physical Quality of Life Index (PQLI) that aimed at avoiding the limitations posed by Gross National Product by measuring literacy, infant mortality, and life expectancy (Morris 1978).

The "Human Development Report" produced by the United Nations Development Programme incorporated the QOL parameter in 1997 in the form of the Human Poverty Index based on figures like access to safe drinking water, and prevalent poverty(Undp 1997). The 1990s saw the indexing of countries based also known as the Human Development Index, which was later adjusted to the Inequality-adjusted Human Development Index from the year 2010 onward. The Human Development Report also measures gender gaps as Gender Development Index calculating the disparities between men and women in the health, knowledge, and living standards dimensions (Unit 2005).

Another such attempt in the same direction was attempted by the Economist Intelligence Unit of The Economist also known as "Quality of Life Index" that tracked the non-monetary social indices through its Life satisfaction survey, that quantified figures like material well-being, health, political freedom, security and stability, family and community life, climatic conditions, job security, and gender equality (International: The lottery of life|The Economist n.d.). This was later renamed as Where to be born Index as the lottery won predetermine the success of an individual based on the favorable dependent factors in a particular place (lessons from Bhutan on the pursuit of happiness above GDP|World Economic Forum n.d.).

The mountain nation of Bhutan is on a marathon run to promote QOL in the form of Gross National Happiness comprising not just the economic factors but also the moral development among its citizens. This was to be achieved riding on the back of good governance, environmental protection, cultural safeguard, sustainable and equitable economic development, and stepping into the future technologies for a holistic overall development (Interrogation 2008; World Happiness Report 2020 | The World Happiness Report n.d.). Good ideas resonate throughout the globe and it gave birth to the concept of measuring world happiness in the shape of the World Happiness Report by the United Nations Sustainable Development Solutions Network that added happiness and well-being along with the economic indicators for measuring the development vector of the country (About the HPI | Happy Planet Index n.d.). Another similar index is the "Happy Planet Index" that combines four parameters namely well-being, life expectancy, inequality and injustice, average ecological footprint, which factors for a long happy life (WVS Database n.d.). A noteworthy mention at this stage would be of the World Values Survey (WVS) that tries to quantify and measure the intangible values like social, economic, religious, cultural, and political change in values of the global fraternity (Gender Development Index (GDI) | Human Development Reports n.d.).

A dimension of "Welfare Economics" popularly referred to as "Happiness Economics" is often vetted against traditional developmental economics as a much more inclusive and appropriate measure for a rounded development (What are the economics of happiness? | Yale Insights n.d.; The Economics of Happiness | Greater Good n.d.; The Economics of Well-Being n.d.; Why we need to choose happiness over economics | World Economic Forum n.d.).

Quality of life is not just about holistic development and economic well-being it has a hidden aspect, the factor of disposable time. The lack of which is referred to in the popular culture as "Time Poverty" which essentially means the individual does not have enough time for the contribution to social and human capital development (Kalenkoski and Hamrick 2014). QOL is also extensively used in the health care sector to measure the success or outcome of various medical procedures. According to the World Health Organization (WHO), QOL is an individual perceptual context based on individual standards, expectations, standards centered around the cultural value systems. WHO has created an instrument known as WHOQOL (WHOQOL-100 and WHOQOL-BREF) to measure the quality of life (WHO 2014).

25.1.3 Socioeconomic Changing Roles and New Dimensions of Governance: Good Governance and E-Governance

The world has been mostly governed by a democratic setup. In a democracy, every citizen of a county can choose their representative and the government is formed with the majority of a particular representative group. The role of the government in democratic orientation was to provide goods and services to its countrymen. This system is perfectly alright when there is no such competition, complexities, or higher commitments toward achieving improvised quality of life in consonance with rapid technological development. The state-controlled machinery has an inherently limited aspiration for higher growth trajectories rather it develops conservative and status quo syndrome with the growth of population it becomes unrealistic to provide goods and services to all its citizens. The momentum of liberalization privatization and globalization (LPG) coupled with the technological revolution had mended all of us into a massively competitive world both at an individual level and in a collective manner. It has made the state untenable to act as a provider of all utility products and services there is a paradigm shift that the state has initiated the role of being facilitator so that new entrepreneurial expeditions can be improvised and innovative startups can be patronized. The notion of good governance has become the focal point not only to maintain law and order rather enhance a vibrant economy and ensure superior quality of life for its countrymen. There are two schools of thought of any good governance ecosystem of which a section of experts advocate for a higher order of human intervention with an increased face of humanistic touch to make the governance system highly acceptable and adaptable. On the contrary, the other school conservatively believes indiscriminate human intervention in the governmental machinery necessarily makes the system biased prejudiced ambiguous, and most importantly a genesis of corruption. Human-centric governance essentially suffers from procrastination syndrome. To address all these challenges and emerging dynamics of societal aspirations good governance has become the minimum condition for gaining and retaining power in the political corridor of the country where a substantial function can be accomplished by the e-governance ecosystem.

25.1.4 Revolution in E-Governance Ecosystem across the World

The United Nations Department of Economic and Social Affairs under the banner of the Division for Public Administration and Development Management devised a tool United Nations E-Government Development Database (UNeGovDD) that tracks the global E-Readiness for the member states of the UN. This database lists Denmark (Rank 1), Republic of Korea (Rank 2), Estonia (Rank 3), Finland (Rank 4), Australia (Rank 5), Sweden (Rank 6), United Kingdom of Great Britain, and Northern Ireland (Rank 7), New Zealand (Rank 8), the United States of America (Rank 9), Netherlands (Rank 10) among the top ten performers for the year 2020 out of 193 participant nations. Countries in the Indian subcontinent is ranked in the same list in the following order Sri Lanka (Rank 85), India (Rank 100), Bhutan (Rank 103), Maldives (Rank 105), Bangladesh (Rank 119), Nepal (Rank 132), Myanmar (Rank 146), Pakistan (Rank 153), and Afghanistan (Rank 169) (UN E-Governance Knowledgebase 2020). These rankings are derived using the United Nations E-Government Development Index (EGDI) that is based on the weighted average of Telecommunications Infrastructure Index (TII) essentially measuring the data connectivity as monitored by the International Telecommunications Union (ITU), Human Capital Index (HCI) measured by the United Nations Educational, Scientific and Cultural Organization (UNESCO), and Online Service Index (OSI) which is an independent survey conducted by United Nations Department of Economic and Social Affairs (UNDESA) in the form of an Online Service Questionnaire (OSQ). Each of the three indices contributes one-third of the value for calculating EGDI of any nation along with a section of Member State Questionnaire (MSQ), all these questionnaires try to assess the online delivery of services, openness of government data, digital divides, use of ICT and implementation in the whole of government.

More countries across the globe are adopting the E-government strategy in various innovative ways to reach the entire society with the whole of government, with formulations like e-participation, data-centric focus revolving around technologies like artificial intelligence and blockchain to reduce the digital divide through digital kiosks in far-flung areas and achieve the development of smart cities/smart villages. Approximately 65% of the member countries now are in the high or very high group of EDGI ranking while 22% have moved up the rank since the year 2018. It is noteworthy to mention that the EDGI ranking has a positive correlation with the income level of the country. In the new normal of COVID19 E-Governance is facing a massive stress test as most of the face-to-face services were disrupted due to the social distancing guideline. It was evident that countries with robust E-Government infrastructure were able to steer clear through this pandemic with much more ease than those which lacked such facilities. However, the hope notation is that around 85% of the countries globally offer at least one transactional online service. Approximately an average of 14 different services can be availed through the E-Governance platforms of the member states, some of the most common services include applying for a birth certificate, registering or applying for a new business, and paying utility bills online (United Nations E-Government Survey 2020).

If one is to look with the magnifying glass to inspect landmark victory for principalities in the journey of human development then perhaps the bicycle nation of Denmark would possibly beat every other candidate black and blue in every parameter with radical margin. This still stands true for E-Governance where the nation has surpassed Estonia which happens to be an early bird to catch the E-Government train. The Danish governments' official website claims that the nation is digital by default hence you can get all your official work done from the comfort zone of your computer screen within a span of 24 hours be it registering a business, reporting a bicycle theft, or even dealing with a health care issue (The key to Denmark's digital success n.d.). None the less tiny nation of Estonia is still a tough competitor in the E-Governance space with 99% of the services offered online 24 by 7 and an astounding 44% of the citizen's vote digitally using the i-voting platform. The tall claim of this nation not only stops at this, they also officially claim to have saved 844 years of working hours and becoming a hassle-free nation (EAS 2019). Some of the key measures that helped this tiny nation achieve this feat are the digitization of registers to support e-services based on a platform called X-Road, provision of digital id's and making digital signature equivalent to physical signatures. This was backed by the early success of e-banking which helped the nation accept the idea of digital governance. Estonia is experimenting with blockchain in creating a digital embassy and promoting e-residency programs (How Estonia became an e-government powerhouse–Tech Republic n.d.).

India launched the "Digital India" campaign back in the year 2015 with the vision not to miss the train of the information age, the outcome of which was to accesses the various strengths for this adventure. The program revolves around three pillars of digital infrastructure, i.e., (i) Digital infrastructure as utility, (ii) On-demand Services and Governance, and (iii) Citizen Empowerment through Digitization.

India is among the top three digital economies with one of the largest unique digital identification program (AADHAAR), the digital divide is also reducing day by day as mobile services are penetrating the remote hinterlands. India also aims to digitize the healthcare services sector, education sector, and built on a robust E-Governance ecosystem for the future (India's Trillion-Dollar Digital Opportunity 2019). Although these are great ambitions for a nation, however, the base of digital society needs to be built on the backbone of digital governance as is evident in the case of Denmark, Estonia, or South Korea.

25.1.5 Recent Experience of E-Governance Protocol for Mitigating COVID Pandemic

The COIVD pandemic struck the global community with shock and awe strategy paralyzing normal life, halting economic activities, even hitting very hard at the core of community life with forced social distancing, quarantining, and strict containment. The pandemic created such an atmosphere that even the most democratic governments of the world were forced to enact dictatorial practices and enforce sudden lockdowns as a choice between life and death. Several silver bullets were fired as a solution to this pandemic however least was achieved. One of the early outcomes of this pandemic was a global frenzy to trace and isolate COVID-positive cases. This phenomenon gave birth to a new trend of contact tracing apps across the world by governments as a measure to tackle COVID as a non-medical intervention in the form of basic E-Governance (Table 25.1).

Most of these applications track positive COVID cases the users' proximity to access a risk profile. Some of the applications are also able to trace the routes used by a user and alert any possible risk outcome to other users. The "Thai Chana" application possibly differs from the rest as it is designed to be a post COVID measure to step out in the world and perform day-to-day activities including commercial activities.

Even local governments used such half-baked applications as a non-medical intervention or even to disperse E-Governance services like the one launched by the Government of Delhi-NCR in India called "Delhi Corona" that was specifically designed to provide information about the availability of hospital resources and reduce the rush and run around in a massive, populated city like National Capital Region (Kejriwal launches 'Delhi Corona' app for real-time information on availability of hospital beds–The Hindu n.d.). It is not that any government could have produced a perfect application in these desperate times but a much more coordinated effort perhaps could have yielded better results.

World Summit on Information Society (WSIS) is a UN forum that keeps a track of E-Governance progress across the world. It is a forum that keeps a track of the growing digital divide in the world. Around 40% of the global population lives in poverty-prone nations, and approximately 1 billion people have no access to ICT. Poverty and the digital divide are correlated. Several solutions are proposed to overcome the digital obstacle policy solutions like easy user-friendly web services,

Table 25.1 Indicative list of contact tracing apps for COVID-19 across the world

SL. No.	Country of origin	Name of the app
1.	Australian	COVID safe (COVIDSafe n.d.)
2.	Austria	Stopp Corona (Stopp Corona n.d.)
3.	Azerbaijan	E-Tabib (Download the "E-DOCTOR" mobile application n.d.)
4.	Bahrain	BeAware Bahrain (Kingdom of Bahrain–eGovernment Apps Store: BeAware Bahrain n.d.)
5.	Bangladesh	Corona tracer BD (Corona Tracer BD–Apps on Google Play n.d.)
6.	Canada	COVID alert (Download COVID Alert today–Canada.ca n.d.)
7.	China	Close contact detector (China launches coronavirus "close contact detector" app–BBC News n.d.)
8.	Colombia	Coronapp (Coronapp n.d.)
9.	Croatia	STOPcovid19 (STOPcovid19–STOPcovid19 n.d.)
10.	Czech Republic	eRouška (eRouška–chránímsebe, chránímtebe n.d.)
11.	Denmark	Smittestop (Stop udbredelsenaf COVID-19–Smittelstop n.d.)
12.	France	StopCovid (StopCovid I economie.gouv.fr n.d.)
13.	Germany	Corona-warn-app (Bundesregierung I Aktuelles I Veröffentlichung der Corona-Warn-App n.d.)
14.	Ghana	GH Covid-19 tracker app (Bawumia launches GH COVID-19 Tracker App n.d.)
15.	Hungary	VírusRadar (VírusRadar–a Koronavíruskövetéséreés a COVID-19 ellenivédekezésre n.d.)
16.	Iceland	Rakning c-19 app (Information about Covid-19 in Iceland n.d.)
17.	India	AarogyaSetu Mobile app (AarogyaSetu Mobile App I MyGov.in n.d.)
18.	Ireland	COVID tracker app (COVID Tracker App–Ireland's Coronavirus Contact Tracing App n.d.)
19.	Israel	HaMagen (HaMagen–The Ministry of Health App for Fighting the Spread of Coronavirus n.d.)
20.	Italy	Immuni (Immuni - Download Immuni n.d.)
21.	Japan	COVID-19 contact-confirming app (New Coronavirus Contact Confirmation Application (COCOA) COVID-19 Contact–Confirming ApplicationIMinistry of Health, Labor and Welfare n.d.)
22.	Jordan	Aman app (Aman n.d.)
23.	Latvia	ApturiCovid (ApturiCovid n.d.)
24.	Malaysia	MyTrace (MyTrace, a Preventive Counter Measure and Contact Tracing Application for COVID-19–KementerianSains, TeknologidanInovasi (MOSTI) n.d.)
25.	Nepal	COVIRA (COVIRA n.d.)
26.	New Zealand	NZ COVID tracer app (NZ COVID Tracer app I Ministry of Health NZ n.d.)
27.	North Macedonia	StopKorona (StopKorona!–tracing of Coronavirus exposure and protection from COVID-19 n.d.)
28.	Norway	Smittestopp (Smittestopp–app–helsenorge.no n.d.)
29.	Qatar	Ehteraz (Qatar makes COVID-19 app mandatory, experts question efficiency I Qatar News I Al Jazeera n.d.)

(continued)

Table 25.1 (continued)

SL. No.	Country of origin	Name of the app
30.	Singapore	TraceTogether (TraceTogether n.d.)
31.	Spain	App radar Covid (App Radar Covid, la aplicación de rastreoparaEspaña I Noticias de Tecnología en Diario de Navarra n.d.)
32.	Switzerland	SwissCovid app (SwissCovid app and contact tracing n.d.)
33.	Thailand	Thai Chana quick response (Thailand wins n.d.)

installation of e-kiosks, or simplified mobile devices like "Simputer." Some of the factors determining good e-governance are:

- Cooperation between national and international governments.
- Normalization of the legal framework.
- Reasonable costing for interoperable services.
- Holistic promotion of digital literacy, e-learning, etc.
- Preparing underprivileged society for the Tsunami of Information Society (e-Readiness).
- Increased transparency and promotion of e-participation, e-services, etc..
- Overall all citizens inclusive public administration process to address bureaucratic deadlocks.
- Access to knowledge through e-services to achieve an overall improvement in "Quality of Life" (Stoiciu 2011).

Open Government Partnership is another popular concept based on the idea of responsive, accessible, accountable, open governance with long-term benefits across a broad range of issues and citizen engagement in overseeing governmental activities using digital tools (Open Government Partnership 2020).

Some unique technology-led solutions were also innovated during this Pandemic. For example delivery of essential goods to the households in containment zones/ risk-prone areas with participation from hand card vendors and reducing footfalls in local markets, preserving social distancing. Time allocation for consumers for the purchase of particular goods from a particular shop was also a prominent e-governance feat achieved in some cities in India (Covid-19: Use of new-age technologies for e-governance, Government News, ET Government n.d.).

25.2 Literature Review

Tons of literature are present about QOL, E-governance and COVID pandemic, however, a focused approach is applied to narrow down available relevant recent literature, specifically that devise assessment models for measuring QOL relevant to the present study.

25.2.1 Measuring Quality of Life

A dimension of QOL is strictly related to the health of an individual, with this view the "World Health Organization" endeavored to build a tool to measure the "Health for All" with a holistic vision of prompting social, mental, and physical well-being. This tool was named WHOQOL that assessed an individual's relative perception about his expectations, goals, and concerns in the context of his/her culture and value system. The scores are derived from a multidimensional profile of six domains and 24 subdomains (WhoQOL Group 1995). A pilot survey was eventually conducted containing 236 questions 29 dimensions of QOL divided across 6 domains. Each dimension had approximately four sets of questions of perceived and self-reported types of questions. This highly standardized questionnaire was administered after prepping the target population for about two weeks (Group 1998).

WHOQOL has been established over the years as a reliable de facto instrument to measure QOL in various cultural, language settings across the globe. An example is that "WHOQOL-l00 Hindi" was found to be a suitable instrument to measure QOL and "WHOQOL-l00 BREF Hindi" was found ideal for measuring results of a drug trial (Saxena et al. 1998). Similarly in the USA and UK, the standard "WHOQOL-l00" was found effective in meaning QOL for medical patients (Skevington 1999; Bonomi et al. 2000). Again the "WHOQOL-l00 BREF" passed its reliability test in both Korea and Brazil in the localized version. (Min et al. 2002; Berlim et al. 2005).

HRQOL is mostly administered in a pen and paper mode in a self-report, in person, or via a telephonic interview. This rudimentary technique is not just old-fashioned but also lacks the impunity ease provided by the much more responsive digital technologies like a computer or mobile-based instruments. Both the administrator and the responder showed a positive inclination to the use of technology in a survey conducted among 134 patients (Crawley et al. 2000). Several organizations both academic and non-academic are working toward developing instruments to measure HRQOL. Among them are the international society for quality of life research (ISOQOL) and the Scientific Advisory Committee (SAC). The Scientific Advisory Committee is focused on the refinement and development of HRQOL instruments reducing gaps and making them culturally inclusive, easily interpretable (Lohr 2002).

The 1960s played a significant role in highlight hidden poverty using social indicators and reporting. A proposed QOL barometer in public administration is to shift the scale of measurement from material welfare to universal common concerns of the society. This mechanism would help determine the how-to and what-to aspects of public administration through public participation in effectively resolving social issues (Johansson 2002). There stands a conflict between the established "Gold-Standard" translation norms and canonical methods in translating HRQOL instruments. This is specifically true when the barrier of language is deep between the original language of the instrument and the translated language. Alternative methods like an adaption of a dual translation pattern may prove to be efficient in such cases of need bases QOL measure (Swaine-Verdier et al. 2004). Asia may be considered as a prominent example in such a scenario where a large chunk population speaks Mandarin followed by Hindi and other languages like Japanese, Malay, and Tamil. As most HRQOL instruments are adaptions designed in the western hemisphere major issues are incorporated in the process of translation that makes it ill fit for the Asian problem given the strict translation protocols. Asian countries are still infested with illiteracy and remoteness leaving these instruments designed to be self-completed infective and useless in the ocean of linguistic dialects that differs very often. The result is a biased reporting of economic developments rather than the ground-level picture of disease burden. These instruments may be designed keeping in purview the cross-cultural adaptions needed specific to Asian provinces (Cheung and Thumboo 2006).

The European Parliament made its intention clear regarding the measurement of holistic development and look beyond the GDP numbers in 2009 with five-point indicators. This was essential at a time when the world was preparing for measuring societal progress beyond economic growth only (Bache 2013). EuroQOL or EQ-5D is an inter-disciplinary EuroQOL instrument developed by the efforts of five countries composed of a three-level five-dimensional approach designed to measure the health status. The instrument ran into several issues with the initial translation in multiple languages the rectification of which further enhanced the instrument. This experience also resulted in two more versions of the instrument EQ-5D-5L and EQ-5D-Y for youth. None the less EuroQOL is also a popular instrument of choice for measuring QOL across the globe (Devlin and Brooks 2017).

Several sociodemographic variables play a vital role in determining the health and happiness of the community. This was evident in the case of the community residing in Juarez, Mexico (Molina-Herrera et al. 2019). To determine the impact of the multifactorial problem on the quality of life of its citizens a multivariable analysis is needed for understanding community-oriented problems (Callejo et al. 2019).

25.2.2 Governance and E-Governance

The United Nations have initiated e-Government measurement initiatives among 178 member states under the "e-Government Readiness Index" and "e-Participation Index" also known as e-PI and e-GRI using an e-Government Readiness Assessment survey based on Innovation Management Measurement Framework. The IMMF is built on Input, Knowledge Management, Innovation Strategy, organization and culture, portfolio Management, project management, and commercialization headers. These 7 constructs and 19 sub-constructs try to measure and promote e-Governance across the globe (Potnis 2010). An evaluation of three countries India, Ethiopia, and Fiji showed positive results that can help reduce corruption in government administration. Limited usage of e-governance in developing countries has deprived those countries of the benefits of e-governance mechanisms. However, an important point is that ICT technologies need to be revamped to effectively incorporate governmental plans and reduce the digital divide (Singh et al. 2010).

The single project, single governance, and single jurisdiction model is not efficient and hampers the applicability of administration and researches in continuation. e- Governance is a probable solution to solve this crisis and improve upon the existing mechanisms (Kaye 2011). Cloud computing as technology has significant application in the e-governance space as it frees the local governments which are often underfunded and has an infrastructure bottleneck for implementing successful online portals or governance sites. This also makes the portals much more reliable and available as the downtime is almost reduced to zero (Tripathi and Parihar 2011).

The Indian e-governance space is still nascent and there is tremendous scope for improvement and it needs to go through several phases of change management before it can achieve a self-sustainable model. Policy gaps exist in every facet of the e-government aspect as ICT is not completely incorporated into the system. The implementation framework is also not as robust and immense opportunities exist for improvement in the sector (Singh and Kiran 2013).

E-governance of e-commerce is still an under-investigated area. Research in a similar direction between government, business, and civil society highlighted severe lacuna and compromise in the system. E-consumer protection needs to be looked at from the angle of all the three dimensions as mentioned earlier rather than just from a marketing perspective. E-consumer protection will ensure the voice of the consumer is protected by the implementation of e-governance in this era of e-retail (Ha and McGregor 2013).

The early 1990s gave rise to the concept of New Public Management and E-Governance. Increased consumer expectations have left have pressurized local governments across the world to improve upon the governance mechanism and promote good governance. Information and communication technology lead governance is much more transparent and accountable and successful implementation of e-governance requires injection of finances, human resource, administrative, and citizen-oriented changes to be effective. India can achieve this feat by developing well-directed leadership intended to achieve these goals (Sapru and Sapru 2014).

The popular blockchain technology is found to be reliable, safe, and anonymous for building e-governance applications and when coupled with technologies like the Internet of Things will help the creation of e-democracy tools powered by automation and minimizing the security risk of exposed or open systems. This blockchain technology is a decentralized information exchange, used by millions across the globe as it shields the common user from the risks of the world wide web by default, which is unprecedented in historical terms (Qi et al. 2017).

Digital technologies have made citizen participation in the arena of public policymaking by reducing the cost of public participation to a minimum level. However, people would still need encouragement for deliberate participation in these processes. The scope of influence digital technologies can exert from lobbying to e-governance, e-participation, citizen budget, etc. (Baxter 2017). E-governance website has the probability to enhance the democratic nature of government (Lee-Geiller and Lee 2019).

It is a known fact that e-governance has improved the government machinery across the globe making it transparent and accountable. Investigations in the literature show a major contribution of information system management, social networks, and open data in building such governments robust and resilient (Bindu et al. 2019).

25.2.3 Changing Scenarios of E-Governance and Digital Divide in COVID Pandemic

One thing that is already clear in this COVID world is no one will come out of this without something or the other. The lessons of COVID-19 will be a renewed struggle among the fittest in postmodernity testing both the strengths and skills of an individual and the nations as well. This is the new reality social freedom will no longer mean the same things that we once perceived. Old world order will have to stand the test of time against the rising powers. Information technology, e-governance, commerce, health, and artificial intelligence will play a much more potent role than earlier. There is a possibility of human rights conflict with extensive technology-driven monitoring systems for the citizens (Sharfuddin 2020). Taiwan is approximately 200 kilometers far away from the COVID-19 epicenter in China. The collaborative model of governance in Taiwan is often hailed for the early success against COVID as cooperation helped mobilize the essentials at a very early stage (Huang 2020).

COVID-19 pandemic quickly shifted hotspot from China to Europe and other Asian countries depending on the sociocultural context and soon hospitals were overwhelmed with resulting tragic stories. However, based on the responses East Asian countries were able to counter and manage this pandemic in a much effective manner, and an exaggerating amount of similarities exists between the responses of these countries (Shaw et al. 2020).

COVID-19 has pushed the digital agenda to the forefront of the society where normal functioning of day-to-day activities was only possible safely in the digital space. Also, governments across the world were using digital tracking and tracing systems (DTTS) or contact tracing apps to trace individual's locations and behavior patterns for risk analysis. This is where the question gets complicated and twisted as it directly hinders the agenda of cyber-surveillance and is perhaps a step closer than what we expect. The question of digital ethics is an unfinished agenda in this pandemic. While digital is the only answer to function as a normal society without interacting with the risks of COVID (Taddeo 2020).

COVID-19 has exposed many of the drawbacks of our public infrastructure and the lack of attention it has been facing for ages. New normal came with reinvented old terms like social distancing and quarantine soon enough the medical community and the medical systems also switched over to technology-infused solutions to safeguard their personal life. This exposed another dimension of our society the existing digital divide in our rural and remote communities where most of the vulnerable and poor people lived. This switchover left these communities deprived of the so-called digital medicine due to accessibility issues. The telemedicine expansion also was gravely hampered by this digital divide (Ramsetty and Adams 2020; Bakhtiar et al. 2020).

25.3 Objectives

1. To explore and devise QOL Barometer as an integral part of the e-governance initiative by a state for achieving the holistic development of the citizens.
2. To leverage the QOL barometer and IoT ecosystem for combating humanitarian crises due to critical emergencies and pandemic disasters.

25.4 Research Methodology

The present study is designed to harness the appropriate e-Governance ecosystem for holistic development for citizens. The paper floats an idea of creating a dedicated Quality of Life Barometer, especially designed mobile application (Web-Enabled as well) which could become instrumental for enhancing QOL of its fellow users. The paper is developed using secondary information like relevant literature, reports, and recent experiments and experiences undertaken by various governments to handle this COVID pandemic. The exploratory model—QOL Barometer app or CARE protocol has been described with appropriate diagrams and algorithms. These recent experiences of the COVID-19 pandemic exhibit that the global communities have adopted a mobile-based app for exploring real-life databases to mitigate the vulnerability of COVID-19 infection. This has improvised the authors to conceptualize an innovative framework of mobile app-based comprehensive mechanisms for enhancing QOL of its citizens on a broader perspective rather than being confined to mitigating disaster or pandemic situations.

25.4.1 Analysis I

The *Comprehensive Automated Result-oriented E-Governance (CARE)* Protocol is illustrated in Fig. 25.1. This would function through the following indicative sequence. A nationwide mobile app "*CARE App*" (web enabled) will be developed which needs to be installed by the citizens of India. This app may be made compulsory for mobile *Original Equipment manufacturers (OEM)* operating in India. The app would be supported by state-controlled dedicated "*Expert Monitoring Team (EMT)*" for its operation, up-gradation, and e-governance manifestations in India. The EMT group would cover various dimensions of governance by providing relevant information asking for citizen's expectations, impediments, aspirations,

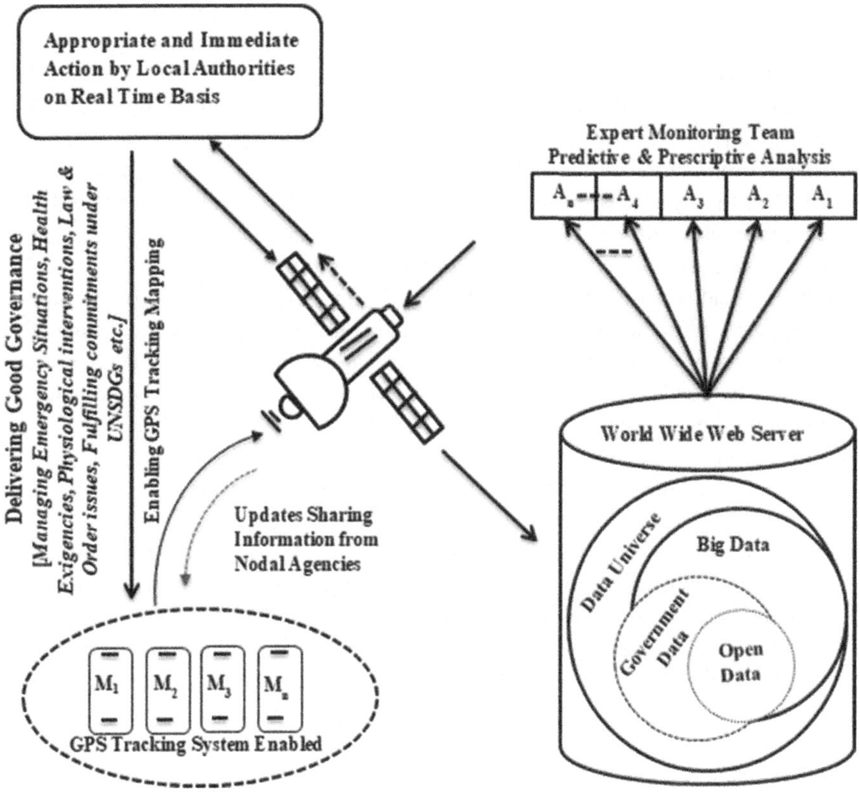

Fig. 25.1 QOL barometer for E-Governance—CARE protocol

quizzes, test batteries, and psychometric analysis so that the state of mind of every citizen can be mapped for predictive and prescriptive analytics along with indicators of holistic development matrix. The indicative list of dimensions may be as follows:

- Real-time information on localized emergency or vulnerability issues (landslide, forest fire, epidemic, endemic, pandemic, etc.)
- Individualized severe ill Health information for providing appropriate and immediate support facilities.
- Various law and order issues.
- Local-level information and feedback on community service on the functioning of respective public health engineer/municipality/Panchayati Raj intuitions.
- Individualized information for household activities, income, and participation in other developmental vectors.
- Various dimensions are covered under UNSDGs, etc.
- Appropriate information from all the citizens of India would be transmitted and stored in World Wide Web servers through dedicated satellites.

- From the server extreme cases in terms of life threats, livelihoods or socioeconomic critical factors both for the individual as well as collective modes would get automatically transferred to the respective expert domain for an appropriate recommendation, advisory, or instructions using Big Data analytics (BDA—Preventative, Predictive and Prescriptive Analytics) or intensive case studies.
- The appropriate recommendation, advisory, or instructions would be communicated to the respective, responsible local authorities (DM/DC, SP, SDO, BDO, etc.) through satellite system for immediate action and successful implementation of e-governance ecosystem using GIS/GPS-based tracking system.
- The local administration would identify the user and provide need-based support available within the jurisdiction of the authorities concerned on a real-time basis so that every citizen of India could be provided improvised good governance system by rendering immediate and need-based support facilities both for an individual level or collective purposes.
- The concerned local authority may also record the action taken report (ATR) they have taken against the corresponding advisory or instructions which would be communicated to all concerned through satellite.
- Through the entire process, the aspirations of data privacy and protection would be maintained until and unless the individualized score on a certain parameter or domain appears to be absurd, exceeding critical ranges. In that case, a GPS tracking system would be operationalized to provide instant/immediate support for the individual or group. Otherwise, the commitment of the state to good governance would be defeated. In general deplorable health hazards accidents emergencies, acute distress threats to life, extreme poverty, absolute hunger, etc. would have to identify for administering instant support by the concerned local authority. However appropriate data privacy and protection protocol as well as ethical practices would be followed without any deviations.

The CARE Protocol if implemented successfully will have several benefits that would percolate in all levels of society. In case of any unprecedented situations like COVID governments and administrative bureaucracy automatically goes to shock and are forced to implement or try policies whose benefits and losses cannot be an estimate or even speculated before the consequence of chain reactions that triggers massive public outrage and media frenzy.

The CARE protocol is a default deterrent and a mechanism to organically connect with the massive population of a country or even a region that may be dispersedly populated. Some of the hypothetical direct benefits transfers for existing situations are listed below:

- In India, approximately 10 people commit suicide every day. Out of which it was found that approximately 12% of victims were illiterate 17% are educated up to primary level, 19% up to middle level, and 23% were educated up to matriculate level (National Crime Records Bureau 2018a). This shows that there is a high degree of stress among the literate population which can be mitigated by examining the psychological status of an individual and application of CARE protocol.

- It was also seen during the same period, out of the number of cases resisted under IPC crimes against women showed that approximately 31% of women faced some kind of domestic violence 26% faced assault on modesty, 22% were kidnapped, and a staggering 10% rape victims (National Crime Records Bureau 2018b). The CARE protocol is a citizen-centric e-Governance model and thus provides direct access to the concerned authority in the virtual sphere with real outcomes. This will drastically reduce the number of attempted assaults as it may be used to trigger alarms in case of perceived threat by the potential victim.
- Out of the total cases registered under IPC Crime against Children were approximately 44% kidnapping, 34% child sexual offense, or child rape (National Crime Records Bureau 2018c). Children are indeed voiceless victims often not in a situation to either assess the attempt to assault or report abuse in case if it was already late thus saving other children from getting victimized. The CARE protocol may be designed to instruct parents and responsible adults to further train children to avoid escape or report such a situation.
- There are various forms of law and order issues (National Crime Records Bureau 2018d; United Nations Department of Economic and Social Affairs (UNDESA). 2015; Chakrabarty 2019) that may be overcome if the care protocol is effectively and efficiently administered.
- The CARE protocol has the potential to fulfill the commitments made to achieve UNSDGs by 2030. It will fulfill Goal 1 (No Poverty), Goal 2 (Zero Hunger), Goal 3 (Good Health and Well-being), Goal 5 (Gender Equality), Goal 10 (Reducing Inequality), Goal 16 (Peace, Justice, and Strong Institutions), Goal 17 (Partnerships for the Goals)(El-nafaty and Bashir n.d.) directly and indirectly if it is implemented and improvised on a continuous cycle.

Hence, the CARE protocol would act as a change agent for the citizen of a country through a comprehensive e-Governance framework. It would bring transformative changes in the lives of the people by elevating QOL or minimizing the impact of social evils and menace. The CARE protocol would embrace with robust, real-time, and responsible manner to accelerate the social developmental process.

The CARE Protocol would ideally be a mobile application as it is targeted at individuals. The installation would form a recognizable flexible, reliable source. The application would be authenticated based on a primary key and two factor or multifactor security authentication mechanisms as illustrated in (Fig. 25.2). The Care Protocol is the QOL Barometer that will help identify, monitor, and fix issues that are concerned with the well-being of a Citizen.

25.4.2 Analysis II

Due to the recent COVID pandemic, it is observed that various countries have taken a digital-based platform for identifying victims. So that appropriate strategies like the social distancing quarantine process and other advisories could be implemented

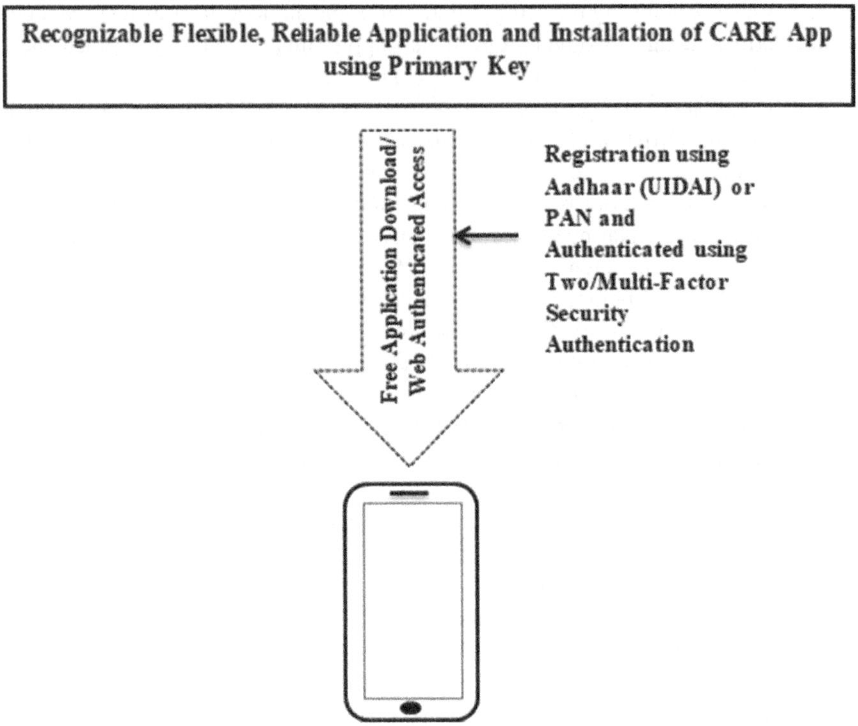

Fig. 25.2 Installation procedure of CARE app using primary key

immediately. The modus operandi of this software or mobile apps essentially have worked on Digital Tracking and Tracing Systems (DTTS). The digital platform enables the user to register their health-related information regarding the COVID pandemic. Based on this the concerned authority can take up steps to facilitate the victims for treatment. On the contrary, the noninfected citizens can get to know that the status and nature of the outbreak in the locality so that they can adopt appropriate defense mechanisms like improvising health hygiene issues and self-induced precautionary measures. However, the state needs to play a better role to improve the QOL of its citizens during pandemic disasters the world has witnessed severe economic downfall leading to loss of jobs, pay cut, struggle to maintain basic needs and amenities. It has a tantamount effect on the physiological state of individuals it has literally challenged the societal ecosystem. The entire discussions, deliberations, thought process and even media broadcasts have been overshadowed by COVID nemesis. The COVID pandemic generally doesn't differentiate in terms of wealth, gender, race, age, or geographic location but the worst victim, of this pandemic, were people belonging to the poor, and people working in the informal sector through long term collateral damage which is anti-thesis of UNSDG's. The developed nations have been supporting the unemployed youth's old age people under its social security schemes but it is a far cry from the people residing in developing and underdeveloped nations. Vaccines, antibiotics, medicines, and other health facilities essentially help the victims to get rid of infections but the long-term effect of an economic pandemic impacts the entire society both the victims and non-victims. Under this critical juncture, the state may provide such financial support to the needy and helpless people saving them from destitution in consonance with the philosophy of Universal Basic Income (National Crime Records Bureau 2018d). The CARE protocol can function as the depth and breadth of the severity that occurred due to critical emergencies and pandemic disasters.

In consonance with the CARE protocol, the state can use the IoT ecosystem in critical emergency and pandemic hotspots. For instance, in identified hotspots, the IoT devices would be installed along with peripheral roads and lanes in the containment zones.

Based on various inputs received from various stakeholders and real-time information received through QOL Barometer critical zones (for emergency crisis) or containment zones (during pandemic disasters) would get identified further to reach the root of the targets. The appropriate IoT ecosystem may be installed with the short vicinity of the target area. The IoT devices would be strengthened by incorporating appropriate sensors. The sensor would support for analyzing image-based analytics, temperature sensor/ heat-sensing cameras (Udgata and Suryadevara n.d.), sound pattern recognition sensors, for instance, the COVID infected patients have certain symptoms like high or mild fever, cough, and cold, and breathing distress. The image sensor would identify the movement of trace passers within and around the containment zone while the thermal sensor would enable the IoT to understand any case of unreported fever, especially at night time within its range. Sound pattern recognizers would help to understand the high frequency of coughing sounds/ nebulization sounds coming out of the containment zones. This would further enable the response team to take appropriate and immediate steps as depicted in Fig. 25.3.

Indicative algorithm of the system:

- First, the containment zone or critical zones would be identified.
- Appropriate IoT ecosystems would be installed which would be embedded with an appropriate set of sensors relevant to the nature of the emergency pandemic management.
- The IoT ecosystem would be connected to the specific portals of local authorities and the response team via satellite.
- The IoT devices would capture all information that is imagery information, sound pattern recognition, and thermal image from the targeted critical/containment zone. Entire information would be passed instantly to the assigned portals through satellite communication. It would also provide the user with approximate Lat/Long of the source location.
- Based on the inputs local authorities and response teams would identify the place, rush to the location, and take appropriate action as per the standard operating procedures.

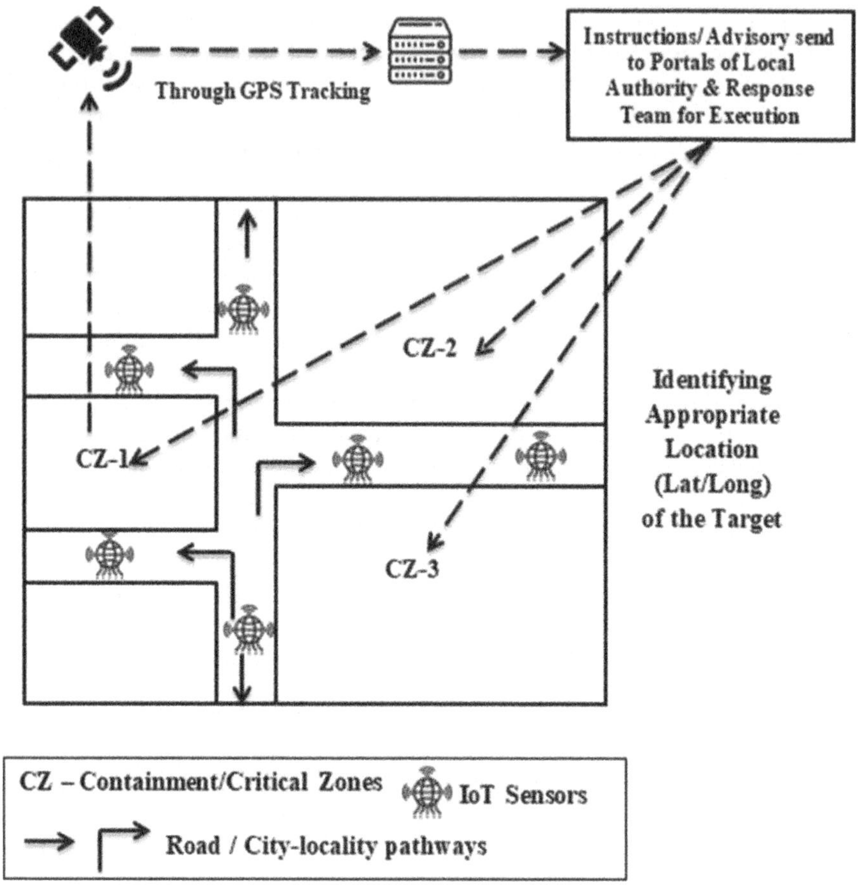

Fig. 25.3 IoT enabled emergency and pandemic management

Apart from this the local authority may also capture information, images by using appropriate unmanned aerial vehicles or drones which may be of additive value for taking a strategic decision.

25.5 Conclusion

In the era of the Fourth Industrial Revolution, it is inevitable for any government to choose an electronic medium for its long-term sustenance. During the issues like critical emergencies and pandemics, disasters are it is so important to sustain the existence of the civilization that the e-governance ecosystem has to be revamped by incorporating modern technological interfaces. This chapter has recommended devising a QOL barometer, a specially designed application known as CARE protocol for a country like India where the system would enable the policymakers, planners as well as local authorities and response teams to act immediately on a real-time basis. There are numerous experiences where various nations adopted similar kinds of approaches during the recent COVID pandemic, but qualitatively this model is somehow different simply because it is not for mitigating only emergency or pandemic issues this can be a weapon for the state to foster holistic development and well-being of the society. Apart from the natural practice this app could be revitalized and leveraged during emergencies and pandemic disasters.

References

AarogyaSetu Mobile App | MyGov.in (n.d.) Retrieved August 20, 2020, from https://www.mygov.in/aarogya-setu-app/

About the HPI | Happy Planet Index (n.d.) Retrieved August 21, 2020, from http://happyplanetindex.org/about

Aman (n.d.) Retrieved August 20, 2020, from https://amanapp.jo/en

App Radar Covid, la aplicación de rastreoparaEspaña | Noticias de Tecnología en Diario de Navarra (n.d.) Retrieved August 20, 2020, from https://www.diariodenavarra.es/noticias/vivir/tecnologia/2020/08/11/app-radar-covid-aplicacion-rastreo-para-espana-698713-3195.html

ApturiCovid (n.d.) Retrieved August 20, 2020, from https://apturicovid.lv/

Bache I (2013) Measuring quality of life for public policy: an idea whose time has come? Agenda-setting dynamics in the European Union. J Eur Publ Policy 20(1):21–38

Bakhtiar M, Elbuluk N, Lipoff JB (2020) The digital divide: how Covid-19's telemedicine expansion could exacerbate disparities. J Am Acad Dermatol

Bawumia launches GH COVID-19 Tracker App (n.d.). Retrieved August 20, 2020, from https://www.ghanaweb.com/GhanaHomePage/NewsArchive/Bawumia-launches-GH-COVID-19-Tracker-App-922885

Baxter DJ (2017) E-governance and e-participation via online citizen budgets and electronic lobbying: promises and challenges. World Affairs 180(4):4–24

Berlim MT, Pavanello DP, Caldieraro MA, Fleck MP (2005) Reliability and validity of the WHOQOL BREF in a sample of Brazilian outpatients with major depression. Qual Life Res 14(2):561–564

Bindu N, Sankar CP, Kumar KS (2019) From conventional governance to e- democracy: tracing the evolution of e-governance research trends using network analysis tools. Gov Inf Q 36(3): 385–399

Bonomi AE, Patrick DL, Bushnell DM, Martin M (2000) Validation of the United States' version of the world health organization quality of life (WHOQOL) instrument. J Clin Epidemiol 53(1): 1–12

Bundesregierung | Aktuelles | Veröffentlichung der Corona-Warn-App (n.d.) Retrieved August 20, 2020, from https://www.bundesregierung.de/breg- de/aktuelles/veroeffentlichung-der-corona-warn-app-1760892

Callejo KED, Sotomayor MR, Ochoa-Zezzatti A (2019) Happiness and its socio-demographic determinants analyzed with datamining: the case of a Community at the North of the border of Mexico. Research in Computing Science 148:135–151

Chakrabarty A (2019) Universal basic income at the bottom of the pyramid: achieving SDGs through financial inclusion in India. Administrative Development A Journal of HIPA, Shimla 6(1):119–141. http://qtanalytics.in/journals/index.php/HIPA/article/view/439

Cheung YB, Thumboo J (2006) Developing health-related quality-of-life instruments for use in Asia. PharmacoEconomics 24(7):643–650

China launches coronavirus "close contact detector" app–BBC News (n.d.) Retrieved August 20, 2020, from https://www.bbc.com/news/technology-51439401
Corona Tracer BD–Apps on Google Play (n.d.) Retrieved August 20, 2020, from https://play.google.com/store/apps/details?id=com.shohoz.tracer&hl=en_US
Coronapp (n.d.) Retrieved August 20, 2020, from https://coronaviruscolombia.gov.co/Covid19/aislamiento-saludable/coronapp.html
COVID Tracker App–Ireland's Coronavirus Contact Tracing App. (n.d.) Retrieved August 20, 2020, from https://covidtracker.gov.ie/
Covid-19: Use of new-age technologies for e-governance, Government News, ET Government (n.d.) Retrieved August 20, 2020, from https://government.economictimes.indiatimes.com/news/digital-india/covid-19-use-of-new-age-technologies-for-e-governance/75993744
COVIDSafe. (n.d.) Retrieved August 20, 2020, from https://www.covidsafe.gov.au/
COVIRA (n.d.) Retrieved August 20, 2020, from https://covira.info/
Crawley JA, Kleinman L, Dominitz J (2000) User preferences for computer administration of quality of life instruments. Drug information journal: DIJ/Drug Information Association 34(1): 137–144
Devlin NJ, Brooks R (2017) EQ-5D and the EuroQOL group: past, present and future. Appl Health Econ Health Policy 15(2):127–137
Download COVID Alert today–Canada.ca (n.d.) Retrieved August 20, 2020, from https://www.canada.ca/en/public-health/services/diseases/coronavirus-disease-covid-19/covid-alert.html
Download the "E-DOCTOR" mobile application (n.d.) Retrieved August 20, 2020, from https://koronavirusinfo.az/az/page/haqqimizda/e-tebib-mobil-tetbiqini-yukleyin
EAS (2019) e-Governance—e-Estonia. https://e-estonia.com/solutions/e-governance/
El-nafaty ASI, Bashir FM (n.d.) Novel vision-based thermal people counting tool for tracking infected people with viruses like COVID-19
Environmental Protection Agency USA (1973) The quality of life concept: a potential new tool for decision-makers. Environmental Protection Agency
eRouška–chránímsebe, chránímtebe (n.d.) Retrieved August 20, 2020, from https://erouska.cz/
Frankl VE (1963) Man's search for meaning: an introduction to logotherapy. Am J Orthopsychiatry 33(2):390–390
Gender Development Index (GDI) | Human Development Reports (n.d.) Retrieved August 21, 2020, from http://hdr.undp.org/en/content/gender-development-index-gdi
Group TW (1998) The World Health Organization quality of life assessment (WHOQOL): development and general psychometric properties. Soc Sci Med 46(12):1569–1585
Ha H, McGregor SL (2013) Role of consumer associations in the governance of e-commerce consumer protection. J Internet Commer 12(1):1–25
HaMagen–The Ministry of Health App for Fighting the Spread of Coronavirus (n.d.) Retrieved August 20, 2020, from https://govextra.gov.il/ministry-of-ealth/hamagen-app/download-en/
How Estonia became an e-government powerhouse–Tech Republic (n.d.) Retrieved August 19, 2020, from https://www.techrepublic.com/article/how-estonia-became-an-e-government-powerhouse/
Huang IYF (2020) Fighting against COVID-19 through government initiatives and collaborative governance: Taiwan experience. Public Administration Review
Immuni - Download Immuni (n.d.) Retrieved August 20, 2020, from https://www.immuni.italia.it/download.html
India's Trillion-Dollar Digital Opportunity, Ministry of Electronics & Information Technology, Government of India. (2019). https://www.meity.gov.in/writereaddata/files/india_trillion-dollar_digital_opportunity.pdf
Information about Covid-19 in Iceland (n.d.) Retrieved August 20, 2020, from https://www.covid.is/
International: The lottery of life|The Economist (n.d.). Retrieved August 20, 2020, from https://www.economist.com/news/2012/11/21/the-lottery-of-life

Interrogation A (2008) Bhutan's gross national happiness index. OPHI l(50):30–33. https://ophi.org.uk/policy/national-policy/gross-national-happiness-index/

Johansson S (2002) Conceptualizing and measuring quality of life for national policy. In: Assessing quality of life and living conditions to guide National Policy. Springer, Dordrecht, pp 13–32

Kalenkoski CM, Hamrick KS (2014) Time poverty thresholds in the USA. In: Michalos AC (ed) Encyclopedia of quality of life and Well-being research. Springer, Dordrecht. https://doi.org/10.1007/978-94-007-0753-5_3704

Kaye J (2011) From single biobanks to international networks: developing e–governance. Hum Genet 130(3):377

Kejriwal launches 'Delhi Corona' app for real-time information on availability of hospital beds–The Hindu (n.d.) Retrieved August 20, 2020, from https://www.thehindu.com/news/cities/Delhi/kejriwal-launches-delhi-corona-app-for-real-time-information-on-availability-of-hospital-beds/article31729239.ece

Kingdom of Bahrain–eGovernment Apps Store: BeAware Bahrain (n.d.) Retrieved August 20, 2020, from https://apps.bahrain.bh/CMSWebApplication/action/ShowAppDetailsAction?selectedAppID=321&appLanguage=en

Lee-Geiller S, Lee TD (2019) Using government websites to enhance democratic E-governance: a conceptual model for evaluation. Gov Inf Q 36(2):208–225

4 lessons from Bhutan on the pursuit of happiness above GDP|World Economic Forum. (n.d.). Retrieved August 21, 2020, from https://www.weforum.org/agenda/2016/05/4-lessons-from-bhutan-on-the-pursuit-of-happiness-above-gdp/

Lohr KN (2002) Assessing health status and quality-of-life instruments: attributes and review criteria. Qual Life Res 11(3):193–205

McCall S (1975) Quality of life. Soc Indic Res 2:229–248. https://doi.org/10.1007/BF00300538

Michalos A, Ferriss AL, Easterlin R, Patrick D, Pavot W (2006) The quality of life (QOL) research movement: past, present, and future. Soc Indic Res 76(3):343–466

Min SK, Kim KI, Lee CI, Jung YC, Suh SY, Kim DK (2002) Development of the Korean versions of WHO quality of life scale and WHOQOL–BREF. Qual Life Res 11(6):593–600

Molina-Herrera E, Ochoa A, Gill T, Ibarra-Mejia G, Herrera C (2019) Conceptualization of a predictive model for analysis of the health outcomes of dust events in a society with Köppen climate classification BW. Research in Computing Science 148:63–89

Morris MD (1978) A physical quality of life index. Urban Ecol 3(3):225–240. https://doi.org/10.1016/0304-4009(78)90015-3

MyTrace, a Preventive Counter Measure and Contact Tracing Application for COVID-19–KementerianSains, TeknologidanInovasi (MOSTI) (n.d.). Retrieved August 20, 2020, from https://www.mosti.gov.my/web/en/mytrace/

National Crime Records Bureau (2018a) Accidental deaths & suicides in India:2018

National Crime Records Bureau (2018b) Crime in India 2018. Vol-I Ministry of Home Affairs

National Crime Records Bureau (2018c) Crime in India 2018. Vol-II Ministry of Home Affairs

National Crime Records Bureau (2018d) Crime in India 2018. Vol-III Ministry of home affairs

New Coronavirus Contact Confirmation Application (COCOA) COVID-19 Contact–Confirming Application|Ministry of Health, Labor and Welfare (n.d.) Retrieved August 20, 2020, from https://www.mhlw.go.jp/stf/seisakunitsuite/bunya/cocoa_00138.html

NZ COVID Tracer app | Ministry of Health NZ (n.d.) Retrieved August 20, 2020, from https://www.health.govt.nz/our-work/diseases-and-conditions/covid-19-novel-coronavirus/covid-19-novel-coronavirus-resources-and-tools/nz-covid-tracer-app

Open Government Partnership (2020) Collecting Open Government Approaches to COVID-19 | Open Government Partnership. In Open Government Partnership. https://www.opengovpartnership.org/collecting-open-government-approaches-to-covid-19/

Potnis DD (2010) Measuring e-governance as an innovation in the public sector. Gov Inf Q 27(1):41–48

Qatar makes COVID-19 app mandatory, experts question efficiency | Qatar News | Al Jazeera (n.d.) Retrieved August 20, 2020, from https://www.aljazeera.com/news/2020/05/qatar-covid-19-app-mandatory-experts-question-efficiency-200524201502130.html

Qi R, Feng C, Liu Z, Mrad N (2017) Blockchain-powered internet of things, e-governance and e-democracy. In: E-democracy for smart cities (pp. 509–520). Springer, Singapore

Quality of Life Research Unit : home page (pp. 1–6). (n.d.) Retrieved August 19, 2020, from http://sites.utoronto.ca/QOL/QOL_model.htm

Ramsetty A, Adams C (2020) Impact of the digital divide in the age of COVID-19. J Am Med Inform Assoc 27(7):1147–1148

Sapru RK, Sapru Y (2014) Good governance through e-governance with special reference to India. Indian Journal of Public Administration 60(2):313–331

Saxena S, Chandiramani K, Bhargava R (1998) WHOQOL-Hindi: a questionnaire for assessing quality of life in health care settings in India. Natl Med J India 11:160–165

Sharfuddin S (2020) The world after covid-19. The Round Table 109(3):247–257

Shaw R, Kim YK, Hua J (2020) Governance, technology and citizen behavior in pandemic: lessons from COVID-19 in East Asia. Progress in disaster science, 100090. Health 5(5):e002655

Singh KN, Kiran R (2013) E-governance success factors: an analysis of E- governance initiatives of ten major states of India. Int J Public Sect Manag 26(4):320–336

Singh G, Pathak RD, Naz R, Belwal R (2010) E-governance for improved public sector service delivery in India. International Journal of Public Sector Management, Ethiopia and Fiji

Skevington SM (1999) Measuring quality of life in Britain: introducing the WHOQOL-100. J Psychosom Res 47(5):449–459

Smittestopp–app–helsenorge.no (n.d.) Retrieved August 20, 2020, from https://helsenorge.no/coronavirus/smittestopp

Stoiciu A (2011) The role of e-governance in bridging the digital divide. UN Chron 48(3):37–39. https://doi.org/10.18356/d51f530b-en

Stop udbredelsenaf COVID-19–Smittelstop (n.d.) Retrieved August 20, 2020, from https://smittestop.dk/

StopCovid | economie.gouv.fr (n.d.) Retrieved August 20, 2020, from https://www.economie.gouv.fr/stopcovid#

STOPcovid19–STOPcovid19 (n.d.) Retrieved August 20, 2020, from https://www.stopcovid19.it/en/

StopKorona!–tracing of Coronavirus exposure and protection from COVID-19 (n.d.) Retrieved August 20, 2020, from https://stop.koronavirus.gov.mk/en

Stopp Corona (n.d.) Retrieved August 20, 2020, from https://www.stopp-corona.at/

Swaine-Verdier A, Doward LC, Hagell P, Thorsen H, McKenna SP (2004) Adapting quality of life instruments. Value Health 7:S27–S30

SwissCovid app and contact tracing (n.d.) Retrieved August 20, 2020, from https://www.bag.admin.ch/bag/en/home/krankheiten/ausbrueche-epidemien-pandemien/aktuelle-ausbrueche-epidemien/novel-cov/swisscovid-app-und-contact-tracing.html

Taddeo, M. (2020). The ethical governance of the digital during and after the COVID-19 pandemic

Thailand wins (n.d.). Retrieved August 20, 2020, from https://www.thaichana.com/

The Economics of Happiness | Greater Good (n.d.) Retrieved August 21, 2020, from https://greatergood.berkeley.edu/article/item/the_economics_of_happiness

The Economics of Well-Being (n.d.) Retrieved August 21, 2020, from https://hbr.org/2012/01/the-economics-of-well-being

The key to Denmark's digital success (n.d.) Retrieved August 21, 2020, from https://denmark.dk/innovation-and-design/denmarks-digital-success

TraceTogether (n.d.) Retrieved August 20, 2020, from https://www.tracetogether.gov.sg/

Tripathi, A., & Parihar, B. (2011). E-governance challenges and cloud benefits. In 2011 IEEE International Conference on Computer Science and Automation Engineering (Vol. 1, pp. 351–354). IEEE

Udgata SK, Suryadevara NK (n.d.) COVID-19, sensors, and internet of medical things (IoMT). In internet of things and sensor network for COVID-19 (pp. 39–53). Springer, Singapore

UN E-Governance Knowledgebase (2020) United Nations, Department of Economic and Social Affairs, Division for Public Institutions and Digital Government. https://publicadministration.un.org/egovkb/Data-Center

Undp (1997) Human development report 1997. Oxford University
Unit EI (2005) The economist intelligence unit's quality of life index. The world in 2005. White Paper. Economist Group, London
United Nations Department of Economic and Social Affairs (UNDESA) (2015) THE 17 GOALS | department of economic and social affairs. Sustain Dev. https://sdgs.un.org/goals
United Nations E-Government Survey 2020 Digital Government in the Decade of Action for Sustainable Development. (2020). https://publicadministration.un.org/egovkb/Portals/egovkb/Documents/un/2020-Survey/2020 UN E-Government Survey (Full Report).pdf
VFI/Logotherapy and Existential Analysis (n.d.) Retrieved August 18, 2020, from https://www.viktorfrankl.org/logotherapy.html#
VírusRadar–a Koronavíruskövetéséreés a COVID-19 ellenivédekezésre (n.d.) Retrieved August 20, 2020, from https://virusradar.hu/
What are the economics of happiness? | Yale Insights (n.d.) Retrieved August 21, 2020, from https://insights.som.yale.edu/insights/what-are-the-economics-of-happiness
WHO. (2014) WHO|WHOQOL: measuring quality of life. Health statistics and information systems (WHO); World Health Organization. http://www.who.int/healthinfo/survey/whoQOL-qualityoflife/en/
WhoQOL Group (1995) The World Health Organization quality of life assessment (WHOQOL): position paper from the World Health Organization. Soc Sci Med 41(10):1403–1409
Why we need to choose happiness over economics | World Economic Forum (n.d.) Retrieved August 21, 2020, from https://www.weforum.org/agenda/2019/06/happiness-may-be-a-choice-except-that-it-s-constrained-by-vested-economic-interests
World Happiness Report 2020 | The World Happiness Report (n.d.) Retrieved August 21, 2020, from https://worldhappiness.report/ed/2020/#read
WVS Database. (n.d.) Retrieved August 21, 2020, from http://www.worldvaluessurvey.org/WVSContents.jsp

The manufacturer's authorised representative in the EU is Springer Nature Customer Service Centre GmbH, Europaplatz 3, 69115 Heidelberg, Germany. If you have any concerns regarding our products, please contact ProductSafety@springernature.com

Printed and bound by CPI Group (UK) Ltd, Croydon, CR0 4YY

22/09/2026

02220954-0010